The Great Power Competition Volume 1

AF848573

Adib Farhadi · Anthony J. Masys
Editors

The Great Power Competition Volume 1

Regional Perspectives on Peace and Security

Springer

Editors
Adib Farhadi
College of Arts and Sciences
University of South Florida
Tampa, FL, USA

Anthony J. Masys
College of Public Health
University of South Florida
Tampa, FL, USA

ISBN 978-3-030-64475-8 ISBN 978-3-030-64473-4 (eBook)
https://doi.org/10.1007/978-3-030-64473-4

© The Editor(s) (if applicable) and The Author(s), under exclusive license to Springer Nature Switzerland AG 2021
This work is subject to copyright. All rights are solely and exclusively licensed by the Publisher, whether the whole or part of the material is concerned, specifically the rights of translation, reprinting, reuse of illustrations, recitation, broadcasting, reproduction on microfilms or in any other physical way, and transmission or information storage and retrieval, electronic adaptation, computer software, or by similar or dissimilar methodology now known or hereafter developed.
The use of general descriptive names, registered names, trademarks, service marks, etc. in this publication does not imply, even in the absence of a specific statement, that such names are exempt from the relevant protective laws and regulations and therefore free for general use.
The publisher, the authors and the editors are safe to assume that the advice and information in this book are believed to be true and accurate at the date of publication. Neither the publisher nor the authors or the editors give a warranty, expressed or implied, with respect to the material contained herein or for any errors or omissions that may have been made. The publisher remains neutral with regard to jurisdictional claims in published maps and institutional affiliations.

This Springer imprint is published by the registered company Springer Nature Switzerland AG
The registered company address is: Gewerbestrasse 11, 6330 Cham, Switzerland

Preface

The Greater Central and South Asia regions have once again become the theater for a complex, high-stakes geostrategic contest. This contest, termed the Great Power Competition (GPC), echoes the "Great Game" of the nineteenth century, in which the Russian and British Empires competed on the "pull-and-tug" buffer zone of Afghanistan. In today's GPC, multiple powers vie for influence, resources, and interests: the Global Powers include the United States, China, and Russia, while the Regional Powers include Iran, India, Pakistan, Turkey, the Central Asian states, and the Gulf countries.

The re-emergence of this new, multidimensional Great Power Competition is reflected in the United States' National Security Strategy, which notes, "After being dismissed as a phenomenon of an earlier century, the great power competition returned." Similarly, the National Defense Strategy (NDS) announced that "inter-state strategic competition, not terrorism, is now the primary concern in U.S. national security."

The Middle East, Horn of Africa, Central Asia, and South Asia form the strategic crossroad known as the Central Region. Here, China is attempting to expand its influence, and Russia is working to regain a place of power internationally. Both China and Russia see opportunities to dominate and influence the Central Region, while the stated U.S. intent is to shift to other regions. The U.S. finds itself at a strategic disadvantage, having been in the middle of a nearly twenty-year conflict in the region. For the U.S. to counter these revisionist powers will require an understanding of their strategic objectives, perspectives, and the challenges that limit their competitive mobilization and the identification of common areas of interest and potential pathways to more holistic approaches, intersections, and/or cooperative efforts.

Those efforts must aim to combat terrorism and transnational organized crime in areas such as Afghanistan, Iraq, and Syria, as well as deter Iranian ambitions and reigning in a nuclear North Korea. A key component of the United States' national security is the global relationship between the U.S. and existing and emerging military and economic powers. These rising powers and their threat to the U.S. status as a Global Power are what have led to the Great Power Competition's re-emergence. More specifically, the NDS asserts that "the central challenge to U.S. prosperity and security is the re-emergence *of long-term, strategic competition* by ... revisionist powers," specifically China, as an economic, technological, and political competitor, and Russia, as an opportunistic military power.

Hal Brands, the Henry A. Kissinger Distinguished Professor of Global Affairs at the John Hopkins University School of Advanced International Studies (Perry World House, 2017) provides further insight into what he sees as the "Six Propositions about Great Power Competition and Revisionism in the twenty first Century":

1. Clashing interests and clashing ideologies have long driven the leading powers in the system to compete with one another; they have often led to intense conflict and war.
2. Great power frictions and competition never fully went away even during the post-Cold War era; they were simply muted by the two defining features of the post-Cold War international system.
3. Great power competition has returned in fuller and sharper form today because the systemic conditions for such competition have become more propitious—and because some of the great hopes of the post-Cold War era have now been dispelled.
4. Great power competition and revisionism are sharper today than at any time since the end of the Cold War. We are seeing that competition in the geopolitical realm, in the sense that Russia and China are increasingly seeking to carve out spheres of dominant influence within their respective "near abroads," to undermine U.S. alliances and partnerships in these areas, and to develop military capabilities needed to achieve regional primacy and project power even further abroad.
5. Intensified great power competition is certainly going to lead to a more dangerous and disorderly international environment, but it need not necessarily lead to a major crack-up of the existing international system. The fact that the major authoritarian great powers are pushing back more strongly against that international order and its defenders—principally the United States and its allies—means that we are likely to see more diplomatic and military crises and a generally higher level of international tensions in the coming years. It is going to make it harder to achieve meaningful multilateral cooperation among the great powers on common security challenges. Just look at how great power rivalry has stymied efforts to resolve the Syrian civil war over the past six years.
6. As nationalism, populism, and retrenchment grow stronger in the coming years, the United States and its friends will generally find it harder to meet the challenge of intensified great power competition. All of these phenomena are broad, and they sometimes pull in opposite directions, so many caveats are attached to this proposition.

With an aim to gain a holistic perspective that better informs decision-making at various levels and to further amplify the vital importance of success in the Central Region to U.S. prosperity and security, the University of South Florida (USF), in collaboration with the National Defense University Near East South Asia (NESA) Center for Strategic Strategies, hosted the inaugural Great Power Competition Conference in Tampa, Florida, on January 29 and 30, 2020. In a novel forum that brought together the industry, military, and academic subject matter experts in one location, the conference facilitated a deep-dive understanding of Great Power and Regional perspectives in the Central Region by advancing the dialogue on identifying key issues for stakeholders within and beyond the Central Region as it relates to regional diplomacy and politics, economic potential and rivalries, security considerations and interests, and the information environment. This Great Power Competition Conference Series is the first in a Great Power Competition Conference Series, with conferences taking place twice per year for nine years at the University of South Florida.

While some publications consider the Great Power Competition topic, none are yet known that explore the Great Power Competition in the Central Region from the collective vantage of military, academic, and industry experts. The Inaugural Great Power Competition Conference, indeed, filled this gap, offering new ideas and lines of inquiry, innovative solutions, and decision support at multiple levels.

This book presents a compilation of papers that emerged from the Great Power Conference held in January 2020 at the University of South Florida. The 20 chapters in this volume cover topic ranging from historical analysis to innovative solutions, changing the dynamics of peace and conflict in the region, and matters pertaining to human security and intervention strategies associated with humanitarian, development, and peace nexus.

Tampa, USA Adib Farhadi

Acknowledgements

The editors would like to extend their sincerest gratitude to the entire Universit of South Florida and National Defense University Near East South Asia Center for Strategic Strategies for their work on the Great Power Competition Conference Series and the resulting edited volume.

They would like to express their appreciation to Dr. Eric Eisenberg, USF Dean of College of Arts and Sciences, for his continued and enthusiastic support for the initiative.

A most heartfelt thanks to Michelle Assaad, Dr. Dianne Donnelly, Michael Peters, Sean Ryan, Andrew Roberts, and the contributing authors to making this book possible.

Dr. Adib Farhadi would like to thank his wife, Elaha, children, Adam, and Sophia, and mother, Maliha, for their support, patience, and love, not only in regard to this project but in everything that life brings.

Contents

About the Editors

Dr. Adib Farhadi is an Assistant Professor and a Faculty Director of Executive Education at the University of South Florida. His research is at the intersection of religion, politics, and conflict with a particular focus on the "Silk Road" Central and South Asia (CASA) Region, which is the subject of his recent book, Countering Violent Extremism by Winning Hearts and Minds. In addition, he is the Director and editor of the Great Power Competition book and conference series. Formerly, Dr. Farhadi served in senior positions in Afghanistan and extensively advised the United States government and various other international organizations. Dr. Farhadi is internationally recognized and a frequent presenter on the topics of Countering Violent Extremism (CVE), Conflict Resolution, Strategic Negotiations and Communication, and Geoeconomics. Dr. Farhadi earned his Ph.D. in Economics from the University of Canberra, M.A. from New York University, and B.S. from East Carolina University.

Dr. Anthony Masys is an Associate Professor and a Director of Global Disaster Management, Humanitarian Assistance and Homeland Security. Former Senior Air Force Officer, Dr. Masys, has a B.Sc. in Physics, an M.Sc. in Underwater Acoustics and Oceanography from the Royal Military College of Canada, and a Ph.D. from the University of Leicester. He is the Editor-in-Chief for Springer Publishing book series: Advanced Sciences and Technologies for Security Applications, and he holds various advisory board positions with academic journals and books series. Dr. Masys is an internationally recognized author, speaker, and facilitator and has held workshops on security, visual thinking, design thinking, and systems thinking in Europe, North America, South America, West Africa, and Asia. He has published extensively in the domains of physics and the social sciences. Dr. Masys supports the University of Leicester (UK) as an associate tutor in their Distance MSc Program on Risk Crisis and Disaster Management.

About the Editors

Surfing the Conflux: Technology, Information Environments and Great Power Competition Convergence

Mark Abdollahian

Abstract The convergence of technology, information environments and great power competition define today's operating environment. Enabled by technology, the sources, speed and scale of change demand agility. To succeed, national security and defense challenges can leverage several arbitrage opportunities: a broader consideration of human behavior, deeper knowledge of complexity dynamics, smarter understanding of others' cognitive bias as well as our own and finally sharper operating environment capabilities. These also necessitate maneuvering the underlying conflux of increasing connectivity, complexity and uncertainty across most human behavior to master operational art. Capitalizing on transformational computing power, big data, AI and computational social sciences are keys to preparing for alternative futures for our defense and security.

Keywords Operational art · Cognitive maneuvering · Technology · Artificial intelligence · Big data · Analytics · Complex Adaptive Systems · Computational Social Science

Introduction

The future operating environment is here. National security and defense challenges stretch and morph into managing complex problem sets. Priorities span a broad and diverse spectrum of issues driven by shifting realities. Lanes in the road merge to include peer competition for US global leadership, economic prosperity, regional multipolarism, extremism, proliferation, state fragility, abusive regimes, forced migration and climate change among others. Welcome to the convergence of technology, information environments and great power competition.

The convergence first begs for situational awareness as technology enables, shapes and accelerates operating environments. As widely recognized, technology is not the problem, it is defining both problem and solution sets in a fourth industrial

M. Abdollahian (✉)
Claremont Graduate University, Claremont, CA 91711, USA
e-mail: Mark.Abdollahian@cgu.edu

© The Author(s), under exclusive license to Springer Nature Switzerland AG 2021
A. Farhadi et al. (eds.), *The Great Power Competition Volume 1*,
https://doi.org/10.1007/978-3-030-64473-4_1

revolution (World Economic Forum 2020). Human computer interactions, big data and the AI revolution present equal tests and prospects. AI performs some remarkable feats already, yet quantum computing and potential supremacy overcoming transistor miniaturization limits is somewhere on the arc ahead.

Technological progress forces us to recalibrate timescales, perspectives and areas of responsibilities for what is possible, often far outside comfort zones. Moving from farming technologies developed during the agricultural revolution to the industrial revolution took over 8,000 years of human progress while moving from the commercialization of the internet to sequencing of the entire human genome only took 9 years (Grossman 2011). Sequencing the human genome was a coordinated 'whole of global industry and governments' approach accomplished in under 15 years and costing billions of dollars. Today this is performed by a single lab in less than 3 days for $1,000 (Forbes 2019). Development of COVID-19 vaccines in record time frames is another recent example.

A vast majority of technology is also now available to almost anyone with access to a computer and the internet. As innovation sources proliferate and scale, they are not confined to elites, nation states and corporations with billion-dollar balance sheets. Whether great leaders or bad actors, extremist recruitment narratives or cultural lifestyle videos, all these compete in the global information marketplace of possibilities. Technology abundance forces us to reconfigure our strategies from seeking often elusive optimality, to hedging our bets to be adaptive to change and resilient to unforeseen shocks. As technology augments very human needs, choices and actions, everyday billions of people use technology to push the envelope of possible operational art.

As an enabler, technology has accelerated modern hybrid warfare. Many great power competitions encompass the conflict-competition-cooperation continuum across physical, information and cognitive environments. Multiple, often conflicting national interests bridge increasingly intertwined political, military, economic, social, information and infrastructure (PMESII) domains. Our shaping tools operate in hyper-competitive, saturated information environments in real time. We cognitively maneuver gray zones with coarse notions of diplomatic, informational, military, economic and social (DIMES) elements of national power addressing symptoms when we need millimeter, neurosurgical precision for attacking root causes.

Narratives and interests can merge and diverge quickly when focused at varying global, regional, national, and local levels. These create rifts, blowback and execution impedance instantly broadcast worldwide. Different local operational realities feed into regional security coordination efforts all juxtaposed against evolving strategic goals. Success often necessitate 'whole of government' approaches orchestrated across multiple domains for impact. Whether we believe it, like it, embrace it or not, the future operating environment is upon us now.

National interests might come and go but the national interest always remains the same—owning the operational environment. Today the convergence of technology, information environments and great power competition necessitate that we operate broader, deeper, smarter and sharper.

While recognizing the convergence, we need to master the undercurrents, the conflux of increasing connectivity, complexity and uncertainty. Enabled by connectivity, the sources of innovation and disruption proliferate faster than we realize. Empowered by complexity, threat surfaces and attack vectors span not just kinetic technology futures but shaping cognitive environments today. Accelerated by uncertainty, disruptive new normals are simply the evolving *status quo* from the resonance of human activities. So why are we here? What do we do? And how can we move ahead, agilely surfing the conflux?

Surfing the Conflux

The convergence lays bare the conflict-cooperation continuum. For example, Turkish military incursions in northern Syria, to minimize perceived terror and sovereignty threats from various Kurdish groups, directly impact US efforts combating Daesh. Risks abound, from combatting violent extremist organization (VEO) social media campaigns to miscalculations pitting Turkish and US forces against each other despite a common NATO alliance. Multiple objectives include domestic political and economic considerations, not the least Erdogan's personal survival and population support.

Other factors only deepen the complexity puzzle: officials smuggling oil funding adversarial efforts, American F-35 versus Russian S-400 security choices, the international community's financial assistance to preempt another refugee crisis while an Iran-Turkey-KSA regional 'soft power' rivalry escalates. All this potential for 6-sided gun fights operate in the shifting context of great power competition: Russian strategic narratives, competitive military and economic investments for regional influence, stress testing alliance portfolios, and a rebalance to Asia due to impending economic security dynamics on the horizon.

Conflict-competition-cooperation continuum objectives include discrediting threat leader legitimacy while deterring military actions, degrading adversaries will to resist while attriting capabilities to exercise such will, persuading neutral populations to support US interests while simultaneously delivering on our own prosperity narratives in an international system with changing order. In response to the convergence, multiple efforts across the community have significantly changed our conversations, thinking and focus.

From thought leadership excellence, revised doctrine and DOD programmatic investments we already know much of what needs to be done. Azimuth checks enterprise wide illuminate what a path forward in the convergence could be: broader consideration of human behavior, deeper knowledge of complexity dynamics, smarter understanding of others' cognitive bias as well as our own and finally sharper operating environment capabilities.

Going broader, deeper, smarter and sharper are guides for possible success. Each one possesses multiple and specific data, model, knowledge and integration arbitrage opportunities. Arbitrage is buying something in one market and simultaneously selling it in another market at a higher price, profiting from temporary price differentials. Exercising arbitrage opportunities is easy money and risk-free profit. Integrating data, models and knowledge both within the enterprise and across the private sector, academia and whole USG is waiting to be realized.

However, the difficulty of us exercising arbitrage opportunities is our own problem, not our adversaries. Our beliefs, feelings, culture, motivational energy, time available, cost envelopes or allocated resources shape and constrain our own objective functions only. Obviously, these create additional threat vectors for exploitation that embolden adversaries. Given we know what to do, how can we do it?

Owning the convergence compels us to surf the conflux—the mixing undercurrents of connectivity, complexity and uncertainty that pervades all our modern operating environments. Surfing the conflux demands we profoundly understand root causes and agilely maneuver elemental convergence forces. A need to swiftly maneuver up surging attack surfaces. Anticipate multiple waves or else drown. Strategically capture the best ones by choice with adaptive work-energy efficiency. Artfully master dynamic tactical, operational and strategic environments, not fixed terrain. While only a simple figurative analogy, surfing the conflux envisions agility of action operating in accord with fluid, multidimensional environments often at the edge of predictability.

The convergence compels us think differently. Broader for better understanding why human behavior is foundational to any mission. Linear and step phased operations plans morph into a simultaneously intertwined PMESII webs necessitating kinetic and non-kinetic responses.

> During the Cold War, the US had one message with many movements to encourage the rejection of communism and influence other peoples and nations to choose capitalism. Now with the Cold War ended, the US is challenged by using cognitive maneuver in more ways than one. It has many more adversaries, is dealing with the increasing speed of human interactions, and facing opponents who are playing by new rules of engagement in the information space. To succeed in this environment, the US must up its cognitive game. Strategic plans must include strategies to influence beliefs, behaviors, and perceptions as well as kinetically coercing them or risk loss. After all, war is won by the "will" of the people (DeGennaro 2017).

Reconceptualization our conflict-competition-cooperation continuum breadth is also necessary: unpacking socio-cultural effects of the human activities to accurately encompass underpinning political, economic, social, environmental, psychological, technological and other relevant scientific fields. The Joint Concept for Human Aspects of Military Operations envisions this broadening.

> The Joint Force will enhance operations by impacting the *will* and influencing the decision making of relevant actors in the environment, shaping their behavior, both active and passive, in a manner that is consistent with U.S. objectives. Human aspects are the *interactions* among humans and between humans and the environment that influence decisions. To be effective at these interactions, the Joint Force must analyze and understand the social, cultural, physical, informational, and psychological elements that influence behavior. Actors perceive these elements over time, mindful of seasons and historical events, and with people having differing notions regarding the passage of time. Relevant actors include individuals, groups, and populations whose behavior has the potential to substantially help or hinder the success of a particular campaign, operation, or tactical action (JC-HAMO 2016).

Deeper from the perspective that defense and security challenges are more tangled, interdependent and complex. Our AORs are not just geographic and multi domain across air, land, sea, cyber and space, but span complexity science: individuals' micro motivations, collective networks of meso organizational behavior and finally the resulting sum of all such behaviors which shape and define macro nations and international structural environments. Of course, many of these scale activities feed-forward to amplify new and often unanticipated behavior and feedback that constrain or incentivize new emergent behavior.

> The actors, actions, and arenas of the emerging global security environment are changing. To navigate these murky waters, the United States requires effective statecraft that relies in equal measure on: (1) resurgent diplomatic application of national will, (2) information and technological overmatch including through artificial intelligence, (3) multi-use conventional and irregular warfighting capabilities, and (4) economic growth fostering national interests across domestic and international private–public partnerships. Such a comprehensive and nuanced approach is needed to achieve strategic success in diverse contested spaces, and address the complex political, economic, social, and ecological challenges that will face the nation (SMA Annual Conference 2019).

Convergence also forces us to embrace the accelerating and evolving nature of operating environments. We need to encode complex adaptive systems thinking and science into our constantly evolving doctrines. Humans are dynamic, strategic and adaptive beings who will continue to do more amazing as well as nefarious things enabled by technology. Several Joint Command efforts already reformulate our doctrinal depth.

> … individuals and groups today have access to more information than entire governments once possessed. They can swiftly organize and act on what they learn, sometimes leading to violent change. When applied to military systems, this diffusion of technology challenges competitive advantages long held by the United States. Our competitors and adversaries are using technology to offset or diminish the physical overmatch of the broad range of U.S. lethal capabilities (JCOIE 2018).

Smarter in the sense of explicitly recognizing cognitive biases, the heuristic short-cuts used by individuals for making sense and meaning of their world, which are already being leveraged by adversaries as both exploits and vulnerabilities.

> … there is growing evidence that US adversaries have recognized the deficiency in US cognitive capabilities and have pursued ways to exploit it to their advantage via gray zone and other technologically-focused tactics … Notably, this was demonstrated by ISIS in the social media campaign that took many in the US defense establishment by surprise; North Korean and Iranian apparent discounting of the credibility of US deterrent threats; Russian activities in Crimea; and Chinese "island building" in the South China Sea (SMA 2017).

Leveraging the benefits of biases, we also need to be smarter in our change processes to achieve what needs to be done. Irreplaceable expert knowledge, wisdom and know-how resides throughout the enterprise. And it is a transformational mission to organize the know-how and know-who, into the enterprise know-what and know-where to deliver across value chains on a national security production function. Learning the tech sector's form-function-feeling approach provides for more organic technology and culture transformation at the convergence.

> Culture = policy + people + process. Once each of these areas is defined, if there are issues, they can be addressed. Once the policy is laid out, the key is to determine the processes needed to enable fulfillment of the policy in order to deliver the desired product or outcome. The people, for example various types of expertise or skill levels, are then assigned to implement a part of the process in line with the policy … In doing so, a culture is created (Haugland 2019).

Sharper from the perspective that most of our tactics, techniques and procedures can benefit from or be disrupted by technology augmentation. There are obvious possible scenarios where lethal autonomous AI weapons simultaneously swarming with "social media, digital and other informational tools will alter, confound, and manipulate facts toward an engineered version of events" (SMA 2019). And there are equally obvious opportunities for greater cognitive shaping, influencing and deterring tools across the conflict-competition-cooperation spectrum.

- Emerging science and technology will continue to disrupt customary characteristics of political and kinetic conflicts among states and non-state actors.
- We cannot reliably predict whether or not human roles will rapidly give way to a more dominant robotic style of war, so we must prepare for a variety of futures, per the Scharre/Horowitz autonomy typologies.
- Humans involved with machines that operate at vastly greater speeds and volumes of data will further create problems of cognitive demand for the human soldier that need to be examined.
- We must investigate this not only in terms of technical performance, but also from a more holistic perspective, to include the social, political, and psychological dimensions of the soldier and of the citizen (SMA 2019).

Sharper also informs our need to capture integration opportunities, so that we can "convert potential challenges into opportunities through foresight, assumption of risk, and initiative" (TRADOC 2019). We need to continue our efforts owning the high ground in computational approaches that fuse different data, theories and models into specific, operationally useful and engineered products.

> Due to the literal complexity of complex adaptive systems, computational tools are inextricably intertwined with their analysis and are able to encode knowledge for more effective coordination across the USG. An improved analytic framework combined with an understanding of existing and emerging artificial intelligence (AI), and a subset of computational tools, allows for an assessment of how computation fits into both the Joint Intelligence Process and Joint Planning Process to support integrated campaigns.
>
> This symbiosis of emerging technology into the everyday processes of military affairs has been the historic catalyst for revolutions in military affairs (Boot 2006), and this integration of technology and perspective through Integrated Campaigning has the potential to not only revolutionize the conduct of military affairs but the whole of foreign policy (Pike 2020).

Much like automated weather forecasting and navigation systems, we need to sharply integrate our broader, deeper and smarter capabilities together to adroitly anticipate and navigate human behavior with the current ease of weather forecasting and geophysical terrain navigation across time, space and distance.

> The ability to provide a Common Operating Picture (COP) from private to president, simultaneously, in real time and the level of insight provided, along with the ability to forecast and predict varied activities, etc. will drive a thirst for covering the globe in such foundational information (Haugland 2019).

The good news is we know how to do what needs to be done by exercising broader, deeper, smarter and sharper arbitrage opportunities. The bad news is we need to do it faster, better and cheaper than peer competitors due in part to technology proliferation and AI. Encompassing these principles, the mission of DOD's Joint Artificial Intelligence Center (JIAC) is to enable transformation across the enterprise disruptive technologies. They also warn us of the urgency and impending gravity of consequences if we do not.

> The costs of not implementing this strategy are clear. Failure to adopt AI will result in legacy systems irrelevant to the defense of our people, eroding cohesion among allies and partners, reduced access to markets that will contribute to a decline in our prosperity and standard of living, and growing challenges to societies that have been built upon individual freedoms (JAIC 2018).

Technology the Enabler

Below are some foundational trends and perhaps insights for where we are going across plausible alternative futures. Obviously, these enabling trends translate into speed, power and capabilities for new shaping, deterring and influencing activities.

1. More computing power in today's smartphones than all of NASA in 1969
2. Computational power grows exponentially
3. 3.5 Billion people have smart phones
4. 60% of the world's population is connected through the internet
5. There are 26 billion devices already connected in the Internet of Things (IOT)
6. More data has been generated in the last few years than all of human history combined

Physicist Micho Kaku remarked there is more computing power in a smartphone today than all of NASA combined had in 1969 to send man to the moon (Kaku 2011). Handheld devices can navigate all the complexities of space travel including trajectories, orbits and keeping astronauts alive. We can easily understand why computational power has increased exponentially while costs have simultaneously plummeted through engineering ingenuity and economies of scale. But what are we doing with all this computational power? As individuals, we have more apps that facilitate daily work productivity and satisfaction in our lives. As organizations, we have enterprise wide software to keep track of specialized information that fuses invaluable human know-how into precise, scheduling know-where processes to deliver on complex, know-what production functions.

Most understand Moore's law of transistor doubling so that over the last 50 years our computing power has increased 3500 times. Yet few comprehend the equal acceleration of algorithmic efficiency, which decreases the amount of computation by half every 16 months. Multiplying doubling hardware speed times with doubling algorithmic efficiency gains equals even more exponential capabilities (Hernandez and Brown 2020).

Given this new power, we have moved to AI based approaches, that can operate even better, faster and at scale. Some AI achieves remarkable results based upon unsupervised, human out-of-the-loop learning, and computational advances. AI researchers' technological singularity hypothesis, popularized by Ray Kurzweil, posits how increasing power coupled with decreasing costs could radically change our world. Just like Monte Carlo methods of computational power led to the Manhattan Project's success before our adversaries, what will emerge as the next weapon of mass destruction from all this computing power? Alternatively, what can be mankind's next moonshot for hope and progress?

Over 3.5 billion people have smart phones today (Stastica 2020a, b) while almost 60% of people in the world have some sort of access to the internet. Given crowd sourcing and multi-party computation, we can donate our smart phones to search for cures to cancer or mine bitcoins. Yet 74% of people in the developing world still do not have smart phones (Statistica 2020a, b). This proliferation of computational power is redistributing equities which creates new 'haves and have nots' in addition to vulnerabilities and exploits. Every individual who can access the internet has the potential to become a cyber shaping target or an attack vector source as our online and physical worlds continue to merge and co-evolve.

Today there are 26 billion devices connected to the internet (Stastica 2020a, b). This is over three times more connected things than people and probably a lot more than the original ARPANET inventors imagined in 1966. Each thing is a sensor, an object, a big data input for creating higher resolution and real time maps of systems from the IOT: smart watches, smart homes and smart cities scale expanding capabilities. Obviously, IOT exponentially increases possible command and control surfaces across physical, informational and human domains for C4I.

More data has been generated in the last few years than all of previous human history combined (IBM 2020). Widely known as the 'big data tsunami', information abundance means many answers to questions are often only a web search away. Big data is often characterized according to four Vs—volume, variety, velocity and veracity. Volume measures the scale of big data with almost 300 billion emails and 5 billion internet searches performed each day. In 2020, this translates into approximately 40 trillion gigabytes of data alone (IDC 2020). Variety measures the span forms of data with both structured data values, KPIs and metrics and unstructured voice, video and sentiment all interconnected in the cloud. Velocity measures the speed of data, with over 500 h of YouTube videos uploaded and 500,000 tweets made per minute (IDC 2020), and on average, every person on the planet is generating 1.7 megabytes of data every second (IDC 2020)—equivalent to 100 pages of text which is more than Sun Tzu's *Art of War*. Veracity measures the inherent value, quality and uncertainty surrounding data, focusing on signal to noise ratios and what insights can we infer for better decisions. Poor information and bad data quality costs the US economy over $3 trillion per year (Redman 2016). Whether the data is meaningful or just gross information pollution depends.

Disruptive Realities

Jim Barksdale, former CEO of AT&T Wireless and Netscape, is credited saying "*If there is data, bring it and we will discuss. If all we have are opinions, we are going with my own.*" (Barksdale 2015) Yet how can we learn to trust data to make better, evidence-based decisions than intuition alone, just like we rely on GPS and smart phones for navigating physical terrain?

On one extreme, there is the story of the Soviet satellite watch officer whose system indicated a US missile launch, yet he chose not to act according to strict protocols. Other sensors and systems did not corroborate information veracity and fortunately his very human expertise and intuition intervened. On the other extreme, there is the private sector mantra describing the data analytics value chain as 'M5' measure, monitor, model, manage and monetize: "If I can measure something I can monitor it, if I can monitor it I can model it, if I can model it I can manage it, and if I can manage it I can monetize it." Obviously the answer lays somewhere in between as we all use GPS navigation to augment driving and yet we are still late for arrival.

Technology re-imagines business models for monetization. The M5 arbitrage opportunities from big data combined with AI and predictive analytics abound while how, where and when we decide to responsibly engage with technology still remains a human choice. Often, we never pay for any of their products or services, happily using free services in apparent win–win service agreements in order to feed into a self-fulfilling 'better customer experience.' Yet they are multi-billion-dollar companies as we and the data are their products.

Technology adoption is facilitated through the seemingly diametrically opposed lens of human emotion. Marketing departments fully leverage anthropomorphism. Devices are giving human names, friendly faces and stylized human–computer interfaces where we can meaningfully connect. They match our mental models and mirror our human perceptions to satisfy our emotions, regardless of the quantum physics that makes them work. *Alexa, Siri* and *Nest* are very carefully thought out to be tech friendly and sold not for the price of the product but for their M5 value chain revenues.

This form-function-feeling approach is the current recipe for technology product success: plug and play task automation driven by scientific engineering, that performs some necessary function which makes us feel good, wrapped with an aesthetically pleasing design served in a 'no brainer' business model.

Technology also births the unanticipated emergence of new market segments. It also enables new influencers that disrupt market dynamics. From humble beginnings, social media has become its own self-reinforcing marketplace of human activities. Social media is not popular solely because it functionally allows us to communicate with more people, but also for how it makes us feel as human beings: sharing our own stories, satisfying deep emotional needs and shaping desires. Strategic narratives, fake news, AI bots and networked illicit actors all operate and cognitively compete in the convergence. Hyper competitive and saturated information environments explicitly target individuals' beliefs, biases and echo chambers to create increasing operational uncertainty.

Broad

We need a broader reconceptualization of the conflict-competition-cooperation continuum. Problem sets facing national defense and security include interconnected, uncertain and complex events across all scales of human behavior Our PMESII optics may or may not be encompassing enough given the conflux.

We need to continue decomposing PMESII domains and socio-cultural human activities to encompass their underlying social science fields from political science, economics, sociology, psychology, anthropology, as well as many others with our physical, environmental and technology sciences. Transdisciplinary or multidisciplinary work focuses on root cause, symptom and solution analysis. Many often start with a simple, yet focused BOGSAT accounting exercise of putting smart people together in a room. In the private sector, cross functional teaming exercises a data or information arbitrage opportunity, connecting the dots and information sets together. Each approach offers valuable insights, experience and evidence that we can leverage into mission success.

Plans and courses of action (COAs) can often become dead on arrival (DOA) as crises morph into an unexpected situations: kinetic actions spawn first order action-reaction political consequences with second order, longer term economic and propagating, higher order social effects. Recent Iranian attacks on Saudi oil production serve as a reminder. As a result, analysis and human decision making is often then overcome by events (OBE) with accelerating developments only inducing increasing uncertainty which necessitates new analysis and decision making for conflict off ramps or cooperation on ramps.

Many Cold War lenses and monolithic strategies for contemporary great power competition are both DOA and OBE now. This is due in large part as the global system has evolved from a relatively disconnected Westphalian system with isolated and ordered elements, towards a massively interconnected, interdependent and more fluid human terrain at the convergence. Our current arbitrage opportunity is to create zoomable optics on our problem sets to master the conflux. Zoom out for better contextual understanding why human behavior is foundational to any mission. Zoom in for addressing root causes as well as associated symptoms for any specific solution.

The Conflux of Connectivity, Complexity and Uncertainty

Economic magnitudes and interdependencies foundational to peace, security and prosperity doctrines have shifted tremendously. Since the collapse of the Soviet

Union, world trade has increased from $3 trillion to over $20 trillion in 2019 (WTO 2020). Enabled by technology proliferation, these impact lives and livelihoods as well as operating environments of choices and constraints across the conflict-cooperation continuum.

Tracking increasing connectivity, each year the World Economic Forum publishes their top global risks (WEF 2019). These risk-trends maps in Fig. 1 provide a PMESII like situational analysis as economists and the business sector use a similar framework called SEEPT—social, economic, environmental, political and technological.

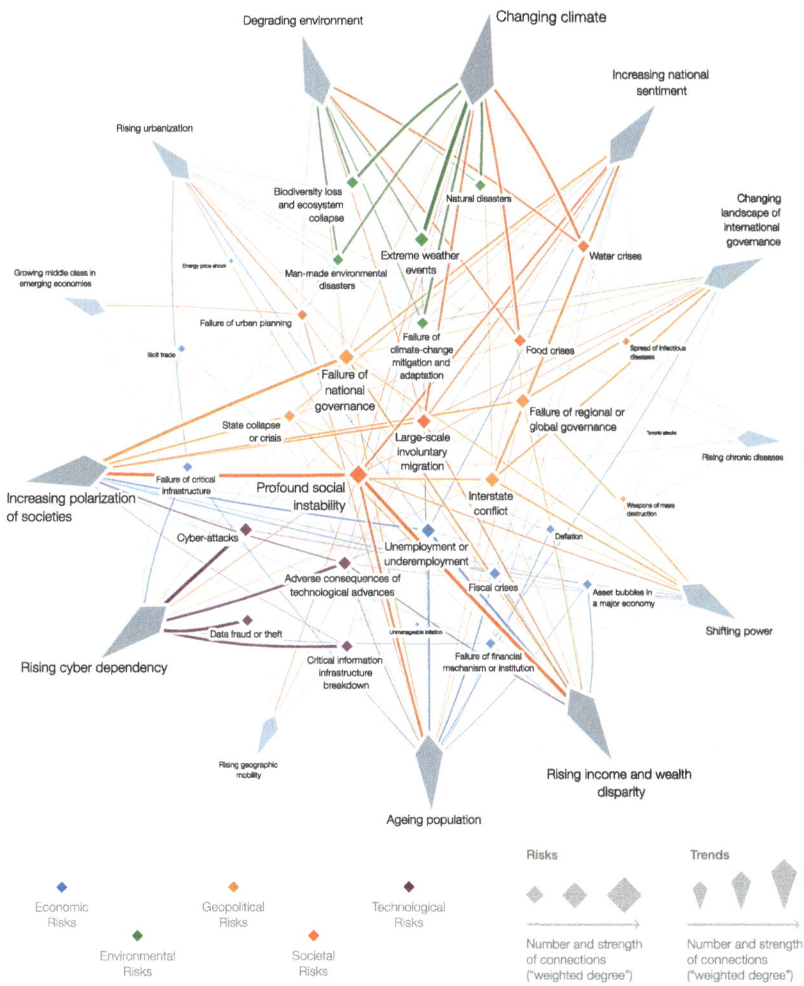

Fig. 1 Risk-Trends Interconnection Map (World Economic Forum Global Risks 2019)

Regardless of frameworks, risk dependency underscores operating environments. It also illuminates trend inter-connectivity emergence.

Risks often span multiple domains, such as violent extremist movements encompassing economics, sociology, political science and cultural studies. This is one important reason why calls for transdisciplinary approaches such as political economy, geo-economics and socio-cultural behavior gain popularity and analytical traction. Uncertainty explicitly shapes and limits cognitive maneuvering and operational action-choice sets. National security operators, analysts and decision makers face the undercurrents of connectivity, complexity and uncertainty challenges daily. Similar to social science researchers, fundamental questions boil down to answering "who does what to whom, how, where, when and ultimately, why" in operating environments.

Arab Spring Case

The Arab Spring illustrates exactly how and why social science data, theories and methods need to be integrated to overcome increasing connectivity, complexity and uncertainty of human behavior. In late 2010, what began as isolated and local protests against Tunisian government repression and poor economic conditions quickly escalated into widespread social movements. These swiftly blossomed regionally, rallying against government autocracy and subsequently many direct attempts at both peaceful and violent regime change occurred throughout the Middle East. All this presented shaping, influencing and deterring challenges, whether met or not, as well as opportunities, whether recognized or not.

Detailed post hoc analysis why the Arab Spring occurred has been performed by thousands of researchers, analysts and experts worldwide. An international relations expert might say that the Arab Spring was the result of increasing parity and dissatisfaction among the government and opposition groups supported by competing interests of various regional and global actors. A comparative political scientist might say ethnic and religious group rivalries coupled with economic inequality drove competition among winning coalitions for and against authoritarian transitions and modernization. A game theorist might say dictatorial regimes responded with mixed repression-liberalization strategies to protestors' coordination behaviors by solving collective action sub-games. A sociologist might say angry, dissatisfied activist networks compressed political spaces which empowered the rise of bolder, more trusted social middle-class movements that broke the wall of fear for change. A political demographer might say that the large, under employed male cohort and youth bulges provided obvious structural political vulnerabilities that government policymakers could not adjust to fast enough by providing public goods and services in individuals' conflict-cooperation choice sets. These are but a few of many excellent social science explanations.

Although none of the above explanations are wrong, not one is right. They all are differing lenses on the same tapestry of historical events. As such, each explanation offers one piece to solve a larger puzzle. Each social science optic, when applied robustly with evidence, sharpens only a portion. Together through integration, they can potentially focus our understanding across complex cases.

Zoomable Optics

We cannot afford any fixed optic on our problem sets anymore: going either wide angle or high-power telephoto regardless of individuals' personal expertise, experience, biases or equities in the room. Gibson's law from American public relations wittily offers 'for every PhD, there is an equal and exactly opposite PhD." They are all important pieces which may or may not solve a particular mission puzzle. We must develop easily zoomable wide angles to capture context, trends and important interdependencies while adroitly zooming into actionable, operational details involved to achieve desired objectives. As seen in Fig. 2, work across research, academia, the private sector and USG communities already integrate our zoomable PMESII plus environmental and technological (+ET) lenses. Human, socio, cultural and behavioral (HSCB) efforts explicitly address root causes, symptoms and solutions throughout security, stability and prosperity value chains.

Today integrated multi domain operations (MDOs) capitalize on shaping, deterring and influencing activities across operational plan phasing. Information environments are a critical conduit to managing beliefs, perceptions and will in root causes as well as expressed symptoms. Creating opportunities and prosperity through tangible and realizable choices for individuals, populations and societies contemplating choices along the conflict-cooperation continuum. Arbitrage opportunities abound across social science.

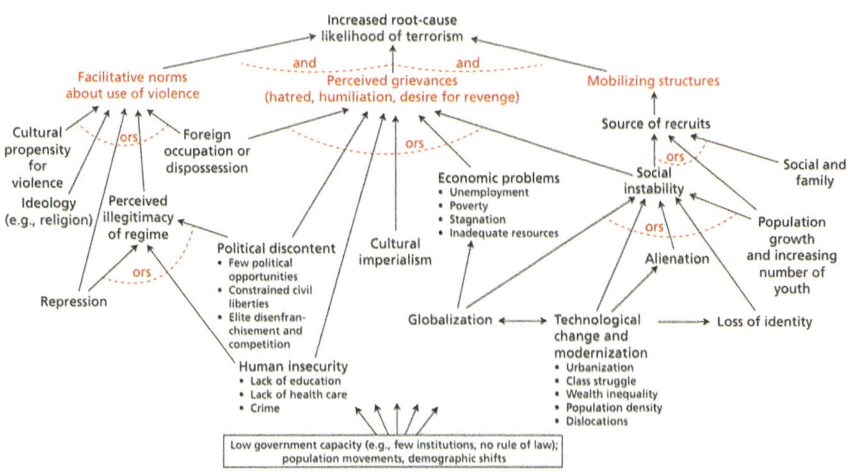

Fig. 2 Root Cause Synthesis (RAND 2018a, 2018b and Norricks 2009)

Deep

Convergence forces us to deeply embrace complex and evolving natures of operating environments. Conflict is inherently about uncertainty, as the payoffs of peace are less than the gamble of war on the conflict-competition-cooperation continuum. Deterrence on the other hand is more about certainty of choices, limiting actions given the likelihood of consequences. This is why notions of force projection capabilities in the physical realm struggles in today's cyber domain. The undercurrent of uncertainty pervades most operational, choice-action sets and is a overwhelming many cognitive information environments.

Humans beings possess agency: the ability to make decisions to solve problems and create opportunities through goal directed behavior. This agency spans individual actors, entities, groups, nation states and the international system itself. Agency and agents operate their own respective production functions to meet desired objectives. Information environments explicitly recognize these means across units of human analysis. This occurs every day on the planet, across all scale activities, with lessons learned through feedback loops that incentivize or constrain choices as well as feed-forward loops that create or deny spaces of possibilities for human action. Convergence pushes doctrinal depth even further, necessitating our frames are never static but must continually evolve and adapt in the face of new uncertainities.

Complex Adaptive Systems

Complex Adaptive Systems (CAS) is a scientific framework for understanding some common and universally connected, uncertain and complex phenomena (Mitchell 2009). Figure 2 demonstrates some principal elements of a complex adaptive system relevant for great power competition operating environments. Here complexity encompasses many elements, dynamically interacting across a number of scales which self-organize into hierarchical orders that produce emergent behavior across multiple domains (Middleton-Kelley 2003).

Complex behavior arises from relationships, interaction and connectivity of elements within a system and between a system and its environment (Kauffman 2000). In human, social, cultural and behavioral systems, connectivity and interdependence means that a decision or action by any individual, group or entity may impact related individuals and/or systems (Holland 1998). Effects are often non-linear and contingent, which might vary with the state of each related individual and system at any particular time (Strogatz 1994, 2013). States of any individual unit, collective element or a system includes its past history and context, including emotional sentiment, stability and whether the unit is near or far from equilibrium (Epstein and Axtel 1996; Axelrod and Cohen 2000).

The first step to decoding complex systems is unit and behavioral classification across levels or scales of activity as seen in Fig. 3. In operating environments, it is

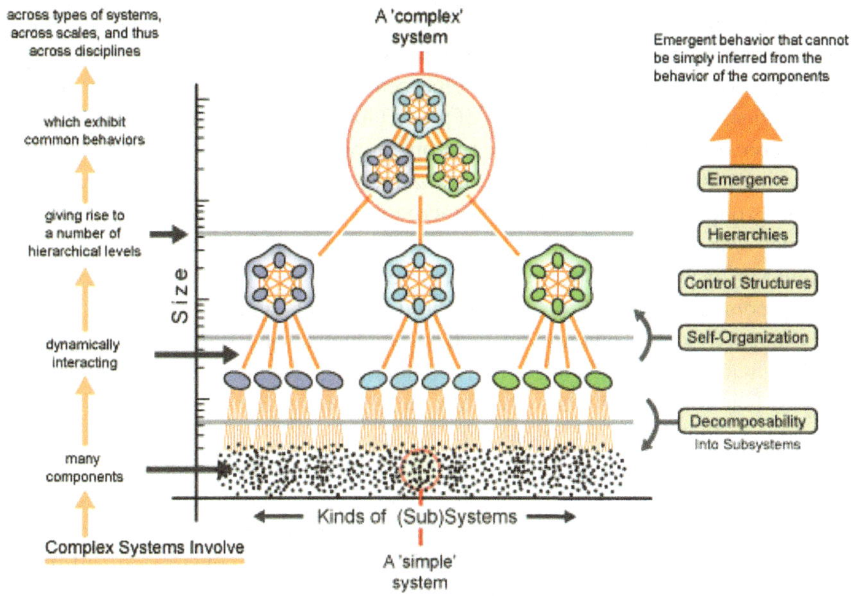

Fig. 3 Some Properties of Complex Adaptive System (New England Institute of Complexity 2011)

important to understand 'who is doing what with or to whom, how, where, when and why'—what micro individual actors do in the context of meso social networks of others reacting and interacting, that create emergent behavioral events that shape and define macro PMESII + ET environments and human behavior.

As humans are strategic adaptive beings, behavior by design will change and evolve. New orders and coherence are created by interactivity and adaptive thinking (Nicolis and Prigogine 1989). Historically warfare has also evolved from eighteenth century land armies, nineteenth century navies, twentieth century air power and now twenty-first century convergence. Novel, behavior or properties can appear when a number of individual entities interact to form more complex behaviors with new patterns, decisions or structural orders. Such new actions provide benefits to some while imposing costs on others and can realize unknown unknowns that lurk in CAS frameworks.

Subsequently, individuals' actions are constrained and incentivized by perceived choice sets shaped by their respective PMESII + ET environments, feedback loops and path dependence (Arthur 1990 and 2002, Axelrod 1999). Sometimes co-evolution occurs where individuals learn new behavior via mimicry or reciprocity and strategically adapt to each other and their environments (Ehrlich and Raven 1964). Arms races, evolving tactics and counter tactics are all foundational to operating environments. Whether negotiating peace agreements in Afghanistan, El Salvador, Cambodia or the Philippines, offering tangible upside choices of prosperity in PMESII + ET environments can help offset the downside risks of individuals' conflict choices.

In both physical and social systems, increasing uncertainty and seemingly complex, stochastic behavior is often generated by some fundamental yet simple processes. Agency, self-organization, emergence, self-similarity, path dependence, catastrophes, co-evolution, the limits of predictability and strategic adaptation are all critical concepts and real phenomena in operating environments. Enabled by technology, increasing connectivity and information flows cause new complex behaviors and groups never anticipated to emerge. Network typologies, as well as their dynamics, can be instrumental (Watts and Strogatz 1998). General Petraeus and General McChrystal's perspectives on insurgency and networked adversaries shifted the way counterinsurgency operates to produce results and save warfighters lives. Unfortunately, VEOs asymmetric tactics, techniques and procedures evolved in response, with the high cost of civilian collateral damage fundamental to their uncertainty of fear conflict calculus. They also enlist competitive information narratives tailored to individuals, populations and countries specific cognitive contexts.

While we intuitively understand complex adaptive systems, we must master it for actionability. Our post WWII leadership created peace through deterrent strength, alliance security organizations and international institutions to facilitate prosperity which created spaces of new choices, which subsequently constrained and incentivized individual and group behavior to be more aligned with prosperity interests. Leaders understand the tradeoff between the supply of economic growth and the demand of political security. This is why sustainable growth is the key buzzword for politicians globally. And why increasing uncertainty is now a new operational cognitive weapon.

If entities and organizations can deliver on the promise of things getting better, ever expanding upside opportunities, who would want to change that vision of the status quo? In fragile states, increased government service delivery and rising market expectations can change actors' conflict-competition-cooperation choice sets from protesting to participation. Macro PMESII + ET environments constrain and incentivize human behavior, not just as individuals, but as networked social groups. What can this tell us about countering adversarial narratives with our own narratives backed by real choices and exercisable opportunities?

Complexity theory does not argue for ever-increasing interconnectivity between related systems or entities. Coherence and order can change over time. Sharp, unexpected catastrophes or transitions from order to disorder and back again can occur (Mitchell 2009). This was explicitly recognized almost a decade ago in the 2011 National Strategic Narrative where complex adaptive systems thinking provide some key insights to understanding the convergence.

Tipping points force disorders and precipitate disintegration in social systems with limited carry capacities, slow institutional response times or inelastic adaptation abilities (Gleick 1987, Waldorp 1992, Çambel 1993). The Black Plague wiped out up to 60% of Europe in the Middle Ages. The American Great Depression led to a 15% decrease in worldwide GDP. Syrian forced migration from civil war or Sudanese climate change all are examples of marked system disorders induced by exogenous shocks and slow adaptation. What sharp V, slow U or structural break L shock-response shapes will COVID-19 propagate across PMESII + ET domains remains to be seen.

Encoding a CAS Doctrine

Through a complex adaptive systems perspective, we can clearly see the need for integrating multiple domains and scale activities to recognize complex behaviors. In the Arab Spring case, Tareq Benaziz's individual experience, emotions and actions were literally the unfortunate spark of greater societal discontent and political activism region wide with globalization repercussions. Many forward leaning military thinkers throughout the community recognize both complexity and constancy of system change (RAND 2019, Lawson 2019).

Channeling Darwin's inspiration among many others, it is not the strongest nor most intelligent species that survives, it is the one most responsive to change. Instead of shying away from complexity's difficulties we need to embrace its totality. Instead of fearing uncertainty, we need to shape it. Instead of finding optimal solutions, we need to add adaptation to operational, tactical and strategic portfolios.

Certain trends, terrains and domains can only be resolved so far ahead with the kind of certainty we desire. Forecasting weather is a chaotic, complex phenomena, yet given high resolution big data, IOT connectivity of weather stations and high-performance computing models, we can accurately forecast out to over a week with known confidence. In understanding chaos, while there are limits to predictability, there are also horizons of useful predictability. For understanding human behavior, we need to move towards organizing and preparing against a range of plausible alternative futures, not just what we possibly believe or probably can estimate. Advances in robust, adaptive decision-making hedges our efforts across deep uncertainty in many spaces and domains (Marchau et al. 2019). This applies equally for our mosaic, distributed warfare and multi domain objective kinetic activities (DARPA Mosaic 2018) as well as to maneuvering the conflict-competition-cooperation continuum.

Smart

Science illuminates how and why individuals make sense and meaning of their world. Cognitive bias are simply heuristic shortcuts we use to navigate our physical and social worlds. As cognitive biases are filters, that subjectively interpret our individual as well as social experiences differently than others, they have important consequences for shaping and influencing actions. These are the many predictable errors in judgement we all can make that generally fall into four categories of 'too much information, not enough meaning, need to act fast and what we remember.' Arbitrage of lessons learned from these behavioral sciences obviously provide actional insights for information operations. Cognitive biases provide exploits for shaping and influencing adversarial beliefs, perceptions and will as well as vulnerabilities for our own.

As crises at the convergence become more complex, frequent and uncertain, the challenges for humans to cognitively cope and develop sufficient responses leaves national policy goals vulnerable. Cognitive biases make us very human, in what we believe, feel and do. Cognitive biases are both errors of fallible judgement as well as deep human intuition and heuristic rules empowering our successful survival. As well as projecting them outwards on others, we also need to reflect inward on how they impact our ourselves and organizations. While these heuristic shortcuts helped our evolutionary survival in the physical environment, today's connectivity, complexity and uncertainty in technology enabled operating environments greatly exposes both their benefits and limits.

Cognitive biases are also key to human progress. Modern societies possess a wealth of knowledge as individual pieces are distributed among various individuals and groups who functionally specialize. The power of economics is that this diversity of know-how, generated specifically with cognitive bias, can be recombined in different ways than any individual imagined, to address new challenges and opportunities (Hausman et al. 2014). This is especially evident in organizational behavior where individual scales of human agency are networked into meso, group scale production functions through collective action. Managerial economics harnesses specialized, individual know-how into productive know-what for any organization. Understanding our institutional capabilities, organizational bias, momentum and new knowledge scalability is critical in the conflux.

Cognitive Biases

Under complex and uncertain circumstances humans are more susceptible to make predictable errors in judgment caused by cognitive biases (HDCDTF US Army 2015). Behavioral scientists have identified hundreds of human cognitive biases that filter how we might perceive and process information and how we perform decision making. Figure 4 shows some of these biases, ranging from the well-known group think, rosy retrospection and confirmation biases to the less recognized gamblers fallacy, loss aversion and to the more humorous Ikea effect.

Too Much Information

- We notice things already primed in memory or repeated often
- We notice when something has changed
- We are drawn to details that confirm our own existing beliefs

Not Enough Meaning

- We tend to find stories and patterns even when looking at sparse data

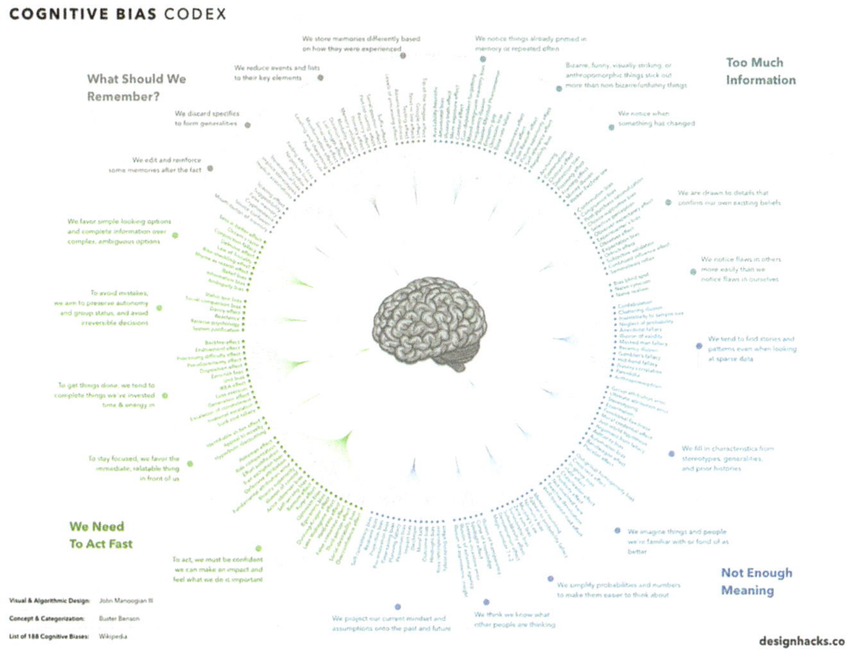

Fig. 4 Cognitive Bias Codex (Wikipedia)

- We fill in characteristics from stereotypes, generalities and prior histories
- We simplify probabilities and numbers to make them easier to think about
- We project our current mindset and assumptions onto the past and future

What We Remember

- We discard specifics to form generalities
- We reduce events and lists to their key elements
- We store memories differently based on how they were experience

Need to Act Fast

- We favor simple-looking options and complete information over complex, ambiguous options
- To avoid mistakes, we aim to preserve autonomy and group status to avoid irreversible decisions
- To get things done, we tend to complete things we have invested in

Across operating environments, each bias is a potential exploit and attack vector and conversely, a vulnerability and attack surface. Fully mining every one is a simple knowledge arbitrage opportunity. They provide the foundations for strategic narratives, fake news propagation, conspiracy theories, and intra-group divisiveness by leveraging anxieties, uncertainty, knowledge gaps and other filtered, individual subjective and group intersubjective realities. They also provide sources of herd immunity. Explicitly recognizing these throughout operational, tactical and strategic spaces is necessary to survive the convergence.

Leveraging Biases for Transformational Change

Cognitive biases also allow us to survive and more importantly, thrive evolutionarily. Biases from too much information permit us to operate in saturated information environments, identifying and digesting the signals that matter the most. Not enough meaning biases forces us beyond our neurological cognitive limits to make inferences about experience. What we remember biases makes sense of lessons learned. Need to act fast biases help us overcome 'paralysis by analysis' for timely agency. While each is an explicit part of our individual and organizational production functions, when harnessed together appropriately, they can help facilitate transformational change.

Expert intuition is the instantaneous, on-demand heuristic lessons individuals have learned by cranking our own observe-orient-decide-act (OODA) loops thousands of times, not only dogfighting at 35,000 feet but operating 5 feet in front of us daily (Gladwell 2008). The lessons learned from failing early and often, savoring success, fast and slow thinking all sharpen up individual production functions. Whether described as 'feel' nuances, 'it depends' subtleties, or 'context' understanding, this is priceless human know-how we cannot afford to lose or ignore.

However, intuition is necessary but not sufficient for addressing the convergence. Unfortunately the velocity and variety of convergence happens faster than any one individual can comprehend. So as individuals, we are left with our default operating assumptions, biases, intuition and experiences when we need to act faster, with more to remember, in the presence of information overload, without enough meaning. Whether we understand quantum mechanics, believe in 'spooky action at a distance', or have quantum field theory dreams, our operating environments are enabled by things often beyond common sense.

Individual, neurological cognitive capabilities are not that much more advanced than early Neolithic man 10,000 years ago. With the advent of written language however, human beings harnessed collective learning (Cristian 2018)—the ability to transfer information across time, space and each other, not condemned to reinventing or rediscovering knowledge individually each time. Proliferation of knowledge across individuals created new orders and novel emergent behavior for harnessing even more sophisticated achievements through coordinated social organization.

"The secret to modernity is that we collectively use large volumes of knowledge, while each one of us holds only a few bits of it. Society functions because its members form webs that allow them to specialize and share their knowledge with others" (Hausman et al. 2014). These are the sophisticated production functions and integrated value chains that determine of current operating environments. But they rest firmly on individual know-how foundations. Daily we happily augment many of our physical activities with technology while the convergence now empowers even more human augmentation possibilities through virtual, augmented and mediated reality. So the management question remains, fusing know-how and know-who, with know-where and know-what together, to focus down to solutions.

Transformation is difficult, not impossible. The DARPA 'special forces' process model for innovation has been employed with great effect across the enterprise. But we still need to embrace the very human elements and age-old management representations of equities under varying degrees of experience, knowledge, information sharing, bureaucratic constraints and professional aspirations.

As operating environments become more complex, enterprise value chains and production functions must become even more sophisticated to continually provide relative value, often in new areas. Expanding definitions of kinetic and non-kinetic MDO, through coordinated whole of government approaches, already capitalizes on valuable lessons learned. Learning the tech sector's form-function-feeling approach, both with individuals and as collective organizations, provides faster and better organic adoption to scale necessary transformation.

Sharp

The need for sharpening our efforts in the convergence is attainable given our enterprise production functions and past success. It is an integration arbitrage opportunity, based upon individual expertise, organizational knowledge, MDO capabilities, systems and engineering prowess. Whether evolutions or revolutions in military affairs, technology combined with human operational, tactical and strategic art are key enablers.

Sharper also informs our need to capture integration opportunities at scale. Most of our tactics, techniques and procedures can benefit from or be disrupted by technology augmentation. We already possess great capabilities of collective know how and know what across the enterprise. Yet wisdom and capabilities can remain siloed with M5 value still left sitting on the table. And there are equally obvious opportunities for more advanced, specialized tools for greater COA shaping, influencing and deterring power and precision.

Many efforts are already capturing these through computational approaches fusing different data sets, theories and models into operationally useful products. Much like automated weather and navigation systems, we need to adroitly anticipate and navigate human behavior across cognitive PMESII + ET domains as well as physical time, space and distance.

From Magellan Maps to Operational Pictures

A simple yet powerful analogy is that we need to move from Magellan's static, sparse and coarse map of the sixteenth century world then to today's real time, dynamic navigation systems for surfing the conflux. Ancient maps are geographically incomplete, with low dimensional representations and even lower resolution actionability. Yet they are the hard-won cumulative bounty of countless explorers. The type of answers that this map provides is to strategically sail west with increasingly vague operational guidance on how and where to tactically succeed. Of course, Magellan's circumnavigation generated more detailed geographical knowledge, created new maps, and inspired countless others based upon his discoveries, stories and journey.

Individual's experiences, worldviews and mental models all create a unique and invaluable map of expertise. This know-how is not limited to the geophysical environment, but across cognitive and PMESII + ET domains. So we leverage Delphic processes, survey communities and crowdsourcing technologies to paint a better operational space. Today's common operational pictures are driven by real time data, sensor and model fusion across broader, deeper and smarter perspectives. Here we currently own the high ground of integrating maps and mosaics given leadership strength and diversity of collaborative academia, research, the private sector and USG communities.

Computational Social Science

How can we push current efforts even further? Advances in computational social science propel our understanding across PMESII + ET domain activities and levels of human activity. Work throughout the community leverages computational social science, by crowdsourcing SMEs invaluable knowledge and fusing big data, AI and predictive analytics with social science theories, methods and models.

The M5 integration opportunity is to federate information, data, models and knowledge together as we already can see the broader, deeper and smarter frameworks to put them into. Operational engineering product end states call for higher resolution and actionable cognitive navigation, avoiding threat vectors and creating opportunity spaces equally. However, applying social science theories in a complex adaptive systems present unique challenges, pushing the limits of scientific explanation and prediction even further. Recently RAND summed up community efforts to embrace such.

1. *Complex adaptive systems*. Social systems are complex adaptive systems (CAS) that need to be modeled and analyzed accordingly—not with naive efforts to achieve accurate and narrow predictions, but to achieve broader understanding, recognition of patterns and phases, limited forms of prediction, and results shown as a function of context and other assumptions. Great advances are needed in

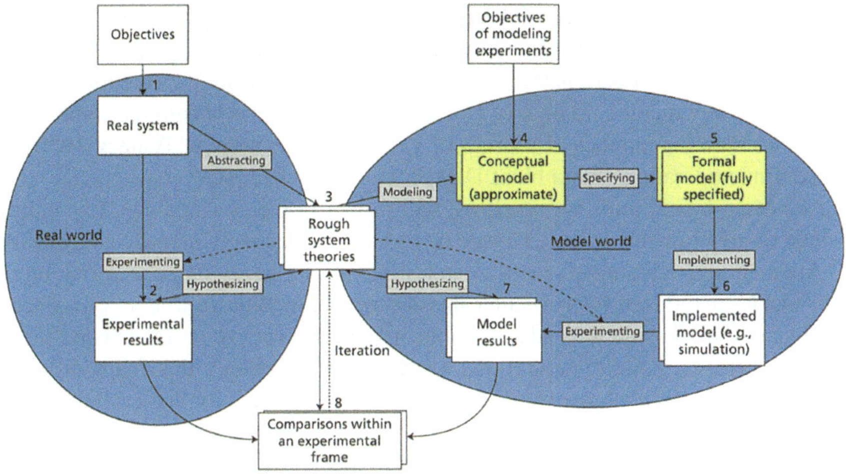

Fig. 5 An Idealized System View of Theory, Modeling and Experimentation (RAND 2019)

understanding the states of complex adaptive systems and their phase spaces and in recognizing both instabilities and opportunities for influence.

2. *Wicked problems.* Many social-behavioral issues arise in the form of so-called wicked problems—i.e., problems with no a priori solutions and with stakeholders that do not have stable objective functions. Solutions, if they are found at all, emerge from human interactions.
3. *Structural dynamics.* The very nature of social systems is often structurally dynamic in that structure changes may emerge after interactions and events. This complicates modeling (RAND 2019).

Despite complex, wicked and dynamic natures, these are hard but not necessarily impossible challenges. We already operate in highly uncertain environments. Figure 5 shows a synthesized system view of theory, modeling and experimentation underlying our integration efforts. Systems are only chaotic at specific times, states and scales of behavior. If we can diagnose and recognize such, prediction and prescription at the other times is both possible and necessary.

Whether navigating physical terrains or cognitive domains, maneuvering relies upon dynamic resource allocation mathematics and engineering: first identifying the right data, theories, models, materials, sensors, tools, second integrating them together to anticipate mapping inputs to expected outputs through any data generating process (DGP) or $f(x)$ in any domain, and then finally optimizing strategies subject to the constraints to solve the problem. We must surf the conflux dynamically, performing continuous M5 navigation on the fly.

Although complicated, any integration effort is exactly about overcoming challenges—automate the difficult, engineer the complex, change what matters and stay focused while doing it. Complexity is not dictated by engineers but necessitated by specifications and design requirements. On one hand, building a high resolution,

PMESII model with detailed and nuanced actionable policy levers that takes six months to gather data and calibrate parameters to accurately populate the model, is often OBE and not actionable. On the other, we know from the scientific law of requisite variety that the number of controllers or levers in a system has to be equal or greater than the number of sources of variance in the system (Ashbey 1956). If we continue operating with a blunt DIMES controller hammer versus precise, root cause neurosurgical tools we need, we will drown in the convergence. Expanding 'whole of government' approaches are necessary to effectively manage complexity.

Augmenting operational, tactical and strategic art necessitates that our products adapt, customized for particular mission needs, modularized in an ecosystem, yielding required resolution and actionable insights to a variety of detailed questions. Figure 6 offers a sample, natural and social science typology that maps different domain theories and behaviors into the scales of human activity that each theory or approach applies to.

For surfing the conflux, combining scientific PMESII, SEEPT and CAS frameworks are powerful. Lanes in the road are being horizontally integrated across disciplines, such as political-economy and environmental politics. Lanes in the road can also be vertically integrated up and down stream within a single field, such as the micro motivations for macro-economic outcomes or pro-social behavior. Neuroscience, evolutionary game theory and quantum biology to name a few, all take advantage of knowledge arbitrages in a complex adaptive systems framework. Pushing ahead in both a transdisciplinary and a cross scale approach present even more obvious opportunities in the matrixed road of the natural and social sciences.

	domains	**SOCIAL**	**ECONOMIC**	**ENVIRONMENTAL**	**POLITICAL**	**TECHNOLOGICAL**
scales	**MACRO**	Human Development Globalization Poverty, Gender & Ethnicity	Global Trade Economic Crises Sustainable Growth	Climate Change Biodiversity Pandemics	International Relations Power Transitions Collective Security	Techno-Science Connectivity AI & Computing Power
	MESO	Movements Culture, Norms & Values Social Spaces	Money & Finance Institutions Collective Action	Environmental Justice Green Movements Ethics	Nation States Governance & Will Organizational Behavior	Innovation Diffusion Internet Social Media
	MICRO	Agency Intersubjectivity Ego, Id & the Individual	Microeconomics Game Theory Decision Theory	Health Consumer Behavior Activism	Positive Political Theory Dispute Resolution Political Bargaining	Inventions Entrepreneurship Consumer Products

Fig. 6 Typology of science domains, theories and approaches across levels of human behavior

Promise Versus Performance Expectations

Despite remarkable success throughout, several challenges exist in realizing promise versus performance for surfing the conflux. Progress can be slow without luxury of controlled experiments, replicability of results and adaptive processes. Explanatory and predictive power from social data can also vary: highly applicable in certain strategic operational contexts while conversely underspecified in other tactical frames.

In engineering our new cognitive navigation tool suites, we need to transform social science orthodoxy into actionable insight heterodoxy. This transformation is occurring now. But as we learned from prior scientific revolutions, such as from astrology to astronomy or from alchemy to chemistry, progress nor acceptance does not happen overnight. Regardless, allies and adversaries continue to press forward in the convergence marketplace. Thus we should first calibrate a few expectations accordingly.

Level set pieces specifically involve data veracity, model accuracy and operational validation. Statistician George Box' admonishment that 'all models are wrong, some are useful' has been both widely quoted and wildly distorted. The point being that instead of arguing endlessly if an answer is the answer, and correct in all cases, models should be evaluated on their usefulness. All models are by design, a simplified but hopefully insightful synthetic version of some limited, narrow slice of reality. The same goes when working with data. Garbage data into any model will always result in garbage results out, known as GIGO. Even worse is when bad data goes into a great model and produces reified outputs, known as garbage in-gospel out.

Debates on model accuracy are also meaningless unless framed in terms of usefulness. We are not going to get the 100% solution given obvious human adaptability nor should we target for that. At 47% odds, blackjack is one of the better player casino games. If we can improve our blackjack system, from learning from experts, using a new mental model or otherwise augmenting our decision making, yielding only 4% more predictive power is great model. Here the astute strategist recognizes that 51% is great accuracy if playing repeatedly and doubling down on bets. Conversely, an additional 4% model uplift in the game of Russian Roulette with 83% odds in our favor to start is probably not going to change most individual's opinions whether to play or not. Sometimes 4% model accuracy uplift on what we are doing currently can make all the difference in the world.

In data science, we constantly calibrate our approaches so they are neither under or overpowered which would mismatch requirements. Not to build a Swiss army knife that answers all generalizable questions poorly but something that is engineered to precisely perform necessary and specific tasks. Thus we need to correctly adjust our return on accuracy investment (ROAI) calculations for any endeavor. Regardless of these well-known challenges, we continue to press ahead across M5 value chains, not leaving any data, model, knowledge or integration arbitrage opportunities on the table.

Alternative Futures

Surfing the conflux necessitates what we need to do at the convergence—agilely capturing the undercurrents of connectivity, complexity and uncertainty. Going broader on our understandings of human behavior foundational to any mission. Going deeper into the complexity of physical and social environments as they continue to morph, merge and co-evolve. Going smarter for informational exploits and vulnerabilities as well as transforming ourselves. Going sharper for integrating the commanding information, knowledge and know-how already across the enterprise.

Why do we need to go broader, deeper, smarter and sharper? Because we can. And if we do not, our adversaries most definitely will. We need to rebalance our risk appetite, from being risk averse, chasing downside losses of a retrospective status quo to becoming more risk acceptant, capturing upside gains in creating new normals. Ultimately our own risk appetite might determine the winner.

> Unless we adjust our future CONOPS and S&T Investments to account for the paradigm shifts that have occurred under our feet, our nation and its intelligence operations will once again awaken too late, to a different reality, which is likely to end badly with significant and long-term impacts to our nation's security and place as world leader.
>
> I project such a negative and reactive outcome to occur either because we lost the cognitive war totally, our adversaries succeed undermining our institutions and democratic foundation to such an extent they are no longer viable, or, because our efforts to counter in the cognitive domain came too late.
>
> If we fail to act in the cognitive domain, we will likely end up in a major kinetic conflict resulting in devastating outcomes, in physical and human toll—recovery is questionable (Haugland 2019).

Yet with technology as the enabler, we have to do everything faster, better and cheaper at scale than competitors. A few prominent AI examples relevant to the conflux serve to recalibrate our timeframes at the convergence. In 2016, Microsoft experimented with a chatbot named Tay for AI to learn from interacting with human users on Twitter. Within 16 h of being released, Tay became a racist, exuding hate speech and Microsoft stopped the service. Tay was targeted by coordinated, extremist groups and trolls on Twitter and began mimicking their offensive behavior as AI has no intrinsic understanding of inappropriate behavior.

In 2017, Facebook researchers created AI chatbots to learn how to effectively negotiate with humans (Forbes 2019) using a generative adversarial network (GAN) approach. A GAN basically turns AI against each other for competitive, offensive versus defensive, strategic action-reaction learning. In doing so, AI performs OODA loops competitively, over and over millions of times, learning faster than humanly possible. Researchers noticed that instead of speaking in proper English, the AI bots began communicating in their own shorthand lingo for more efficient results. This new language, which humans could not understand, also led to better negotiation outcomes.

Creating AI to play the ancient game of go, Google revealed Alpha Zero in 2017. Go is a 2,500 year old board game where opponents surround territory to capture stones. It much more complex than chess with 10^{361} possible moves. Using conceptually simple but operationally complex Monte Carlo tree search methods—thinking out several moves ahead to see what the best likely move is now currently by leveraging computational probabilities and game theory—Alpha Zero taught itself to play the game of go. The result? Alpha Zero learned by itself how to beat thousands of years of collective human go wisdom, in 70 h and at a cost of only $25 million. It also devised new strategies previously unknown to go masters, called *joseki* moves. How it accomplished this feat are lessons from both our OODA loop and evolutionary biology playbooks. Alpha Zero used reinforcement learning to continuously improve its own strategy execution via feedbacks, not missing any lessons learned in a systematic, complete and tireless manner. It also did this learning experimentation millions of times over, as the speed of playing computational games far outperforms that of physical world constraints.

Obviously, AI capitalizes not just on OODA loop speed, but on the exponential capability increases in each step—billions of IOT sensors to observe, relentless deep learning for new orientation unconstrained by our beliefs or biases, expansive searches in spaces where we not dare to decide, and obvious milliseconds to act. Yet from action-reaction AI competition, new behavior emerges that we cannot exactly anticipate but somehow we must prepare for. This forces us to not only plan for what we probably can see on the horizon but against what portfolio of possibilities are out there.

Thus we must recalibrate how we think about alternative future possibilities. In silico experimentation far outpaces in vivo real-world evolution as AI demonstrates. Over 500 million year ago simple, individual cell organisms transitioned to multicellular life through the networked effects of heart, brain, bone and blood cell specialization, known as the Cambrian explosion. The effectiveness of multicellular organisms allowed for more complex, life forms and new behavior. Diversity exploded with different organisms, orders and production functions, propelling new strategies, predator–prey ecological dynamics and complex food webs. Whether or not we are at the tipping point of another explosion in the anthropocene era is not a question of if, but when.

Human strategy will continue to co-evolve with technology to define new defense and security challenges. Here we may or may not choose to explore, operate and own competitive environments of new ideas, orders and wicked problems. At the convergence, surfing the conflux, the future operating environment is upon us. Going broader, deeper, smarter and sharper is not impossible, it is necessary.

References

B.W. Arthur, *Positive Feedbacks in the Economy* (Scientific American, 1990)
B.W. Arthur, *Increasing Returns and Path Dependence in the Economy* (Michigan, 1995)
B.W. Arthur, Is the information revolution over? If history is a guide, it is not, in *Business 2.0* (2002)
W.R. Ashbey, *An Introduction to Cybernetics* (Chapman and Hall, London, 1956).

R. Axelrod, M.D. Cohen, *Harnessing Complexity: Organisational Implications of a Scientific Frontier* (Free Press, 2000)
J. Barksdale, https://www.goodreads.com/quotes/655987-if-we-have-data-let-s-look-at-data-if-all (2015)
A.B. Çambel, *Applied Chaos Theory: A Paradigm for Complexity* (Academic Press, 1993)
D. Christian, *Origin Story: A Big History of Everything* (Little, Brown and Company, New York, 2018).
P. DeGennaro, The power of cognitive maneuver: don't underestimate its value. Small Wars J. (2017)
Department of Defense, Joint concept for human aspects of military operations (JC-HAMO) (2016)
Department of Defense, Joint concepts for operating in the information environment (JCOIE) (2018)
Department of Defense, Summary of the 2018 Department of Defense artificial intelligence strategy: harnessing AI to advance our security and prosperity. JAIC (2018)
J.M. Epstein, R. Axtel, Growing artificial societies: social science from the bottom up (Brookings Institution Press, 1996)
P.R. Ehrlich, P.H. Raven, Butterflies and plants: a study in co-evolution. Evolution **18**, 586–608 (1964)
Forbes India, The first human genome project took 12 years, technology can now do it in 3 days (2019). https://www.forbesindia.com/article/one-ceo-club/the-first-human-genome-project-took-12-years-technology-can-now-do-it-in-3-days/58961/1
M. Gell-Mann, *What is Complexity?* (Complexity Journal, 1995)
M. Gladwell, *Outliers: The Story of Success* (Little, Brown and Company, New York, 2008).
J. Gleick, *Chaos: Making a New Science* (McDonald & Co, Cardinal, 1987).
L. Grossman, 2045 the year man becomes immortal. Time Magazine (2011)
E.L. Haugland, Future military intelligence CONOPS and S&T investment roadmap 2035–2050: the Cognitive War (NSI Team, 2019)
R. Hausmann et al., *The Atlas of Economic Complexity* (MIT Press, Cambridge, 2014).
D. Hernandez, T. Brown, AI and efficiency. OpenAI (2020).https://openai.com/blog/ai-and-efficiency/#rf3
J. Holland, *Hidden Order: How Adaptation Builds Complexity* (Addison Wesley, Boston, 1995).
J. Holland, *Emergence: From Chaos to Order* (Addison Wesley, Boston, 1998).
Human Dimension Capabilities Development Task Force (HDCDTF), Cross-cultural competence: overview of cross-cultural training theory and practice for the army. US Army (2015)
International Business Machines, Six reasons to update your data science (2020). https://www.ibm.com/analytics/data-science
International Data Corporation, The digital universe of opportunities (2014). https://www.emc.com/leadership/digital-universe/2014iview/index.htm
S. Kauffman, *Investigations* (Oxford University Press, Oxford, 2000).
V.A.W.J. Marchau et al., *Decision Making Under Deep Uncertainty from Theory to Practice* (Springer, New York, 2019).
B. McKelvey, Self-organization, complexity catastrophe, and microstate models at the edge of chaos, in *Variations in Organization Science: In Honor of Donald T. Campbell*, ed. by J.A.C. Baum, B. McKelvey (Sage, 1999a)
B. McKelvey, Visionary leadership vs distributed intelligence: strategy, microcoevolution, complexity, in *Proceedings of EIASM Workshop*, Brussels, June 1999 (1999b)
M. Mitchell, *Complexity: A Guided Tour* (Oxford University Press, Oxford, 2009).
E. Mitleton-Kelly, Ten principles of complexity & enabling infrastructures in complex systems and evolutionary perspectives on organizations: the application of complexity theory to organizations. Elsevier (2003)
G. Nicolis, I. Prigogine, *Exploring Complexity* (W.H. Freeman, New York, 1989).
T. Pike, Analysis and artificial intelligence in integrated campaigning. SMA (2020)
RAND, Priority challenges for social and behavioral research and its modeling (2018a)
RAND, Discontinuities and distractions—rethinking security for the year 2040(2018b)

T.C. Redman, Bad data costs the U.S. $3 trillion per year. Harvard Bus. Rev. (2016)

D. Silver et al., Mastering the game of Go without human knowledge. Nature **550**, 354–359 (2017)

Statistica, Number of mobile phone users worldwide from 2015 to 2020 (2020a). https://www.statista.com/statistics/274774/forecast-of-mobile-phone-users-worldwide/

Statistica, Internet of Things (IoT) connected devices installed base worldwide from 2015 to 2025 (2020b). https://www.statista.com/statistics/471264/iot-number-of-connected-devices-worldwide/

Strategic Multilayer Assessment White Paper, A cognitive capabilities agenda a multi-step approach for closing DoD's cognitive capability gap. NSI (2017)

Strategic Multilayer Assessment Annual Conference, The evolving anatomy of conflict in a dynamically changing world. NSI (2019)

Strategic Multilayer Assessment, *Complex Adaptive Systems Interview with Dr* (Sean Lawson, NSI, 2019).

Strategic Multilayer Assessment Periodic Publication, On the horizon: security challenges at the Nexus of state and non-state actors and emerging/disruptive technologies. NSI (2019)

S.H. Stogatz, *Nonlinear Dynamics and Chaos: With Applications to Physics, Biology, Chemistry, and Engineering* (Westview Press, Nashville, 2015).

United States Army Training and Doctrine Command (TRADOC) and the Army Futures Command (AFC), *Mad Scientist: Disruption and the Future Operational Environment Conference* (2019)

M.M. Waldrop, *Complexity: The Emerging Science at the Edge of Order and Chaos* (Penguin, London, 1994).

D.J. Watts, S.H. Strogatz, Collective dynamics of 'small world' networks. Nature **393**(6684), 440–442 (1998)

World Economic Forum, *The Global Risks Report*, 14th edn (2019)

World Economic Forum, *The Global Risks Report*, 15th edn (2020)

World Trade Organization, World trade developments, 2018–2019 (2020). https://www.wto.org/english/res_e/statis_e/wts2019_e/wts2019chapter03_e.pdf

Conceptualizing the Great Power Competition and U.S. Geoeconomic Strategy for the Central and South Asia (CASA) Region

Adib Farhadi

Abstract Once a central point of competition in the 19th-century "Great Game" between the Russian and British Empires, with Afghanistan as the "pull-and-tug" buffer zone, the Greater Central and South Asia regions have once again become the theater for a Great Power Competition (GPC). Vying for influence, resources, and interests are the modern Great Powers, namely the United States, China, Russia, and the Regional Powers, namely Iran, India, Pakistan, Turkey, and the Central Asian and Gulf countries. The rival powers of Russia and China, in particular, continue to compete with the United States for hegemony through economics and other non-military soft-power means. Despite the high-stakes complexity and global implications, no consensus has yet been reached on a conceptual underpinning for the GPC that might effectively guide the U.S. strategic engagement in the region. This chapter attempts to fill this gap by defining and conceptualizing GPC with a focus on geoeconomics in the Central and South Asia region. The chapter furthermore recommends the U.S. transition to the New Silk Road Initiative's soft-power strategy in the CASA region to preserve the hard-fought gains of the last two decades in Afghanistan and advance U.S. relevance and influence in this critical Great Power battleground. Geoeconomic and New Economic Geography theories support the recommendation.

Keywords Great power competition · United States · Afghanistan · Russia · China · Belt and road initiative · New silk road initiative · New economic geography · Geoeconomics · Rare earth minerals · Critical minerals · Lithium

Introduction

Over the past decade, the international political system has come to be characterized as a Great Power Competition (GPC) in which multiple would-be hegemons compete for power and influence. Instead of a global climate of unchallenged U.S. dominance,

A. Farhadi (✉)
University of South Florida, Tampa, FL, USA
e-mail: farhadi@usf.edu

© The Author(s), under exclusive license to Springer Nature Switzerland AG 2021

A. Farhadi et al. (eds.), *The Great Power Competition Volume 1*,
https://doi.org/10.1007/978-3-030-64473-4_2

revisionist powers, notably the People's Republic of China and the Russian Federation, are vying for dominance through military and economic means. Also engaged in the GPC are regional powers such as India, Iran, Saudi Arabia, and Turkey; rogue regimes such as North Korea; and violent extremist organizations such as ISIS and Al-Qaeda.

A critical GPC battleground is the Central and South Asia (CASA) region. The high-stakes, multi-player GPC unfolding in this region is a natural extension of the Great Game of the nineteenth and twentieth centuries, in which the British and Russian empires competed for control of the territory among their respective empires. However, the 21st-century GPC is a more complex contest involving a greater number of players and global implications. Moreover, its battles are as much economic as they are political or military.

A major geostrategic initiative emerging in the GPC context is the modern Silk Road, a massive collaborative infrastructure project that promises to integrate the trade and economies of over 70 countries, ranging from Europe to China and beyond. For all of CASA's economic potential, it remains one of the least economically integrated areas of the world. The modern Silk Road, set to re-unify the region, represents a revival of the ancient Silk Road, a network of shifting land and sea trade routes that knit together Asia, Europe, and Africa for millennia until its disintegration in the nineteenth century. Although the U.S. has been entrenched in the CASA region for nearly 20 years, fighting the War on Terror in Afghanistan, it is China who is the leader in rebuilding the modern Silk Road network of land and sea trade corridors through its Belt and Road Initiative (BRI).

To date, the U.S. has not engaged any meaningful Silk Road projects, despite launching its own New Silk Road Initiative (NSRI) in September of 2011. China's BRI, a reaction to the U.S.'s NSRI, was launched in 2013 by President Xi Jinping, and thus far, more than $200 billion has been invested into the project. The estimated total cost of the BRI by 2027 is $1.2–$1.3 trillion (Chatzky and McBride 2020). China's action and reaction demonstrate the vying for power and influence, which is the essence of the GPC. With the planned military withdrawal from Afghanistan in late 2021, the U.S. has stated its intention of shifting attention away from the CASA in favor of a more isolationist foreign policy. This withdrawal leaves room for both China and Russia to seek greater dominance and influence in the region at the cost of U.S. global power and influence. Suppose China and Russia control access to Central and South Asia's natural resources. In that case, the U.S. stands to lose not only its economic hegemony but also its autonomy in future efforts to counter violent extremism, fostered in the poverty and isolation of the region that led to the tragic events of 9/11. The U.S.'s potential disengagement from, and isolationist approach to, Afghanistan and the CASA region may perpetuate the formation of ineffective national security policies and jeopardize the competitiveness of the American private sector in the global marketplace. Such an initiative underpins the GPC framework outlined in the U.S.'s National Security Strategy (NSS) and National Defense Strategy (NDS).

While the U.S.'s competitive position as a global power and influencer is a major tenet posited in the NSS and the NDS, the documents leave the Great Power Competition undefined, likely because the term "GPC" was still evolving and barely applied at the time of the documents' construction. Nevertheless, in recent years, the term has come to shape international relations and actions and is now "firmly entrenched in conventional defense thinking" (Boroff 2020). Once an arcane concept and term, the Great Power Competition has become linked with every aspect of defense, strategy, and national security (Boroff 2020). As such, "GPC" must be uniformly understood so that policies and actions can be shaped accordingly.

Due to the relative newness of the term "Great Power Competition" and the overall lack of a universal definition and understanding, many gaps remain in its conceptual underpinnings. Thus, this chapter presents a conceptual perspective on the Great Power Competition, focused on geoeconomics relevant to current and future affairs. The chapter proceeds in two major sections. The first section of the chapter provides background on the Great Power Competition, including the origins and definition of the term, and explores the geoeconomic role of two major U.S. rivals in the GPC: Russia and China. Following the GPC conceptual framework, the chapter argues that the U.S.'s best interest is to revive the NSRI and establish the states' relevance and economic power in this critical GPC battleground. This second section posits that the U.S. role in the GPC must support Afghanistan's economic development, which is at the heart of CASA. This argument is further supported by theoretical frameworks of geoeconomics and New Economic Geography.

Background and Definition of the Great Power Competition

The Great Power Competition era is one in which the international political system is defined by competition for power and influence through military and economic means. Since post-WW II, the GPC has been defined as an environment "marked not by unchallenged U.S. dominance" or by close cooperation among world powers but, rather, by competition for military and economic power and influence among the Great Powers of the United States, China, and Russia (Colby and Mitchell). While these modes of modern power have had various materializations, economic rivalries have increasingly prevailed over military power and presence in foreign lands (Csurgai 2017).

Like the Cold War, the Great Power Competition is a quiet manifestation of power. Its theatre has shifted away from displays of hard power to one in which "cyber espionage, defense planning, 5G spectrum, international aid, intellectual property theft, supply chains, coalition-building, and election interference" dominate (MITRE 2020). Increasingly, the Great Power Competition has led to competition for the natural resources that will become critical in the future. Lithium, for example, is one such critical resource found in abundance in CASA, especially in Afghanistan, and its presence creates a desire to control, compete, influence, and possess. The GPC economic rivalries play out in regional projects aimed at increasing economic power and influence. As such, the potential spheres of state influence and theatres of war are endless.

The 21st-century GPC is a complex contest involving a significant number of players and global implications. Its battles are as much economic as they are political or military. The GPC great powers have developed economic projects, and political and military alliances intended to shape the international order in their favor and shift power away from their competitors. As in past power competitions, economic projects in the CASA region are positioned as vital for the facilitation of hard and soft power projections and economic growth. Joseph Nye, Harvard University political scientist, observed that soft power, or a country's reputation and influence, is just as crucial as statecraft as a military power. While the NSS states that perception of strength, specifically military strength, and the vitality of alliances can affect rivals' willingness to abandon or forgo aggression, soft power can attract allies on the state level and foster an international environment that is distinctly pro-American (Musgrave 2020).

Though the Great Power Competition is a distinctly modern era, the GPC has been often regarded as an extension of the Great Game, a period of expansionist competition during the nineteenth century between Russia and the United Kingdom, and as a continuation of the Cold War between the United States and the Soviet Union. In a later chapter of this volume, Mitch Shivers argues that "the Great Power competition is anything but new to Central Asia. Since the early Nineteenth Century, great powers have jostled and gamed with one another, diplomatically and politically, using the mountains, deserts, and steppes of Central Asia as a 'pitch' (Shivers 2020). Indeed, in the historical periods and the current period, the crossroad to the competition is the CASA region. As with the historical Great Game, Afghanistan is once again geographically situated at the intersection of the modern GPC.

In *The Tragedy of Great Power Politics*, Mearsheimer (2001) argues that the international system exists in a state of anarchy, which is to say that no stable or neutral government body can in fact oversee states and state actions. As such, this state of anarchy places competing states in a constant power competition and fear of each other's actions and intentions. Mearsheimer further argues that it is this fear and uncertainty that leads to offensive economic and military measures. He contends, "states recognize that the more powerful they are relative to their rivals, the better their survival chances. Indeed, the best guarantee of survival is to be a hegemon, because no other state can seriously threaten such a mighty power" (Mearsheimer 2001). Following from this argument, the goal of individual states competing in the Great Power Competition is to achieve hegemon status and, thus, exist in a state of permanent security.

The term "Great Power Competition" first emerged on the world stage in 2008 when Robert Haas, then-president of the Council on Foreign Relations, used the term to refer to a reality that was assumed to have already passed. He concluded that the "challenges derived from globalization will dominate the century," noting that "great-power competition and conflict is no longer the driving force of international relations" (Friedman 2019). This sentiment echoes officials in Washington at the time, including former President Barak Obama, who, in his 2006 book, *The Audacity of*

Hope, wrote that expansionist states of the past and great power competitions "no longer exist" (Friedman 2019). The mood in Washington was that the state of world affairs and its challenges, such as terrorism, climate change, and pandemic disease, called for global cooperation rather than global competition (Friedman 2019). As this chapter explores, many instances exist where U.S. national interests are indeed better served through collaboration than competition. Although the temperature in Washington has shifted away from collaboration and towards competition, these two strategies should not be firmly held as competing ideologies, but rather seen as complementary and fluid, each useful in different circumstances to further U.S. national interests.

By 2008, the utopian mindset of collaboration had shifted, with Chinese and Russian objectives becoming very clear in the South China Sea and neighboring countries, respectively. President Obama responded to these actions with an attention pivot to the Pacific. Meanwhile, Russia's actions in the Middle East, a region vital to the U.S., also signaled a growing assertiveness that many now see as a "prelude to a new phase of Western depreciation" (Kempe and Massolo 2019). In 2008, Defense Secretary Chuck Hagel stated that "'state-on-state aggression' in Europe and 'competition between rising powers' in the Asia–Pacific region," as well as "'enduring and emerging powers, are challenging the world order that American leadership helped build after WWII'" (Friedman 2019). Such international competitions began to arise because of state dissatisfaction with the status quo, leading to economic warfare, as seen in the trade wars between the United States and China, as well as political warfare and military conflicts (Wohlforth 2009).

As the world transitioned to the second decade of the twenty-first century, liberalism, utopianism, and international cooperation gave way to the age of conservatism, realism, and global competition. Though state competition has always existed, the globalization age, alongside growing technological advances in warfare and trade, signify an age of competition such as has not yet been seen, with consequences throughout the world.

The shift was tied to a perceived loss of power by the U.S. According to MacDonald and Parent (2011), "For some observers, the wars in Iraq and Afghanistan represent the ill-advised last gasps of a declining hegemon seeking to bolster its plummeting position." These same observers might note that the sudden change in foreign policy from President Obama to President Trump, from global cooperation to global competition, is also reflective of a declining power. States like China and Russia thus see an opportunity to increase their own global prowess. International actors such as Russia and China claim the diminishing of U.S. global power in order to shift global politics in their own favor. Yet, many thinkers describe this diminishing of the U.S. as perception and not reality (Colby and Mitchell 2020). These perceptions have been driven not by questions of capability, but rather by, for example, shifting commitments to the peace and stability of Afghanistan (Kempe and Massolo 2019).

The concept of the Great Power Competition, still relatively undeveloped, has its official origins for U.S. national policy in the NSS and NDS, released in 2017. The NSS details national security concerns as well as presenting the administration's plans for addressing them, while the NDS presents new frameworks and priorities for the

United States. Both the NSS and the NDS shifted focus away from counterterrorism and counterinsurgency and, instead, placed an increasing emphasis on "great power competition" in regions such as CASA. According to the NSS and the NDS, the Great Power Competition's main competitors, known as the Great Powers, are the United States, Russia, and China. The NDS states, "The central challenge to U.S. prosperity and security is the reemergence of long-term, strategic competition by … revisionist powers." Defeating these rivals, should deterrence fail, will require far-reaching changes in what the American military purchases and how it fights.

Despite an increased focus on containing the expansionist ambitions of China and Russia in the CASA region, the strategy does offer prospects for collaboration and compromise, committing the U.S. to "seek areas of cooperation with competitors from a position of strength, foremost by ensuring our military power is second to none and fully integrated with our allies" (NSS 2017).

The NDS and the NSS describe the Great Power Competition as a complex environment in which the United States has shifted its focus from counterinsurgency and counterterrorism to the Great Power Competition with Russia, China, North Korea, Iran, and violent extremist organizations. However, policy documents provide no further explanation or universal understanding. This lack of discernment has wide-reaching issues for the military, the defense industry, the diplomatic arm, and U.S. policymakers (Boroff 2020). An unclear definition creates uncertainty, wherein development and training are not focused on the immediate strategic environment. Without a universal understanding of the Great Power Competition, a defense strategy is challenging to develop and implement. In 2019, it was reported that the Pentagon has not yet ascertained how to "compete" with Russia and China, nor even settled on a definition for the "competition" in "great power competition" (Williams 2019).

The GPC strategy has been evolving quietly in the U.S. No longer is the U.S. military focused only on combating rogue states, terrorist groups, and other deadly, albeit relatively weak, enemies. Instead, the U.S. Defense Department has prioritized its primary focus on China and Russia as the great power rivals contesting American military advantages and threatening to reorder the world. In 2018, Ali reported on the unveiling of the strategy, stating that the U.S. military had "put countering China and Russia at the center of a new national defense strategy," in "the latest sign of shifting priorities after more than a decade and a half of focusing on the fight against Islamist militants" (Ali 2018). In the same report, Ali quoted then-U.S. Secretary of Defense James Mattis's speech, noting, "We will continue to prosecute the campaign against terrorists that we are engaged in today, but great power competition, not terrorism, is now the primary focus of U.S. national security" (Ali 2018).

This new focus requires a restructuring of the U.S. military and new approaches to force planning. Since the end of the Gulf War, the U.S. has maintained a two-war approach. This approach proved to be "essential for meeting the ongoing demands of forward presence, crisis response, regional deterrence, humanitarian assistance, building partnership capacity, homeland defense, and support to civil authorities" (Goure 2013). In their analysis, Brand and Montgomery (2020) note that the U.S. Defense Department has departed from the two-war approach and has shifted toward a one-war standard.

Brand and Montgomery (2020) continue their analysis by arguing that the shift "is a recognition that defeating a great-power adversary would be far more difficult than anything the U.S. military has done in decades and that losing a great-power war would be devastating to America's global interests." Given these potentially devastating effects, an "all-hands-on-deck" approach is required for the U.S. to compete adequately in the Great Power Competition.

Russia and China as Major Players in the GPC

The United States' global competition with emerging military and great economic powers has become a key focus of U.S. national security strategy. The NSS and NDS policy documents singled out China and Russia as the United States' two main competitors for global power. The economic rivalries of the GPC are revealed in the regional projects that China and Russia, among others, have undertaken to increase their economic power and influence. Marine Corps Gen. Joe Dunford concurs that "China and Russia are doing what they can to challenge the U.S. and target American capabilities. This proactive approach means the two nations are working to subvert America's network of allies and partners and are actively seeking to negate the American military's ability to project forces when and where needed and sustain them" (Garamone 2020). With a perspective that the United States is weakened not only by domestic political competition but also decades of war in Iraq and Afghanistan, China and Russia are positioned to shift the status quo away from the era of American dominance.

Though all three Great Powers play important roles in the CASA region through economic, cultural, political, and military means, all three have opportunities to dominate and influence the region. It is Russia and China—not the U.S.—who are actively positioned as key regional players. This less attuned posture on the part of the U.S. is a costly mistake.

Lt. Gen. Thomas W. Bergeson, Deputy Commander of U.S. Central Command, addressed Russia and China's pursuit of their interests in CASA at the inaugural Great Power Competition Conference in January 2020 at the University of South Florida (USF). These interests, which shift historical alliances and strengthen power and influence, take the shape of regional economic and infrastructure projects, such as China's Belt and Road Initiative soft power projections, hard-power displays, and partnerships with regional powers and players. China and Russia's distinct approach has progressively shifted the status quo of the past thirty years.

Russia's Interplay in the GPC

The U.S. competition with Russia is hardly new. Lovotti (2019) argues that the tension between the U.S. and Russia has dominated the world stage since World War

II. It was only with the fall of the Soviet Union in 1990 that the United States was regarded as the world's sole Great Power. However, with Russia subverting international treaties, such as the North Atlantic Treaty Organization (NATO) and using its "veto authority over nations on its periphery," the U.S. (NDS 2018) position faltered in 2001. These actions have had the power to "shatter NATO and change European and Middle Eastern security and economic structures to its favor," through technological advances, sovereignty violations, and a growing and modernizing military NDS 2018).

Russia has also made gains in international diplomacy; this is especially true in the Middle East as well as in Central and South Asia. This global energy-producing region has been significant to the U.S. for decades. This importance is reflected in the United States' four national security objectives that revolve around the free flow of energy resources and "preventing the growth of state or non-state actors antagonistic to the U.S." (RAND 2020). Russia works at interrupting these factors by using its soft power and its insertion in Middle Eastern and CASA region politics. Russia's capacities extend beyond that of the U.S. For example, Russia opens discussions with opposing sides of conflicts (i.e., Syria and Afghanistan) and facilitates peace talks such as the Astana Process, a Russian-led forum established in 2017 to negotiate peace in Syria along with parallel Afghan peace talks (Lovotti 2019; Thepaut 2020). As one of the most turbulent areas of the world, CASA often calls for international cooperation to end conflicts, and the increasing Russian (and Chinese) presence in the region is cause for concern among American planners and policymakers (Wechsler 2019).

In addition to Russia's increasing diplomatic presence, its military power is growing exponentially as well. Dunford points to concentrated Russian defense spending on upgrading their military weaponry and capabilities and modernizing their nuclear enterprise. Notably, Russian military expenditure has decreased significantly and consistently since 2016, and while the Russian economy still cannot compete with that of the U.S., Russia has been effective in complicating the U.S. position abroad. Sean Ryan (2020) insists the Great Power Competition is not solely about action, but also about the willingness to act. He notes in his address at the January 2020 GPC Conference at USF, "When we look at the possibility of escalation—consider that it is not just about objectives, methods, and resources. It is about the will to do it. Russians have demonstrated that they are willing to conduct disruptive operations." These disruptive operations are demonstrated by Russian cyberwarfare during the American elections of 2016, where it exerted its soft power and diminished the United States' soft power. Indeed, as Marcus concurred, Russia has been "waging a 'grey' or undeclared war against the West in recent years. This has many elements: cyber-attacks; disinformation campaigns; electoral interference; the funding of extremists in Western countries and so on" (Marcus 2020).

It is widely acknowledged that rare earth and critical minerals are vital in the Great Power Competition context. Rare earth minerals are used for military and consumer technologies such as electric vehicles and cellphones (Burke and Scott 2019). China has an overwhelming monopoly on rare earth minerals, supplying 67% of global rare earth mineral production (Kiggins 2015). Russia, who holds the world's fourth-largest rare earth mineral reserves, imports 90% of its rare earth minerals from China (Cottig et al. 2019). This dependence on China for rare earth mineral importation and reserve creates an enormous risk given the supply insecurity (Kiggins 2015; Lyrichikova and Stolyarov 2020). Like Russia, the United States imported 80% of its rare earth mineral supply from China (Reuters 2019). In the Great Power Competition era, dependence on competing countries for minerals used for critical, defense, and civilian purposes can result in an economic and geopolitical weakness (Burke and Scott 2019).

However, Russia plans to break its dependence on China by actively seeking foreign investments of $1.5 billion to develop further and expand its rare earth mineral industry (Lyrichikova and Stolyarov 2020). It also plans to offer to back foreign investments, reduce mining taxes, and give cheaper loans to investors (Lyrichikova and Stolyarov 2020), according to Lyrichikova and Stolyarov (2020). With the proper incentives to investors, Russia hopes that it will become the second-largest producer of rare earth minerals after China by 2030. This plan places further stress on the United States to keep pace in the Great Power Competition by developing its independent earth minerals source.

China's Interplay in the GPC

China proves to be an even greater rival to the U.S. in the GPC than Russia as China's economy is much larger and more robust than that of Russia. The country is described as economically and militarily ascendant, with a growing global presence.

China has embarked on a far-reaching military modernization that Colby has characterized as being in "deep contravention to our [U.S.] interests" (Ali 2018). Writing in a later chapter of this volume, Michael Singh of the Washington Institute for Near East Policy argues that the United States has singled out China as "its foremost peer and rival" in the CASA region. Both the United States and China compete for the same resources—energy, economic presence, and influence. Speaking at the 2020 GPC Conference at the University of South Florida, Singh argued for the importance of considering China's competitive focus of displacing the U.S. as hegemon, particularly in Asia, and in attaining global influence over the international order.

China's designs on hegemony have been especially evident in and around the South China Sea. In early April, for example, a Chinese aircraft carrier was deployed close to Japan and Taiwan in a move that seemed opportunistic and calculated, given the plight of another nearby carrier, the U.S.S. Theodore Roosevelt, which was pulled off operations (and its captain relieved of his post) after hundreds of the crew had contracted COVID-19. The U.S. Navy also had a powerful amphibious task group at sea in the region at the time, reinforced by an Australian frigate, and was "maintaining a drumbeat of other presence and freedom-of-navigation operations" (Childs 2020). According to Childs (2020), "Washington has explicitly accused Beijing of seeking to exploit the pandemic, while China has fired back that it is the U.S. that is creating instability." Tensions between the two powers show no signs of abating soon.

Over the past decade, China's key project in the Great Power Competition has been the Belt and Road Initiative (BRI), an infrastructure network of railroads, roads, pipelines, and electricity projects that directly passes, and links, 60 countries, engaging 60% of the world's population across Asia and Europe, though it also involves many other states (Cai 2017; Shivers 2020). The BRI has served its purpose of increasing the Chinese presence throughout Asia and into Europe by creating the infrastructure that links countries by land and sea.

The BRI is the crux of Chinese president Xi's foreign policy. It is an economic and diplomatic strategy designed to promote Chinese influence in the region and beyond and weaken U.S. dominance in the regional and global economy, including minimizing the U.S.'s policy effects of containing China (Cai 2017). Shivers (2020) argues that while the BRI has often been compared to the post-World War II U.S. Marshall Plan, it is actually "12 times larger than the Marshall Plan (in constant dollars), [and will] be stretched over a 3.5 times longer time period. Bigger in scale, the BRI nevertheless shares a common objective with America's earlier Marshall Plan." The BRI is estimated to cost upwards of USD 8 trillion upon completion. Alongside the infrastructural projects, the Asian Infrastructure Investment Bank (AIIB) directly funds projects in the countries that line the BRI, eliminating the need for funding from institutions aligned with the U.S. and its allies.

Russia–China Alliance and the Need for U.S. Allies

Russia and China have historically not enjoyed a particularly friendly relationship. Sharing a large border, they have had distinct tensions over territorial expansion and naval designs on the Artic (Hill et al. 2020). However, despite these tensions, Russia and China share various interests, and most importantly, they share what Hill et al. (2020) refer to as "a mutual distaste" for the United States' operating in their backyards. This "mutual distaste" has resulted in various policies and alliances between Russia and China designed to shift the world order away from the United States.

The U.S.' economic war with China has resulted in a trade war, which has lasted for over 18 months and has resulted in billions of dollars' worth of tariffs on the imports of the United States and China (Smith 2019). Moreover, this trade war has resulted in the recent formation of a partnership between Russia and China, with Chinese President Xi Jinping promising Russian President Vladimir Putin that China was "ready to go hand in hand" with Russia (Smith 2018) in this trade initiative. According to Smith (2019), "Russian and Chinese news agencies reported recently that the two nations planned to almost double their trade over the next five years, hitting $200 billion by 2024 compared to $107 billion in 2018 by implementing collaborative projects in energy, industry, and agriculture." Given this alliance between Russia and China, the U.S. requires stronger alliances of its own.

While not always translating to action, such alliances are alluded to in U.S. strategy documents. The NSS states that "the United States will seek areas of cooperation with competitors from a position of strength, foremost by ensuring our military power is second to none and fully integrated with our allies and all of our instruments of power" (NSS 2017). The same document details the United States' plans to enhance missile defense; protect intellectual property, which it claims has long been a victim of Chinese thievery; tighten visa procedures; and protect data and underlying infrastructure. The NDS recommended "a more lethal, resilient, and rapidly innovating Joint Force, combined with a robust constellation of allies and partners." The document stated a plan to maintain American power and preserve peace through strength and acknowledged the costs of not implementing the strategy: U.S. global influence could decrease cohesion amongst allies, and partners would erode, costing the U.S. access to markets; this, in turn, would cause a decline in prosperity and American standard of living (NSS 2017). According to Ali (2018), the NDS identified international alliances as "critical" for the U.S. military, emphasizing "a need for burden-sharing," but also noted the contrast to President Trump's "public criticism of allies who he says unfairly take advantage of U.S. security guarantees." Carpenter demonstrates the United States' tendency to turn away from alliances in recent years, citing President Trump's choice to withdraw troops from Germany and his "questioning of the U.S. commitment to defend our allies." Carpenter notes, "The Trump administration has taken a go-it-alone approach and neglected our alliances, which has greatly weakened our position vis-a-vis Russia and China" (Tucker 2020). In Carpenter's view, a Biden presidency would be more likely than a second Trump presidency to "put our democratic allies and partners at the forefront of its foreign policy" (Tucker 2020).

As Twining and Quirk (2020) recently noted, for the U.S. "to prevail in any great power competition," it needs "allies and partners." The nation's "strategic purpose is served by helping friendly democracies build the capabilities to meet their people's needs while also strengthening their resilience to protect themselves from Russian and Chinese predation." The measure of success will be "how effectively we are enabling global partners to better serve their citizens and repel opportunistic attempts by Russia and China from extending their illiberal influence" (Twining and Quirk 2020).

Russia and China have ancient relationships with Afghanistan. Writing in this collection, Professor Martin Weinbaum (2020) details how Afghanistan has been a site where the historic rivalry of the United States and the Soviet Union has played out since 1950 (albeit, the pre-1990 Soviet version of the competition differs from today's). Weinbaum (2020) explains that the competition over Afghanistan "went only so far as their mutual concern that the other be kept from exercising a commanding influence in the country." While not discounting Russia as a U.S. GPC rival, China is the U.S.'s main competition. China's relationship with Afghanistan dates to the seventh century due to the Silk Road, which established a friendly, mutually prosperous relationship between the two countries (Rahajay 2019). China was one of the first Great Powers to issue a statement on "the Agreement for Bringing Peace to Afghanistan," expressing its support for the deal and calling for the "orderly and responsible withdrawal of foreign troops to avoid a power vacuum and possible terrorist resurgence" (Burns 2020). Like all other countries sharing a border with Afghanistan, China has active worries about the resurgence of terrorism in Afghanistan and instability spilling into its territory. Recognizing that cooperation is the key to peace and stability in Afghanistan, China also hosted several regional meetings and dialogues between the Taliban and the Afghan government and increased its diplomatic and cultural presence in Afghanistan (Rahajay 2019). China's Belt and Road Initiative would bring in economic prosperity and connectivity into Afghanistan, which Professors Frederick Starr and Barnett Rubin argue is the key to peace and stability in Afghanistan. Rubin (2020) argues that China has begun to develop a conception of their national interests through the Belt and Road Initiative, which is focused on creating international order as a public good that it will benefit from in a way similar to how the United States developed an interest in Europe through the Marshall Plan after the two World Wars. The enduring relationships between U.S. rivals Russia and China lend further support to the claim that the U.S. must remain engaged in Afghanistan.

U.S. GPC Strategy in Afghanistan

After nearly two decades of fighting the War on Terror in Afghanistan, the U.S., following a Peace Agreement with the Taliban, has agreed to withdraw its military forces by late 2021. Ambassador Zalmay Khalilzad (2020 GPC Inaugural Conference, USF) attributed this withdrawal agreement to the need to shift attention and resources to the Great Power Competition in other theatres at the very moment that both China and Russia are seeking to dominate and influence the Central and South Asia region. This planned withdrawal can be expected to cost the U.S. its presence in the region and its standing in the GPC, thereby potentially rendering it irrelevant and threatening U.S. national security. For the U.S. to compete effectively in the GPC, it must maintain an active soft-powers presence in the CASA region, and, specifically, in Afghanistan. This centrally located, landlocked nation bordering six other countries sits between East and West and has the potential to become a bridge that connects and stabilizes the entire region. The U.S. will effectively compete with China's BRI in the region if it ensures a durable peace and economic stability in Afghanistan through the NSRI.

The key to Afghanistan's stabilization, peace, and regional integration lies in continued U.S. economic engagement and its estimated untapped USD 1–3 trillion of natural resources, including rare earth and critical minerals such as lithium. In 2018, lithium was classified as a critical mineral by U.S. Executive Order 13,817 (National Archives 2020). The American Geosciences Institute (2020) describes critical minerals that are deemed vital to the economy and whose supply may be disrupted. Lithium and has been dubbed "white petroleum" (Sheraz 2014), and many scholars have noted that future resource conflicts will not focus on oil, but, rather, minerals such as lithium (Gulley et al. 2018). As such, the United States must maintain unfettered access to Afghanistan's vast lithium wealth. If the U.S. military withdrawal results in a complete disengagement from Afghanistan, China and Russia and other regional rivals will engage and gain the upper hand in the great power resource competition.

Afghanistan's wealth has not yet been tapped because the country lacks the trade and transit infrastructure to bring the resources to market. For decades, Afghanistan has been in a precarious security situation as an aid-dependent rentier state suffering under a presumed "resource curse." Now, Afghanistan is even further destabilized by the withdrawal of the U.S. military and other NATO countries and major reductions in international foreign assistance. The loss of aggregate economic benefits from military presence leaves Afghanistan in danger of collapse, which would return the country to its failed state status of the 1990s. Afghanistan has already become a seedbed for global terrorism, insurgent-fueled violent conflict, and drugs, concerns that threaten peace and security worldwide. The global COVID-19 pandemic has brought even greater precarity to the region, putting a halt to its evolving peace talks. The increasingly precarious state of Afghanistan, coupled with pending U.S. disengagement by late 2021, would concern anyone familiar with recent history. Recall that it was in the aftermath of the U.S.'s disengagement from Afghanistan following the Russian military's 1989 withdrawal that led to the tragic events of 9/11.

However, should Afghanistan overcome its aid dependency as a rentier state and replace the lost revenues through sustainable economic growth, it can achieve durable peace and prosperity that, in turn, would help stabilize the entire region. Through U.S. support, Afghanistan has the potential to monetize its mineral wealth and become economically integrated with its CASA neighbors. The external factors that will provide Afghanistan with the necessary socio-cultural transformation from a culture of war to a culture of peace and from a war-based economy to a peace-based economy is an autonomous economy dependent on its natural resources. This NSRI transformation calls for cooperation, not competition, among the great powers.

The New Silk Road Initiative Vision

After four decades of continuous war and an economy dependent on foreign aid, it seems difficult to imagine a peaceful Afghanistan with a sustainable economy in a prosperous region. Continuing to depend on U.S. military and NATO forces to safeguard its security and foreign aid as a chief driver of economic development are no longer viable options. Afghanistan must shift to a new economic growth model based on the country's $1–3 trillion natural resources to ensure peace and economic sustainability for its people. For the land-locked war-torn country Afghanistan to succeed in this transformation from war-economy to peace-economy, it must (1) break down regional trade and transit barriers, (2) invest in the transportation and energy infrastructure to facilitate the growth of transnational economic corridors, and (3) attract foreign investment from near and far on a scale previously unforeseen in Afghanistan. The New Silk Road Initiative aims to establish Afghanistan as an efficient trade and transit hub that connects China with Europe and beyond. The NSRI seeks to revive Afghanistan's historic role as an essential land bridge for Asia, facilitating the exchange of commerce, culture, people, and ideas. Many have claimed that its geographic location has determined Afghanistan's fate, that Afghanistan could just as easily be an economic hub or a battleground.

The NSRI represents a shared commitment to promote private-sector investment, increase regional trade and transit, and foster a network of economic linkages throughout the region. With these components in place, Afghanistan and the region can maximize the value of natural resources, build human capacity, create jobs, generate revenue to pay for needed services, and capitalize on the region's economic potential in the years ahead. The NSRI vision is intended to make these objectives a reality for South and Central Asian people by advancing a comprehensive, long-term economic strategy centered on a peaceful Afghanistan.

The NSRI shared vision goes beyond Afghanistan and comprises the following elements. (1) The U.S. vision of a Silk Road connecting India, Pakistan, and Afghanistan with Central Asia and then Europe is not just a large-scale trade, transportation, and infrastructure project but also a metaphor for a vast and growing network of relationships among Central Asian states, Pakistan, and India, encompassing extraction of natural resources, power generation, railroads, and gas pipelines linking together Central and South Asian countries. (2) For several years, many scholars, experts, and well-informed observers of the CASA region have advocated a policy involving a coherent and coordinated infrastructure, trade, and transport that would reintegrate Afghanistan with its Central and South Asian neighbors. Scholars such as Frederick Starr, Andrew Kuchins, Reuel Hanks, Leif Rosenberger, Richard Ponzio, Gregory Gleason, Vladimir Paramonov, Alexei Strokov, and others have called for a serious "Silk Road" policy that would restore this long-lost regional integration. The policy aims to stabilize Afghanistan economically and geopolitically, strengthening Central Asia against great power threats to its real independence and tying it more to South Asia than to Russia or China or Iran. Such a plan will provide an economic basis for Afghanistan to recover from and possibly even terminate the war, resulting in a coherent and established economic base going forward. The NSRI vision, reflecting growing regional support for closer cooperation and deeper integration, was made possible by the Afghanistan reintegration into the region's fabric, the resumption of traditional trading routes, and the reconstruction of significant infrastructure links, broken by decades of conflict. The state of Afghanistan's current war-centered economy is a direct reflection of four decades of devastation that subjugated the country and its people.

The prerequisite for this visionary transformation is durable peace. Seemingly endless Columbia and Northern Ireland wars have resulted in lasting peace agreements. The NSRI provides the only viable vision for peace, prosperity, and cooperation within Afghanistan, the region, and the U.S. Since its inception in mid-2011, the NSRI was designed to represent and advance the core U.S. values of economic and political freedom, respect for one's neighbors, and increasing regional and global prosperity by encouraging government-to-government, business-to-business, and government-to-business cooperation across borders. The NSRI's primary aim is for the U.S. and its allies to bolster peace and stability in the region by supporting a transition to trade and opening new markets that connect Afghanistan to Central Asia, Pakistan, India, and beyond. Countries in the region know they have more to gain economically by working together than by being isolated.

The NSRI, a comprehensive geoeconomic strategy for achieving peace, stability, and sustainable economic development in Afghanistan and the CASA region, would reestablish continental corridors where Afghanistan has not played a viable part for quite a while. Afghanistan has increasingly powerful neighbors with some of the fastest-growing economies in the world. For the past 40 years, Afghanistan has been the "black hole," yet with the help of regional actors and the United States, who already has a firm foothold in the country, Afghanistan can achieve peace and prosperity. Afghanistan can indeed be stabilized through its own economy. The NSRI would be an international web and network of economic, trade, transit, and people-to-people connections that link Central and South Asia, with Afghanistan at its heart, which would allow Afghanistan to attract new foreign sources of private-sector investment for its vast minerals and connect to markets abroad. Afghanistan is the pivot and, as of now, the missing link in such a Eurasian network. It is the confluence of cultures and interests that propels the vision of the New Silk Road Initiative. This initiative can once again revive the shared prosperity that the Ancient Silk Road once provided across Asia. This geographic connection would expand the U.S. influence in the region while containing China and Russia's influences, who are currently developing ambitious projects to expand the reach of their hard and soft power.

The U.S. Role in the New Silk Road Initiative

The United States, having maintained a military and economic foothold in Afghanistan for nearly two decades, is the only global power that can establish and properly champion the NSRI to succeed. As the NSRI is designed to rebuild the trust and relationships that have been lost after decades of war, not only within Afghanistan but also within the region itself, bringing a durable peace and stability in Afghanistan is vital for the U.S. to compete in the Great Power Competition effectively. In the complex and volatile environment of the GPC, the U.S. can exert leadership and "convening power," supporting and facilitating coordination of governments, businesses, and non-governmental organizations interested in aiding the region through collaborative economic projects that will improve stability in the region. The U.S. can convene a broader community of stakeholders around the shared geoeconomic goal of integrating Afghanistan and the region and securing investments for its vast natural resources. Starr and Farhadi (2012) argue that the U.S. should put aside its costly "big-ticket infrastructure projects" and instead embrace "initiatives that require the U.S. to exercise its convening power and 'soft power' leadership." By working to facilitate a durable peace and economic stability in Afghanistan, the U.S.

can secure its presence, relevance, and strength in relation to global and regional powers competing in the GPC.

The alternative is bleak. If the U.S. and the West were to choose to abandon Afghanistan or even give the perception of abandonment at this critical moment, the U.S. would lose its credibility and influence in the CASA region, and the vulnerable Afghan state would deteriorate into failed state status once again. Rival great and regional powers would carve out Afghanistan and wage proxy wars over its vast minerals, particularly its rare earth and critical minerals. In such a situation, the U.S. would face even greater national security threats than at present, further weakening its position in the GPC. Notably, should the United States not integrate Afghanistan into an economic corridor, China most certainly will. This dire scenario would be a devastating loss of power, leadership, and opportunity for the United States, specifically as it relates to mineral extraction. As Farhadi and Bekdash (2020) argue, the era of the Great Power Competition is essentially a competition for mineral resources.

In February of 2020, the United States led by Ambassador Zalmay Khalilzad, Special Representative for Afghanistan Reconciliation at the State Department (and keynote speaker at the 2020 GPC, USF) signed a peace deal with the Taliban that sets conditions for a full military withdrawal from the country. Afghanistan is now approaching the Intra-Afghan negotiation stage, between the Taliban and the Kabul government. Khalilzad insisted that in order to focus on the Great Power Competition, peace in Afghanistan must be established, and the resources freed from the Afghan conflict must shift per the National Defense and National Security Strategies. However, to have a durable peace, the United States should remain engaged in Afghanistan through soft-power economics as opposed to hard-power military power. This engagement would be advantageous for the United States in the Great Power Competition era, allowing for the U.S. private sector to invest in Afghanistan's minerals and maintain its relevance and influence in the region.

The Great Power Competition era has spurred the United States to a peace deal with the Taliban in which the U.S. will militarily withdraw from Afghanistan by late 2021. This military withdrawal from Afghanistan should not mean abandonment, but rather a transformation from military engagement to NSRI. In the GPC era, the NSRI should be the U.S. response to China's BRI as well as its plan for sustained peace in Afghanistan. Michael Singh (2020), whose chapter appears later in this volume, argues that the BRI was itself a response to U.S. policy shifts and action. Singh (2020), citing Nadege Rolland, states that the "BRI emerged as a result of two developments that led the Chinese leadership to rethink their approach to achieving China's rise: the global financial crisis of 2008, which hampered Chinese economic growth, and the U.S. 'rebalance' to Asia, viewed in Beijing as an effort to contain China." Indeed, the BRI is now an overt attempt to contain the U.S. in the region. If the U.S. intends to maintain its role in the international order in the Great Power Competition era, it must effectively compete with the BRI.

Theoretical Basis for Recommendations: Geoeconomics and New Economic Geography Theory

As the Great Power Competition's defining feature is its economic character, China is widely accepted as the U.S.'s main competitor in the Great Power Competition. It is for this reason that this chapter recommends that the U.S.'s strategy in the GPC encompass economic influence. This chapter recommends that the U.S. strengthen its engagement in Afghanistan's economic development and integration, specifically through support for Afghanistan's transportation and trade infrastructure, underpinned by geoeconomics and New Economic Geography (NEG) theoretical frameworks.

Geoeconomics

Through economic dominance, states can act geostrategically in the world space through hard power projections. This economic power, described by Jaeger and Brites (2020) as a "mechanism for obtaining resources of power," is known as geoeconomics. Geoeconomics theory explains that economic intervention, such as that by the U.S., may exercise its soft and convening power in Afghanistan as an effective way to influence world power distribution structurally. According to Fußstetter (2016), "the aim of geoeconomics is to gain a sustainable competitive advantage over other countries and regions through geographical, cultural, or resource-related aspects." This advantage is accomplished by gaining control over foreign economies' strategic sectors (Csurgai 2017). The use of these economic means by international actors can structurally influence world power distribution in their favor. Geoeconomics can create spheres of influence while also reaping economic benefits, which is the goal of both China's Belt and Road Initiative and the U.S.'s New Silk Road Initiative.

In the current Great Power Competition era, economic and military power are interconnected, particularly in natural resources. Two prominent thinkers of geoeconomics, Harris and Blackwell (2016), argue that the U.S. has been losing ground internationally because it has continued to rely on military power while other nations "increasingly carry out geopolitical combat through economic means. Policies governing everything from trade and investment to energy and exchange rates are wielded as tools to win diplomatic allies, punish adversaries, and coerce those in between" (Harris and Blackwell 2016). Jaeger and Brites (2020) assert that geoeconomics is, in fact, the only economic power that can promote and sustain national interests and sustain the power of the Great Powers through the maintenance of military, diplomatic, and intelligence services. They characterizes the Great Power Competition era as one of the international political economies in which economic factors determine international relations. This rationale can be seen in the race to procure the vaccine for COVID-19 through development or economic power.

Edward N. Luttwak, the authority on geoeconomics, predicted in the 1990s that economic rivalries would replace the ideological rivalries between the Western liberal and communist collectivist models of society. In other words, rather than a focus on military power, trade, finance, and important military technology, mastery of soft-power would prevail (Chacko 2016). This shift from military power to economic power is what Luttwak terms *geoeconomics*, which he defines as the "interstate rivalry and conflict conducted through commerce methods" (Chacko 2016). The prioritization of economic growth over military displays would, Luttwak argues, leads to competition for trade routes, access to natural resources, and conquering markets (Csurgai 2017). This shift away from military power has indeed been the dominant characteristic of the Great Power Competition, though military power does remain important. Csurgai argues that military confrontations and direct control of foreign lands, a methodology that has recently been given priority over economic power flexing, is now seen as less advantageous than the use of indirect strategies and soft power. However, such economically focused state behavior has been seen previously in the mercantilism era, which was closely associated with the nation state's rise. Mercantilism led to Venice becoming the most powerful player of the Ancient Silk Road. The machination of the modern-day Great Power Competition, like those of the mercantilist era, has the potential to yield changes in the global stage and shift the concentration of power.

Many authors in this volume suggest that the United States should remain engaged in Afghanistan to aid in its socio-economic development. The United States should remain strategically engaged in Afghanistan to develop the extractive industry and establish an infrastructure that allows for the creation of the New Silk Road Corridors to provide market access for Afghanistan's minerals. This will, in turn, benefit the United States. The U.S.'s geoeconomic strategy for Afghanistan's minerals will put the U.S. in the role of entrepreneur. Moiso and Paasi (2013) echo this sentiment, characterizing the move as a privatization of the state.

According to geoeconomic theory, future conflicts will take the form of resource competition (Fußstetter 2016). Both the United States and China have a great need for the minerals and extractives that are abundant within the CASA region. Though China's consumption of minerals and petroleum falls far below the levels of other industrialized countries, its government is setting the stage for future consumption levels, fully expecting that as China continues to industrialize and its economy continues to grow, its need for energy will continue to rise. The United States Energy Information Administration (2020) projects that U.S. energy consumption levels will also continue to rise. With this competition for resources, it becomes vital that the United States gain the upper hand in the resource competition and firmly establish the New Silk Road. With the establishment of the New Silk Road comes many opportunities. First, the United States would find itself in a position of economic power and security through a stable stream of resources. China currently has a near-monopoly on rare earth minerals and is unafraid to use this monopoly politically. This political power translates into economic power. Secondly, with the establishment of the New Silk Road, the United States would have immense influence in the region. This soft-power influence is vital in the creation of economic systems through cultural understanding. This economic stability, especially in lands that have long suffered under the yoke of wars in Afghanistan, translates into sustainable peace. Lastly, the NSRI is the only viable option to successfully end the "endless war" and still preserve the gains of the last two decades for the U.S. and the Afghan people.

New Economic Geography

New economic geography (NEG) illuminates why Afghanistan's transportation infrastructure, of all potential investment sites, is a critical economic intervention. For landlocked countries such as Afghanistan, transport connectivity is the most pressing priority for economic development. Landlocked Least Developed Countries, in particular, must devote significantly more resources to a functioning and effective transportation sector than countries with access to the sea. Because the traditional development aid model has proved itself neither viable nor sustainable in Afghanistan, the NEG offers the most appropriate economic growth theory for this landlocked, fragile, conflicted state. NEG states that once an economic belt is established, production, growth, development, and consumerism will flow, and the economy will grow (Krugman 1991).

As defined by Paul Krugman, the theoretical framework of NEG bears directly on Afghanistan's economic growth challenges because of the country's landlocked geography. Thus, unlike "neo-classic" or "endogenous" traditional economic growth theories, NEG recognizes the importance of geography and transportation costs as a major enabler or hindrance to economic growth. Gallup et al. (1999) argue that "geography matters" to economic growth due to high and difficult transport costs and procedures. NEG economic growth theory asserts that lowering transportation costs will have a direct positive effect on economies of scale and "agglomeration," called the "home market effect," required for sustainable economic growth (Hoaby 2005). NEG also asserts that "transportation costs will decline due to such things as economies of scale, technological advances that increase efficiency, better transport infrastructure, and removing procedural barriers" (Hoaby 2005). Also, the traditional economic theory, with its emphasis on "static allocative efficiency," has a limited capacity to explain how economic growth and development can be achieved through such mechanisms as research and development and technological diffusions, especially in fragile, conflict-affected states (Agalewatte 2004). NEG is much more capable than the traditional economic theory of explaining how such dynamic factors influence growth and enable a country and international community to use resources more effectively to achieve a sustainable economic development model (Agalewatte 2004).

Summary

This chapter has presented a general conceptualization of the Great Power Competition, a proposed U.S. strategy for engagement, and a brief theoretical basis for that proposed strategy. The relative newness and extreme complexity of the GPC has prevented such a conceptualization from being cogently presented before now. The chapter described the origins of this geopolitical situation and of the GPC terminology itself, gave an overview of the major powers and their strategies and objectives to date, and proposed a role for the U.S., centered on Afghanistan, which is a precarious, but resource-rich country at the heart of Asia, the main battleground of the GPC. Finally, the chapter offered a brief theoretical basis for this proposed strategy, founded in geoeconomic and New Economic Geography theory. As this chapter recommends, it is in the United States' best interests to support Afghanistan in becoming a hub of the New Silk Road initiative in Central and South Asia, a move that will preserve U.S. relevance, influence, and two-decade gains in Afghanistan in this critical GPC battleground.

Countering the revisionist powers in the GPC requires the U.S. not only to understand its rivals' strategic objectives, perspectives, and challenges, but also to identify common areas of interest and potential pathways to more holistic approaches, intersections, and/or cooperative efforts in combatting terrorism and transnational organized crime in areas such as Afghanistan and Syria, deterring aggressive Iranian affronts, and reigning in a nuclear North Korea. There is a need for diverse perspectives in this increasingly complex and adaptive environment. If the United States does not remain engaged in Afghanistan as a partner in its extractive industry, the country will surely be engaged by competitors, such as Russia and China, the latter of which already has a firm foothold in the strategic rare earth and critical mineral sector.

References

T. Agalewatte, Competitive industry policy for economic development in Sri Lanka: lessons from East Asia. University of Wollongong Thesis Collection (2004)

American Geosciences Institute, Critical mineral basics. AGI.org (2020), https://www.americangeosciences.org/critical-issues/critical-mineral-basics

I. Ali, *U.S. Military Puts 'Great Power Competition' at Heart of Strategy: Mattis* (Reuters, 2018), https://www.reuters.com/article/us-usa-military-china-russia/u-s-military-puts-great-power-competition-at-heart-of-strategy-mattis-idUSKBN1F81TR

A. Boroff, What is great power competition, anyway? Modern War Institute (2020), https://mwi.usma.edu/great-power-competition-anyway/

H. Brand, E.B. Montgomery, One war is not enough: strategy and force planning for great power competition. Texas Natl. Secur. Rev. **3**(2), 80–92 (2020)

R. Burns, Pentagon sees Taliban deal as allowing fuller focus on China. AP News (2020), https://apnews.com/21926f8afb7ccf3599f735d6fb966854

S.E. Burke, W. Scott, US needs a strategy for minerals competition: it's not just rare earths. The Hill (2019), https://thehill.com/opinion/national-security/449031-us-needs-a-strategy-for-minerals-competition-its-not-just-rare

P. Cai, *Understanding China's Belt and Road Initiative* (Lowy Institute, 2017), www.lowyinstitute.org/publications/understanding-belt-and-road-initiative#_ednref1

P. Chacko, India and the Indo-Pacific from Singh to Modi: geopolitical and geoeconomic entanglements, in *New Regional Geopolitics in the Indo-Pacific: Drivers, Dynamics and Consequences*, ed. by P. Chacko (Routledge, New York, 2016)

A. Chatzky, J. McBride, China's massive belt and road initiative. Council Foreign Relations (2020), https://www.cfr.org/backgrounder/chinas-massive-belt-and-road-initiative

N. Childs, Great-power competition and Covid-19. Military Balance Blog (2020), https://www.iiss.org/blogs/military-balance/2020/05/great-power-competition-covid-19

E.A. Colby, A.W. Mitchell, The age of Great Power Competition. Foreign Affairs (2020), https://www.foreignaffairs.com/articles/2019-12-10/age-great-power-competition

A. Cottig, T. Woodall, J.M.T. Javier, Russia struggling to capitalize on rare earth reserves. SP Global (2019), https://www.spglobal.com/marketintelligence/en/news-insights/latest-news-headlines/russia-struggling-to-capitalize-on-rare-earth-reserves-52525919

G. Csurgai, The increasing importance of geoeconomics in power rivalries in the twenty-first century. Geopolitics **23**(1), 38–46 (2017)

A. Farhadi, A. Bekdash, Afghanistan's Lithium as strategic U.S. focus in the great power competition, in *Great Power Competition: Regional Perspectives in the Central Region*, ed. by A. Farhadi, A.J. Masys (Springer, New York, 2020)

U. Friedman, The new concept that everyone in Washington is talking about. The Atlantic (2019), https://www.theatlantic.com/politics/archive/2019/08/what-genesis-great-power-competition/595405/

A. Fußstetter, Geoeconomic implications of border conflicts in Asia based on resource competition. Georg-August-Universität (2016)

J.L. Gallup, A. Sachs, A.D. Mellinger, Geography and economic development. Int. Reg. Sci. Rev. **22**(2), 179–232 (1999)

J. Garamone, Operation warp speed on track for end-of-year vaccine delivery. United States Department of Defense (2020), https://www.defense.gov/Explore/News/Article/Article/2311177/operation-warp-speed-on-track-for-end-of-year-vaccine-delivery/

D. Goure, The measure of a superpower: a two major regional contingency military for the 21st century. The Heritage Foundation (2013), https://www.heritage.org/defense/report/the-measure-superpower-two-major-regional-contingency-military-the-21st-century

A.L. Gulley, N.T. Nassar, S. Xun, China, the United States, and competition for resources that enable emerging technologies. PNAS **115**(16), 4111–4115 (2018)

R.D. Harris, J.M. Blackwell, *War by Other Means: Geoeconomics and Statecraft* (Belknap Press, Cambridge, 2016)

F. Hill, T. Madan, A. Sloat et al., Balancing act: major powers and the global response to U.S.-China great power competition. Brookings Institute (2020), https://www.brookings.edu/research/balancing-act-major-powers-and-the-global-response-to-us-china-great-power-competition/

S. Hoaby, Can the new economic geography explain inequality between the great plains and the great lakes? (2005)

B.C. Jaeger, P.V.P. Brites, Geoeconomics in the light of international political economy: a theoretical discussion. Braz. J. Polit. Econ. **40**(1), 22–26 (2020)

F. Kempe, G. Massolo, Preface, in *The MENA Region: A Great Power Competition*, ed. by K. Mezran, A. Vervelli (The Atlantic Council, Washington DC, 2019), https://www.atlanticcouncil.org/wp-content/uploads/2019/10/MENA-Preface.pdf

R.D. Kiggins, The strategic and security implications of rare earths, in *The Political Economy of Rare Earth Elements: Rising Powers and Technological Change*, ed. by R.D. Kiggins (Palgrave Macmillan, London, 2015)

P. Krugman, Increasing returns and economic geography. J. Polit. Econ. **99**(3), 483–499 (1991)

C. Lovotti, Redistribution of power in the Middle East: Moscow's return to Syria, in *The MENA Region: A Great Power Competition*, ed. by K. Mezran, A. Vervelli (The Atlantic Council,

Washington, DC, 2019), https://www.atlanticcouncil.org/wp-content/uploads/2019/10/MENA-Preface.pdf

A. Lyrichikova, G. Stolyarov, Russia has $1.5 billion plan to dent China's rare earth dominance. Reuters (2020), https://www.reuters.com/article/russia-rareearths/russia-has-15-billion-plan-to-dent-chinas-rare-earth-dominance-idUSL8N2F73F4

MITRE, Great power competition. Mitre.org (2020), https://www.mitre.org/news/focal-points/great-power-competition

P.K. Macdonald, J.M. Parent, Graceful decline? The surprising success of great power retrenchment. Int. Secur. **35**(4), 7–44 (2011)

J. Marcus, What's going on between Russia, US, Afghanistan? BBC News (2020), https://www.bbc.com/news/world-asia-53220163

J.J. Mearsheimer, *The Tragedy of Great Power Politics* (W.W. Norton and Company, New York, 2001)

S. Moiso, A. Paasi, From geopolitical to geoeconomic? The changing political rationalities of state space. Geopolitics **18**, 267–283 (2013)

P. Musgrave, Trump's Covid-19 response shows he hasn't given up on competition with China that doesn't involve guns and bombs. Business Insider (2020), https://www.businessinsider.com/covid-19-us-inability-to-compete-beyond-military-power-2020-7

National Archives, A federal strategy to ensure secure and reliable supplies of critical minerals. Federal Register (2020), https://www.federalregister.gov/documents/2017/12/26/2017-27899/a-federal-strategy-to-ensure-secure-and-reliable-supplies-of-critical-minerals

National Defense Strategy of the United States of America, United States Department of Defense (2018), https://dod.defense.gov/Portals/1/Documents/pubs/2018-National-Defense-Strategy-Summary.pdf

National Security Strategy of the United States of America, The White House (2017), https://www.whitehouse.gov/wp-content/uploads/2017/12/NSS-Final-12-18-2017-0905.pdf

E. Rahajay, China-Afghanistan relations in the post-Taliban period. Himal. Cent. Asian Stud. **23**(4), 124 (2019)

Reuters Business News, U.S. dependence on China's rare earth: trade war vulnerability. Reuters (2019), https://www.reuters.com/article/us-usa-trade-china-rareearth-explainer/u-s-dependence-on-chinas-rare-earth-trade-war-vulnerability-idUSKCN1TS3AQ

B.R. Rubin, Humanitarian imperative over great power competition in Afghanistan, in *Great Power Competition: Regional Perspectives in the Central Region*, ed. by A. Farhadi, A.J. Masys (Springer, New York, 2020)

S. Ryan, Great power competition in information environment in the CASA Region, in *Great Power Competition: Regional Perspectives in the Central Region*, ed. by A. Farhadi, A.J. Masys (Springer, New York, 2020)

U. Sheraz, Foresight as a tool for sustainable development in natural resources: the case of mineral extraction in Afghanistan. Resour. Policy **39**, 92–100 (2014)

M. Shivers, Central Asia great power competition: a 200-year history continues, in *Great Power Competition: Regional Perspectives in the Central Region*, ed. by A. Farhadi, A.J. Masys (Springer, New York, 2020)

M.K. Singh, China and the United States in the Middle East: between dependency and rivalry, in *Great Power Competition: Regional Perspectives in the Central Region*, ed. by A. Farhadi, A.J. Masys (Springer, New York, 2020)

E. Smith, Russia does not want to exploit US-China trade war, finance minister says. CNBC (2019), https://www.cnbc.com/2019/10/19/siluanov-russia-does-not-want-to-exploit-us-china-tensions.html

F.S. Starr, A. Farhadi, Finish the job: jump start Afghanistan's economy. Central Asia-Caucus Institute Silk Road Studies Program (2012), https://www.silkroadstudies.org/resources/pdf/SilkRoadPapers/2012_11_SRP_StarrFarhadi_Afghanistan-Economy.pdf

C. Thepaut, The Astana process: a flexible but fragile showcase for Russia. The Washington Institute (2020), https://www.washingtoninstitute.org/policy-analysis/view/the-astana-process-a-flexible-but-fragile-showcase-for-russia
P. Tucker, How would Biden wage great power competition? Defense One Magazine (2020), https://www.defenseone.com/technology/2020/07/how-biden-will-wage-great-power-competition/166570/
D. Twining, P. Quirk, Winning the great power competition post pandemic. The American Interest (2020), https://www.the-american-interest.com/2020/05/11/winning-the-great-power-competition-post-pandemic/
United States Energy Information Administration, Total Energy (2020), https://www.eia.gov/totalenergy/
W.T. Wechsler, U.S. withdrawal from the Middle East: perceptions and reality, in *The MENA Region: A Great Power Competition*, ed. by K. Mezran, A. Vervelli (The Atlantic Council, Washington DC, 2019). https://www.atlanticcouncil.org/wp-content/uploads/2019/10/MENA-Preface.pdf
M.G. Weinbaum, Afghanistan and the great powers in regional geopolitics, economics and security, in *Great Power Competition: Regional Perspectives in the Central Region*, ed. by A. Farhadi, A.J. Masys (Springer, New York, 2020)
K.B. Williams What's the Great Power Competition? No one really knows. Defense One Magazine (2019), https://www.defenseone.com/threats/2019/05/whats-great-power-competition-no-one-really-knows/156969/
W.C. Wohlforth, Unipolarity, status competition, and great power war. World Polit. **61**(1), 28–57 (2009)

Afghanistan's Lithium as Strategic U.S. Focus in the Great Power Competition

Adib Farhadi and Ayman Bekdash

Abstract The contemporary geoeconomic Great Power Competition can be understood as global competition for access to a finite supply of necessary rare earth and critical minerals such as lithium. China's current near-monopoly on these minerals presents a potential national security risk to the U.S. as access to these vital resources depends on the two powers' unpredictable relationship. The critical mineral lithium is used in a wide range of technologies and is expected to increase in demand in the coming years significantly. As an alternative to Chinese-controlled lithium sources, Afghanistan presents a promising option, rendering this fragile, centrally located country a key strategic site for the U.S. in the modern Great Power Competition. However, even though vast mineral wealth has been discovered in Afghanistan in recent years, the country is missing the essential mining market access infrastructure to turn these resources into actual wealth to sustain itself financially. Therefore, this chapter argues that as the U.S. plans to withdraw militarily from Afghanistan, it should transition to a strategic role of helping Afghanistan develop its vast mineral wealth. Such a strategy will ensure Afghanistan does not return to a failed state and ensure the U.S. a reliable source of critical mineral lithium.

Keywords Afghanistan · Great power competition · Lithium · Rare earth minerals · China · Economy · Central Asia

A. Farhadi
University of South Florida, Tampa, FL, USA
e-mail: farhadi@usf.edu

A. Bekdash (✉)
DGC International, 7950 Jones Branch Dr, McLean, VA 22102, USA
e-mail: abekdash@dgci.com

© The Author(s), under exclusive license to Springer Nature Switzerland AG 2021

A. Farhadi et al. (eds.), *The Great Power Competition Volume 1*,
https://doi.org/10.1007/978-3-030-64473-4_3

Introduction

The contemporary geoeconomic Great Power Competition can be understood as a global competition for access to a finite supply of necessary rare earth and critical minerals. The primary distinguishing feature of this era is a competition for resource security with pipeline routes and natural resources "as the center of gravity of the New Great Game" (Rahman 2014). This competition for resource security has taken on many forms and is fought on many battlegrounds, both in the diplomatic and physical spheres. At present, a near-monopoly on certain rare earth and critical minerals belongs to China, whose exclusive access to mining and distribution of these minerals gives it power over those countries who depend on access to those resources, including the United States. China's dominance of the rare earth and critical minerals market has expanded to the point that it currently supplies 95% of the world's rare earth (Bernard and Sugarman 2012). Among U.S. imports of minerals, China currently accounts for approximately 80% of the supply due to its near-monopoly control of all the mineral processing facilities. While on an official tour of southern China in 1992, then-Premier of China, Deng Xiaoping, is reputed to have claimed, "The Middle East has its oil; China has rare earth" (Kiggins 2015). Chinese control of the market is largely attributable to China's mines, as well as its relaxed environmental regulations and low labor costs (Gholz 2014).

For China to have such power over such a strategically important product is a major concern not only for the United States, but also for Russia, who is actively seeking foreign investments of $1.5 billion to further develop and expand their rare earth mineral industry (Scheyder 2019; Lyrichkova and Stolyarov 2020). According to Lyrichkova and Stolyarov (2020), Russia aims to become the second-largest producer of rare earth after China by 2030. This ambition places further stress on the United States, particularly so in the era of the Great Power Competition, to develop its own independent source of rare earth and critical minerals.

China's overwhelming monopoly on rare earth, amounting to 67% of global rare earth mineral production, constitutes a security risk for the U.S. because of the insecurity of supply and overwhelming dependence on China's production and reserve (Kiggins 2015; Lyrichkova and Stolyarov 2020). As Dixit explains, strategic minerals are those that are imperative to the security of a country but are obtained largely from foreign sources because the supplies available within the country concerned would not be adequate in a time of national emergency. The three criteria defining strategic minerals are that they are essential for national defense, essential for industry and civilian uses, and do not have any suitable substitutes (Dixit 2015). Butt and Thomas explain that strategic minerals are those considered to be essential for critical civilian and military needs in quantities not available from either domestic sources or secure foreign sources, and for which short-term substitutes are also not available (Dixit 2015). In the case of the GPC, China has demonstrated that it is unafraid to weaponize its monopoly on strategic rare earth and critical minerals against other states, a manipulation of power that has been described as the "new geopolitics of minerals" (Kiggins 2015; Dixit 2015).

Given that the Great Power Competition is essentially a competition for rare, necessary minerals, the United States could find itself in a position of weakness should it enter a military conflict in the coming years. China understands the strategic importance of rare minerals and has used their dominance of the rare earth market to manipulate political events in their favor. China's understanding of the power bestowed by its rare earth access, and its corresponding flexing of political power, is clearly seen in, for example, the diplomatic entanglement between China and Japan in late 2010 (Kiggins 2015). China has repeatedly used rare earths to coerce modification of policy on the part of another state, and its unilateral reduction of rare earth exports to all rare earth consumers by 35% jolted policymakers in rare earth consuming states (Kiggins 2015). For these reasons, rare earth mineral sourcing has become a major concern for the U.S. in the context of the GPC. As James Litinsky, co-chair of the United States' only rare earth mine, has repeatedly stated, the U.S. is in dire need of a "sustainable supermajor for the Western supply of these minerals" (Scheyder 2019).

One of the critical minerals at play in the GPC is lithium. A member of the alkali metal group, a group of metals that are lightweight and very reactive to oxygen, lithium is a powerful superconductor in small quantities and has found its way into many technologies. The uses of lithium have historically been centered around the production of glass and ceramics, but the mineral has increasingly been used for medicinal purposes, as well as in warfare as an active ingredient for nuclear weapons. Its power as a superconductor lies in the fact that, when the element comes into contact with water, it forms into an alkali; according to Kavanagh et al. (2018), it "has the highest specific heat capacity (at 25 °C) of any solid element" and "is the most polarizing of all the alkali metals and more electronegative than H." Due to its reactive nature to oxygen, lithium is not found as pure metal in nature but rather is in various salts and minerals. Today, lithium is used in technologies ranging from cell phones, televisions, and computers, to fiber optics and pharmaceutical products, as well as many other important technologies that are used in mass throughout the world.

Lithium's lightweight conductivity makes it a key ingredient in a highly desirable type of battery. Lithium-based batteries are currently preferred for electric cars and other uses because they can carry far more energy than other batteries in a small and lightweight form (Rapier 2019). It is estimated that by 2040, over half the cars in the world will be electrically powered and contain a lithium battery pack (Rapier 2019). The rise in green energy, such as the use of lithium-powered electric vehicles, is increasing demand for lithium. The global sales of electric cars have increased tenfold in the last five years, and with more and more countries placing bans and restrictions on petroleum powered cars, sales will only grow in the future (Rapier 2019). According to Peiro et al. (2013), the production of lithium secondary batteries grew by 25% between 2000 and 2007, which is the single largest area of growth in lithium use. As the manufacturing and popularity of electric vehicles increases, demand for lithium will as well. Piero describes the way that many countries and government bodies have created initiatives for the manufacturing and consumption of electric vehicles, with the EU publishing several directives to promote electric vehicles, and several European countries such as France, Britain, Spain, and Denmark launching national programs to promote electric car use, creating initiatives such as tax cuts and free parking for electric cars, as well as investing in the infrastructure needed for electric cars to be manufactured as well as used (Peiro et al. 2013). Yet, reports by the British Geological Institute state that even if the electric vehicle market does not take off in the manner in which it has been predicted, lithium demand will continue to increase at a rate of 8% per year due to its important role in daily technologies (British Geological Survey 2016).

Lithium is expected to soon be classified as a critical raw material (CRM), defined by the British Geological Survey as "a material which forms a strong industrial base, producing a broad range of goods and applications used in everyday life and modern technologies" in 2020 (British Geological Survey 2016). As such, lithium is now integral to modern life. Yet, according to Kavanagh et al. (2018), lithium is "far less abundant in the universe than it has been predicted to be." Given the importance and rarity of lithium, as a mineral whose supply is currently dominated by China, the U.S. needs to establish its own sources for minerals such as lithium for the sake of global security.

This paper argues that the U.S.'s strategy for accessing rare earth minerals should center on the country of Afghanistan, which is the most promising alternative source of such minerals and is thus a key strategic site in the Great Power Competition. Vast mineral wealth has been discovered in Afghanistan in recent years, and all that is missing to turn these resources into actual wealth and prosperity are the requisite mining and market access infrastructure. The present moment happens to be a historic one for Afghanistan, as the era of U.S. military presence in Afghanistan comes to a close in May 2021. For the first time in forty years, warring parties are willing to make peace: the Taliban, working in concert with the U.S. and its allies, have negotiated a peace agreement in which the U.S. agreed to withdraw its military presence in exchange for the Taliban's assurance that they will prevent protection of terrorist groups within their borders.

Even as the U.S. withdraws militarily, it has an active role to play in establishing and maintaining a durable peace. Such peace depends on the stabilization of Afghanistan's economy around a profitable commodity; thus, Afghanistan must explore which economic avenues it has available to it. The clearest path to economic autonomy, a byproduct of which would a sustained peace for the region, is in the extraction of Afghanistan's natural resources. On the other hand, a failure to create opportunities for prosperity would result in Afghanistan reverting to its past state of poverty, societal regression, insurgency, and a safe haven to violent extremism. It is the stance of this paper that the U.S. must shift to a form of developmental presence and economic engagement to support Afghanistan in monetizing its vast mineral wealth. Afghanistan's most promising route to a stable peacetime economy is to create the infrastructure necessary to mine, transport, and trade its mineral wealth, including lithium. Because the United States already has a firm foot in Afghanistan, the U.S. is a natural partner for Afghanistan in developing its vast rare earth mineral wealth. Not only would this development allow for other countries to circumvent China, but it would also allow for the United States to maintain independent access to these strategic minerals. Afghanistan presents itself as a strategic country in which the U.S. can maintain a presence to counter the problem of the Chinese monopoly.

This paper elaborates on how the monetization of Afghanistan's lithium can stabilize its economy and integrate it with its neighbors of the Central Region. The paper also discusses the strategic role that the United States should play in Afghanistan in order to maintain a durable peace through economic development. The United States should invest in Afghanistan's lithium as well as investing in its infrastructure and extractive industry. First, we explain Afghanistan's various challenges related to its geography and political climate, the need for continued U.S. engagement in the region, and the promise of Afghanistan's minerals as the key to its economic development and integration. Second, we explore how the development of Afghanistan's trade and transportation infrastructure can enable the country to monetize its mineral wealth. Finally, we discuss Afghanistan's neighbors in the region to illustrate how the development of Afghanistan's infrastructure for bringing its mineral wealth to market can contribute to economic integration of the region and stabilize the region via shared prosperity.

Afghanistan's Lithium as Strategic U.S. Focus in the GPC

The Central Region has been described as the most important battleground of the Great Power Competition, with China, Russia, and the United States all maintaining a firm and growing presence in the region. Afghanistan, in particular, has proven itself to be vital to the Great Power Competition due to its central location and precarious condition. Since the nineteenth century, Afghanistan has been dependent on foreign aid, beginning with the large subsidies provided by the British to Afghan Emirs (Fayez 2012). This dependence on foreign funds continued into the Cold War, the Afghan Civil Wars, and then again, albeit on a larger scale, upon the United States' occupation of Afghanistan (Fayez 2012). Since 2001, the U.S. has provided billions in foreign aid to fund social programs, basic government functions, infrastructure projects, and security. Beginning in spring 2020, however, the United States has begun to withdraw its funding from Afghanistan due to the COVID-19 pandemic and the February 29, 2020, peace deal between the United States and the Taliban. The imminent withdrawal of foreign aid leaves Afghanistan in urgent need of a strategy for rapidly stabilizing a decentralized peacetime economy.

The country's longtime dependence on foreign aid has created a rentier state that Fryklund (2013) describes as "fueling a culture of corruption never previously seen in Afghanistan" due to an influx of cash and a lack of "accountability to its citizens as well as a lack of oversight from its donors." The Afghan economy has been in ruin for decades and has largely been based on opium production and narcotics trafficking (Risen 2010). Additionally, the Taliban's presence in the country still presents a security concern. While major population centers are largely not in danger of being overrun by the Taliban, high-profile attacks still remain a persistent threat. In order to achieve a durable peace, Afghanistan requires a sustainable economy independent of foreign aid. Even though 30% of Afghanistan's people currently live under the poverty line and 40% of the 6-million-strong labor force is unemployed, such a self-sufficient economy is possible: Afghanistan is a very rich country (Fayez 2012), merely requiring the means to mine its mineral wealth and bring it to market.

Yet Afghanistan faces major obstacles in tapping into its mineral wealth. Beyond the problem of needing adequate infrastructure to extract the minerals and transport them, Afghanistan must also contend with the overwhelming narrative of the resource curse. The thesis of the resource curse first came about in 1995 when Sachs and Warner (1995) studied 97 developing countries and observed that countries with a high ratio of natural resource exports in 1971 had lower economic growth rates by 1989 than countries that were not endowed with natural resources. The resource curse only became a working hypothesis in the aftermath of the 1970s when resource-rich countries showed slower rates of economic growth (Bruenecker et al. 2014). Yet, much evidence demonstrates that the resource curse is not a given, but rather a mere hypothesis that benefits former colonial powers by creating dependency on foreign aid and imports in "traditional" countries. It is the position of this paper that the key to Afghanistan's peace and autonomous prosperity lives under its surface and within its lands in the form of its natural resources. Thus, the country's best hope is to return to its historic focus on mining and trading natural resources such as rare earth minerals, critical minerals, and gemstones.

In 2010, the United States and the Afghan Geological Survey teams discovered vast mineral deposits throughout Afghanistan, comprising copious amounts of iron, copper, cobalt, and gold. Sheraz (2014), a Senior Policy Analyst for COMSTECH, places the estimates of the discovery of this extractable resource at around $1–3 trillion. Among the 14,000 minerals discovered in Afghanistan are its vast lithium reserves, a material that is vital to the age of the technology-dependent global economy (Reeves 2012; Dowd 2013). According to Kavanagh et al. (2018), lithium deposits scattered throughout the country exist at concentration levels between 41 and 99 mg/L. Kavanagh et al. (2018) say that such lithium concentrations would place Afghanistan among the top producers of lithium, alongside Australia, Bolivia, Argentina, and Chile.

Lithium is a critical resource material that is becoming increasingly important in the modern global economy. Such new wealth could not only permanently alter the Afghan economy, which currently has a gross domestic product (GDP) of about $12 billion per year, but can also alter the global standing of the U.S. with immediate access to Afghanistan's lithium. According to Rahman (2014), the discovery of lithium in Afghanistan could provide the country a much-needed economic boost and create thousands of jobs for Afghans. It could also encourage political stability through licit exports and taxation, as well as help to eradicate violent extremism. Lithium in Afghanistan could create a new Saudi Arabia, a state that has benefited greatly from its natural resources.

The discovery of lithium, referred to as "white petroleum," in Afghanistan comes at a crucial moment (Chazan 2019). China, which currently dominates the lithium product manufacturing industry due to its ability to produce and sell lithium products cheaply and without environmental restraints, has been actively investing in lithium mining projects to keep up with its increasing internal demand and has also actively been limiting exports of lithium, which could give rise to a supply security issue. Chossudovsky (2019) says that China, already a major investor and trading partner to Afghanistan, has its sights set on Afghan lithium. The Chinese state-owned company, the Metallurgical Corporation of China, already has a firm footing in Afghanistan.

Afghanistan therefore finds itself in a unique position at an opportune moment. Lithium is in high demand, and that demand will only continue to grow in the years to come. The vast Afghan deposits of lithium could be the answer to an increasingly China-dominated lithium market as well as the increasing global demand for lithium. According to Jeffrey Reeves of the Griffith Asia Institute, demand is currently equal to supply; however, by 2020, demand will begin to greatly outpace supply (Reeves 2012), with estimates of demand being placed at 900,000 tons per year by 2025 by the British Geological Survey (2016). This untouched resource could not only make Afghanistan the New Saudi Arabia of lithium, but also place Afghanistan's strategic partner in the venture in a position of power in the new technology-driven global economy.

Clearly, the key to maintaining a durable peace in Afghanistan is economic development. Afghanistan, rich in vital minerals such as lithium, can achieve economic autonomy with the development of a sustainable extractive industry and its accompanying infrastructure. These minerals can and should be the instrument of economic development, providing necessary jobs for the youthful population of Afghanistan as well as for former Taliban fighters. Holding an estimated $1–3 trillion in minerals, Afghanistan should neither be a poor country, nor should it be a failed state. Given Afghanistan's internal challenges and the larger geostrategic context of the Great Power Competition, the U.S. has a critical role to play in diplomatically facilitating the peace process and ensuring a sustained peace in Afghanistan through shared prosperity.

Building Afghanistan's Capacity to Monetize Its Minerals

Afghanistan's challenges in extracting mineral resources are largely centered on security concerns, issues of corruption, and lack of infrastructure. In a 2019 article for *Al Jazeera*, Pikulicka-Wilczewska corroborates these factors as roadblocks that have stymied Afghanistan's development and also notes a general absence of legal and organizational frameworks that could help facilitate resource extraction and reduce corruption. Given these challenges, how might the United States support Afghanistan's mineral extraction? The answer lies in engaging soft power to galvanize investment by the private sector, which can in turn foster the emergence of a long-term peacetime economy driven by Afghan citizens' own interest in the private sector alongside foreign private actors. Although the Afghan government is reluctant to award foreign investors with contracts due to their belief that the profits of mineral extraction would be distributed outside of the country—and would thus prefer the nationalization of its resources—a compromise here would be to engage in public–private partnerships with the government. In this way, Afghanistan can transition towards a developmental state with heavy macroeconomic planning along-side a market economy. To navigate this challenge, the Afghan government and its citizens need the necessary technical skills and expertise to properly emerge as an autonomous stakeholder in the region, one that is empowered by foreign private sector investment rather than subservient to it.

The theoretical basis for this recommended course of action is that of geoeconomics. Luttwak defines geoeconomics as "interstate rivalry and conflict conducted through the methods of commerce" (Chacko 2016). According to Fußstetter (2016), the aim of geoeconomics is to gain "a sustainable competitive advantage over other countries and regions through geographical, cultural, or resource-related aspects." The use of these economic means by international actors can structurally influence world power distribution. Geoeconomics can be used to create spheres of influence while also reaping economic benefits. It goes without saying that the Great Power Competition has geoeconomic dimensions as multiple entities compete for influence and world power distribution using often economic means. In the current era of the Great Power Competition, economic and military power are tied to each other, and nowhere is this clearer than in the geoeconomics of natural resources.

If the United States does not remain engaged in Afghanistan as a partner in its extractive industry, competitors such as China, who already have a firm foothold in the extractive market, or Russia, who would like to, most certainly will. China has already demonstrated their willingness to act geopolitically and geoeconomically for minerals and energy. This has been demonstrated multiple times and in multiple ways: China's pivot strategy to Central and South Asia and the Middle East, China's One Belt One Road initiative, the geostrategic strategy of territory acquisition in the South China Sea and the Strait of Malaka. China, a monopoly power in the minerals market, has manipulated the minerals market after an international crisis with Japan. Moiso and Paasi (2013) characterize this as a geoeconomic social stance, or "the process by which states seek to accumulate wealth through market control rather than through acquisition and control of territory." To allow either Russia or China into Afghanistan to develop the extractive industry would weaken the position of the United States in The international system from an economic, military, and influential standpoint.

In developing Afghanistan's capacity to monetize its mineral wealth, new economic geography (NEG) theory points to the importance of focusing on transportation infrastructure, in particular, due to Afghanistan's landlocked geography. According to NEG theory, landlocked countries such as Afghanistan must devote significantly more resources to the transportation sector, and they must take advantage of telecommunications and information technology advances to circumvent the additional burden on the transportation sector. As NEG economic growth theory asserts, lowering transportation costs will have a direct positive effect on economies of scale and "agglomeration," called the "home market effect," required for sustainable economic growth (Hoaby 2005). NEG also asserts that "transportation costs will decline due to such things as economies of scale, technological advances that increase efficiency, better transport infrastructure and removing procedural barriers" (Hoaby 2005). Traditional economic theory, with its emphasis on "static allocative efficiency," has a limited capacity to explain how economic growth and development can be achieved through such mechanisms as research and development and technological diffusions, especially in fragile, conflict-affected states (Agalewatte 2004). NEG is much more capable than traditional economic theory of explaining how such dynamic factors influence growth and enable a country and international community to use resources more effectively to achieve a sustainable economic development model (Agalewatte 2004). Thus, the NEG offers the most appropriate economic growth theory for a landlocked, fragile, conflicted state. NEG states that once an economic belt is established, production, growth, development, and consumerism will flow, and the economy will grow (Krugman 1991).

At present, the poor infrastructure in Afghanistan makes transport difficult and adds to the overall logistical cost overhead necessary for investment in mineral resources. Due to this multitude of factors, mining only comprises "between 7 and 10%" of the nation's GDP, according to Pikulicka-Wilczewska (2019). To achieve sustainable economic growth, one of the top priorities of Afghanistan is to make the country into a regional trade and transit hub. Just as peace is required for Afghanistan's economic autonomy, infrastructure is required as well. This is particularly so given that Afghanistan is a landlocked country. Recognizing the significant contribution of this sector in terms of creating employment opportunities and reducing poverty, Afghanistan intends to initiate several major projects to facilitate the country in becoming one of the most open trade and transit economies in the region.

Improving Afghanistan's transportation infrastructure capacity is key to building its autonomous economy. The next several pages will describe the current Afghan transportation network, the current bureaucratic system around trade across borders, and the weaknesses and strengths of these systems, noting concrete regional and modal differences and how these delimit the transportation environment. Particular attention will be paid to the specifics of lithium extraction and export, such as the fact that lithium's water reactivity limits the viability of various transportation modalities.

It must be noted that several factors influence both the export and transportation infrastructure environments, but that we have determined to fall outside the scope of the present paper. Political relationships, for example, constrain the range of Afghan export possibilities, as well as the potential gains in the national security of the United States. Given the American national security interest in lithium extraction, this paper focuses on U.S. as well as Afghan political relationships with those countries neighboring Afghanistan that would prove essential to the export process, as well as the current supply chain for key lithium imports to the United States. Special consideration is given to China, Iran, and Russia, particularly these countries' geopolitical roles in countering U.S. interests both regionally and globally.

Rail linkages, dry ports, and multimodal airport hubs will be the most effective solution for getting mined goods to market with minimal corruption and hands in the process. The potential benefits of establishing such infrastructure in Afghanistan are numerous. Along with its associated office space and technology, this infrastructure can support the increased trade/exports and transit required for economic sustainability and long-term stability in Afghanistan and the region. In addition, it can attract significant amounts Foreign Direct Investment (FDI) to accelerate local and regional economic activity. Because of the inherent infrastructure and strategic locations, transferring U.S. bases to entities in Afghanistan can provide a critical jump-start to the establishment of these hubs.

Being a landlocked country, Afghanistan is becoming more dependent on aviation as a form of transportation, both for individuals and businesses. Air is increasingly becoming an alternative way to transport road, rail, and sea cargo due to security concerns, unfavorable political developments, and changes in customs laws that result in extra charges being borne by Afghan traders. Various air corridors to and from Afghanistan have been established in the past few years, which have provided direct trade routes where Afghan goods, often perishable, can arrive safely and in a timely manner. These air routes have connected Afghanistan with several Asian countries, as well as with London.

Efforts focusing on trade facilitation in Afghanistan as propagated by the United States government are nothing new and have been seen as a key strategy for rebuilding the Afghan economy during conflict transition. Efforts taken by USG bodies, such as the Agency for International Development, have focused on assisting Afghanistan's ascension to the World Trade Organization, facilitating bilateral and regional trade agreements with large Afghan partners (including Pakistan), and improving customs efficiencies and laws (USAID 2019). These efforts, done in conjunction as the Trade and Accession Facilitation for Afghanistan project (TAFA) and Afghanistan Trade and Revenue Project, attempted many of the tried-and-true methods that have proven successful globally in facilitating trade for LDCs and other nations (TAFA 2019). Despite these efforts, Afghanistan still has an abysmal rating for Trading Across Borders by the World Bank's Ease of Doing Business metric, ranking 177th globally (Doing Business 2020).

Afghanistan's trade rating is hampered by numerous challenges. In every single measurable factor, Afghanistan lags behind the rest of South Asia in cost and time for import and export. The most abysmal of these metrics lies in the cost for imports, with the cost of border and documentary compliance vastly larger than the regional average and exponentially larger than the OECD High Income Average. The only metric out of the eight in which Afghanistan is regionally competitive is the time it takes to achieve export border compliance, which is more efficient than the regional average, but still four times more challenging than the OECD high-income average. Border compliance is actually the most effective aspect of trading across borders in Afghanistan, with both the import and export processes being the closest to the regional norm (Doing Business 2020).

Another key issue that plagues Afghanistan's export viability is the status of the state's infrastructure, which is an issue of significantly higher importance due to Afghanistan's landlocked nature and its lack of a seaport. Quite understandably, Afghanistan's current level of infrastructural quality is low, with the mileage of paved roads roughly half the sum of unpaved roads, and the ongoing civil war severely degrading the quality of both (SIGAR 2016). Exacerbating this problem is the limited nature of the Afghan rail system, which remains extremely hard to expand due to the topographical features of the state (Glassner 1983). Furthermore, the multi-faceted civil war continues to wreak havoc on the infrastructural system in multiple ways, including checkpoints across the country, damage to infrastructure from explosives, and kidnapping and ancillary human security threats. Air travel, while safest, is prohibitively expensive for most aspects of trade, and the transportation of hazardous materials (such as lithium) falls under even stricter regulations. While efforts to expand Afghan air trade capabilities have been relatively successful—including the creation of an air freight corridor between India and Kabul—the heavy export of an extremely volatile element by air is untested, and without improvements to the capacity of Hamid Karzai International and the other airports across Afghanistan, it remains to be seen whether this is a viable solution (Indian Government 2017).

Rail provides an intriguing option for Afghan export capacity. The current rail system is incredibly limited, however, with three short linkages: from Mazar-e-Sharif to Uzbekistan, Torghundi to Turkmenistan, and Aqina to Turkmenistan (Kakar 2017). Rail, on paper, provides the best resolution for the logistical issues plaguing large-scale exports, as rail is more cost-effective and able to transport in higher quantities than air freight or trucking, as well as less vulnerable to the unique logistical issues of transporting lithium (e.g. volatility when in contact with water). Yet, the limited nature of Afghanistan's rail system severely hampers this a viable solution, with the currently known extractable lithium naturally occurring in Daykundi and Nuristan, which are far removed from the current skeletal rail system (Cocker 2011). Expansion of the current system within the state would alleviate this problem; however, the extreme level of investment that would be required to expand this system enough to make rail transport from these locations economically viable is prohibitive.

Due to this infrastructural restriction, the only viable logistical solution, despite the challenges that befall it, would be a multi-modal solution, utilizing all logistical aspects to avoid reliance on a single one. This would entail the utilization of air freight with embarkation and debarkation predominantly occurring at the two largest and most capable airports: Bagram AFB, which is currently being utilized by the United States and allied nations in support of Operation Resolute Support and Operation Freedom's Sentinel, and Hamid Karzai International Airport, which is Afghanistan's largest. The international airports in Kandahar, Mazar-e-Sharif, and Herat can also provide logistical support and solutions. Of these, the international airport in Mazar-e-Sharif provides the most intriguing solution. As mentioned previously, Mazar-e-Sharif currently has a rail linkage with Uzbekistan, which can be utilized to haul large volumes across the border and further onwards. To augment these solutions, trucking will be essential for overcoming logistical and topographical challenges.

In order for these solutions to be viable, key investments must be made in administrative solutions and infrastructure, both in terms of Afghan development and United States national security. Investment would take two forms. The first is trade facilitation measures specifically focused on improving Afghan capacity in handling large-scale natural resource exports as well as driving down the cost and man-hours to achieve documentary compliance as well as border compliance. The second form of investment would be a more traditional form of investment, targeted at improving infrastructural capacity by investing in expanding rail capacity and air freight capacity, including storage and ancillary services as necessary. These investments, focused on improving the underlying factors that currently limit extraction opportunities, will create a more favorable investment environment for lithium extraction and, by proxy, will help bulwark the national security of the United States by ensuring a supply of lithium that is not at the mercy of Chinese central decisions. To fully achieve this objective, however, the United States must prioritize incentivizing investors who are friendly to the interests of the United States to lead this extraction effort, in collaboration and at the behest of the Afghan government.

Expanding trade promotion investments as well as rail and air freight capacity will ensure a multi-modal approach while bypassing erstwhile troublesome neighbors. We believe the missing link is the lack of a unified rail gauge as well as airport cargo infrastructure, which severely limits potential trade and forces Afghanistan to rely on neighbors to an unhealthy degree. By eliminating the need to transport lithium for long road stretches and deal with potential security hazards, government interference (internal and external), and infrastructure gaps, this approach will ensure long-term Afghan security and provide a means for the government to demonstrate the legitimacy of these sorts of large-scale investments to a wary global investor community.

Regionalism: Engaging Afghanistan's Neighbors

Afghanistan's landlocked geography means that its export concerns necessarily call for cooperation with other regional actors, which include Pakistan, India, Iran, China, and the Central Asian states. These nations play an important role in Afghanistan's domestic affairs, have a vested interest in peace in Afghanistan, and play a part in the Great Power Competition. This section of the paper highlights regional partnerships that the United States must to facilitate. A large part of the United States' strategy for achieving and then maintaining a durable peace in Afghanistan involves using its convening power with these regional actors. These diplomatic and economic relationships are necessary to develop and maintain in order to provide an environment where Afghanistan can thrive.

Pakistan

Perhaps the single largest foreign influence on Afghanistan, aside from the United States, is Pakistan. This neighboring country will play a large role in any facilitation of increased export capabilities in Afghanistan. Pakistan is currently Afghanistan's second-largest export market and its largest import partner, per OEC, and is within two percent of Afghanistan's largest partner, India (Observatory of Economic

Complexity 2020). Pakistan shares the largest contiguous border with Afghanistan, of all of the neighboring states (the Durand Line), and it remains a key player in Afghan politics and regional considerations. Currently, the Pakistan–Afghan relationship is nuanced. The two states share heavy ethnic ties, with Pashtuns comprising a significant percentage of both states' populations, and Pashto being a recognized language in both states. Also, both states have agreed to the Afghan–Pakistan Trade Transit Agreement. However, that agreement was in place for only about five years before the Afghan government pulled out in 2015 due to various disagreements, mainly focusing around the prospect of the additions of India and Uzbekistan as parties to the agreement (Husain and Elahi 2016). Furthermore, political and economic ties have been strained between Afghanistan and Pakistan due to the ongoing civil war within Afghanistan. Afghanistan has repeatedly accused Pakistan of providing direct and indirect support to various anti-government forces within the state of Afghanistan in order to further their own political aims, as well as providing a safe haven for those waging war against the internationally recognized government in the mountainous border region between the two states, where the rule of law has consistently been hard to enforce (Azamy 2015). The accusations by Afghanistan have been consistently rebuked by the Pakistani government and ISI; however, theories regarding the links between Pakistan and these groups have been supported by outside actors, including the United States. In 2011, then-Joint Chief of Staff Admiral Mullen called out Pakistan's actions thusly: "In choosing to use violent extremism as an instrument of policy, the government of Pakistan … jeopardizes not only the prospect of our strategic partnership but Pakistan's opportunity to be a respected nation with legitimate regional influence. They may believe that by using these proxies, they are hedging their bets or redressing what they feel is an imbalance in regional power" (Shahzad 2011). In 2015, the Afghan government and the Pakistani government signed a Memorandum of Understanding to jointly combat terrorism and divulge key information to one another, which was followed by another in 2018, this time with China as a member.

While relations between the United States and Pakistan are generally positive, they have historically differed in their positions on Afghanistan. The United States and Pakistan are long-term allies, dating back to the Cold War, where the United States viewed Pakistan as a bulwark against Communism and as an ally in South Asia to offset India's declared non-alignment (Gartenstein-Ross and Vassefi 2012). This continued post-9/11, with then-President Pervez Musharraf working closely with the United States in the Global War on Terror, culminating in Pakistan's designation as an American major non-NATO ally in 2002. However, Pakistan's continued "double game" of nominally supporting counter-terror measures while continuing to fund groups that Pakistan viewed as amenable to Pakistani national security, including the safe harbor of Al-Qaeda mastermind Osama bin Laden in Abbottabad, Pakistan, caused rifts in the Pakistani–U.S. relationship. This led to the United States freezing all military aid flows to Pakistan in early 2018, although they resumed in late 2019 (Landler and Harris 2018). Despite Pakistan's history in aiding and abetting extremist groups in Afghanistan, both United States policy and the current ground truths necessitate the inclusion of Pakistan in the ongoing peace processes, with Prime Minister Imran Khan publicly committing to utilizing Pakistan's resources and gravitas to aid the talks.

India

One of the most influential geopolitical partners of Afghanistan is India, which views Afghanistan as a natural geopolitical ally and a natural counterweight to Pakistan. India has had longstanding ties with multiple Afghan governments and has invested heavily in the state. As the largest purchaser of Afghan goods globally, India is also the second-largest originator of Afghan imports (Gartenstein-Ross and Vassefi 2012). These ties predate the current Afghan state, with friendly relations between the two existing since the signing of a Friendship Agreement in January 1950 between the then-Prime Minister of India Jawaharlal Nehru and the Ambassador to India Najibullah Torwayana (not to be confused with the later President of Afghanistan, Mohammad Najibullah) (Government of India 2017). India was the only regional actor to recognize the Democratic Republic of Afghanistan, led by the aforementioned Mohammad Najibullah, and the two states currently enjoy a strong relationship. This relationship has, both currently and historically, been buttressed by long-term flows of Indian aid into the Afghan state, absent the era of Taliban control of Afghanistan.

Aid from India to Afghanistan has funded the construction of schools and public services buildings, including hospitals, educational exchange programs, and military aid. In 2017, then-Ambassador to India Shaida Mohammad Abdali, in a speech highlighting India's involvement in Afghanistan, noted that India "is the biggest regional donor to Afghanistan and fifth largest donor globally with over $3 billion in assistance. India has built over 200 public and private schools, sponsors over 1,000 scholarships, [and] hosts over 16,000 Afghan students" (Godbole 2017). Afghanistan and India signed a strategic partnership agreement in 2011, and projects such as the Afghan–Indian Friendship Dam, also known as the Salma Dam, have continued to strengthen the two states' relations. Most importantly for the purposes of this paper, India and Afghanistan have facilitated the creation of two air freight corridors to alleviate Afghanistan's reliance on cross-border traffic due to its landlocked nature (Government of India 2017). These freight corridors provide a key opportunity to expand Afghanistan's export capacity and enjoy increased access to the global market, due to India's influential role regionally and globally, as well as its ports.

Iran

Another state that borders Afghanistan and shares a complicated past and present with the Afghan state is Iran. Iran shares Afghanistan's third-largest border, and it has remained actively involved in the Afghan state since each state recognized the other in 1935. Linguistically, the predominantly spoken language in Afghanistan is Dari, a dialect of Farsi. Historically, Afghanistan and Iran have found themselves in political contention since the Islamic Revolution in Iran in 1979 and the Russian invasion of Afghanistan that same year (Worden 2018). Tensions between the two states have long been exacerbated by disputes over the water rights of the Helmand River, which was the subject of an unratified deal between Iran and Afghanistan in 1973. Key political differences have also strained their relations, particularly the close relationship between Afghanistan and the United States, which maintains an exceptionally adversarial stance against the Iranian government. While both parties (Iran and the United States) nominally backed the same actors pre- and post-Taliban ouster, with Iran backing the Northern Alliance and opposed to the Sunni-sectarian Taliban, who have a history of vicious attacks and massacres against the Shia populations in Afghanistan, Iran is highly opposed to the United States' presence and relationship inside Afghanistan and with the current Afghan government (Milani 2020). Due to this, Iran has been accused of promoting and abetting sectarian groups inside Afghanistan, including Gulbuddin Hekmatyar's Hezb-e-Islami, as well as providing safe haven for the Taliban due to their ongoing conflict with the United States. While these relationships, particularly the latter, seem counterintuitive due to the sectarian odds between the two parties, multiple parties, including the Saudis, the United States military, ISAF (International Security Assistance Force), and key Afghan officials insist on it, with the former commander of all U.S. forces in Afghanistan, Gen. David Petraeus, stating that:

> [ISAF] did interdict a shipment, without question [from] the…Quds Force, through a known Taliban facilitator… Iranians certainly view as making life more difficult for us if Afghanistan is unstable. We don't have that kind of relationship with the Iranians. That's why I am particularly troubled by the interception of weapons coming from Iran. But we know that it's more than weapons; it's money; it's also according to some reports, training at Iranian camps as well (The Situation in Afghanistan 2011).

Despite these differences, Iran and Afghanistan have made some overtures toward improving their relations and economic linkages. The largest of these efforts center around the Chabahar Port, located in Iran, and a tripartite agreement among Iran, India, and Afghanistan to expand capacity and utilization there (Nader et al. 2014). Chabahar provides an intriguing option for Afghanistan to improve its export location diversity and lessen its reliance on Pakistan for global trade; however, the continued animosity between Iran and the United States, which is still Afghanistan's largest foreign backer, has rendered this option non-viable for the time being.

Iran has found itself at the center of the region's export capabilities. An examination of these capabilities demonstrates that Iran, Pakistan, and India are key for Afghanistan's global connectivity. However, Pakistan does not have nearly as many trading routes, ports, roads, or pipelines as its immediate neighbors, India and Iran. Pakistan's historically strained relationship with India had led to Pakistan being bypassed (Times of India 2018). Pakistan does have the Gwardar Port; however, it had succeeded in isolating itself by not allowing India, the region's most populated country, land access to Central Asia and Afghanistan, specifically. It is widely considered by the region's states that an economically integrated and strong Afghanistan means a strong and stable region.

Iran's Chabahar Port provides various commercial and also strategic advantages since Iran acts as a regional connection hub. Due to its centralized location, Iran is vital for the International North–South Transport Corridor (INSTC), which runs from India to Europe. Iran provides vital connections to countries in Central Asia that are landlocked, and it also is the shortest, fastest, and most cost-effective route through the INSTC (Times of India 2018). The Chabahar Port project has led to a trilateral trade and transit corridor which has greatly benefitted all the countries involved (Times of India 2018). Between 2018 and 2019 alone, trade between India and Afghanistan, facilitated by Iran, has jumped 40% and can be placed at a value of $1 billion. Chabahar Port can be seen as the door to the landlocked countries of Central Asia and the key to Afghanistan's vital export trade, which would free it from the culture of war that has so long gripped the country. Iran has also committed itself to rebuilding and stabilizing Afghanistan through the Chabahar Port. It has reduced the port fees for Afghanistan as well charging reduced tariffs for goods from Afghanistan (Farooq 2019). This is in the hopes that it would stimulate exports and allow for the maximum amount of profit to enter the country.

The Chabahar Port is still in phase one, with many additional expansions in the works. India plans on further building a railroad that would connect Chabahar to the Bamiyan province in Afghanistan (Farooq 2019). There is also going to be an air freight corridor as well as a $16 billion free trade zone around the city of Chabahar (Times of India 2018; Bahgat 2017). However, renewed sanctions on Iran have provided a heavy roadblock to its success and have led to diminished enthusiasm and funding from India. For example, Indian and Afghan banks have not been able to transfer money through the port because of US sanctions against Iran. In May of 2019, At least 50 containers of Afghan goods to be exported to India and China through the Chabahar Port were blocked from doing so due to United States sanctions against Iran (Chaudhury 2019). India also slashed the development budget for Chabahar Port by a third (Kapoor 2019). India's waver to import Iranian oil, was also not renewed (Iyengar and Defterior 2019). Interestingly, China has continued ignored U.S. sanctions against Iran and has chosen to continue importing Iranian oil (Iyengar and Defterior 2019). Governor of Sistant-and-Baluchestan, Ali Oset-Hashemi, told a Chinese delegation in 2015 that "Iran stands ready to provide lucrative business opportunities to the countries that stood by it during hard times" (The Iran Project 2015). Surely, this positioning with Iran against the United States, only further serves to heighten the current trade war between China and the United States, as well as to challenge the hegemony of the United States.

In 2016, the European Union released their "EU Strategy for Relations with Iran After the Nuclear Deal" in which they refer to their future relationship with Iran as "not indispensable" but certainly as "unavoidable" (European Parliament 2016). The document further expands upon this sentiment:

> the cost for the EU to pursue its policies vis-à-vis the region will be much higher and less effective with Iran absent from the table. Neither does the absence of an EU–Iran relationship mean that such a relationship void will remain empty. If Europe is not present on the Iranian scene, be it in trade or politics, other actors (China, Japan, India, etc.) will claim that space. As a result, in order for the EU to make any headway in addressing issues of concern and build a more stable relationship with Iran, the EU must devise a medium to long-term strategy for regular, sustained dialogue with Iran. In other words, the EU must have a clear notion of what a structured and strategic relationship with Iran can and should look like. Such a rethink is not about rewarding or punishing the Islamic Republic of Iran, but rather, about the

role and position of the EU in the Middle East in general and how it can pursue its interests most effectively (European Parliament 2016).

The European Union recognizes the role that Iran plays in the region, and the importance of that role. If the United States does not do so as well, it risks losing its own position in the post-WWII order to another contending state, and contributing to Afghanistan's loss of its link to the world.

China

China also plays a key geopolitical and geostrategic role in Afghanistan. By maintaining a small, remote border (the Wakhjir Pass) between the two states, it has essentially locked down all potential travel between the two states, which abuts two natural reserves in both states. China limits access at this crossing due to the belief that the violence in Afghanistan could spill over into the predominantly Muslim, heavily repressed Xinjiang Province (Malik 2014). While there have been proposals to open a trade corridor and improve infrastructure at this pass, the geopolitical considerations as well as the topography have rendered this proposition moot. Despite the lockdown in cross-border transit, China views Afghanistan as critical to its regional aims and the propagation of its Belt and Road policy, and it has heavily invested in Afghanistan, particularly in the extractives sector, where projects such as the extraction of oil and natural gas products in the Amu Darya basin have been awarded to Chinese firms (Jin 2016). Furthermore, the Afghan rail system that begins in Hairaton terminates in China, providing the opportunity to increase rail freight trade between the two nations. Nevertheless, this paper focuses on the possibility of incentivizing Afghan lithium extraction to loosen the Chinese stranglehold on lithium. That stranglehold poses a potential national security threat to the United States.

China's Grand Strategy rejects the notion of the nation-state and, according to Araya (2019), instead reimagines "the world as a single complex of supply chains and trade arteries" wherein China can leverage the current geotechnological shift and develop new markets for its advanced technologies. These technologies include electric vehicles (EV), telecommunications, robotics, artificial intelligence (AI), semiconductors, clean energy technology, advanced electrical equipment, rail infrastructure and maritime engineering (Araya 2019). For all of these technologies, lithium is vital. In short, China plans to grow and exert its power and influence through economic expansion. Rahman (2014) argues that China's grand strategy is aimed at displacing the current world system in which financial, economic, and strategic dominance center on America. The signature project of China's Grand Strategy is its Belt and Road Initiative (BRI). In 2011, China's President Xi described the initiative as a "vast network of railways, energy pipelines, highways, and streamlined border crossings" that travel from East Asia to Europe by land and sea. This project, described by many as the New Silk Road, is funded by the National Development Bank (NDB), which according to the bank's constitution will "support public or private projects through loans, guarantees, equity participation and other financial institutions," as well as cooperating "with international organizations and other financial entities, and provide technical assistance for projects to be supported by the Bank" (New Development Bank 2020). As such, China has gained the power to influence the development of its neighboring countries, but also allow those countries to contribute to the growth of China through increased trade and shared prosperity.

With technology at the center of China's Grand Strategy, the importance of Afghanistan to China cannot be overstated. Indeed, China has been the country most active in Afghanistan's extractive industries (Bernard and Sugarman 2012). China has hardly hidden its intent to crown Afghanistan as the center of the BRI or to occult its intentions in regard to Afghanistan's extractive industries. China's relationship with Afghanistan is focused not on reconstruction of security, but rather, on commerce (Bernard and Sugarman 2012). China has therefore been quietly benefitting from the fruits of the United States' labor to reconstruct Afghanistan and build peace.

China's Grand Strategy calls for it to control Afghanistan's vast natural resources in order to meet the Chinese demand. It routinely makes geopolitically driven resource investments through state-owned companies and outbids its commercially driven Western competitors (Bernard and Sugarman 2012). Afghanistan's first contract, a 30-year lease, went to the Metallurgical Corporation of China (MCC), a Chinese mining consortium, giving them the rights to one of the largest copper deposits in the world at Aynak, 25 miles south of Kabul in Logar Province. This $3 billion project is the largest foreign investment and business venture in Afghanistan's history and spearheads China's ambitious effort to monopolize Afghanistan's natural resources (Bernard and Sugarman 2012).

Rare earth and critical minerals have both political and economic importance, giving significant power to those states or entities that can control the production and/or distribution of a given strategic commodity. This is especially true if those states are willing to flout global trade rules and agreements. If China is allowed to dominate Afghanistan's rare earth minerals, the United States, as China's main competitor, will find itself locked out of the market. This alone may be a decisive factor in future confrontations (Butler 2014). China has demonstrated that it is unafraid to manipulate its political power through natural resources.

Gulf Cooperation Council (GCC) States

Afghan relations with the GCC states, notably the United Arab Emirates, Saudi Arabia, and Qatar, shape the geopolitical situation in which Afghanistan currently finds itself. Historically large backers of the Afghan government and heavy investors into the infrastructure there, GCC states have sent troop detachments in support of the current conflict against the Taliban. However, these states have also historically been friendly with the Taliban. During the U.S.-backed insurgency against the Soviet-backed government, most of the foreign fighters involved in this conflict came from these states, including the predecessors to al Qaeda and key individuals such as Osama bin Laden. Support for the *mujahideen* and "the Islamic struggle" in Afghanistan came predominantly from these states, including routing fighters and funding. Saudi Arabia and the UAE were also two of the only states that recognized the Taliban as the government of Afghanistan, and the current political office of the Taliban resides in another GCC state, Qatar, which has had strained relations with the rest of the Gulf over the last few years (Gannon 2018). Despite these historical links, the GCC nations formally recognize the current government in Afghanistan as the legitimate one, and they have invested in key areas such as road infrastructure, housing, and ancillary public services to support efforts there. As key partners of the United States, these states can play a role in expanding Afghan logistical and exploitative capacity.

Central Asian States

Lastly, one of the most important geopolitical relationships and opportunities to expand Afghanistan's global export capacity lie with the other Central Asian states with which Afghanistan shares a border, namely Tajikistan, Turkmenistan, and Uzbekistan. These states, which share with Afghanistan a history of colonization by the Soviet Union and Russian Empire, maintain large borders and combined logistical networks with Afghanistan, including road and rail. These logistical networks, utilized by the United States government in their current efforts against the Taliban, are less politically fraught than those with Pakistan, and allow for transport into Russia and the rest of Europe. These states view peace inside Afghanistan as an important policy objective to curtail cross-border violence, and have begun small-scale but noticeable investments into Afghanistan, stating multiple times their interest in pursuing more concrete efforts (Government of Afghanistan 2019). The United States has attempted to promote continued growth in relationships between Afghanistan and these states and will continue to do so (United States Strategy for Central Asia 2020). Expansion of these networks as an alternative to air freight and road logistics could promote and diversify Afghanistan's trade options, allowing for greater export opportunities.

These dynamics in international relations limit the potential transportation solutions we may consider. Namely, export frameworks which include Iran should not be onsidered within the realm of possibility. Other countries like India, UAE (and other GCC countries), Pakistan, and particularly the northern states (Tajikistan, Uzbekistan, Turkmenistan), should be the focus of potential transportation solutions. That is not to say that political implications are associated with those areas (from regional hegemons like Russia and China) which may interfere with supply chains and political stability.

Conclusion

To date, the United States has spent over $2 trillion USD in Afghanistan, of which $714 billion USD has been on war and reconstruction (Almukhtar and Nordland 2019; Amini 2017). Yet, Afghanistan is missing key trade and transport infrastructure that could enable it to monetize its vast mineral wealth and free it from over-reliance on neighbors. If the United States supports Afghanistan in expanding its trade promotion investments, including unified rail gauge and airport cargo infrastructure, Afghanistan's vast wealth in lithium and other minerals can be monetized using a multi-modal approach that bypasses erstwhile troublesome neighbors. By eliminating the need to transport lithium for long stretches by road, with the attendant potential for security hazards, government interference (internal and external), and delays caused by infrastructure gaps, Afghanistan's long-term security can be ensured, and the government can have a means of demonstrating to a wary global investor community the legitimacy of large-scale investments in Afghanistan.

The 2008 Afghan National Development Strategy put forth the vision of a peaceful Afghanistan that is "a stable Islamic constitutional democracy," "a tolerant, united, and pluralistic nation that honors its Islamic heritage," and "a society of hope and prosperity based on a strong, private-sector led market economy, social equity, and environmental stability." Such a vision can be achieved in part through continued U.S. engagement after the military drawdown. The alternative is to allow Afghanistan to regress into a narco-mafia state that harbors terrorists who threaten the security of the United States and its allies. Thus, continued U.S. engagement in Afghanistan is critical to ensuring a just and durable peace that stabilizes the region through shared prosperity. In return, this continued engagement will preserve U.S. National Security—a critical goal in the environment of the Great Power Competition.

Acknowledgements The authors would like to extend their sincerest gratitude to Major General Edward M. Reeder for his key insights and valuable feedback. They would also like to thank Joseph Wagner, and Peter Brodzik, for their research contributions and a special thanks to Michelle Assaad for her tireless work of editing and research.

References

T. Agalewatte, Competitive industry policy for economic development in Sri Lanka: lessons from East Asis. University of Wollongong Thesis Collection (2004)

S. Almukhtar, R. Nordland, What did the US get for $2 trillion in Afghanistan? The New York Times (2019), https://www.nytimes.com/interactive/2019/12/09/world/middleeast/afghanistan-war-cost.html

M. Amini, At stake in US military efforts to stabilize Afghanistan: at least $3 trillion in natural resources. CNBC News (2017), https://www.cnbc.com/2017/08/18/trumps-afghanistan-strategy-may-unlock-3-trillion-in-natural-resources.html

D. Araya, China's grand strategy. Forbes (2019), https://www.forbes.com/sites/danielaraya/2019/01/14/chinas grand strategy/#2d4051c1f180

H. Azamy, It's complicated: the relationship between Afghanistan, Pakistan, and the Taliban. Foreign Policy (2015, March 2), https://foreignpolicy.com/2015/03/02/its-complicated-the-relationship-between-afghanistan-pakistan-and-the-taliban/
G. Bahgat, US-Iran relations under the Trump administration. Mediter. Q. **28**(3) (2017)
C. Bernard, E. Sugarman, Afghanistan's copper conundrum. Cauc. Int. **2**(3) (2012)
British Geological Survey, Mineral profile: Lithium (2016), ftp://ftp.bgs.ac.uk/pubload/MineralsUK/Lithium_Profile/Lithium_profile_July2016.pdf
M. Bruenecker et al. (eds.), *Resource Curse or Cure?* (Springer, New York, 2014)
C.J. Butler, Rare earth elements: China's monopoly and implications for US national security. Fletcher Forum World Aff. **38**(1) (2014)
P. Chacko, India and the Indo-Pacific from Singh to Modi: geopolitical and geoeconomic entanglements, in *New Regional Geopolitics in the Indo-Pacific: Drivers, Dynamics and Consequences*, ed. by P. Chacko (Routledge, New York, 2016)
D.R. Chaudhury, Iran sanctions: Indo-Afghan banking via Chabahar Port hit. The Economic Times (2019), https://economictimes.indiatimes.com/news/economy/foreign-trade/iran-sanctions-indo-afghan-banking-via-chabahar-port-hit/articleshow/69348517.cms
Y. Chazan, China rushes to dominate global supply of lithium. The Diplomat (2019), https://thediplomat.com/2019/02/china-rushes-to-dominate-global-supply-of-lithium/
M. Chossudovsky, More American troops to Afghanistan, to keep the Chinese out? Lithium and the battle for Afghanistan's mineral resources. Global Research (2019), https://www.globalresearch.ca/more-american-troops-to-afghanistan-to-keep-the-chinese-out-lithium-and-the-battle-for-afghanistans-mineral-riches-2/5605456
M. Cocker, Chapter 24A. Summary for the mineral information package for the Nuristan rare metal pegmatite area of interest (2011), https://pubs.usgs.gov/of/2011/1204/pdf/24A.pdf
G. Dixit, Strategic mineral resources: concept and significance. IRJSMH **6**(5) (2015)
Doing Business, Afghanistan ease of doing business (2020), https://www.doingbusiness.org/content/dam/doingBusiness/country/a/afghanistan/AFG-LITE.pdf
A.W. Dowd, Silver lining in Afghanistan? Fraser Forum **26** (2013), https://search.ebscohost.com/login.aspx?direct=true&db=edb&AN=87722605&site=eds-live
European Parliament, An EU strategy for relations with Iran after the nuclear deal (2016), https://www.europarl.europa.eu/RegData/etudes/IDAN/2016/578005/EXPO_IDA(2016)578005_EN.pdf
K.O. Farooq, Chabahar port: a step toward connectivity for India and Afghanistan. The Diplomat (2019), https://thediplomat.com/2019/07/chabahar-port-a-step-toward-connectivity-for-india-and-afghanistan/
H. Fayez, The role of foreign aid in Afghanistan's reconstruction: a critical assessment. Econ. Polit. Wkly. **47**(39), 65–70 (2012)
I. Fryklund, Foreign aid is Afghanistan's resource curse. Foreign Policy in Focus (2013), https://fpif.org/foreign_aid_is_afghanistans_resource_curse/
A. Fußstetter, Geoeconomic implications of border conflicts in Asia based on resource competition. Georg-August-Universität (2016)
K. Gannon, Pakistan, Saudis, UAE join US-Taliban talks. AP News (2018, December 17), https://apnews.com/182f8532e5864459b80f3a868239d9f3
D. Gartenstein-Ross, T. Vassefi, The forgotten history of Afghanistan–Pakistan relations. Yale J. Int. Aff. 38–45 (2012), https://yalejournal.org/wp-content/uploads/2012/04/Article-Gartenstein_Ross-and-Vassefi.pdf
E. Gholz, Rare earth elements and national security. Council on Foreign Relations (2014), https://www.google.com/url?sa=t&rct=j&q=&esrc=s&source=web&cd=&cad=rja&uact=8&ved=2ahUKEwj5kJGQnOHqAhVkmeAKHU5JCNAQFjAAegQIAxAC&url=%3A%2F%2Fwww.cfr.org%2Fsites%2Fdefault%2Ffiles%2Fpdf%2F2014%2F10%2FEnergy%2520Report_Gholz.pdf&usg=AOvVaw3n4alD8mw486RuoDwQIzDx

M.I. Glassner, *Transit Problems of Three Asian Land-Locked Countries: Afghanistan, Nepal, and Laos*. Contemporary Asian Studies Series, vol. 4 (1983), https://digitalcommons.law.umaryland.edu/cgi/viewcontent.cgi?article=1056&context=mscas

S. Godbole, *Strengthening India–Afghanistan Strategic Relations in an Uncertain World*. Address by H.E. Shaida Abdali, Ambassador of Afghanistan to India (2017, May 1), https://www.brookings.edu/events/india-afghanistan-and-connectivity-in-south-asia-address-by-h-e-shaida-abdali-ambassador-of-afghanistan-to-india/#cancel

Government of India, India and Afghanistan establish direct air freight corridor. Ministry of External Affairs (2017, June 19), https://mea.gov.in/press-releases.htm?dtl/28546/India_and_Afghanistan_establish_Direct_Air_Freight_Corridor

S. Hoaby, Can the new economic geography explain inequality between the great plains and the great lakes? (2005)

I. Husain, M.A. Elahi, The future of Afghanistan–Pakistan trade relations. USIP (2016, October 11), https://www.usip.org/publications/2015/08/future-afghanistan-pakistan-trade-relations

R. Iyengar, J. Defterior, India is buying more US and Saudi oil because of sanctions on Iran. CNN Business (2019), https://www.cnn.com/2019/06/26/business/indian-oil-us-iran-sanctions/index.html

W. Jin, The US withdrawal and one belt one road: Chinese concerns and challenges in Afghanistan. Strateg. Assess. **19**(3), 69–79 (2016)

J.H. Kakar, Construction of Turkmen-Afghan railroad begins (2017, December 1), https://www.pajhwok.com/en/2017/12/01/construction-turkmen-afghan-railroad-begins

S. Kapoor, The Chabahar angle to Iranian sanctions. The Hindu Business Line (2019), https://www.thehindubusinessline.com/opinion/the-chabahar-angle-to-iran-sanctions/article28334469.ece.

L. Kavanagh, J. Keohane, G. Garcia, A.L. Cabellos, J. Cleary, Global lithium sources—industrial use and future in the electric vehicle industry: a review. Resources (3), 57 (2018). https://doi.org/10.3390/resources7030057

R.D. Kiggins, *The Strategic and Security Implications of Rare Earths* (Springer, New York, 2015).

P. Krugman, Increasing returns and economic geography. J. Polit. Econ. **99**(3), 483–499 (1991)

M. Landler, G. Harris, Trump, citing Pakistan as a 'Safe Haven' for terrorists, freezes aid. The New York Times (2018, January 4), https://www.nytimes.com/2018/01/04/us/politics/trump-pakistan-aid.html

A. Lyrichkova, G. Stolyarov, Russia has $1.5 billion plan to dent China's rare earth dominance. Reuters (2020), https://www.reuters.com/article/russia-rareearths/russia-has-15-billion-plan-to-dent-chinas-rare-earth-dominance-idUSL8N2F73F4

H.Y. Malik, Geo-political significance of the Wakhan corridor for China. Fudan J. Humanit. Soc. Sci. **7**(2), 307–323 (2014). https://doi.org/10.1007/s40647-014-0017-z

M. Milani, Iran and Afghanistan Primer (2020), https://iranprimer.usip.org/sites/default/files/IranandAfghanistan.pdf

S. Moiso, A. Paasi, From geopolitical to geoeconomic? The changing political rationalities of state space. Geopolitics **18**, 267–283 (2013)

A. Nader, A. Scotten, A. Rahmani, R. Stewart, L. Mahnad, Iran and Afghanistan: a complicated relationship, in *Iran's Influence in Afghanistan: Implications for the U.S. Drawdown* (RAND Corporation, 2014, May 7), pp. 5–22. https://www.jstor.org/stable/10.7249/j.ctt1287mjf.8

New Development Bank, Our work (2020), https://www.ndb.int/about-us/essence/our-work/

Observatory of Economic Complexity, Afghanistan. OEC World (2020), https://oec.world/en/profile/country/afg/

L.T. Peiro, G.V. Mendez, R.U. Ayres, Lithium: sources, production, uses, and recovery outlook. J. Min. Metals Min. Soc. **65** (2013)

A. Pikulicka-Wilczewska, Why is Afghanistan unable to extract its vast mineral wealth? Al Jazeera (2019), https://www.aljazeera.com/indepth/features/afghanistan-unable-extract-vast-mineral-wealth-190527111748895.html

K. Rahman, The new great game: a strategic analysis. The Dialogue (2014), https://www.google.com/url?sa=t&rct=j&q=&esrc=s&source=web&cd=&ved=2ahUKEwj-hZHKnOHqAhXpnuA

KHV5TARcQFjAAegQIBBAC&url=http%3A%2F%2Fwww.qurtuba.edu.pk%2Fthedialogue%2FThe%2520Dialogue%2F9_1%2FDialogue_January_March2014_1-26.pdf&usg=AOvVaw2tNqX1ySSACVGERzg43Mat

D. Rank, Leveraging US-China cooperation to build a regional consensus on Afghanistan. United States Institute of Peace (2018)

R. Rapier, Why China is dominating lithium-ion battery production. Forbes (2019), https://www.forbes.com/sites/rrapier/2019/08/04/why-china-is-dominating-lithium-ion-battery-production/#2751b8813786

J. Reeves, Mining in Afghanistan. Issues (99), 20–22 (2012), https://search.ebscohost.com/login.aspx?direct=true&db=eue&AN=77652504&site=eds-live

J. Risen, U.S. identifies vast mineral riches in Afghanistan. The New York Times (2010), https://www.nytimes.com/2010/06/14/world/asia/14minerals.html

J.D. Sachs, A.M. Warner, Natural resource abundance and economic growth. NBER Working Paper Series (1995), https://www.nber.org/papers/w5398.pdf

E. Scheyder, Exclusive: pentagon races to track U.S. rare earths output amid China trade dispute. Reuters (2019), https://www.reuters.com/article/us-usa-rareearths-pentagon-exclusive/exclusive-pentagon-races-to-track-us-rare-earths-output-amid-china-trade-dispute-idUSKCN1U727N

A. Shahzad, Pakistan warns U.S. "Can't Live Without Us". CBS News (2011, September 23), https://www.cbsnews.com/news/pakistan-warns-us-cant-live-without-us/

U. Sheraz, Foresight as a tool for sustainable development in natural resources: the case of mineral extraction in Afghanistan. Resour. Policy **39**, 92–100 (2014)

SIGAR, Afghanistan's Road Infrastructure: Sustainment Challenges and Lack of Repairs Put U.S. Investment at Risk. SIGAR (2016), https://www.sigar.mil/pdf/audits/SIGAR-17-11-AR.pdf

The Iran Project, Iran wants China in Chabahar development (2015), https://theiranproject.com/blog/2015/10/08/iran-wants-china-in-chabahar-development/

The Situation in Afghanistan (2011), https://www.govinfo.gov/content/pkg/CHRG-112shrg72295/html/CHRG-112shrg72295.htm

Times of India (2018), India to get control of key port in Iran for 18 months. https://timesofindia.indiatimes.com/articleshow/62963561.cms?utm_source=contentofinterest&utm_medium=text&utm_campaign=cppst

Trade and Accession Facilitation for Afghanistan (TAFA) (2019, May 7), https://www.usaid.gov/news-information/fact-sheets/trade-and-accession-facilitation-afghanistan-tafa

Treaty of Friendship Between the Government of India and the Royal Government of Afghanistan (2019), https://www.mea.gov.in/bilateral.documents.htm?dtl/6584/Treaty+of+Friendship

United States Strategy for Central Asia 2019–2025: Advancing Sovereignty and Economic Prosperity, United States Strategy for Central Asia 2019–2025: Advancing Sovereignty and Economic Prosperity (2020), https://www.state.gov/united-states-strategy-for-central-asia-2019-2025-advancing-sovereignty-and-economic-prosperity/

USAID Support to Afghanistan's WTO Accession (2019, May 7), https://www.usaid.gov/news-information/fact-sheets/usaid-support-afghanistans-wto-accession

S. Worden, Iran and Afghanistan's Long, Complicated History (2018, August 13), https://www.usip.org/publications/2018/06/iran-and-afghanistans-long-complicated-history

Central Asia, Great Power Competition, and Achieving Peace in Afghanistan

Humayun Hamidzada and Richard Ponzio

Abstract Building on its August 2017 South Asia Strategy, the Trump Administration, in February 2020, unveiled a new Central Asia Strategy. Similar to the important roles performed by India and Pakistan to foster peace in and greater connectivity with Afghanistan, the countries of Central Asia—namely, Kazakhstan, Kyrgyzstan, Tajikistan, Turkmenistan, and Uzbekistan—are also poised to make significant contributions in these areas as the United States transitions in Afghanistan following the signing, on February 29, 2020, of its historic deal with the Taliban movement. In this chapter, the authors explore key political, economic, and other aspects of the recent past and potential further support the Central Asian Republics can offer to increase stability and prosperity in Afghanistan, within the broader context of Great Power competition involving the United States, Russia, China, and other major countries.

Keywords Central Asia · Afghanistan · Great power competition · U.S. central Asia strategy · Peace agreement · Taliban · Regional economic connectivity

H. Hamidzada
Stimson Center, 82 Summitcrest Drive, L4S 1 A8, Richmond Hill, ON, Canada
e-mail: hamidzada@gmail.com

R. Ponzio (✉)
Stimson Center, 2138 California Street, NW Apt. 106, Washington, DC 20008, USA
e-mail: rponzio@stimson.org

© The Author(s), under exclusive license to Springer Nature Switzerland AG 2021
A. Farhadi et al. (eds.), *The Great Power Competition Volume 1*,
https://doi.org/10.1007/978-3-030-64473-4_4

Introduction

The United States and the Taliban, the Islamic Emirate of Afghanistan,[1] signed a historic peace agreement on February 29, 2020 (Hansler 2020). Signed in Doha, Qatar and witnessed by U.S. Secretary of State Michael Pompeo, it concluded over eighteen months of intensive negotiations, led by the U.S. Special Representative for Afghan Reconciliation Ambassador Zalmay Khalilzad. The peace deal was preceded by a weeklong "reduction in violence" period. The "Agreement for Bringing Peace to Afghanistan" outlined a series of commitments from the U.S. and the Taliban on such issues as troop levels, counter-terrorism, and the intra-Afghan dialogue, with the ultimate goal of paving the way for "a permanent and comprehensive ceasefire" (U.S. Department of State 2020a).

Ever since the signing of the agreement, the Covid-19 virus has impacted the entire world, violence has continued in Afghanistan, and both developments are complicated further an escalation of public tension between the United States and Afghan leaders, as both President Ashraf Ghani and former Chief Executive Abdullah Abdullah have declared themselves president having failed to resolve their differences (Neuman 2020). The political impasse over the presidential election results have emboldened the Taliban, who are making advances both politically and militarily. A consensus is building, internally and internationally, that Abdullah Abdullah could lead the peace process on behalf of the Afghan state (Allen and O'Hanlon 2020). Despite Secretary Pompeo's warning and Americas' decision to cut $1 billion in economic aid, the United States and the international community have acknowledged Ashraf Ghani as the winner of the presidential election, as declared, in December 2019, by the Afghan Independent Elections Commission.

The U.S. strategy of seeking a political settlement to the conflict in Afghanistan has had significant repercussions among Afghanistan's neighbors and in the region more broadly. Especially with the signing of the recent U.S.-Taliban agreement, many countries in the region continue to fear a precipitous withdrawal. The new peace-focused strategy for now is aimed at an orderly withdrawal that aims to leave behind a stable Afghanistan, albeit one with a Taliban-infected political order. Nonetheless, assuming that the key U.S. strategic objective of ensuring that Afghanistan does not become a haven for international terrorists is concretized, a stable Afghanistan would offer unprecedented opportunities for regional cooperation. Under this scenario, the five Central Asian countries would have the most to gain.

Even before the U.S. initiated its peace-oriented strategy several developments in Central Asia had led to a shift in perspective with regard to Afghanistan. Gradually,

---

[1]Humayun Hamidzada is a nonresident fellow at the Stimson Center in Washington, DC, and director of the Afghanistan Peace Research Project at the York University Centre for Asian Research, Toronto. Richard Ponzio is director of the Stimson Center's Just Security 2020 Program. From 2010 to 2014, he was a senior adviser to the Office of the Special Representative for Afghanistan and Pakistan in the U.S. Department of State. They wish to thank the United States Institute of Peace for granting permission to update an earlier version of their August 2019 USIP Special Report, "Central Asia's Growing Role in Building Peace and Regional Connectivity with Afghanistan", for this chapter.

the region's *cordon sanitaire* approach to Afghanistan, which sought to prevent the southern neighbor's perceived pathologies from infiltrating the region, had begun to evolve towards an approach of greater engagement, based on the assessment that a more active role in trying to solve Afghanistan's problems was a better long-term strategy than seeking to insulate themselves from these problems. This was accompanied by an attitude and rhetoric that sought to emphasize Afghanistan's historic Central Asian identity.

While Central Asia for now is not a major player in the future of Afghanistan, U.S. policy should not ignore the region. Central Asia can play a positive role in peace-making but especially in sustaining an eventual peace. As crucial as India and Pakistan are to sustainable peace in Afghanistan, as acknowledged by America's 2017 South Asia strategy, similarly persuasive arguments can be made about Afghanistan's northern neighbors. This is why it is promising that the "United States Strategy for Central Asia 2019–2025: Advancing Sovereignty and Economic Prosperity", unveiled on February 5, 2020, bodes well in signaling to the countries of this vital region that they too have a central role to play in supporting the achievement of peace in Afghanistan (U.S. Department of State 2020a). Two of the Strategy's six main objectives, in fact, emphasize America and the five Central Asian Republic's commitment to:

1. **Expand and maintain support for stability in Afghanistan**. The Central Asian nations will become stable, secure, and engaged partners of the United States and continue support for international efforts to stabilize Afghanistan.
2. **Encourage connectivity between Central Asia and Afghanistan**. The Central Asian states will develop closer ties with Afghanistan across energy, economic, cultural, trade, and security lines that directly contribute to regional stability (U.S. Department of State 2020b).

This signal is critically important at a time when most "regional" initiatives seem to overlook Central Asia, with talk of a new "6 + 1" (involving major states such as Iran, Pakistan, India, China, Russia, and the U.S. plus Afghanistan) configuration or a revived Quadrilateral (involving China, the U.S., Afghanistan, and Pakistan) platform. One exception may be a "C5 + 1 + 1" (The Central Asian Republics plus Afghanistan and the United States), building on the historic "C5 + 1" attended by U.S. Secretary of State Michael Pompeo, on 3 February 2020 in Tashkent—*the first time this high-level configuration was convened in the Central Asian region*. The two strategies (one on Central Asia alongside the current South Asia strategy) wield the potential now to complement and reinforce one another, as well as ensure that the implementing agents of American foreign policy—whether from State, the Department of Defense, USAID, or elsewhere—are sufficiently coordinated to promote a "unity of purpose". Working together, a synched-up U.S. Strategy for South *and* Central Asia could create more favorable conditions to enable the U.S. to finally reduce its military presence and transition to a more normal, long-term relationship with Afghanistan and its wider region.

In this chapter, we provide a historical background to the major but insufficiently acknowledged policy shifts that have brought about a new geopolitical configuration in Central Asia that has created this opportunity. In addition, the chapter provides a current account of the many distinct ways the five Central Asian Republics—Kazakhstan, Kyrgyzstan, Tajikistan, Turkmenistan, and Uzbekistan—individually and collectively contribute to efforts to build a more stable and prosperous Afghanistan in a more stable and prosperous Central Asia. With a critical lens, we ask whether recent developments genuinely represent a new era of Central Asian openness, in which a hopefully peaceful Afghanistan can acquire a greater Central Asian identity.

Dismantling the Cordon Sanitaire

It is easy to overlook the historical fact that the modern five countries of Central Asia were born into an unstable region. As the Soviet Union fell the five Soviet Socialist Republics that became independent countries had to deal with a southern neighbor that was moving from insurgency to civil war. In the ensuing three decades, Central Asian nations have gone through their own crises but have largely emerged from them somewhat stronger. Until very recently, they had collectively adopted a defensive position against Afghanistan as a means of addressing their fears that religious extremism, terrorism, and the narco-economy could overwhelm their fragile political systems. These fears varied depending on how far countries were from Afghan borders and perceptions of their own vulnerabilities. As they have gained in stability, in the past few years all Central Asian states have tended to adopt more engaged policies on the theory that such policies may contribute more effectively to Afghan and regional stability than defensive disengagement. Nonetheless, despite the US-led state-building attempt in Afghanistan, the pathologies that Central Asian states once feared have not disappeared and, in many cases, have grown worse. Islamic extremism, drug-trafficking, and terrorism remain factors in Central Asian policy thinking and will certainly dominate again if the current opportunity for peace, to which they are all invested, fails.

Historically, Uzbek, Tajik, and other Central Asian citizens have found refuge in Northern Afghanistan during periods of Mujahedeen and Taliban rule. The cross-border traffic of Islamist groups goes back almosst three decades when Central Asians would cross Afghanistan to attend Madrassas and terrorist training camps in Pakistan. However, the increase in ISIS Khorasan-claimed attacks inside Afghanistan have precipitated growing Russian allegations of ISIS making inroads through northern Afghanistan into Central Asia. Russian officials are also on the record claiming that the United States supports ISIS infiltration of Central Asia, citing "mystery helicopters" transporting men and supplies (Ramani 2018). While this narrative, rejected by Afghan and U.S. government officials, has been played to a domestic Russian audience, little evidence exists to support a dramatic surge of ISIS in Afghanistan (Ramani 2018). While the Central Asian Republics share Russia's concern about the spread of extremist ideologies from Afghanistan, they have not publicly expressed concerns that the U.S. has facilitated the movement of extremist groups from Afghanistan.

Besides serving as a steady source of revenue for the Taliban movement and other militia groups battling the Afghan government and its international allies,[2] the cultivation of opium and the production and trafficking of heroin remains a multi-billion dollar industry and a chief concern to Afghanistan's Central Asian neighbors. According to the United Nations Office on Drugs and Crime (UNODC) Afghanistan Opium Survey for 2018, 2630,000 hectares were under cultivation. This was a drop from the unprecedented 328,000 hectares reported in 2017 (UNODC and GoIRA 2018). The drop, however, was due to severe drought in Afghanistan rather than effective counter-narcotics policies or improved rule-of-law. Afghanistan remains the largest supplier of opium, accounting for about three-quarters of global production (UNODC 2018a). While yields and prices fell in 2018 due to the drought and a supply gut, in 2017, heroin generated $4.1 to $6.6 billion in exports for the Afghan economy, and the taxing of local farms involved in poppy cultivation contributes annually between $76 to $121 million to the Taliban movement and another $40 to $63 million to other anti-government groups (UNODC 2018a).

Opiates produced in Afghanistan, mainly in the form of heroin, find its way to Central Asia and Russia in large quantities. Between 2011 and 2015 an estimated 42.5 to 74.5 tons of pure heroin annually crossed into the northern route from Afghanistan (UNODC 2018b). The new connectivity projects involving road and railway links between Afghanistan and Central Asia have raised concerns about the drug trade potentially benefiting from transport and trade facilitation improvements. However, scholars, such as Dr. Sebastien Peyrouse from George Washington University, have noted that corruption, the complicity of security forces, and the economic and social crisis in some Central Asian states may be larger factors in fueling Central Asia's drug problem than improved infrastructure and trade relations with Afghanistan (Peyrouse 2017).

While Central Asian states have frequently evoked their fear of extremism migrating from Afghanistan to Central Asia, they are also aware of home-grown extremism that colors their international reputation, particularly reports of Central Asian nationals fighting alongside ISIS in Syria and Iraq. The primary drivers of extremism in Central Asia as a whole have been linked to economic stagnation, repressive regimes, and to some degree the rediscovery of political identities (Plekhanov 2018). The Central Asian States have provided a fertile ground for Islamic State (ISIS) recruitment, particularly in the Fergana Valley traversing Uzbekistan, Tajikistan, and Kyrgyzstan. It is estimated that close to 5,000 Central Asian foreign fighters have traveled to Iraq and Syria, of which 1,500 are from Uzbekistan alone (Barrett 2017). Those who have survived and return home bring back extremist ideologies and firsthand combat experience. The punitive actions of Central Asian governments often go beyond the targeting of ex-combatants and affect religious groups and politically marginalized populations. The crackdown on religion and political activities further deepens resentment toward the state.

[2]Widespread opium cultivation and trafficking is also known to support corrupt Afghan government officials and local militia commanders supported by the Afghan government.

Alarm bells have gone off long ago in Central Asia, Moscow, and European capitals on the rapid expansion of local and ISIS-inspired extremism in Uzbekistan and beyond. Collective measures, such as the Collective Security Treaty Organization (CSTO) and the conventions of the Shanghai Cooperation Organization (SCO) on combating extremism, have rightly called for "unified and concerted efforts" to focus on terrorism, extremism, and transnational organized crime (Russian Embassy U.K. 2018). However, these regional frameworks and declarations fail time and again when it comes to the implementation of coordinated counter-terrorism policies. Mariya Omelicheva, Associate Professor in the Department of Political Science at the University of Kansas argues that, "They routinely fail to address the multiple and complex ways that these activities intersect. This is a serious oversight. It makes collective security initiatives incapable of appropriately countering local and regional security threats" (Global Affairs 2018).

The Central Asian approach to countering radicalization and terrorism is described by some experts, with the common terminology of "the good, the bad, and the ugly" (Global Affairs 2018). While there is greater recognition and willingness for good forms of cooperation across Central Asia on the grave threats these phenomena present, when it comes to meaningfully understanding the root causes and adopting workable national and regional solutions, bad and sometimes outright ugly failures come to light. Police action, repression, and punitive measures alone have failed to produce meaningful progress. However, on the whole, Central Asia has fared far better in preventing ISIS sponsored and inspired attacks compared to other regions, such as the Middle East and even Europe. This is perhaps due more to a relatively smaller ISIS presence in Central Asia than necessarily better policies per se (Lister 2015).

Major Policy Shifts

Three major strategic shifts, if leveraged properly, could help to finally address the pathologies from Afghanistan that have prevented the consolidation of regional stability based on economic cooperation. These are the implementation of a peace-centered strategies by the United States and Afghanistan beginning in 2018, and the "opening" of Uzbekistan following the accession to the presidency of Shavkat Mirziyoyev in 2016.

U.S. South Asia Strategy

In August 2017, the United States introduced its new strategy for South Asia. The strategy called for "a political settlement that includes elements of the Taliban in Afghanistan" but made this contingent on military gains (The White House 2019). The regional element of the policy focused on putting pressure on Pakistan so that it would in turn reduce support for the Taliban and encourage them towards negotiations. The strategy had little to say about Central Asia—a feature noted by Central Asian policy makers (The White House 2019).

Perhaps, the most significant outcome of the Trump administration's South Asia Strategy is the signing of the U.S.–Taliban peace agreement, providing for a withdrawal plan that was concluded on February 29, 2020 in Doha, Qatar, as detailed below.

A year after the adoption of the South Asia Strategy, following mixed results on the battlefield, the U.S. broke a longstanding policy of not talking directly to the Taliban without the participation of the Afghan government and began unilateral discussions with the movement's political office in Doha. In September 2018, Ambassador Zalmay Khalilzad was appointed as Special Representative for Afghan Reconciliation and given a mandate to negotiate the terms of a possible military withdrawal from Afghanistan with the Taliban in exchange for a Taliban commitment to ensure that Afghanistan would not become a haven for international terrorists. Khalilzad's mandate also included urging the Taliban and the Afghan government to seek agreement on a ceasefire and begin negotiating a future political settlement.

While advancing on the bilateral issues between the United States and the Taliban, Khalilzad urged, with little success, the Afghan Government and the Taliban to form negotiating teams and begin to engage with each other (Tolo TV News 2018a). While this has not yet happened, perhaps the most significant meeting of this nature occurred in Moscow, from February 4–5, 2019, where Taliban Representatives sat around the same table exchanging views with a wide array of Afghan political actors, except for Afghan government representatives (Radio Free Europe 2019). A meeting that would have included government representatives was scheduled to take place in Doha in late April 2019, but broke down when an agreement was not reached on the composition of Kabul delegation.

The apparent determination of the United States to withdraw responsibly from Afghanistan has helped to align the region. It is increasingly becoming clear that the United States, Russia, and China are slowly finding common ground in their view of the Afghan conflict (Xinhua 2019). Tensions between the United States and Russia over Afghanistan have, for example, been greatly lowered (Khalilzad 2019). But Central Asia as a region does not appear to be a coherent part of the strategic thinking about Afghanistan's future—though Khalilzad did visit Tashkent and other Central Asian capitals in the context of his peace efforts. Absent that clarity, it is difficult for regional countries to adjust their own strategies. Central Asian countries have expressed doubts as to whether the United States has clear aims and the staying power to allow its longest war to transition to a peaceful conclusion (Rubin 2018).

President Ghani's Bid for Peace

In February 2018, at the second Kabul Process conference, President Ghani announced that he was prepared for unconditional talks with the Taliban. Despite the Taliban not responding, and following the beginning of U.S. unilateral talks with the Taliban, in November 2018, President Ghani began taking some institutional and symbolic measures to "focus attention on the peace process" (Office of the President of Afghanistan 2018). The measures included re-establishing the High Peace Council, multiple advisory boards with broad membership, government negotiating teams, a consultative Loya Jirga, and a large Ulema meeting.

While the U.S.–Taliban track of negotiations culminated in a recent landmark formal agreement, the Afghan Government-Taliban track, formulated as part of the larger intra-Afghan dialogue, has been riddled with disagreements, delays, and setbacks.

A principle issue has been the parallel nature of the two tracks, where the U.S. held talks with each side separately making commitments to the Taliban, without the Afghan government's consent that yet required its action, A prime example is the provision to release over 5000 Taliban prisoners (and 1000 from Afghan government side) as a pre-condition touted as a "confidence building measure" before the start of the intra-Afghan dialogue. This was to begin on March 10 but has witnessed delays over disagreements. Further brinkmanship by Ambassador Khalilzad has at last facilitated the start of a series of "virtual" meetings between the working groups from both sides over prisoner releases, currently aimed to begin on March 31, 2020 (Tolo TV News 2020).

In addition, the Afghan Government has also delayed formation of a broad-based national negotiating team to engage the Taliban in the intra-Afghan dialogue. A widely criticized list headed by Masoom Stanikzai, former NDS Chief, was first leaked to the press by Palace officials, in late March 2020, and later rejected by the Ministry of Peace spokeswoman on Twitter (Anwari 2020).

The Afghan Government is also delaying a closer vetting of Taliban prisoners and pushing for a permanent ceasefire while talks are underway, as opposed to advocating a reduction in violence measure (Aljazeera English 2020). Given the Afghan Government's failure to overcome internal rifts (including during its "National Unity Government" initiated in 2014) and coming-up short lately in forming a broad national consensus for peace, the Afghan Government is unable, at present, to unite major political opposition groups behind these demands and its larger peace agenda. As a result, the first round of public interactions between Afghan leaders and Taliban has lacked a full and coherent level of Afghan Government participation.

Uzbekistan's Opening

Uzbekistan President Shavkat Mirziyoyev's ascension to power in Tashkent in 2016 following the death of President Islam Karimov opened prospects for important shifts in Uzbekistan's domestic and foreign policies. Mirziyoyev initiated a policy of opening up to the world and immediate neighbors, both politically and economically, as well as initiating governance reforms and granting new freedoms to Uzbekistan's citizens. Some have described these waves of changes the "Uzbek Spring," leading to increased citizen participation in public life and national governance (Kramer 2017).

Uzbekistan, which borders all four other Central Asian countries, declared Central Asia to be the priority of its foreign policy. Mirziyoyev has sought to increase trade with all these neighbors and reduce tensions, in particular with Tajikistan. This central Asian orientation has set the stage for Uzbekistan's renewed interest, over the past several years, in Afghanistan's peace process and greater economic connectivity with Afghanistan and its other Central Asia neighbors. Uzbekistan sees Afghanistan (ranking among its top eight trade partners) as an extension of Central Asia and an opportunity to expand trade and commerce with South Asia. The March 2018 Tashkent Conference on Afghanistan, which occurred on the heels of the February Kabul Process conference, cemented Uzbekistan's leading role in the promotion of deepened political and economic links between Afghanistan and Central Asia and indicated a strong interest in supporting Afghanistan's peace process (see below).

The various regional forums to which the country belongs, such as Asian Development Bank-facilitated Central Asia Regional Economic Cooperation (CAREC) program, the Shanghai Cooperation Organization (SCO), C5 + 1 (Central Asia 5 + United States) potentially becoming C6 + 1 (to include Afghanistan), and the Istanbul "Heart of Asia" process, have provided Uzbekistan with an opportunity to project leadership in areas of cooperation and mutual benefit. Explored further below, these vehicles for Uzbekistan to exert regional leadership build on the country's experience, from the mid-1990s under the late President Karimov, when the Afghanistan focused 6 + 2 grouping of countries was formed under UN auspices.

With the United States, Uzbekistan has developed a multifaceted bilateral relationship, culminating in an official visit by President Mirziyoyev, to the White House in May 2018. The United States and Uzbekistan successfully concluded their Annual Bilateral Consultation (ABC) for 2019. President Trump highlighted the importance of new military, trade and investment relations between the two countries (PBS NewHour 2018). Military cooperation, potentially opening further assistance to U.S. operations in Afghanistan, is key to these renewed relations.

General Joseph Votel, U.S. Central Command (CENTCOM) Commander, visited Uzbekistan days before the official visit of the Uzbek President to Washington to finalize details on military cooperation. Uzbekistan continues to play an important role in America's strategic calculus concerning Afghanistan.

Uzbekistan is well-placed to advance a Central Asian initiative to coordinate with other peace and security-related forums concerning Afghanistan. Tashkent could also be a convenient location, both geographically and politically, for follow on meetings from Doha and Moscow and potentially merging these U.S. and Russian-led tracks to the Afghan peace process. Further, the integration of Afghanistan with its immediate northern neighbors—in the U.S. and European calculus for Central Asia, by focusing on economic connectivity in addition to security cooperation—would incentivize the Central Asian Republics to view Afghanistan as a partner for reaching new markets rather than as a threat and source of problems (Umarov 2019).

A Potential Strategic Convergence

On March 27, 2018 the Tashkent Conference on Afghanistan: Peace Process, Security Cooperation and Regional Connectivity highlighted a potentially new era of regional and international cooperation with Afghanistan. The conference was attended by the United States, Russia, China, the United Nations, and twenty other countries and international organizations. It endorsed the need for a political settlement to bring an end to the conflict in Afghanistan (Government of Uzbekistan 2018). Co-chaired by President Mirziyoyev and his Afghan counterpart Ashraf Ghani, the forum brought together regional countries directly involved in the conflict (through proxies and other clients), as well as regional and global powers (who are often at odds with each other) to address the sources and solutions to the long-standing and violent conflict. Though largely symbolic, the conference culminated in a clear international consensus (the Tashkent Declaration), promising to pave the way for a regional agreement with specific security and economic cooperation goals in order to achieve greater regional peace and stability.

The Tashkent Conference also triggered a series of follow-up high-level visits and exchanges between Kabul and Tashkent. This included, in October 2018, meetings in Kabul between Uzbek Foreign Minister Abdulaziz Kamilov, and both Afghan President Ashraf Ghani and former Afghan President Hamid Karzai (Government of Uzbekistan 2018). Further, on March 5, 2019, Afghan National Security Advisor Hamdullah Mohib visited Tashkent for a meeting with Viktor Makhmudov, Secretary of the National Security Council of Uzbekistan (ONSC 2019).

Such meetings are essential in terms of following up on the declaration points that require bilateral actions. Historically, high-level meetings have often resulted in limited outcomes as the machinery of governments often fails to follow through on the decisions of their political leaders with concrete actions.

Given the changing dynamics of the Afghan conflict and the United States' willingness to work with the region, the Tashkent and to some extent recent Moscow meetings (see below) could compliment United States' efforts. The Afghan government sees Tashkent as central to its peace process approach, hoping to leverage cooperation from its northern neighbors to reduce Pakistani influence. In a recent meeting with the Uzbek foreign minister, President Ghani underlined that the Tashkent Conference was "aligned with Afghan government's peace initiative" and was a "model of regional cooperation" (Government of Afghanistan 2019).

While the Tashkent Conference reflected an international and regional consensus on the desire for a political settlement in Afghanistan, it did not commit the parties to take concrete and measurable actions towards achieving peace and reconciliation in Afghanistan. In such conferences, the United States' policy position sets the tone, as it contributes by far and away the largest share of international security and economic assistance to Afghanistan.

Uzbekistan, at the same time, has continued to play a constructive and complimentary role to the United States' strategy in Afghanistan, as both the Uzbek President and Foreign Minister assured Secretary Pompeo during his March 3, 2020 visit with the five Central Asian nations in Tashkent. Mr. Pompeo thanked President Shavkat Mirziyoyev of Uzbekistan for "forging a closer, strategic partnership" with Washington in trying to advance the Afghanistan peace talks (Wong 2020).

A Region of Opportunity

Contributions of Individual Central Asian States to Building Peace and Regional Connectivity with Afghanistan

As part of the engagement strategy, each Central Asian country has sought ways to contribute to stability in Afghanistan while positioning themselves to take advantage of the huge benefits that would be reaped should a peaceful Afghanistan allow these isolated countries easier access to the world. This section describes these for each country.

As noted earlier, the reforms and openness advanced by its President, Shavkat Mirziyoyev, has enabled Uzbekistan to fast emerge as a leading regional player, exerting influence in various regional forums to which the country belongs, such as CAREC, the SCO, and the Istanbul "Heart of Asia" process. Uzbekistan's history of regional leadership extends back to the mid-1990s under the late President Karimov, when the Afghanistan focused 6 + 2 grouping of countries was formed under UN auspices.

A regional platform to address Afghanistan-centered issues with cooperation from the country's six immediate neighbors, plus the United States and Russia, is critically needed. Given the ongoing complexities of the U.S. and Russian peace tracks, perhaps a Central Asian venue, such as Samarkand (with its rich historical significance for the wide region), could provide both a venue and necessary Uzbek/Central Asian regional leadership to the peace process in Afghanistan. Both 6 + 2 and C5 + 1 (plus Afghanistan for both frameworks) could be good candidates. A 6 + 2 regional forum configuration, co-chaired by the United Nations, could be revived, among several other options. What brings hope to Uzbekistan's ability to be an effective regional broker is its renewed relationship and credibility with Afghanistan as well as the great powers, such as the United States and Russia, in addition to its more immediate neighbors. For instance, when President Mirziyoyev made an official visit to meet with President Trump, in May 2018 in Washington, D.C., he succeeded in warming relations with the U.S. without undercutting its relations with Moscow and Beijing (The Diplomat 2018).

Alongside Uzbekistan, Kazakhstan is the other relatively large major Central Asian Republic that has flexed its political muscles in support of Afghanistan and regional stability. Shortly after Afghan President Ghani's visit in December 2017 to Tashkent, Kazakhstan—the first Central Asian state to land a non-permanent seat on the UN Security Council—utilized part of its one-month presidency of the Security Council in January 2018 to highlight the importance of building a regional partnership between Afghanistan and Central Asia, and to strengthen the linkages between security and economic development in this pivotal relationship. According to Kairat Umarov, Permanent Representative of Afghanistan to the United Nations, "Until we made it a top priority again [on the Security Council], Afghanistan was merely a second or third order concern of the Council" (Umarov 2019).

In early January 2018, Ambassador Umarov, led a Security Council visit to Kabul (the body's first since 2010) to learn how the Council could further assist efforts on the ground. The delegation also used the opportunity to convey in Umarov's words "… the need to view Afghanistan not as a threat to security in the region but as an important partner." He added, "It is critical to integrate the economies of Afghanistan and neighboring countries, including the Central Asian States, through enhanced interaction and connectivity, by implementing regional infrastructure, trade, investment, transit and transportation projects" (UN News 2018). In short, Umarov stressed that the Council's visit achieved a "breakthrough" by recognizing that "you cannot fence off Afghanistan", and that, on the contrary, "achieving greater regional connectivity was essential to building sustainable peace" (Umarov 2019).

Shortly after the visit, Kazakhstan convened a special session of the Security Council to confirm a new heightened level of cooperation on political, security, and economic matters between Afghanistan and Central Asia. "Regional cooperation offers opportunities to address common concerns, including counter-terrorist financing, improving border security, fostering dialogue with religious institutions and leaders, and countering human trafficking and drug smuggling", explained UN Secretary-General António Guterres in his remarks to the Council (Guterres 2018). Kazakhstan's Foreign Minister, Kairat Abdrakhmanov, further argued in his country's presidential statement against the impracticality of progressing on border management, counter-narcotics, economic development, and so on without Central Asia's involvement (United Nations 2018).

The session also reiterated the support of the Security Council's fifteen members for earlier UN General Assembly and Security Council resolutions on Afghanistan. Both the Security Council visit to Kabul and its special session on Afghanistan demonstrated heightened Kazakh political leadership toward the Afghan peace process, buttressing its longer-standing economic and development assistance engagement vis-à-vis Afghanistan (see below).

Tajikistan convened two significant international meetings in 2018 that address issues of central concern to Afghanistan and its wider region. The first, a "High-Level International Conference on 'Countering Terrorism and Preventing Violent

Extremism'" held from May 3–4, 2018 in Dushanbe, brought together some four hundred delegates from forty countries and thirty international organizations. In referencing the meeting's "Dushanbe Declaration," Mr. Farhod Salim, Ambassador of the Republic of Tajikistan to the United States, recalled the "Joint Action Plan for Central Asia" (November 30, 2011) as the first comprehensive regional framework within the UN's broader Global Counter-Terrorism Strategy (GCTS) (Salim 2018). Closely related to the themes addressed in Dushanbe, Kairat Abdrakhmanov, Foreign Minister of Kazakhstan, introduced signatures by foreign ministers and representatives of more than 70 nations, on September 28, 2018 at United Nations Headquarters a new "Code of Conduct Towards Achieving a World Free of Terrorism (Government of Kazakhstan 2018). Unlike some 19 existing international conventions designed to combat terrorism, the Code of Conduct concisely presents ten easy to understand measures that countries can commit to in the fight against terrorism (Government of Kazakhstan 2018).

The second Tajikistan-hosted meeting was hosted from June 20–21, 2018 in Dushanbe and called the "High-Level International Conference on the International Decade for Action, 'Water for Sustainable Development, 2018–2028′." It brought together diverse government, international organization, business, and civil society representatives, as well as scientists and experts, to promote integrated water resource management, and the implementation of the 2030 Agenda for Sustainable Development's water-related Sustainable Development Goals (SDGs) (High Level International Conference on the International Decade for Action 2018). Given Tajikistan's shared use of the massive Amu Darya river basin with Afghanistan, Uzbekistan, and Turkmenistan, as well as common hydro-power transmission lines project (CASA-1000) with Afghanistan, Pakistan, and Kyrgyzstan, its demonstrated leadership, both regionally and beyond, is timely. Alongside the preceding month's counter-terrorism conference, these two back-to-back diplomatic-expert forums in Dushanbe have placed greater regional and global attention on—and possibly translating into enhanced resources and political support for—how the Central Asian States can better contribute further to building peace and regional connectivity with Afghanistan.

Through the Asian Development Bank (ADB) facilitated CAREC program—with mutually reinforcing support through the Regional Economic Cooperation Conference on Afghanistan (RECCA) and "Heart of Asia" forums as noted earlier—Central Asian countries have individually and collectively expanded their trade and transport ties with Afghanistan. In particular, the "hard" investments in transport infrastructure along CAREC Corridors #3 (involving Afghanistan and all five Central Asian Republics), #5 (connecting Afghanistan, Tajikistan, and western China), and #6 (connecting Afghanistan, Tajikistan, Kazakhstan, and Uzbekistan), along with new "soft" instruments (e.g., the ADB-facilitated Cross-Border Transport Agreement between Kyrgyzstan, Tajikistan, and Afghanistan), have accelerated cross-border economic exchanges in one of the still least economically integrated regions of the world. In January of 2018, Eklil Hakimi, then Afghan Minister of Finance, announced that as part of a new trade and transit agreement between Afghanistan and Uzbekistan, Uzbekistan would reduce transit fees for Afghan goods by 50%; now charging only $1,250 per container full of goods that travels through Uzbekistan (Tolo TV News 2018b).

Turkmenistan and Tajikistan have also exerted regional and broader international leadership vis-à-vis Afghanistan by hosting, in November 2017 and March 2012 in Ashgabat and Dushanbe, respective RECCA conferences. Both gave particular focus to regional energy cooperation, whereby Afghanistan serves as a major conduit for flows from low population and energy-rich Central Asia into densely populated and energy-poor South Asia. Beginning in December 2005, successive rounds of the Regional Economic Cooperation Conference on Afghanistan have provided a regional platform for Central, South, and Southwest Asian Foreign and Economic Ministers to engage their counterparts in Kabul in order to forge a consensus around concrete policies and investment project opportunities that deepen cross-border economic relations centering on Afghanistan. At RECCA-VII in Ashgabat, for instance, a ministerial statement was signed for the Turkmenistan-Afghanistan-Pakistan 500 kV Line (TAP-500). While the Turkmenistan, Afghanistan, Pakistan, India Natural Gas (TAPI) Pipeline Company Limited reported on its recent contract with an engineering firm to commence design and survey work. This latter mega-project (estimated to cost $22.5 billion), once fully implemented, will accelerate regional economic interdependence and demonstrate the real value of regional cooperation.

Earlier, the RECCA-V gathering in Dushanbe lent support to the Central Asia-South Asia Electricity Transmission and Trade Project's (CASA-1000) newly established national Working Groups dedicated to the project's implementation. CASA-1000—which will bring hydropower from Tajikistan and Kyrgyzstan via new electrical transmission lines to Afghanistan and Pakistan—is estimated to net Afghanistan up to an estimated $175 million annually. It will also generate indirect employment opportunities for over 100,000 people, and link the economies of Afghanistan, Kyrgyzstan, Pakistan, and Tajikistan.[3] Similar tangible benefits are projected for other priority trade facilitation, natural resource extraction, and transport initiatives, including:

- A Cross-Border Transport Agreement between Afghanistan, Tajikistan, and Kyrgyzstan.
- Efforts to promote customs harmonization and cross-border economic zones since RECCA-V.
- The signing last year, at RECCA-VII, of the Lapis-Lazuli Transit, Trade & Transport Route connecting Afghan traders with regional and markets further afield via Turkmenistan, Azerbaijan, Georgia, and Turkey.

[3]See strategies presented at the fifth (March 2012, Dushanbe), sixth (September 2015, Kabul), and seventh (November 2017, Ashgabat) editions of the Regional Economic Cooperation Conference on Afghanistan (visit: https://recca.af/).

Alongside and reinforcing the 11-member nation Central Asia Regional Economic Cooperation Program-CAREC,[4] and the14-member nation Istanbul "Heart of Asia" Process,[5] the Regional Economic Cooperation Conference on Afghanistan has succeeded in sustaining a dialogue on Afghan regional economic priorities that matter to the states of Central Asia and other regional development partners on seven successive occasions since 2005. Moreover, at successive RECCA gatherings, the forum's inclusive approach to regional cooperation has harnessed the talents and resources of regional governments and international agencies. Increasingly, scholars, the business community, and civil society organizations have become involved too. Finally, through private sector exhibitions and steps to engage business leaders in policy dialogue, RECCA has created new opportunities for facilitating private investment and innovative regional public–private-partnerships. At the same time, if this critical regional platform is to cement the economic cooperation gains achieved in recent years—signaling a new era of openness and cross-border commercial exchange—a far more serious commitment and skillful follow-through to regional investment projects and policy priorities agreed to at RECCA forums are needed urgently. Rigorous economic impact analyses for all regional investment projects and closer collaboration between RECCA, CAREC, the Heart of Asia, and other regional bodies and informal platforms would help to enhance Afghanistan's economic and political ties with its northern neighbors and beyond.

Recommendations

First, the **Central Asian Republics should band together collectively** and encourage something along the lines of C5 + 1 + 1 or perhaps a revamped version of the earlier 6 + 2 framework (plus the inclusion of Afghanistan in both), where Central Asian States, Uzbekistan in particular, provide growing political leadership in support of the Afghan peace process and development. Such an effort should directly reinforce the U.S.-led "Doha Track" and Russia-led "Moscow Track", as well as a more inclusive and credible intra-Afghan peace process. Especially for the proposed C5 + 1 + 1, conditions are highly favorable, building on the first C5 + 1 meeting in the Central Asian region at the level of Foreign Ministers with U.S. Secretary of State Pompeo, on February 3, 2020 in Tashkent, which reaffirmed commitments by the Central Asian Republics "to contribute to the peace process in Afghanistan, practical next steps in joint border security work, and regional efforts to improve economic and energy connectivity" (U.S. Department of State 2020c).

Here, the five Central Asian States have much to offer in assisting these concurrent new initiatives in Afghanistan's peace process: by hosting regular follow-up

[4]Note: The five Central Asian Republics were founding members in 1997 and Afghanistan joined in 2005.

[5]Note: The five Central Asian Republics were founding members and Turkmenistan co-leads, along with Azerbaijan and Afghanistan, the regional infrastructure technical working group.

meetings, giving regular political support, and encouraging regional confidence-building through cross-border economic cooperation with Afghanistan. Collectively, the Central Asian States stand to benefit from engaging Afghanistan on larger-scale economic integration projects, from the Five Nations Railway Corridor to the TAPI and CASA-1000 mega energy projects. These enable Afghanistan to be seen as a partner in development, rather than as a threatening breeding ground for violent extremist ideologies.

Just as the January 2019 "India-Central Asia Dialogue", convened in Samarkand, Uzbekistan, brought Afghanistan into the traditional "Central Asia 5" formulation, it is encouraging that the newly energized "C5 + 1" ministerial with the United States—underway since November 2015—is now entertaining the idea of including Afghanistan too (possibly expanding into a "C6 + 1" ministerial forum). Similarly, the U.S.-Central Asia Trade and Investment Framework Agreement (TIFA) should integrate Afghanistan, given the potential mutual benefits derived by all parties from treating Afghanistan as an integral part of Central Asia.

Second, and individually (but still in direct support of a stronger regional approach), each of the Central Asia States has a further contribution to make to building peace and greater connectivity between Central Asia and Afghanistan. For one, **Uzbekistan** is encouraged to build on the momentum created by recent U.S. peace efforts and the Tashkent Conference in March by pressing ahead with an institutional follow-up to the conference's commitments outlined in the Tashkent Declaration. Specifically, it could formally propose co-hosting, with the United Nations, a revamped 6 + 2 forum to coordinate peace and regional cooperation efforts vis-à-vis Afghanistan, that could, initially, build on the recent U.S.-Russia-China tripartite and expected (first) expansion to include Pakistan, India, and Iran. In consultation with the Afghan Government and the UN, Uzbekistan is also encouraged to establish a secretariat for potentially headquartering the revamped 6 + 2 forum in Tashkent. In cementing its internal economic and political reforms, it could further invest in efforts that harness ongoing regional cooperation programs (e.g., through CAREC, and the Belt-and-Road Initiative, and it is promising to see that it will hose the next RECCA meeting). Uzbekistan should be encouraged to increase bilateral trade and commerce with Afghanistan and the United States.

Kazakhstan should continue to exert influence politically in Central Asia, in major regional and global bodies (e.g., the SCO and UN), and vis-à-vis major powers (e.g., Russia and China) in support of concrete measures to build peace and greater regional connectivity between Central Asia and Afghanistan. Meanwhile, **Kyrgyzstan** should continue to assist completion of the CASA-1000 project, improved implementation of the ADB-facilitated Cross-Border Transport Agreement with Tajikistan and Afghanistan and strengthening of CAREC Corridor #3.

Building on its 2018 hosted international conference on countering terrorism and water for sustainable development, **Tajikistan** is well positioned to promote innovative new measures for improved Afghan-Tajikistan bilateral cooperation and Afghan-Central Asian regional cooperation to combat violent extremism and improve opportunities for transboundary water sharing. As co-chair of the Istanbul-Heart of Asia conference in 2019, Tajikistan should work with Afghanistan to maintain this important regional cooperation framework. **Turkmenistan** should continue to focus on accelerating implementation of the two major energy projects that traverse Afghanistan, TAPI and TAP-500, as well as the Lapis-Lazuli Transit, Trade & Transport Route connecting Afghan traders with regional and markets further afield via Turkmenistan, Azerbaijan, Georgia, and Turkey.

Third, **Afghanistan,** seizing on the momentum of the Tashkent Conference and renewed U.S. and Russian efforts, is encouraged to build on the recently concluded Consultative Loya Jirga. Specifically, it should engage the majority of political actors from across the board and launch a national and highly inclusive (to include both opposition groups and rival presidential teams) consultation process leveraging existing structures, including the Afghan President's new Advisory Board for Peace. The goal, in short, must be to bring the country's political leadership and population on board for a negotiated peace with the Taliban. To safeguard the gains of the past eighteen years, the Afghan Government must ensure the rights and liberties of all Afghans, men and women, and that the Taliban become part of the new Afghanistan (and not the other way around). Integral to this should be a continuous dialogue between Afghanistan and its northern neighbors, one that regularly pinpoints concrete political and economic measures the Central Asian Republics and Afghanistan can undertake to both overcome obstacles and further the causes of peace and greater economic connectivity within their region.

Fourth, and equally important, building on its recent historic agreement with the Taliban, in February 2020, the **United States** should commit to ensuring the ultimate goal of bringing peace and justice to the Afghan people and not just a withdrawal plan. Recognizing that the Afghan war can end with cooperation from *both* global and regional powers, the United States, Russia, and China should agree to convene a Presidential Summit involving the U.S., Russia, China, and the Afghan Government, as well as the Central Asian Republics and other important regional stakeholders, to agree on a peaceful resolution of the Afghan war based on regional and wider international cooperation. In its Annual Bilateral Consultations (ABCs) with Kazakhstan, Kyrgyzstan, Tajikistan, Turkmenistan, and Uzbekistan, the United States should encourage the inclusion of at least one standing agenda item dedicated to strengthening collaboration on cross-border economic ties with Afghanistan and the wider Central Asia region.

In addition to agreeing to the proposed Presidential Summit, both **Russia** and the **U.S.** should support the proposed revival of the 6 + 2 forum to coordinate peace and regional cooperation efforts vis-à-vis Afghanistan. In establishing stronger State-to-State relations with the Afghan Government, Russia should also cease providing the Taliban with political and other types of support and welcome greater Central Asian support for peacebuilding and economic connectivity vis-à-vis Afghanistan.

Fifth and finally, beyond intergovernmental relations, **Business-to-Business Cooperation** between Central Asia and Afghanistan should be enhanced through a new RECCA Chamber of Commerce & Industries and measures to empower women entrepreneurs through joint training, regional trade fairs, and other cross-border economic cooperation activities. **People-to-People and Educational Exchanges**, including joint artistic and cultural performances, should also be expanded across the region as a means of facilitating greater social and cultural understanding in Greater Central Asia.

The authors wish to thank Professor Sergei Plekhanov (York Center for Asian Research at York University) for sharing key insights and resources on Central Asia, Robert Kiel (The Stimson Center) for research assistance and Robert Kiel, Samuel Koralnik (The Stimson Center), and Colin Cookman (USIP) for editorial assistance. In particular, they wish to thank Scott Smith (USIP) for his thoughtful substantive guidance and editorial advice and the feedback received from participants attending the University of South Florida's Inaugural Great Power Competition Conference from January 29–30, 2020 in Tampa, Florida.

Interviews for this study were conducted with (on background) Mr. Kadambay Sultanov, Deputy Chief of Mission, Embassy of Uzbekistan to the United States (March 5, 2019);H.E. Kairat Umarov, Permanent Representative of Kazakhstan to the United Nations and former Ambassador to the United States (February 27, 2019); (on background) Ambassador of Afghanistan to Pakistan and former Afghan Finance Minister H.E. Dr. Omar Zakhilwal (on several occasions in October and November 2018); Former U.S. Assistant Secretary of State for South and Central Asia Ambassador Robert Blake (August 30, 2018); (on background) Mr. Nadir Naim, Deputy Chair of the High Peace Council of Afghanistan (May 9, 2018); (on background) Mr. Bashir Farooq, Deputy CEO of Hight Peace Council of Afghanistan (May 7, 2018); Ambassador Omar Samad, Senior Fellow, Atlantic Council (January 5, 2019); Javid Ahmad, Non-Resident Fellow, West Point/Atlantic Council (December 2018); Professor Sergei Plekhanov, Political Science Department, York University, and former Russian MFA official (January 15, 2019); David Raikow, Head of Office, Faryab Provincial Field Office, UNAMA (November 6, 2019); Claudia Schneider, Political Affairs Officer, Embassy of Germany to the United States (September 26, 2018); and Dr. Fredrick Starr, Founding Chairman of the Central Asia-Caucasus Institute (October 18, 2018).

References

Aljazeera English, Afghan government delays Taliban prisoner release, endangering deal. Aljazeera English Website. (2020).https://www.aljazeera.com/news/2020/03/afghan-gov-delays-taliban-prisoner-release-endangering-deal-200314153003123.html. Accessed 25 Mar 2020

J.A. Allen, M.E. O'Hanlon, Resolving the Ghani-Abdullah impasse in Afghanistan. The Brookings Institution. (2020). https://www.brookings.edu/president/resolving-the-ghani-abdullah-impasse-in-afghanistan/. Accessed 26 Mar 2020

N. Anwari, @7Anajia, Najia Anwari. Spokesperson State Ministry for Peace, Government of Afghanistan (2020)

R. Barrett, Beyond the caliphate: foreign fighters and the threat of returnees. The Soufan Group. (2017). https://thesoufancenter.org/wp-content/uploads/2017/11/Beyond-the-Caliphate-Foreign-Fighters-and-the-Threat-of-Returnees-TSC-Report-October-2017-v3.pdf. Accessed 25 Oct 2018

The Diplomat, Uzbekistan does not drift among great powers. The Diplomat. (2018).https://thediplomat.com/tag/russia-uzbekistan-relations/. Accessed 15 Oct 2018

Global Affairs, Russia, Counter radicalization policies in Central Asia: the good, the bad and the ugly. Global Affairs, Russia. (2018).https://eng.globalaffairs.ru/PONARS-Eurasia/Counter-Radicalization-Policies-in-Central-Asia-The-Good-the-Bad-and-the-Ugly-18507

Government of Afghanistan, President Ghani discusses economic cooperation, peace process with Uzbek Foreign Minister. Office of the President of Afghanistan website. (2019). https://president.gov.af/en/news/2/15/2019/9. Accessed 19 Feb 2019
Government of Kazakhstan, Another international initiative of President Nazarbayev implemented: code of conduct towards achieving world free of terrorism signed. The Ministry of Foreign Affairs, Republic of Kazakhstan. (2018).https://mfa.gov.kz/en/content-view/realizovana-ese-odna-mezdunarodnaa-iniciativa-prezidenta-n-nazarbaeva-podpisan-kodeks-povedenia-po-dostizeniu-mira-svobodnogo-ot-terrorizma. Accessed 15 Oct 2018
Government of Tajikistan, Dushanbe Declaration. (2018). https://dushanbeconf2018.tj/declaration. Accessed 30 Sept 2018
Government of Uzbekistan, Declaration of the Tashkent Conference on Afghanistan: peace process, security cooperation and regional connectivity. Ministry of Foreign Affairs Website. (2018a).https://uza.uz/en/politics/declaration-of-the-tashkent-conference-on-afghanistan-peace--28-03-2018?m=y&ELEMENT_CODE=declaration-of-the-tashkent-conference-on-afghanistan-peace--28-03-2018&SECTION_CODE=politics. Accessed 30 Sept 2018
Government of Uzbekistan, Uzbekistan government delegation visited Afghanistan. Ministry of Foreign Affairs website. (2018b). https://uza.uz/en/politics/uzbekistan-government-delegation-visited-afghanistan-10-07-2018?m=y&ELEMENT_CODE=uzbekistan-government-delegation-visited-afghanistan-10-07-2018&SECTION_CODE=politics. Accessed 30 Sept 2018
Guterres, A, Remarks to the Security Council on building regional partnership in Afghanistan and Central Asia, to link security and development. U.N. Secretary General's Office. (2018). https://www.un.org/sg/en/content/sg/speeches/2018-01-19/building-regional-partnership-afghanistan-and-central-asia-remarks. Accessed 16 Aug 2018
J. Hansler, U.S. Taliban sign historic agreement. CNN. (2020). Accessed 20 Mar 2020
High-Level International Conference on the International Decade for Action, Water for sustainable development, 2018–2028. Conference website. (2018). https://wsdconf2018.org/. Accessed 30 Sept 2018
Z. Khalilzad, Special interview with Zalmay Khalilzad. Tolo News Website. (2019).https://www.youtube.com/watch?v=PAfNHTYnbaA. Accessed 30 April 2019
A.E. Kramer, Once closed and repressive, Uzbekistan is opening up. The New York Times. (2017). https://www.nytimes.com/2017/10/25/world/asia/uzbekistan-politics.html. Accessed 15 July 2018
T. Lister, ISIS goes global: 143 attacks in 29 countries have killed 2,043. CNN. (2015). https://www.cnn.com/2015/12/17/world/mapping-isis-attacks-around-the-world/index.html. Accessed 20 Nov 2018
S. Neuman, U.S. 'disappointed' in Afghan leadership, will slash $1 billion in aid, Pompeo Says. National Public Radio (NPR). (2020). https://www.npr.org/2020/03/24/820550175/u-s-disappointed-in-afghan-leadership-will-slash-1-billion-in-aid-pompeo-says. Accessed 23 Mar 2020
Office of the President of Afghanistan, News release. (2018). https://president.gov.af/fa/2/11/5/18. Accessed 10 Nov 2018
ONSC, NSA @hmohib met with Dr. Viktor Mahmudov, Uzbekistan's NSA in Tashkent today. The two sides discussed a range of issues, including partnerships on counter-terrorism, counter-narcotics, and expanding infrastructure, economic, and cultural cooperation Tweet from Office of National Security Council (ONSC), Afghanistan. (2019).https://twitter.com/NSCAfghan/status/1102925895344549889.Accessed 4 April 2019
S. Peyrouse, Diagnosing Central Asia's drug problem: the real drivers are poverty and corruption, not infrastructure. Reconnecting Asia web portal, Center for Strategic and International Studies. (2017). https://reconnectingasia.csis.org/analysis/entries/diagnosing-central-asias-drug-problem/. Accessed 3 March 2019
S. Ramani, Why Russia exaggerates Islamic States presence in Afghanistan. The Diplomat. (2018). https://thediplomat.com/2018/04/why-russia-exaggerates-islamic-states-presence-in-afghanistan/

RFE/RL, Taliban, Afghan delegations hail 'successful' Moscow meeting, pledge to meet again. Radio Free Europe/Radio Liberty. (2019). https://www.rferl.org/a/taliban-afghan-delegation-say-talks-to-continue/29756223.html. Accessed 19 Feb 2019

B. Rubin, Is Afghanistan ready for peace? How great powers can end the war. Foreign Affairs J. (2018). https://www.foreignaffairs.com/articles/afghanistan/2018-07-30/afghanistan-ready-peace. Accessed 15 Oct 2018

Russian Embassy U.K., Convention of the Shanghai Cooperation Organization on combating extremism. Embassy Wesbite. (2018).https://www.rusemb.org.uk/fnapr/6271

F. Salim, Briefing by Ambassador of the Republic of Tajikistan. United States Institute of Peace (USIP). (2018). Accessed 8 June 2018

Tolo TV News, Govt-Taliban VTC held over prisoner release, date set. Tolo News website. (2020).https://tolonews.com/afghanistan/govt-taliban-vtc-held-over-prisoner-release-date-set. Accessed 25 March 2020

Tolo TV News, Khalilzad calls on Taliban, Afghan government to form negotiating teams. Tolo News website. (2018a).https://www.tolonews.com/afghanistan/khalilzad-calls-govt-and-taliban-form-negotiating-teams. Accessed 30 Oct 2018

Tolo TV News, Uzbekistan drops transit fees for Afghan goods by 50%. Tolo News Website. (2018b). https://www.tolonews.com/business/uzbekistan-drops-transit-fees-afghan-goods-50. Accessed 15 Oct 2018

UN News, In Afghanistan, Security Council reiterates support for efforts to restore peace and progress. (2018). https://news.un.org/en/story/2018/01/1000442#.Wmc4aWhSxPY. Accessed16 Aug 2018

United Nations, Security Council presidential statement calls for action to avert threats against security, stability in Afghanistan ahead of debate on pressing challenges. U.N. Press Release. (2018). https://www.un.org/press/en/2018/sc13170.doc.htm. Accessed 20 Aug 2018

UNODC, *Afghan Opiate Trafficking along the Northern Route* (UNODC, Vienna, 2018a).

UNODC, World Drug Report 2018, booklet 3, analysis of drug markets: opiates, cocaine, cannabis, synthetic drugs. United Nations Office on Drugs and Crime (UNODC). (2018b). https://www.unodc.org/wdr2018/prelaunch/WDR18_Booklet_3_DRUG_MARKETS.pdf. Accessed 8 May 2019

UNODC and GoIRA, Afghanistan opium survey 2017: challenges to sustainable development, peace and security. United Nations Office on Drugs and Crime (UNODC) and Islamic Republic of Afghanistan Ministry of Counter Narcotics. (2018).https://www.unodc.org/documents/crop-monitoring/Afghanistan/Afghanistan_opium_survey_2018.pdf. Accessed 8 May 2019

U.S. Department of State, Agreement for bringing peace to Afghanistan between the Islamic Emirate of Afghanistan which is not recognized by the United States as a state and is known as the Taliban and the United States of America.Official Text of the U.S.-Taliban agreement posted on State Department website. (2020a). Available at https://www.state.gov/wp-content/uploads/2020/02/Agreement-For-Bringing-Peace-to-Afghanistan-02.29.20.pdf. Accessed 20 Feb 2020

U.S. Department of State, Secretary Pompeo's participation in the Tashkent C5+1 ministerial. (2020b). https://www.state.gov/secretary-pompeos-participation-in-the-tashkent-c51-ministerial/

U.S. Department of State, United States Strategy for Central Asia 2019–2025: advancing sovereignty and economic prosperity. (2020c). https://www.state.gov/united-states-strategy-for-central-asia-2019-2025-advancing-sovereignty-and-economic-prosperity/

U.S. White House, Remarks by President Trump on the strategy in Afghanistan and South Asia. (2019). https://www.whitehouse.gov/briefings-statements/remarks-president-trump-strategy-afghanistan-south-asia/. Accessed 15 April 2019

E. Wong, Violence must stop before Afghanistan peace agreement, Pompeo says. The New York Times. (2020). https://www.nytimes.com/2020/02/03/us/politics/Afghanistan-Pompeo-Taliban-Peace.html. Accessed 25 March 2020

Xinhua, China, U.S., Russia meet on Afghanistan issue, agree on further talks. Xinhua News Agency. (2019). https://www.xinhuanet.com/english/2019-03/23/c_137918082.htm. Accessed 1 May 2019

A Failure to See: The Strategic Error of Peacetime Deployments as a Direct Cause for Failure in Afghanistan

Andre Hollis

Abstract In the weeks immediately following the September 11, 2001, senior U.S. leaders in the Departments of Defense (DoD), State, Justice, the U.S. intelligence community and others met repeatedly to plan and discuss the immediate steps that the U.S. would undertake to respond to Al Qaeda's attack. Many in global media, academia and think tanks have and continue to opine with regard to decision-making at that time—namely the decision to rely mainly upon Afghan forces to find Osama Bin Laden and the later decision to overthrow Iraqi President Saddam Hussein—but few have focused upon an even greater strategic error made at that time. This strategic error—the decision not to deploy U.S. military and civilian personnel to Afghanistan until the strategic objectives had been met—directly led to failures in information collection, analysis, force deployment decision making and the inability of senior leaders to understand the nature of the conflict that they were undertaking. This Chapter describes the reasons that this failure was strategic, its consequences and offers future senior policymakers several recommendations to avoid future catastrophes.

Keywords Strategic error · Failure in afghanistan

Introduction

In the weeks immediately following the September 11, 2001, senior U.S. leaders in the Departments of Defense (DoD), State, Justice, the U.S. intelligence community and others met repeatedly to plan and discuss the immediate steps that the U.S. would undertake to respond to Al Qaeda's attack. Many in global media, academia and think tanks have and continue to opine with regard to decision-making at that time—namely the decision to rely mainly upon Afghan forces to find Osama Bin Laden and the later decision to overthrow Iraqi President Saddam Hussein—but few have focused upon an even greater strategic error made at that time. This strategic error—the decision

A. Hollis (✉)
Tigris International, 344 W. Maple Avenue Ste 279, Vienne, VA 22180, USA
e-mail: ahollis@tigrisinternational.us

© The Author(s), under exclusive license to Springer Nature Switzerland AG 2021

A. Farhadi et al. (eds.), *The Great Power Competition Volume 1*,
https://doi.org/10.1007/978-3-030-64473-4_5

not to deploy U.S. military and civilian personnel to Afghanistan until the strategic objectives had been met—directly led to failures in information collection, analysis, force deployment decision making and the inability of senior leaders to understand the nature of the conflict that they were undertaking. This Chapter describes the reasons that this failure was strategic, its consequences and offers future senior policymakers several recommendations to avoid future catastrophes.

Rationale for Establishment of Long-Term Deployments to Afghanistan Starting in 2002

Persistent Situational Awareness

"Persistent Situational Awareness" refers to the concept that, by observing and analyzing all kinds of information—including through interaction with individuals—one can more accurately identify trends, current or potential future obstacles and identify solutions before those problems manifest or worsen. While many think of such awareness in terms of "intelligence" collection (e.g., communications and other technical means), many tend to overlook or underestimate the value of "HUMINT"—information gained from individuals whose credibility is constantly assessed through interaction and independent verification. In such cases, an actor such as the U.S. Government can influence local or Host Nation decision-making by establishing trusted relationships with leaders or actors by earning their trust and confidence.

This situational awareness, once developed, can be useful on many strategic levels. By establishing interpersonal levels of trust, one can better determine the credibility, willingness and limitations of the person or persons that one seeks to influence. Correspondingly, the person to be influenced will more easily accept one's advice if they believe that the influencer understands their challenges and is willing to assist as long as the need exists. When an influencer is unable or unwilling to commit throughout the duration of the problem, the decision-maker is naturally less likely to accept advice from someone who will not assist them as long as needed.

Thus, U.S. Government decisions—continuing as of the publication of this Chapter—to limit tours of service in Afghanistan to as short as six months (U.S. Marine Corps), continue to limit and, in most cases, prevent the development of persistent situational awareness developed from interpersonal relationships. First, by indicating to Afghan national, regional and local leaders that westerners advising or assisting them in military, civilian or humanitarian aid will only do so for limited periods, we prevent those seeking to provide assistance from developing the initially required trust needed to first establish credibility. Second, we thereby prevent the influencer from gaining sufficient knowledge to determine whether the person to be influenced possesses the necessary competence, willingness or integrity needed to affect a solution. Finally, the influencer fails to gain the needed strategic understanding of the local or regional problem necessary to recommend to senior U.S.

leaders in the Host Nation or higher headquarters in Washington, D.C. actions that might advance our strategic interests. Without this persistent situational awareness, U.S. leaders have and will continue to develop and implement policies lacking in local or regional understanding or credibility.

Creation and Development of Key Personal Relationships

This trust and confidence, moreover, can neither be assumed simply because the seeker has an official position or can affect constructive benefits through their official duties. Trust and confidence necessary to influence another to act (or not act) in a manner sought first requires that the parties establish and grow inter-personal credibility through actions taken over time. Only after that inter-personal relationship has been established over time can the seeker thereafter influence the person to effect changes that will last.

An excellent example from popular U.S. movie culture is the role of "Tom Hagen" in the movies "Godfather" and "Godfather II." In these movies, Tom Hagen, played by actor Robert Duvall, is a trusted advisor to the leaders of the Corleone crime family. In the accepted stereotype of Italian-American crime families, trusted members or advisors are normally selected from longtime family members or other close associates whose family lineage could be traced back to Sicily. Tom Hagen, however, was an American of European descent who had been adopted by the Corleone family. Tom was not Sicilian but, yet, was "intimately involved in all the family business." The Hagen character achieved this level of access and trust over years of personal relationship, credibility and proven trust. While Hagen could never be "made" (accepted formally as a member of the crime family), he earned that access and trust and was often sought for his counsel on criminal and other activities.

Likewise, I personally found that becoming a trusted advisor to leaders of normally closed societies was possible as long as I earned their trust through credible actions. From 2011 to 2012, I served as the Advisor to the Honorable Zarar Osmani—then the Minister for Counternarcotics for the Government of the Islamic Republic of Afghanistan (GIROA). As one might imagine, contrary to many usual western (and, particularly, American) assumptions, my prior U.S. Government and international experience in counternarcotics,[1] while enabling me to begin that relationship as experienced, did not create an automatic level of trust on the part of Minister Osmani, senior Afghan staff or international counterparts (including U.S. military and civilian leaders in Afghanistan). From my first day in Afghanistan, I had to constantly earn their trust through actions.

Additionally, I began that experience with several disadvantages. First, I arrived in Kabul, Afghanistan with no ability to speak Dari or Pashto—the primary languages

[1]From 2001 to 2003, I served as the U.S. Department of Defense Deputy Assistant Secretary of Defense for Counternarcotics—an unusual office with its own $1.3 billion annual budget and global and domestic statutory authorities.

spoken by Afghans. Indeed, many of the Afghans spoke four or five languages—including English—as a result of education that they received in school or refugee camps. Thus, while I began with some knowledge of strategic Afghan CN issues (as a result of my previous U.S. Government experiences), I was initially unable to advise change because I lacked the ability to articulate to Afghans in the languages that they spoke.[2] Thanks to the Minister's vision and graciousness, I eventually became conversant in Dari—in exchange for assisting him in his mastery of English.

Second, I began that experience competent in CN issues from a U.S. perspective but ignorant of the issues from an Afghan and, even, a Central Asian strategic, operational or cultural perspective. My prior experience involved U.S. strategic CN policy and program development—with knowledge of issues impacting South America and East Asian—but, like most Americans who became involved in the issues post-9/11, I lacked substantive understanding of how CN impacted Afghans or the region.

Third, I quickly realized that the Minister and his staff did not need or seek an American view of their problems but, rather, someone who might provide advice that would be of practical value to their Ministry and Afghanistan CN problems. Solutions that I knew worked in the U.S., Colombia or other places where we had successfully assisted other Governments were not necessarily practical or useful in Afghanistan.[3] Thus, the more likely route of success was to first listen and learn from the Afghans what they thought were their challenges and thereafter slowly suggest potential solutions.

Thus, it became quickly obvious that, to be effective, I first needed to earn and grow the confidence of these Afghan decision-makers and influencers before offering counsel. Indeed, I needed to wait until asked, first before offering suggestions. Doing otherwise—offering a western perspective on solutions before understanding their particular needs or gaining their initial trust—would only confirm their stereotypes of western arrogance.

Establishing that necessary credibility therefore could not be assumed simply because I was a U.S. citizen and former U.S. Government official. I had to earn that trust (and keep it) and, in order to do so, I had to spend time earning their confidence. I could not earn this trust if they believed that I only intended to serve for an arbitrary period of time. Whether six months, nine months or even two years, if I indicated that my time in service to the Minister was finite, my value would diminish each day from the start. Thus, when asked by Minister Osmani on my first day "how long

[2]Ironically, the State Department had refused my pre-deployment request to take Dari language classes at the State Department's language school in northern Virginia because "the Afghans speak English, you won't need it". This faulty assumption hampered my ability to become effective.

[3]For example, the employment of aerial or ground-based spraying of crops, such as successfully done in Colombia, were not useful in Afghanistan. Cocoa crops in Colombia were largely multi-hectare fields in jungle or otherwise rural areas where aerial spraying largely did not impact civilians, other agriculture or livestock. In contrast, many opium fields in Afghanistan were small-hectare plots located adjacent to farmers' homes or the livestock that they relied upon for sustenance. Additionally, the concept of spraying chemicals upon Muslim populations had its own potential strategic drawbacks, as one might imagine.

would I serve as his Advisor," I told him that I would stay as long as I was useful to him.[4]

The value of this open-ended utility cannot be overstated. By making clear to the Minister and his senior staff that I served as long as needed, they gave me the opportunity to earn their personal trust. After several months of service, including partaking of many professional, social and cultural activities, traveling with the Minister into areas that U.S. citizens were not allowed to travel due to risk and, eventually, moving into the Minister's home and meeting his female family members, we established a level or trust that caused the Minister to consider and, many times, act as I advised.

Sadly, U.S. military and civilian leaders continue to fail to establish and maintain the necessary interpersonal relationships necessary to acquire or maintain credibility with Afghan local, regional and national leaders. Continuous rotations of key personnel prevent the establishment of long-term relationships needed to influence Afghan leaders in ways that may advance both Afghan and U.S. strategic interests.

Development of Subject-and Area-Expertise

Sadly, as discovered after the initial post-9/11 response into Afghanistan, the United States and its allies lacked substantive area- or subject-matter expertise in the issues that continue to bedevil Afghanistan today. Initial U.S. representatives—military and intelligence personnel—lacked even basic maps, language comprehension or cultural awareness.[5] Soldiers and intelligence operators initially entering Afghanistan relied upon Fodor's, Lonely Planet, National Geographic and other commercially-available map information because the U.S. lacked any useful, current intelligence.[6] The U.S. lacked knowledge of key Afghan family, tribal, regional and ethnic relationships that could have aligned in concert with U.S. interests. The U.S. also lacked interpersonal relationships that would have alerted us to the fact that many from whom we sought assistance had a well-documented history of betrayal, criminality and human rights abuse. While these and many other deficiencies can fairly be placed into context (i.e.,

[4]To be clear, I want to thank several mentors of my own—many of whom are prior U.S. Special Forces veterans trained in working with Host Nationals—who advised me prior to deployment to do exactly that. As a result of my statement, the Minister told me that he would give me the chance to earn his trust.

[5]The U.S. military primarily relies upon U.S. Special Forces (SF) and Foreign Area Officers (FAOs) who receive regional language and cultural training theoretically enhanced by 179 day deployments (SF) or three year Embassy tours (FAOs) that, it is hoped, will maintain their skills. The problem on 9/11, however, was that the regionally focused SF Group—Fifth Group—focused on the Middle East, as a whole. Thus, over 90% of the Fifth Group operators in service on 9/11 possessed language and cultural skills relating to Arab nations and mastered the Arabic language. Afghanistan is not an Arab nation and its two main languages are Dari and Pashtu. Finally, since the U.S. maintained no diplomatic presence in Afghanistan, there were no FAOs skilled in Dari, Pashtu or Afghan political, cultural and other unique attributes.

[6]While my sources are individuals who were some of the first Americans in Afghanistan the six months subsequent to 9/11, the reader can find dozens of similar, unclassified reports via the internet.

necessity), the failure to rectify many of those deficiencies after the initial invasion cannot be ignored.

Several subject matters became manifestly relevant to post-9/11 Afghanistan—areas that the U.S. Government was, at the time, incompetent.[7] An obvious area was (and remains) post-conflict governance. There are many academics who have written countless treatises and opined at numerous conferences but few had practical experience assisting an unfamiliar culture with re-establishing governance. Issues such as law enforcement, security, establishment of government institutions, anti-corruption, tax and legislation all sound familiar in a classroom but in an Islamic region with ties based upon family and clan rather than lines on a map, they are as foreign to U.S. "experts' as brain surgery is to the author.

Second, the post-9/11 return of narcotics trafficking as a threat to future Afghan governance was apparent to many in the U.S. Government immediately after September 11. Indeed, in the first (of many) emergency supplemental appropriations enacted by the U.S. Congress just days after 9/11, elements of the Department of Defense requested—and received—$68 million for "counter-drug activities" in Afghanistan and in support of the U.S. State Department's Office of International Narcotics and Law Enforcement (INL). The ability to see this upcoming challenge to U.S. strategic goals (i.e., withdraw from Afghanistan with a functioning, capable Government in place), however, was, sadly, a minority view.

Most senior leaders in the Department of Defense—notably the U.S. Central Command[8] and senior officials reporting to Secretary of Defense Donald Rumsfeld—consciously and actively opposed any effort by the Department to proactively counter the return of narcotics. General Officers perceived the narcotics issue as not a military priority and, even if it were, something for someone else to address. Senior civilian appointees likewise believed that the Secretary had no interest in employing the Department's authorities, money or experience in counternarcotics. In fact, the author was publicly threatened with firing for sending a written memorandum suggesting limited CN efforts in Afghanistan—in support of INL—in November of 2001. The Department of Defense—the overwhelmingly majority actor in U.S. interagency efforts relating to Afghanistan from 2001 to 2005—was not going to develop any expertise in Afghan-related CN.

In hindsight, it is not clear that active, willing Defense Department CN efforts in Afghanistan would have resulted in a materially different outcome. From 1989, the U.S. Government, with substantial Defense Department assistance (albeit unwilling for many years leading to 2001), developed CN subject-matter expertise through its efforts to support Colombia. In fact, as of 2003, the U.S. can proudly claim that, in support of the Colombian Government, it was successful in combatting a narcotics-funded insurgency. For example, while many Defense Department

[7]To be clear, I use the word "incompetent" not in a pejorative sense but, rather, as a statement of fact. The author has never attended medical school. If the reader needed immediate brain surgery and relied upon the author, they would most certainly die. The author is incompetent with respect to surgery of all kinds.

[8]The geographically-focused U.S. military command responsible for the most of the Middle East stretching east to Afghanistan (but not Pakistan).

uniformed personnel deployed for 179-day period to Colombia, they made many such deployments. While back at home station, they continued to develop and grow their language, cultural and Colombia-specific knowledge. Many developed interpersonal relationships that grew over multiple returning deployments. They developed personal, geographic and institutional knowledge relating to Colombia and its challenges. Likewise, other. Agency employees served multi-year tours in Colombia. State, Justice (particularly the Drug Enforcement Administration) deployed their best employees—with incentives of pay and attractive career advancement—who developed subject and area expertise vital to U.S. policy and program development and support to the Colombian Government. Programs such as persistent vetting of Colombian law enforcement, military and Governmental officials helped to identify potential corrupt actors and mitigate their effects. These programs, coupled with these subject and area expertise developments, provided U.S. senior policymakers with reliable, objective sources of information upon which to develop, implement and oversee U.S. policies and programs.

The development of this subject-matter expertise, however, was limited in scope and benefit. The expertise was useful in Colombia—as a result of Colombian will[9] and limits upon Defense Department activity[10] but was of limited utility with respect to Afghanistan's narcotics challenges. First, the Defense Department was materially the sole U.S. Agency with presence or ability in Afghanistan. Other Agencies provided limited personnel and expertise but themselves were limited by their inability to deploy expeditionary personnel. INL, the State Department Office responsible for civilian programs to combat narcotics and development Afghan law enforcement, did not place a single Foreign Service Officer into the U.S. Embassy in Kabul until 2005. Thus, the Defense Department had no interagency partner with the authority or resources to address narcotics issues in Afghanistan even if it could convince Defense Department counterparts to assist in the issue.

Second, Defense Department capabilities were regionally aligned and therefore not competent in Afghan CN issues. The Special Forces personnel experienced in counter-narco terrorism, for example—the Seventh Special Forces Group—were limited by policy and tradition to Central and Latin America. Even when senior Defense Department officials suggested to Central Command that it consult with

[9]Unlike Afghanistan and other locations, Colombia had and maintained an ability to govern (albeit limited and damaged by narcotics and civil war). What Colombia possessed, however, was a collective national identity (most Colombians self-identified as Colombians and as Catholics) and a collective national will. Colombia, for example, levied a one-time war tax in 2002 equivalent to $1 billion. The Colombian people gave $1.3 billion in response. The author is unable to identify any historical effort—including in the United States—when the population donated more money to the Government to combat threats than that which the Government requested.

[10]The Congress authorized the Defense Department to provide specific unclassified and classified support to the Colombian Government but limited its role by placing a ceiling upon the numbers of uniformed and civilian Defense Department personnel who could be deployed to or visit Colombia. As an unintended consequence, the Defense Department had to operate as a true partner to other U.S. Agencies (notably, the State Department) and devise collective interagency programs and policies. No similar limits existed upon the Defense Department with respect to Afghanistan in 2001 or afterwards.

its Southern Command[11] and Seventh Special Forces Group counterparts, Central Command officials rejected the idea.

Finally, even after some of the U.S. Government's Colombia-focused expertise were invited to assist with Afghan CN issues, they found their expertise largely inapplicable to Afghanistan. Unlike Colombia, Afghanistan had no functioning Government. Unlike Colombia, Afghan identity was (and, sadly, remains) largely non-existent. Those with Afghan passports are more likely to self-identify based upon ethnicity (Pashtun, Tajik or Hazara), tribal affiliation and family lineage. Finally, skills and interagency relationships necessary to develop and implement coordinated strategic policy and programs did not exist since not all Agencies participated—as in U.S. Colombian policy development—as relative equals.

The U.S. also lacked area expertise. As noted earlier, the Defense Department lacked experienced FAOs with Afghan-specific language, cultural and political competence. The State Department—and notably INL and the Agency for International Development (AID) lacked any Afghan expertise since the U.S. had no diplomatic presence in Afghanistan for decades. What little initial area expertise the U.S. gained in the initial eighteen months after 9/11 came from tactical-level interpersonal relationships with local leaders whose trustworthiness, ability and will to implement goals consistent with U.S. priorities were, to be diplomatic, unreliable. Finally, the well-known inability of U.S. non-Defense Agencies to deploy personnel—in an expeditionary capacity—prevented these Agencies from developing a true partnership with their Afghan or Defense Department counterparts. Many might convince a few intrepid employees to deploy to Afghanistan for 90 or 180 days but they could not order or recruit willing employees to deploy and develop area expertise for the necessary multi-year period. When Agencies (including the Defense Department) did send employees on temporary assignment to develop expertise in Afghan issues such as CN, enemy reintegration, police training, anti-corruption and other governance-related issues, they often sent those employees unwanted by other offices and, in most cases, clearly incompetent to the mission. The author personally knew dozens of employees sent by the Departments of Defense, State, Justice and Homeland Security who were manifestly incompetent and, in many cases, unwilling to learn. Thus, the U.S. Government was unable to develop the necessary expertise to provide Washington-based policymakers the needed information and recommendations to effect beneficial policies or programs.

[11] U.S. Southern Command is the Defense Department Command responsible for Defense activities in Central and Latin America. Thus, any expertise in combating narcotics within the Defense Department was largely limited to Southern Command and its component units such as Seventh Special Forces Group.

Unintended Consequences

To be sure, the U.S. Government leaders of the past fifty years, including those leading the Department of Defense from 2001 to today, did not foresee the consequences of their decisions to rely upon short-duration deployments of less-than-competent personnel. The failures cited earlier were, of course, evident as well during the Vietnam conflict when U.S. forces deployed for one-year engagements and our civilian Agencies were not directed to deploy expeditionary personnel to Southeast Asia, in significant numbers, in order to bolster the South Vietnamese Governments. We can be reasonably sure that many of these leaders simply lacked the foresight and competent advice that perhaps would have led to them to better deployment decisions.

The consequences of these failures were as apparent in 1973 as they are now. First, the continual rotation of military and civilian personnel eliminated the ability to develop, analyze and collect key tactical, operational and strategic information that could have informed leaders at all those levels. Thus, when new leaders assumed responsibility, many arrived bereft of key information that might have prevented them from recreating past mistakes or committing mistakes heretofore avoided. Second, continual rotation naturally created opportunities for the less than competent (and, in many cases, unwilling) to take tactical, operational or strategic responsibilities. Since there were few opportunities to "weed out" these less-than-stellar performers, capability that might have built upon previously successful efforts often did not occur.[12] The rewarding (or lack of immediate removal) of incompetent leaders was not lost on local leaders at all levels of Afghan Government. Many saw the replacement of good (or even excellent) American leaders due to arbitrary timelines as a key sign that the U.S. and its allies were not serious in their commitment to assist in rebuilding Afghanistan. Finally, the revolving door approach to American assistance prevented the development of key interpersonal relationships with Afghan leaders at all levels and Afghans, in general. When each successive western leader arrived, promising great change, many Afghans began to ignore (and even oppose) those efforts. We will examine these three consequences and their fatal impact.

Absence of Institutional Knowledge Passed on to Successors

The revolving door approach to personnel deployments—military but, in particular, civilian—continues to prevent the collection and growth of any institutional memory. As a direct result, successively deployed personnel continue to repeat past mistakes,

[12]In fact there are numerous, verifiable accounts during the Vietnam, Afghan and Iraq conflicts where incompetent leaders were promoted or allowed to retire rather than publicly shamed for their incompetence. Former Coalition Provisional Authority leader Ambassador Paul Bremer and retired Army Lieutenant General Ricardo Sanchez are only two, of many, examples.

ignore demonstrated past successes and miss opportunities to improve assistance to Afghanistan.

From the earliest deployments after September 11, 2001, U.S. policymakers made critical, strategic mistakes. One of the greatest of these mistakes is the continuing policy of limiting deployments of personnel—military and civilian—to arbitrary time-limited periods. Whether it was six months (many European nations), nine months (U.S. Marines) or one year (U.S. civilians), this policy ignored the simple fact that almost all of these personnel were incompetent in Afghan versions of issues for which they were supposed to be responsible.

To be sure, many of those deployed were experienced. Military personnel (in many but not all cases) were competent in low-intensity conflict (in contrast to set piece battles against nation states). Civilian experts in agriculture, finance, legal systems, transportation, counternarcotics, healthcare and other critical fields possessed academic or practical experience and truly wished to assist the Afghan people. None, however, were competent in understanding or finding solutions to Afghan versions of those problems. To those of us who actually worked directly and regularly with Afghans—as colleagues rather than abstract figures—we quickly realized that it would take significant time, effort, relationships and, frankly, luck, for us to become competent in Afghan versions of these difficult and complex problems.

I, for example, arrived in Kabul in 2012 relatively competent in counternarcotics—including Afghan counternarcotics. I have served three years as the lead civilian in the Department of Defense for counternarcotics. I led successful efforts to support Colombian counternarcotics policies, programs and operations. From my first trip to Afghanistan in December 2001, I at least knew something of Afghan narcotics challenges and obstacles. If anything, my limited experience prepared me for the realization that I had much to learn in order to be useful to the Afghan Minister for Counternarcotics—an official for whom I had volunteered to serve as his Senior Advisor.

Even with that experience and humility, however, I was immediately overwhelmed by the depth of what I did not know or understand. Issues of eradication to which I felt comfortable in the Colombian or U.S. domestic context were completely useless in addressing Afghan eradication needs. The absence of an extradition mechanism for those who trafficked in Afghan-produced narcotics rendered traditional U.S. law enforcement impotent. European production of precursors necessary to process raw opium into heroin became a major factor that was unknown in the Latin American context. These are just a few of the newly experienced legal, cultural and ethical experiences for which historical U.S. and western expertise were useless.

As a result, it was necessary to spend the time to develop knowledge—knowledge only accessible through trusted relationships with knowledgeable and trustworthy Afghans. These relationships would grow as a result of the determined, repetitive and heartfelt efforts of western "experts" who understood that arbitrary time limits imposed on deployments had to be ignored. If one wanted to become competent in a particular subject matter, one would have to invest the necessary time, language and culture education, patience and love of chai before earning the trust of a knowledgeable Afghan. Needless to say, this could never be done if the Afghan believed that you would leave in six, nine or twelve months.

As a result, western "experts" continued to deploy to Afghanistan, introduce themselves to Afghan counterparts with promises of goodwill and determination and fail in gaining basic understanding of the problems for which they were supposed to assist.[13]

Untrained and Incompetent Successors Lead to Loss of Gains

Second, the west, led by the U.S. opted to prioritize the deployment of personnel—not necessarily the right personnel—to satisfy the public relations goal of showing action rather than taking the time to select and deploy the right kind of people. While perhaps excusable in the initial twelve to eighteen months of the conflict, the continued deployment of bodies—particularly those not wanted by others—served to ensure that valuable lessons were neither learned nor preserved.

One cannot emphasize enough that the quality of deployed personnel—particularly civilian—was overall mediocre and oftentimes poor. In addition to the reasons stated above—deployed personnel who lacked competence in Afghan versions of commonly understood problems—western and U.S. leaders often sent civilians who lacked even basic understanding of issues. Many governments—particularly the U.S.' State Department—could not or would not require their best and brightest to deploy to Afghan. Many lacked an expeditionary capability and, therefore, could only produce volunteers. While many of these volunteers genuinely wished to be helpful, a sizable number deployed for other reasons including bonus pay enticements, personal crises (as if deploying to a war zone could or would ease those personal issues) or to maintain their employment.

Many of these Government agencies therefore sought help in the form of personnel provided by government contractors. Companies such as Dyncorp, CACI, Lockheed Martin and hundreds of others made enormous profit by placing people in temporary positions even though those contractors were not satisfactory for the assignment. The goals of the contracting company (profit without controversy) were very different than the goals of the government customer. It should not be a surprise, therefore, that "police trainers" were hired—based mostly on their past experience as local police

---

[13]Many in the U.S. would counter that those deployed served multiple deployments and, therefore, gained the necessary expertise. This is absolutely false. Issues, relationships and knowledge differed greatly based on the local or regional parts of Afghanistan in which the person deployed. Narcotics in Balkh or Herat were (and remain) significantly different than those in Helmand, Kabul or Nangahar. Relationships, culture—even language—also were vastly different. Thus, those who made more than one deployment but to different parts of Afghanistan were additional hindered by their assumption that what they learned one year earlier in one part of Afghanistan was relevant to their successive deployment. They were sadly mistaken.

officers in the United States—who lacked language, culture and even experience in training.

Operational and Strategic Afghan Leaders Lacked Personal Relationships and Foundation for Trust

As a result of these strategic errors, Afghan leaders (many of whom privately sought and welcomed U.S. advisors) lacked the trusted, personal relationships with westerners upon which they could develop long-term trust and rely upon for long-term strategic advice. Numerous Afghan leaders worked with multiple "advisors" sent for temporary tours who either lacked Afghan-specific expertise or were unable to recognize that their U.S.-centric ideas were inapposite.

Many in the U.S. Government did not (and still do not) appreciate the political, cultural and physical risk Afghan leaders assumed when they were seen as publicly relying upon western "advisors". For some, the risk was direct and simple. If they had family residing in parts of Afghanistan in which Taliban, Al Qaeda, or other anti-government elements operated, overt cooperation with westerners often placed their families at direct risk. For other Afghan leaders whose ambition directed them to seek higher, future office, reliance upon westerners was perceived as a political weakness.

Once I gained the trust of the Minister to whom I served, he included me in all his dealings—domestic and international. I attended Afghan National Security Council meetings, dinners at the Iranian and Russian Embassies and meetings with Afghan political and religious leaders to discuss non-counternarcotics issues. After a time, however, the Minister asked me to take a less obvious role (including not attending certain events) because his family had been threatened. Furthermore, President Karzai—his political mentor—did not wish for his allies to be seen as publicly relying upon American advisors. I readily accepted his wishes and continued to serve him—albeit from a "behind the curtain" perspective. Complying with his wishes further increased his trust in me. Many American advisors, however, could not or would not accept these Afghan realities and, therefore, remained outside trusted Afghan circles.

Lessons Learned

There are several lessons to be learned from the failures to develop long-term interpersonal relationships and subject- and area-expertise relating to Afghanistan. While these lessons are likely not useful to the U.S. with respect to Afghanistan today given the lack of will to continue to assist the GIROA in its narcotics and other challenges, they can serve as useful lessons to future policymakers prior to decisions to assist weak or non-existent governments in the future.

First and foremost, the highest priority should be the determination that, if the U.S. is willing to intervene to replace or assist in development of a Host Nation government, it should do so only with the collective, long-term will to identify, deploy and support interagency personnel who will remain until its strategic goals are largely met. Similar to the deployment of U.S. military personnel during World War Two, the U.S. must send personnel who will remain in the relevant area to develop interpersonal relationships and subject area expertise. Limited-term deployments, while attractive and certainly of benefit to families of those deployed, cannot meet the national goals of developing reliable, objective information that U.S. senior policymakers will need.

Will such expeditionary deployments cause tremendous sacrifice? Absolutely. Like World War Two, personnel deployed may not return home to families for periods of several years. Even if programs are established to allow convalescent and other temporary leave to spend time at home, these personnel must remain and continue to develop the critical knowledge bases that inform policymakers. As seen during the Vietnam, Afghanistan and Iraq conflicts, revolving door deployments fail to develop, maintain or increase the information needed to identify tactical or operational changes and thereby the strategic solutions necessary to achieve strategic goals as soon as possible. An unintended benefit, perhaps, is that such a known sacrifice will likely cause U.S. policymakers and the voting public to consider more fully the consequences of such decisions to intervene.

Second, it is imperative that U.S. policymakers recognize that it must develop subject and area expertise across all parts of the U.S. Government that is more than just regionally based. Government employees with regional responsibilities (e.g., Middle East, Africa or Asia writ-large) are not sufficient for potential challenges that are unique to locations within those regions. A Middle East focus is insufficient to understand Palestinian, Iranian or Afghan issues. Those issues are unique to those peoples. Arabic is not the only language in that region just as Spanish is not the only language of South America and Chinese is one of many languages in East Asia. Counternarcotics expertise in Latin America is largely unhelpful to understanding or combating CN issues in Afghanistan, Iran, Malaysia or Turkey. Third, the development of these areas of expertise cannot occur in the United States. U.S. Government employees must deploy and remain in their specific regions in order to develop the necessary area and subject matter expertise. Changes must be made in order to attract the right kinds or personnel and provide new career incentives in exchange for these long-term deployments.

Third, the U.S. continues to largely ignore the wealth of subject- and area-expertise that is available within the international commercial and private sectors. For years, commercial and other private actors of U.S. and other friendly nationalities have traveled to, resided, conducted business and developed long-term personal relationships with counterparts in parts of the world where the U.S. Government has maintained limited or even non-existent access or influence. Many possess the necessary language, cultural personal expertise that allows them to work with, for and influence key local actors. Many would volunteer to assist the U.S. government—particularly when doing so would likewise benefit their own personal or commercial goals. For example, there are prior-service U.S. military and civilian personnel who established credible and beneficial interpersonal relationships while deployed temporarily in parts of Afghanistan. Those actors—now retired from U.S. Government service—might still assist if asked, particularly if the request related to a subject or area in which they maintain expertise. U.S. policymakers should enact programs that identify this collective knowledge and utilize it in ways that could assist U.S. policymakers.

Conclusion

The United States Government and its allies will continue to face international security challenges well into the future that require persistent local area knowledge, expertise and, most importantly, long-term personal relationships. Current governmental structures—most notably diplomatic, military-to-military and short-term business relationships—will continue to fail to provide the quality of knowledge and influence necessary to mitigate or obviate international security risks.

While our post-September 11, 2001 intentions for Afghanistan may never be realized, we can, if we choose, learn from our strategic mistakes and benefit in the future. We must develop, maintain and then mature a true public/private partnership of subject matter experts whose subject matter expertise and, more importantly, subject area expertise is assessed, recruited and retained for long-term use. We must break down historical barriers between public and private sector strengths and weaknesses to meld capabilities that will serve U.S. interests on a global, long-term level. Finally, we must be willing to constantly assess how we are performing these tasks and break with traditional personal policy when and where it benefits overall U.S. interests. Doing these things will not be easy but, in the long run, will save lives and resources.

Andre Hollis The author served as the Deputy Assistant Secretary of Defense for Counternarcotics from 2001 to 2003. Additionally, he deployed to Afghanistan from 2012 to 2013 as the Principal Advisor to the Minister of Counternarcotics for the Government of the Islamic Republic of Afghanistan. The author highlights the failure of senior U.S. leaders to strategically deploy personnel who could have identified, analyzed and solved many strategic errors made as a result of a lack of institutional and personal relationships with Afghan officials. The conclusions and opinions included herein are solely those of the author and do not reflect or represent any U.S. Governmental position or policy.

Winning the Great Game in Afghanistan: Bottom Up, Not Top Down

Andrea Jackson

Abstract Great powers have often tried and failed to control Afghanistan. They failed to understand Afghanistan's powerful social structures and work by with and through them to reach their goals. Instead, each great power has attempted to impose a social and governmental system on Afghanistan from the top down. Each has met resistance from Afghan society, which is organized from the bottom up around patrilineal families that compete with one another for honor. Afghans are motivated by an effort to accrue honor for their patrilineal family. The fathers of Afghan families come together to solve shared problems in their patrilineal villages or in mixed, urban communities. They select a leader called a Wakil, or representative, who executes their decisions by organizing coercive force, collecting money, executing projects and making and enforcing rules. Wakils can form alliances or choose to fight outside organizations, like great powers, or their adversaries. When brought together in alliances, Wakils can generate unified action across thousands of villages and communities. This chapter describes the values and authority structures that motivate Afghans; how the Soviet Union and the United States failed to work with them; and how any great power can motivate Afghans to do what they want. It describes what a great power would need to do to motivate Afghans to deny safe haven to terrorist organizations or pier competitors, to facilitate the extraction of valuable natural resources, to allow traffic on a road or deny it, or to facilitate or prevent opium production.

Keywords Afghanistan local governance · Afghanistan social structure · Afghan culture · Afghan values · Afghanistan Wakil · Patrilineal familism · Consensus based decision making · Honor · Islam · Great power competition · Cold War · Operation Enduring Freedom · Irregular warfare · Pakistani foreign policy · Soviet policy in Afghanistan · US policy in Afghanistan · Afghan government · Social · Economic and political development Afghanistan · Anti-Soviet *jihad* Afghanistan · International terrorism

---

A. Jackson (✉)
240 8th Street SE, Washington, DC 20003, USA
e-mail: andreavjackson@gmail.com

© The Author(s), under exclusive license to Springer Nature Switzerland AG 2021
A. Farhadi et al. (eds.), *The Great Power Competition Volume 1*,
https://doi.org/10.1007/978-3-030-64473-4_6

Introduction

Great powers have often sought to control Afghanistan or prevent their rivals from doing so. During the 19th Century, Russia and Great Britain struggled for control over Afghanistan during the Great Game. Russia sought to gain access to the warm waters of the Persian Gulf and Indian Ocean. It sought to access the vast resources of India's population, its natural resources and economy. It also worked to prevent Great Britain from accessing those resources or expanding into Russia's sphere of influence. Great Britain, for its part, wanted to use Afghanistan as a buffer to protect the riches of colonial India from Russian expansion.

During the Cold War, the Soviet Union and the United States clashed over Afghanistan. In the 1970's, the Central Asian Republics contained roughly 20% of the Soviet population. from the Islamist influence of Pakistan. Pakistan threatened to pull the Muslim populations of these republics away from communism and toward Islam and a political union with Pakistan. The Soviets therefore sought to establish a buffer zone, a communist, Pashtun, nationalist Afghanistan. The Soviets hoped that their support might enable Afghanistan's Pashtun population to fragment Pakistan by annexing Pashtun and Baluch inhabited Pakistani territory, creating a Greater Pashtunistan. The creation of a Greater Pashtunistan would fragment Pakistan, blunting the Islamist threat to Soviet unity. It would also allow the Soviets to build a naval base in Baluchistan, from which they could threaten oil transiting the Persian Gulf and influence politics in the greater Middle East (Goldman 1984). The Soviets hoped that they might be able to export their ideology to the impoverished Afghan population and put in place a Soviet style of government in Afghanistan. Further, under the soil and rock of Afghanistan lay valuable reserves of gas and oil, along with what is now valued at a trillion dollars in mineral and other geological resources that could support Soviet industrialization in Central Asia. For all these reasons, the Soviets supported communist student proxy groups in Kabul and provided military and economic assistance to the Pashtun nationalist government of Daud Khan. When the communist groups overthrew Daud Khan's government and were incapable of managing the backlash against their ideology, the Soviets invaded Afghanistan.

In the late 1970's and 1980's, the US, for its part, sought to check the expansion of communism and block the Soviets from extracting Afghan resources and accessing the strategic ports of the Indian subcontinent. The Reagan Administration also relished the idea of tying the Soviets down in an exhausting counterinsurgency (Jalali 2017).

In 2001, the US was attacked from Afghanistan by the international terrorist organizations nurtured there during the anti-Soviet *jihad*. In response, it expelled Al Qaeda and overthrew the Taliban government that had provided them sanctuary. By applying a European model for structuring society, the US and its NATO allies sought to build a centralized Afghan state that would be capable of denying terrorist organizations sanctuary.

In their efforts to control Afghanistan, each great power has confronted an Afghan society that steadfastly resists the imposition of foreign social systems and repels efforts to extract its resources. Afghan society itself is the most powerful but forgotten actor in the great power competition in Afghanistan. Afghan society has its own structure, its own values and its own rules. Great powers that have, wittingly or unwittingly, worked with this indigenous social structure have succeeded, while those who ignored the local social structure and its rules have failed.

A Matter of Perspective: Bottom Up, Not Top Down

Most great powers that have competed for control in Afghanistan have sought to impose a system or a leader from the top down. They have tried to identify and work with national or at least ethnic group or tribal level interlocutors who can assist them in controlling some aspect of Afghan behavior. But, that is not how Afghanistan is organized.

There is no Afghan nation around which one can build a centralized state. Afghans use the term *qawm* to describe solidarity groups ranging from nation to ethnic group to tribe.[1] Although "Afghan" is a synonym for Pashtun, the Pashtuns make up only 42% of the population (Anderson 1983; Brown University). Afghanistan contains three other major ethnic groups, Tajik (27%), Hazara (9%) and Uzbek (8%), and there a total of 22 ethnic groups in Afghanistan (Brown University; Dupree 1980). There is no single leader of any of these ethnic groups, which are subdivided into tribal confederations, tribes, clans and subclans.

[1] A *qawm* is a solidarity group that can be as large as a nation or ethnic group or as small as a tribe. It can be based on kinship, occupation or location of residence. Rubin (2002), p. 25.

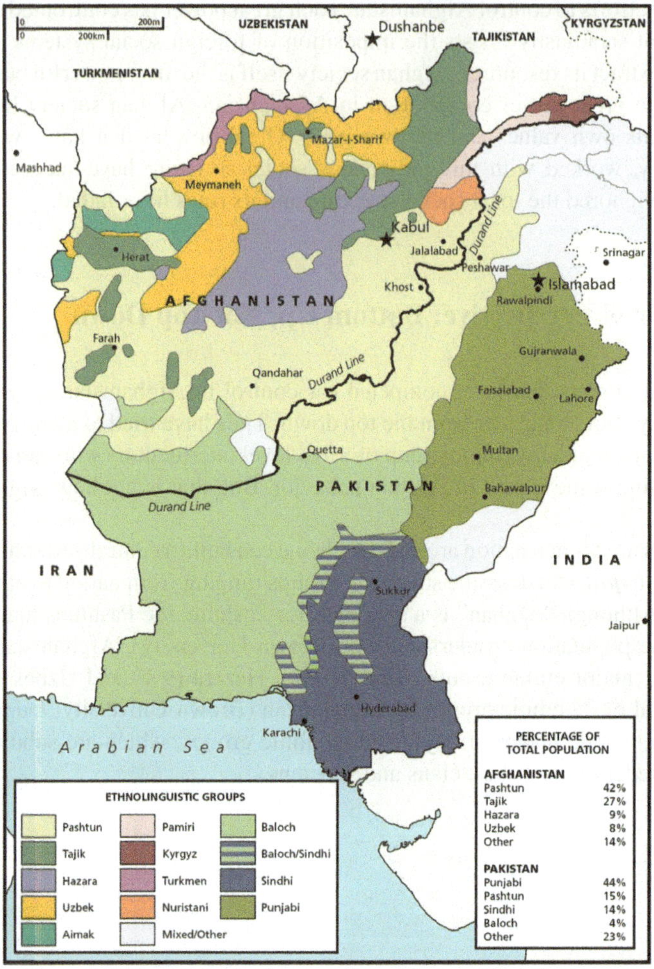

Map of Afghan Ethnic Groups. Providence: The Choices for the 21st Century Program, Brown University, 2013.

Additionally, there is a set of warlords who are generally associated with a particular ethnic group and geographic area. These warlords grew out of the anti-Soviet *jihad*. Rather than concentrating on governing populations, warlords are normally focused on maximizing external resources to provide to their narrow patronage group, comprised of *mujihadeen* who were part of their fighting groups.[2] Warlords are therefore not effective partners for great powers seeking to deny sanctuary to an irregular adversary, like Al Qaeda or the Islamic State, or to prevent another great power from accessing Afghan territory. Warlords are also not effective partners in providing or denying access to a road system, or imposing a system of government, like communism or liberal democracy (Jackson 2017). However, they can be effective in securing specific scarce, geographically concentrated resources like the Amu Daria gas fields, which Rashid Dostum's Uzbek *Jumbaish* militia secured for the Soviets (Rubin 2002).

While great powers, as centralized states, have often sought top down solutions, Afghanistan is organized from the bottom up. Afghans have never lived in a world where a social organization larger than an extended patrilineal family of several generations could be relied upon to assist them in addressing the frequent, unpredictable series of calamities they face. The state in Kabul is often separate from and a threat to the local system upon which Afghans rely. This disconnect between the state and society has characterized all Afghan governments across the 19th, 20th and 21st centuries, from modernizing monarchy, to military dictatorship, to the communist system, to the Taliban and to the hyper-centralized democracy the US and its allies installed after the Bonn Conference of 2001. Afghans rely upon themselves and their close patrilineal kin, and not the state, because the country, "in which they have survived has been a harsh and pitiless world, and so have been the valleys and villages into which the individual Afghan has been born. Neither the state nor your neighbor could be counted on to defend your life and your rights if you were weak. Any property or privilege an Afghan claimed, he would have to secure and defend for himself; and any assistance he obtained from others would depend on what he could offer in return. The strong person, the one who had honor and self-respect, was best able to defend himself and had most to offer as a friend and ally. Such have been the formative life experiences which have, over the centuries, shaped Afghan values and self-reliance: in the last instance, your life and security—and that of your family—depend squarely on your own force and standards" (Barth 1987).

Afghans are born into patrilineal families in which all members are subordinate to the father of the family. Most Afghans live in multigenerational homes, where the father of the family has authority over the actions of his wife, unmarried daughters, sons and their wives and children, and sometimes younger brothers and their wives and children. This patrilineal family is the most important level of social organization throughout Afghanistan. The members of a patrilineal family share physical and economic security. In traditional agricultural villages, the father of the family owns

[2]For example, the commanders on the road between Spin Boldak and Herat during the civil war between 1992 and the Taliban seizure of power in the area in 1994 demanded tolls, harassed, raped and murdered travelers, even those from the area. These mini-warlords were not distributing their booty to communities in Kandahar. They were using it to show their own power, not to govern. This behavior continued after the US and NATO occupation, when militias who had fought during the war used their positions to accrue money and power with little relationship to the population in the areas where they were supposed to serve as commanders of the local police. Jackson (2017), pp. 193–7, 436, 445–8.

the home, land and livestock, and controls the wealth of the family. Even in families where the sons have found jobs in the government or in a service industry (tailor, shopkeeper, taxi driver, etc.), the father of the family collects their wages and decides what to do with them. In the long run, even young men who move to Kabul and join the Afghan National Army or some other government organization, still rely upon their father, brothers, uncles and cousins for their long term physical and economic security, in case of state collapse and/or civil war (Jackson 2017).

Afghan families have long lived in a social world where there is no central authority to impose its will on individuals or groups who would take land, women or other possessions that do not belong to them, and where personal reputation is paramount in protecting those possessions. A father who heads a household, must build a reputation that will deter individuals from taking his belongings (Barth 1987). An honorable man demonstrates that no one can tell him how to live, that he can determine what he does and what the members of his household do (Jackson 2017). To have honor, he must be able to defend his possessions, which Pashtuns[3] call *namus*, in the form of women, gold and land, or *zan, zar, zamin* (Tomsen 2011). Honor, or *nang*, is the source of a man's value to his family, and to his extended family. Without honor, he cannot deter attacks on his *namus*, or possessions, and without his *namus*, he cannot contribute to his wider, extended family's *nang*. *Nang* is the totality of the group's honor (Dupree 1980).

In general, the more sons a man has, the greater his perceived ability to defend his possessions, and the greater his honor (Dupree 1980). This is only true, however, if he can control their behavior. If he cannot, he is seen as weak and dishonorable (Jackson 2017). Each family competes with each and every other family for greater honor. This honor, this reputation of being powerful and capable of defending one's possessions, discourages people from attempting to take them. It is the greatest shared asset of a patrilineal family because without it, all other assets are in danger, and alliances with others impossible.[4]

An extended patrilineal family normally resides in a village where the other residents are patrilineal cousins of the father of the family. The family owns a plot of land and a residence, surrounded by mud walls, and possibly comprised of several homes. The family's residence and fields are surrounded by those of patrilineal cousins who compete with one another for honor, but unite to face shared problems and external adversaries (Jackson 2017). Alliances, depending on the problems or threats they form to solve, can extend across hundreds or even thousands of villages, as they did during the *jihad* against the Afghan communists and their Soviet allies, mobilizing hundreds of thousands of fighters from patrilineal villages (Edwards 2002). Families from these villages who had migrated to Kabul and other cities before the communist coup in 1978 often returned to join their cousins as *mujihadeen* fighting on behalf of their village (Jalali 2017).

---

[3]Although this describes the Pashtun system of patrilineal honor, the same held true for Tajik rural communities and urban communities with multiple ethnic groups.

[4]While the terms used in this description are Pashto, the same system of patrilineal honor governs behavior in Tajik, Hazara and Uzbek communities.

The description of village life above may appear anachronistic in the 21st Century, but roughly 75% of Afghans live in rural villages, where they grow wheat, vegetables, walnuts, and, if they are blessed with plentiful water, pomegranates and grapes (Italian Agency for International Cooperation 2018). They also raise and graze cattle, sheep, goats and chickens. Many families augment their agricultural income with income from a government salary, often earned by a son who is a member of the Afghan military or police. Still others augment their income by running a shop at the district *bazaar* or operating a taxi. The same system of patriarchal authority within a competition between families for honor governs nearly all aspects of life even in Afghanistan's most internationally connected city, Kabul (Jackson 2017).

From the perspective of Afghan society, power does not reside at the top, in the state. Instead, it builds from the bottom up, village by village, on a topic by topic basis that is often impermanent. In Afghanistan, villages aggregate to generate power based on the honor of each patrilineal family. This honor is rooted in patrilineal family and patrilineal honor, rather than a sense of national, or ethnic, or tribal, or even clan or village, honor.

Any power seeking to motivate Afghans to do anything, to guard a road, to prevent an adversary from moving through the population, to assist or allow the extraction of natural resources, is very likely to fail if they do not do so from the bottom up. It is only through the aggregation of patrilineal families, first into villages that act in unison, then into valleys or groups of villages, that a great power can generate the unified action it desires. There is a mechanism for aggregating competing patrilines to act in unison. That system is built around the consensus of the fathers of patrilines to act together, and a leader called the Wakil, or representative, who executes tasks based on their consensus (Jackson 2017).

Meet the Wakil

The competition between patrilineal families for honor exists in the context of a system for building consensus among the fathers of families, who gather as equals to discuss shared problems and agree on how to solve them. It is through this consensus system that decisions about any actions a great power might want the population to take are made. It is also through this system that communities choose to ally with one another or the government or any other organization, or choose to fight them.

Nearly every community in Afghanistan is centered on a mosque which the families who reside there use and maintain. The fathers of the families gather in the mosque to select, by consensus, one from among them to serve as their Wakil, meaning representative.[5] The job of the Wakil is to solve the problems of the people of his community. He works to understand the problems of residents and to develop

[5] Wakil is Arabic for representative. This term is also used for representatives elected by voters in districts throughout Afghanistan to participate in the Afghan parliament, or Wolesi Jirga. But, if you ask an Afghan who his Wakil is, the person he will name is the Wakil in his home community. There are a small number of very wealthy communities in Kabul, which are inhabited by very wealthy

solutions to them. He does this by constantly interacting with them, walking around the community, asking people how they are, and attending prayers five times a day in the community mosque (Jackson 2017).

When a Wakil identifies a problem, he calls a meeting of the fathers of the families to develop consensus on how to solve the problem. Once they have reached consensus, the fathers of the families provide the Wakil with the resources he needs to solve their shared problems. With the consent of the fathers of the families, the Wakil can collect funds, organize projects, organize labor, raise and command security forces, purchase supplies, hire outside specialized labor, resolve disputes between families, and shun or expel residents from the village or community. The Wakil also proposes and implements alliances with other villages and communities, the government and insurgent organizations. He implements community decisions about organizing and executing violence against neighboring communities, the government or insurgent or terrorist organizations.[6] The Wakil wields the power to motivate the families in his community to do the things great powers might want them to do (Jackson 2017). And, in the aggregate, Wakils can deliver—or prevent—what great powers want. A series of examples below of what Wakils are, how they are chosen and the functions they perform describes the mechanism through which actions and resources are coordinated within and across communities.

On an empty, gravel strewn hillside near the Kabul airport, several families began building houses illegally on government land during 2003. The families were originally from a rural community in Parwan, but they had been displaced by Pashtun families during the war between Ahmad Shah Massoud's Northern Alliance and the Taliban (1996–2001). The families took refuge in Massoud's Panjshir Valley stronghold. In 2003, they attempted to return to their village in Parwan, but discovered that the Pashtuns had destroyed their homes and farms before fleeing. The group had decided they would have better luck in Kabul, moved there and began building their homes. Within two months, there were eight families and they began to have problems that required unified solutions. They needed someone to resolve property disputes between residents and to represent them to the government and other communities. They needed water and electricity. Also, they were praying in their houses and they wanted to build a mosque where they could pray together.[7]

people often associated with warlords or the government, where there is no Wakil. Afghans often say that in these areas, everyone is his own Wakil. This is also how they refer to the chaos of the civil war (1992–6) that Afghans consider to be the worst period in their history. Jackson (2017), pp. 238–315, 339–362, 487–497.

[6]Because the fathers of the families decide whether or not to provide resources to the Wakil, the Wakils are generally accountable to their communities. However, in some cases, Wakils have taken advantage of their positions. When this happens, the elders, some of whom were appointed by previous Wakils and continue to serve the community, call a meeting without the Wakil in order to select a new one. When this happens, the old Wakil generally runs away. Jackson (2017), pp. 238–315, 339–362, 487–497.

[7]Afghan males over the age of 16 pray between 2 and 5 times per day in their community mosque. The mosque lies at the center of a community so that residents can easily reach it. The borders of urban communities are based on where the residents attend mosque. In Afghanistan, Mullahs teach Muslims that Allah made men to pray and that on Judgement Day, he will count up their good

The fathers of the eight families met in one of their homes to select someone with the education and experience to solve the community's problems. They agreed that they needed to choose from the five educated fathers. Each person wrote the name of the candidate he wanted to be Wakil on a piece of paper. The man they selected was a 67-year-old Tajik from Parwan who had completed the 6th grade[8] and could therefore read and write. His son had fought for Massoud and he was an honorable man. When the Wakil was selected, he told the people,

> I do not want to be the Wakil, but thank you for trusting me. I will be responsible in this area. I will give a good position to all of the other candidates. I will make them the elders because maybe they know more than me. I don't want them to be unhappy. If you agree with them, then they will be my elders.

In choosing the other candidates as his partners, as the community elders, and in saying that he did not want to be in charge, the Wakil showed that he did not intend to impose his will upon his fellow heads of household, that he intended to rule by consensus. The elders and the fathers of the families accepted him and felt they now had a real community, not just a set of houses on a rocky hillside.[9]

When representatives of the government saw the houses, they approached the people and said, "You made houses here? This is the government's land."

The Wakil told the government personnel, "The Taliban destroyed our houses in Parwan and there was a desert here with feral dogs and we built houses here. We know the houses are illegal, but maybe every Afghan is illegal. Everyone can build a house in the desert. It had no benefit for the government."

The fathers of all of the families went to the District Police Station together. The head of the police station was a Tajik from northern Afghanistan and he was a good man. The Wakil told him, "I want to be Wakil because I want to be responsible for my area."

The elders explained to the Afghan National Police (ANP) Commander how and why they had chosen this man as their Wakil. The police commander said he was very happy about how the community had chosen their Wakil. He said, "If the people are happy, I am happy. Maybe if they build more houses, we will send police there to provide security" (Jackson 2017).

By selecting a Wakil, the people legitimated their illegal settlement in the eyes of the government (Jackson 2017). Within seven years, there were hundreds of homes

---

actions and bad actions to determine who will go to Paradise and who will go to Hell. When he counts up these actions, prayers in the mosque will be worth five times the rewards of praying at home. Jackson (2017), pp. 322–62, 460–78.

[8]Until the expansion of education after the post-2001 government, the highest level of education available to children outside of the major cities was the sixth grade. Some wealthy families could afford to send their children to high school and college in Kabul, or the combined high school and military academy there. Jackson (2017), p. 415.

[9]Wakils are selected for life, or until they are too old to work. They can be removed if they If the Wakil misuses funds, abuses his power in any other way, or behaves inappropriately. The elders call a meeting with the fathers of the families of the community without the offending Wakil, decide whether or not to replace him and select a new Wakil. When this happens, the old Wakil runs away. Jackson (2017), pp. 267–9.

in the illegal settlement, housing Tajiks, Pashtuns and Hazaras. Despite the differing ethnic identities and places of origin of the families, and the lack of patrilineal relationships between them, life in the community was orderly. People could expect that they and their possessions would be safe because of the consensus-based system of the Wakil (Jackson 2017).

A Wakil constantly tends to the unity of his community, to generating agreement and a shared pattern of beliefs and behavior that allows residents to come to consensus. Wakils and elders tend this unity by constantly interacting with the members of the community, and by organizing a practice of Islam that advocates unity with one's family and neighbors. The Wakil and the elders he selects walk around the community greeting people and asking how they are. They attend prayers five times a day in the community mosque and are continuously available to residents to assist them in solving their problems. They provide advice. They accept complaints about people violating the rules to which the community has agreed. They assist residents in interacting with the government, Taliban and other external organizations (Jackson 2017).

The Wakils also collect and manage funds and projects at the community mosque, which is built and maintained using community funds. The mosque is a shared asset of the community. The Wakil and elders identify possible candidates to be Mullah. They are often from outside the community so that the Mullah is not a member of the council of fathers of the families who make decisions and so that he cannot be the Wakil. The Mullah lives in the mosque and is provided food, water and electricity by the community. Each family pays him a portion of their produce or their wages (Jackson 2017).

Mullahs are hired labor and are not full members of the community. They provide their guidance on Islam, but the rules of the community are generated through the consensus of the fathers of the families and executed by the Wakil and elders. By selecting Mullahs from outside the community and managing their wages and accommodation, the Wakils ensure that Islamic guidance from the Mullah does not generate rebellion against them and instead reinforces the patrilineal system, consensus among the patrilines and the legitimacy that confers on the Wakil. What the Mullahs provide to the Wakils is an ideology that emphasizes unity and support for one's neighbors that constantly bolsters the consensus of the fathers of the families and undercuts the pressure Afghan culture places on patrilines to seek honor at the expense of their neighbors (Jackson 2017).

Afghan Mullahs teach people in their communities that the seemingly random suffering and pain in their lives are simply a passing problem in this world. This world is only Allah's test, that he uses to determine each person's disposition in the next one, which is all that really matters. The Mullahs say that on Judgement Day, people, and even an individual's own hands, other body parts, or belongings, will testify to Allah in front of the assembled masses about an individual's *halal* (good or legal in the *sharia)* and *haram* (sinful or illegal in the *sharia*) actions in this world. Allah will then use this information to determine whether or not that person should be admitted to Paradise or consigned to Hell (Jackson 2017).

For example, when discussing whether or not it was right for tractor and thresher operators, who come to all the villages in a Tajik inhabited valley from the neighboring Pashtun valleys, to charge even the poorest villagers for their services in the summer of 2014, a village Mullah described Allah's test, saying,

> The thresher earns *halal* money. He does not take bribes from anyone or steal. It is a good job for him. On Judgement Day, all of the thresher's money will be in front of him. The money will say to Allah, 'This man found me from *halal* actions.' When a person gets money from *haram* actions, on Judgement Day, the money will come in front of him and say, 'This man is very bad and found me from *haram* actions.' Allah will punish him in front of all of the people. His family will be ashamed. Allah will order the angels to take a person who earns money from *haram* actions to put him in Hell. If someone finds *halal* money, Allah will tell the angels to put him in Paradise and make a good place for him. He will be relaxed. Allah will say, 'I promised the people who did not break my rules, who did not kill, steal, take bribes or do something bad in the world that I would make them a good place in Paradise.' Allah said, 'Those who did not remember me and obey my rules in this world, I will make a very bad place for them.' A lucky man obeys the rules of Allah and does good things in this world (Jackson 2017).

The widespread belief among Afghans in a proximate Judgement Day provides Mullahs with the ability to affect their behavior. The rules Afghan Mullahs advocate promote mutual aid and unity among neighbors. Afghan Mullahs teach that it is incumbent upon neighbors to greet one another, attend one another's weddings, parties and funerals and assist one anther whenever requested. Mullahs admonish their followers to respect and help one another. They say that on Judgement Day, if your neighbors are happy with you, then Allah will be happy with you. Throughout Afghanistan, people lend things to their neighbors, assist them in times of need, even at financial cost or personal risk, because they want Allah to be happy with them. This mutual aid reinforces the unity which underpins the Wakil's authority and diminishes fragmentation within communities that results from the competition between patrilines for honor.

Mullahs also reinforce the importance of generating a shared understanding of problems facing a community by telling the truth. As one rural Mullah said at Friday prayers,

> Lying is the darkness of faith, and Allah will never forgive a liar because lying creates disunity that is harmful for the people and society. If you want the community to be strong, it must be unified. Do not lie because lies divide the people. Lies make people disagree and then the people of our community will not be united to help one another. If someone lies, the people from the area will not be united and they will not follow the rules of the Wakil because they will not agree on what happened. People who lie might say, 'We are all humans and we do not need to listen to the Wakil or elders of the area.' They say one thing in the face of the Wakil, and they say something else behind his back. These men create disorder and disunity, and Allah will never forgive them. When you do your work, you should not lie. All money earned through work based on lies is *haram*,[10] and Allah will not forgive the person who makes money from lies (Jackson 2017).

[10]Forbidden in Islam.

The fathers of the families in most Afghan communities require all male residents over 16 years of age to pray in the mosque five times a day, with exceptions for their work, school and health requirements. The Wakils say that they must ensure men are able to provide for their families and that they do not want inflexible requirements for Islamic practice to cause men to hate Islam because Mohammad said you cannot force a man to pray. In practice, this means men pray in their community mosque between two and three times per day. This rule greatly enhances the power of the Wakil because it allows him to interact with all adult male residents of the community regularly. Because they are constantly together, the male members of the community develop a shared understanding of the threats from thieves, predatory government officials, organized criminal gangs, insurgents, and terrorists and from anything else that might challenge the members of the community. They also discuss problems and how to solve them. Additionally, the simple act of praying together, standing as one, prostrating themselves as one and reciting the same rhythmic Arabic words as one single, unified, ordered congregation, enhances the cohesion of the community (Jackson 2017). Further, on Friday, the Mullah's sermons reinforce Allah's requirement that residents think of each other as humans and care for one another and make one another happy by greeting one another and caring about and assisting one another (Jackson 2017).

Most residents of these communities agree with the rule about praying as many times as possible at the community mosque and comply with it, both because the families agreed with it and because many believe Allah made men to pray and they will suffer in the next world if they do not accumulate as many *halal* actions as possible in this life. A Wakil has near perfect information about who lives in each house in his community and how often they attend prayers. The example below describes how one Wakil enforced this rule in his community (Jackson 2017).

A young man and his wife moved out of his father's house in another neighborhood and into a new house in a traditional market community in downtown Kabul. The young man, freed of his father's authority, decided he did not want to attend prayers. The Wakil realized the young man was not attending prayers and thereby undermining unity. The Wakil could not get to know this young man and his problems and the young man would not feel compelled to help his neighbors if he did not attend prayers. The Wakil, elders and Mullah visited the new resident in his home to advise him. They asked him why he did not attend prayers and whether there was some problem that was preventing him from attending that they could help him solve. They then informed the young man that regular attendance at prayers in the mosque was a rule in his new community. The Wakil further informed him that if he did not begin attending prayers in the mosque, the people of the community would force their Wakil to throw him and his family out of their home. The young man was embarrassed and promised the Wakil, Mullah and elders that he would attend prayers at the mosque regularly from then on, but that his work meant he could only attend prayers at dawn, at dusk and at night. From that day forward, the young man attended prayers at the mosque three times per day on work days and five times every Friday (Jackson 2017).

Wakils can organize residents to throw someone out of their house for breaking a rule in the community. In most cases, the family in question hears that the fathers of the families and the Wakil have agreed to organize to expel them and leaves before they do so. A young man from the illegal community in northern Kabul described earlier wanted to buy a cell phone. But, his father told him they did not have money for one. The young man stole their neighbor's hen in order to sell it and use the proceeds to buy the cell phone. When the neighbor realized his hen was missing, he suspected his neighbor's son had stolen it and waited, perched in a tree, for the young man to leave his father's home, to confront him. The young man opened the front gate to his father's home and walked out carrying a case. The owner of the missing hen jumped out of the tree and confronted him, demanding that he open the case. Inside, he found the missing hen. The father took the hen home and walked to the mosque to report what the hen thief had done. The Mullah was the only one at the mosque and the father reported the incident to him. It was then incumbent upon the Mullah to advise the father of the hen thief that his son had stolen the hen. If the Mullah did not advise him that theft was *haram*, and that he needed to control his children, then on Judgment Day, Allah would ask the Mullah why, when he had known the rules of Islam, he had not shared that knowledge. Allah would punish the Mullah for this (Jackson 2017).

The Mullah stood in front of the people after Friday prayers and told them that the son of one of the community members, whom he identified by name, had stolen his neighbor's hen. He asked the people gathered in the mosque what the community should do about this bad action. The Mullah told them that it was his opinion that if the boy continued to live in the community, then all of the people would want to leave to live somewhere else because he made the place so chaotic. He continued, saying that if the community residents remained in the area, then the boy would want to run away because they would want to punish him. The Mullah concluded that the boy should leave because he repeatedly behaved so badly in the community. Some of the residents stood and said they would support doing whatever the Wakil, elders and Mullah decided to do about this boy (Jackson 2017).

The Mullah asked the father of the boy, who was in the congregation, "Why do you not want to control your son?"

The father said, "I have tried many times, and I told him not to steal, but he continues to do this behind my back. Whatever you want to do, I will agree with it."

People in the congregation told the Mullah that they wanted him to make a decision based on Islam about what to do. The Mullah said, "The boy should leave our community, or we should cut off his hand."

The Mullah told the father to throw the boy out of his house, so that he would never do this bad action again. The father of the boy stood in front of all the people gathered at the mosque and said, "From when I was a young until now, I have not done any bad actions in this community. You know that I am a good person. I have tried to advise and discipline my son, but he will not listen. I throw my son out of my house. If he comes back, I will kill him."

The people agreed with the decision of the father of the hen thief. The Wakil told the people, "The father of the hen thief has thrown his son out of his house in our presence. Now, we must throw him out."

The Wakil said that he would punish anyone who helped the hen thief because the entire community had expelled him. The Wakil said, "We will kill him if he comes back."

The hen thief was not in the mosque during the meeting. Later that day, someone who was present told the hen thief that his father and the community had cast him out; that if anyone helped the hen thief, he would be punished; and that if the hen thief returned, his father would kill him. The hen thief was frightened that his father or the people of the community might hurt or kill him if he did not leave, and he fled to Iran (Jackson 2017).

When problems arise for members of an Afghan community, Wakils can propose rules to solve them. If they consent to the rules and punishments, residents are for the most part willing to assist the Wakil in enforcing rules by providing information to the Wakil when people violate the rule. In the illegal community in northern Kabul, there had been a number of fights after people hired bands to come and play at wedding parties. When a young man gets married, the wedding party is a celebration and a demonstration of the honor of the patrilineal family. The father of the groom hires a band and a singer and the men take great pride in causing the singer to play the song they want him to play.[11] It shows their power when they can cause the singer to play their song. People dance and some people drink alcohol. At one of these weddings, there was an incident that caused the Wakil to propose that bands not be allowed to play at parties in the community. Many of the men in the community were dancing at the wedding party and some of them appeared to have been drinking. One of the older, wealthy wedding guests was dancing with a young, handsome man from the community. The older man took out some money while the boy was dancing and waved the money in a circle around the boy's head. This action indicates to people that the boy is his "playboy" or boyfriend. In some circles in Afghanistan, wealthy, powerful older men have young, handsome boyfriends. These "playboys" are almost like the property of the older man, like a wife. Having a "playboy" shows how powerful and wealthy the older man is. The older man is demonstrating that money is nothing to him because he has so much of it and that this handsome boy would rather be with him than have a wife. While these relationships demonstrate the honor of the older man, they bring great shame on the family of the boy because he is like a woman. His family was too weak to raise him and take care of him, so he belongs to a man who has sex with him. It is a great shame for a family when one of its sons is a "playboy".

The cousin of the young man who was dancing was an enemy of the older man. He was very angry that his cousin was bringing shame on the family by appearing to be the playboy of an older man and especially this older man. So, the cousin got a gun and shot his dancing cousin in the leg because he had allowed this older man to wave money around him while he danced. Both the enemy and the cousin ran away after the shooting and began fighting one another with knives. The Wakil took the injured dancing boy to the hospital and no one called the police because they would have made the situation worse.

[11] Women attend a separate happiness party that is normally held in a separate house.

Many of the men from the community had been at the party and had witnessed the violence. In the week that followed, the people of the community had talked about how horrible this incident had been. On the following Friday, the Wakil called a meeting in the mosque to discuss it. The Wakil and Mullah told the people that in Islam, you are not supposed to listen to music and that this incident demonstrated why. Listening to music leads to other bad behaviors, like drinking and fighting with guns and knives. The people in the mosque voted to not allow bands to come to parties in the neighborhood anymore. They agreed that people could listen to whatever they wanted in their own homes, but that music at parties caused people to fight. In this case, the Wakil and the Mullah organized people to solve the problem of drinking and fighting at parties in the neighborhood by proposing a rule to the people that they accepted (Jackson 2017).

Wakils can also organize units of young men to address threats to the honor and therefore the security of residents, and partner with fellow Wakils to solve security problems. The high school girls from the illegal community in Northern Kabul, like girls throughout Kabul City, walk a long distance to school. They pass through a community where there is a group of young men and boys who wait for them every afternoon. The boys tell the girls they are beautiful and try to touch them and hand the girls their cell phone numbers. When they do this, the boys are questioning the honor of the girls' fathers and brothers and uncles and cousins. The boys would never do this in their own neighborhood because if they did they would suffer retribution. For weeks, the girls went home and told their families about the harassment by the boys from that community. Some of their fathers approached the Wakil and Mullah about how to determine who these boys were who were and stop them from harassing their daughters.

The fathers, the Wakil and the Mullah gathered at the community mosque to discuss the problem and come up with a solution. The Wakil said, "As a community, we should not get into a fight with this community. We should not go directly to the boys to tell them not to do this and fight with them. We should go through their Mullah."

The Wakil selected two boys from the community who were familiar with the area where the boys were gathering, were trusted by the Wakil and Mullah and were physically imposing. The Wakil instructed the boys not to get in a fight with the boys, but to identify them and the community they came from and return. The boys executed their task and the Wakil, Mullah and elders went to the mosque of the other area and told the Mullah their concern. The Mullah there called in his Wakil. The Wakil asked the boy's Mullah and Wakil whether there was any law in their community or if everyone just did as they pleased. The Wakil of the boy's community replied, "We have laws, but some people in the area do not listen or do not know. Maybe some of these people are proud they can tease the girls. We will try as much as possible to prevent this action from happening again."

The Wakil, Mullah and elders from the community on illegal land left the area where the teasing had occurred. From that day forward, the daughters of their community were never teased by the boys from the offending community again. Based on this initial cooperation, the Wakils of the two areas met again to discuss problems with organized crime. They invited several other Wakils and they all began to cooperate by ensuring that young men who lived in their communities did not commit crimes in any of the communities who were party to their agreement. The Wakils were able to organize their communities to prevent boys from harassing girls walking home from school and to prevent organized crime in their communities. They did so by identifying shared problems and working together to ensure their residents assisted in implementing shared solutions (Jackson 2017).

Without the Wakils, the Afghan social system would tend toward a war of all against all driven by each patriline's effort to gain honor at the expense of other patrilines. But, the Wakil system allows patrilines to come together, to unify to solve shared problems and to reach out to other communities and their Wakils to cooperate. The Wakil mediates disputes, solves problems and nurtures a shared understanding of problems and a consensus on solutions, not only to internal problems, but to threats and opportunities from governments, insurgent and other organizations. Wakils can unify and aggregate the drive for honor and the desire to pass Allah's test to motivate their communities and to form alliances with communities facing similar problems (Jackson 2017).

Great Powers Meet the Wakil

Great powers have often failed to recognize the existence of this powerful system for motivating behavior and aggregating the power of many patrilines to achieve their ends. Instead, most have sought to impose a foreign social system on Afghan society from the top down, which the basic organizing units of Afghan society, its patrilines, have rejected. If the great powers had identified more narrow aims, instead of seeking wholesale revolutionary social change in Afghanistan, and worked to build alliances from the bottom up to achieve their aims, they might have succeeded.

As it was, each of the great powers moved into Afghanistan and often unwittingly encountered the power of the consensus based organizations of honor-accruing patrilines, refused to consult them, and failed to achieve their ends.

The narrow Soviet aim was to defend their Central Asian Republics from Pakistan and potentially expand their influence into it, causing Pakistan to fragment and gaining access to warm water ports in proximity to the Persian Gulf. This did not require a wholesale Sovietization of Afghan society, but that is what the communist government of Afghanistan pursued.[12] The communists who seized power in 1978 immediately sought land reform, cancelation of all debts, secularization, indoctrination in communist ideology, and women's liberation. Also, during 1978 and 1979,

[12]Some might argue that the Soviets did not intend for these reforms and repression to occur. However, beginning in the early 1950s, the KGB recruited Afghan assets, including the founders of the Afghan communist party. While the exact sequence and timing of actions of the Afghan communist government may not have been what the Soviets intended, they were consistent with Soviet doctrine and implemented by a party the Soviets had caused to come into being. Jalali (2017), pp. 346–8.

the communist government murdered as many as 27,000, or some estimate 50,000, people and jailed between 14,000 and 20,000 more (Jalali 2017).

In so doing, the Afghan government drew the ire of the Wakils and the patrilines they represented, who rose up against reforms that disrupted their way of life and brought shame upon them by telling them how to live. The original uprisings were organized by *khans*[13] who formed *jirgas*[14] and approached villages to whose families they were related to request that the village join their rebellion against the communists. Each village provided fighters to a *lashkar*[15] that then attacked the district government center from which the communist policies were emanating (Edwards 2002). The United States, in its effort to tie the Soviets down in an irregular war against guerrilla fighters in Afghanistan's formidable mountains and deserts, unwittingly provided funding, training and equipment to patrilines that had aggregated themselves into a massive rural uprising that prevented the Afghan communists and their Soviets sponsors from achieving any of their goals (Jalali 2017). US funding greatly improved the technology used by the *mujihadeen*, who began the conflict with Lee Enfield rifles, but soon had Kaleshnikovs, RPG-7s, mortars, rockets, anti-tank weapons and Stinger anti-aircraft missiles. The US covertly provided Pakistan with $12 billion in aid and tens of thousands of tons of equipment.[16] The US achieved its narrow aims by bolstering the firepower of the patrilineal fighting groups that comprised the *mujihadeen.*

But, also unwittingly, the US backed a Pakistani effort to generate revolutionary change in Afghanistan and failed to encourage or allow an alliance between patrilineal groups that could have generated an Afghan government that handled the shared interests of the patrilineal villages and communities. Pakistan's narrow aim was to defeat the Soviets, and, if possible, reduce the threat from India and the threat by Afghan Pashtuns to create a Greater Pashtunistan out of their Pashtun and Baluch-inhabited territory. Instead of working directly with the patrilines and their Wakils to reach these goals, the Pakistanis sought to Islamicize Afghanistan's tribal and patrilineal rebellion against the communists and their Soviet sponsors. Their effort to establish revolutionary, modernizing Islamic state that would transform Afghanistan's patrilineal society from the top down met the honor based patrilines and was as unsuccessful as Soviet efforts to eliminate them (Edwards 2002).

---

[13] *Khans* were large land owners who were members of a clan and subclan of a Pashtun tribe who took it upon themselves to use their largess to provide loans and hospitality, as well as assistance interacting with the government, to members of their patrilineal tribe. In return, they used their influence to "tie the knot of the tribe", reducing infighting between patrilines and thereby enhancing everyone's honor by unifying them. A unified, as opposed to a fragmented, tribe presents a greater deterrent to the government, and other tribes or organizations that might want to take any member of the tribe's belongings. Anderson (1983), pp. 124–130.

[14] *Jirga* means circle. It is a council of fathers of families or of their Wakils and elders.

[15] A *lashkar* is a tribal army.

[16] By 1983, the CIA supplied roughly 10,000 tons of equipment, shipping it through the port in Karachi and then transporting it by land to Rawalpindi and Quetta to ISI depots for further distribution to the Islamic parties and subsequently to commanders. By 1986, the amount had increased to 65,000 tons. Jalali (2017), p. 280.

Pakistan used the Afghan *jihad* to establish an Islamist *jihadist* force generation system along the Afghan-Pakistani frontier to train Muslims from Afghanistan, Pakistan and other countries to wage *jihad* in Afghanistan, Central Asia and elsewhere (Tomsen 2011; Brown and Rassler 2013). It consisted of a religious-military complex of *madrassas*, mosques and militant training camps. The *madrassas* taught fundamentalist Deobandi Islam with a focus on anti-Western, anti-Hindu *jihad*. The *madrassas* were and continue to be the source of the Taliban and the militant camps were the training ground for foreign, especially Arab, fighters who became Al Qaeda (Tomsen 2011; Brown and Rassler 2013). Pakistan sough to use this force generation system to establish an Islamist, fundamentalist regime in Kabul which would form an Islamic Union with Pakistan under the leadership of Gulbadin Hekmatyar. The union would be buttressed by Saudi oil money and would expand into Soviet Central Asia and Indian Kashmir through their strategy of guerrilla *jihad* (Tomsen 2011).

When Al Qaeda attacked the US from Afghanistan in 2001, the US intervened with a narrow aim. It sought to destroy Al Qaeda, remove the Taliban government that had hosted it and establish some governing structure that could prevent the return of the terrorists to Afghanistan. Instead of just focusing on these narrow aims, the US and its NATO allies overreached. They ignored the existing systems of patrilineal power that could have been aggregated to achieve these limited and realistic goals. Instead, the US and its NATO allies sought to alter Afghan society, replacing its patrilineal, village level governance with a hyper-centralized government that would defend individual rights and establish democratic elections and a capitalist market economy. In order to remove the Taliban and Al Qaeda quickly, using as few personnel as possible, the US allied itself with the warlords who had reigned during Afghanistan's chaotic civil war (1992–6), most of whom had been exiled by the Taliban. This decision meant that the government ministries, and particularly the security ministries, were dominated by the warlords, who used their positions to enrich themselves and their former fighters (Jackson 2017). The corruption of the warlords, coupled with the alien system the US and its allies were attempting to impose, led to failure and the narrow goals the US sought have yet to be achieved.

Both the Soviets and the US attempted to build centralized governments that were mirror images of their own and they did not use either patrilineal honor or Islam to motivate personnel. Both governments were characterized by corruption, as personnel used their positions to acquire honor for their partiline, as opposed to performing the tasks the government was intended to perform. The Soviets and the US both faced irregular adversaries, but built Afghan partner militaries that were trained and organized to conduct quick, decisive operations and failed to hold territory. Both also suffered from high rates of desertion and difficulty motivating troops to take the risks required to patrol and engage the enemy outside of large-scale clearing operations (Jalali 2017). The partner Afghan military and security organizations failed to make it in the interests of patrilineal village leaders to assist them in expelling the irregular adversaries they faced. Villagers are the only people who are constantly present, with access to information on the presence of irregular adversaries, and the ability to attack them without risking ambush *en route* to their location. For that reason, both lost control of the rural areas that are home to the bulk of the Afghan population.

These great powers had the opportunity to ally with the existing patrilineal system of honor, to achieve their narrow ends. When they instead confronted the patrilineal system with plans to eliminate it and build a mirror image of their own system, the great powers failed to achieve any of their goals.

But, How Could a Great Power Work with a Wakil?

There are several obvious challenges for a great power attempting to work with the consensus based decision making structures of extended Afghan patrilineal families. The first is the distrust Wakils and communities have of the intentions of all outsiders. After many iterations of outsiders descending upon Afghanistan to impose foreign systems upon them, often through coercive force, the patrilines and the Wakils who represent them are understandably suspicious of the intentions of outsiders. The second challenge is that real power is concentrated not at the national or provincial or even valley level, but at the village or community level. The potential partners who hold the authority to motivate action by Afghans are dispersed across tens of thousands of villages and urban communities across the country.

Only a very clear strategy to leverage the existing authority of consensus based leaders at the village or community level can motivate Afghans to do things a great power wants them to do. To be successful, a great power would need to consult directly with *jirgas* of the fathers of families, laying its basic interests out and describing how these interests align with those of the community. A great power would need to work with communities to arrive at a consensus about how to solve the shared problems. While this sounds unwieldy, the system of patrilineal honor governs most of the communities in Afghanistan and the problems and solutions will be similar across them (Jackson 2017).

For example, the vast majority of communities in Afghanistan, which are rooted in the system of patrilineal honor described in this chapter, want nothing to do with the *jihadist* organizations that threaten to attack the United States and its NATO allies. Al Qaeda, the Islamic State, the myriad other terrorist organizations in Afghanistan, and the Taliban, want to violently impose rules on Afghan patrilineal villages and consensus based urban communities. Young Taliban or other fighters patrolling through communities, beating people for breaking rules to which no one in the community has agreed, brings shame on every family in that community. The weakness displayed by not being able to protect their family members diminishes their honor, their deterrent against people attacking them and taking their belongings (Jackson 2017).

The story of a village at the back of a remote mountain valley in Kapisa Province demonstrates this age old paradigm. The village was beset by what the villagers called "Chichini Taliban", who were Chechen members of an international *jihadist* terrorist group, probably affiliated with Al Qaeda. The village demonstrated its honor by rising up against them in September of 2014. The village Wakil is a famous *mujihadeen* commander from the wars against the Soviets, communists and Taliban. The Chechens built a hardstand fort on the mountain above the village. They abducted the Wakil, beat him severely and told him the rules they would begin violently enforcing in the village. They left the Wakil's body in front of his house. When he recovered from his wounds, the Wakil found that the Chichini Taliban had also prevented two villagers from taking a third villager with an appendicitis to the district clinic, causing the man to die by the side of the road. The Wakil called the villagers to a meeting where they discussed the problem and decided it was either the Chechens or them and planned an assault on the Chechen fort. The Wakil met with the ANP Commander, informed him that the villagers were going to assault the Chichini Taliban fort, and told him he would appreciate some assistance, either in the form of additional ammunition and/or in the form of ANP personnel. The next morning, 500 villagers stormed up the mountain, followed by 25 ANP and 20 Afghan Local Police. They attacked the Chichini Taliban, killing two of them and driving the remaining fighters out, over the top of the mountain. Two villagers were killed in the assault. The Wakil requested ammunition so his villagers could maintain security in the village and on the mountain behind it and the ANP Commander agreed (Jackson 2017). As of the end of 2015, the Chechens had not returned.

The village described above had 1,000 households and it was threatened by a group of less than 20 terrorists (Jackson 2017). But, most villages in eastern Afghanistan have no more than 150 households and are beset by Pakistani Taliban units numbering more than a thousand, and sometimes as many as 10,000, fighters, deployed from the *madrassas* in Pakistan. These villages cannot muster enough fighters to repel the Taliban's larger force or a potential future surge of Taliban seeking to make an example of a single, rebellious village. The problem the villages face is not one of will to repel Islamic terrorist organizations. It is one of collective action. Each of hundreds of villages in an Afghan valley or thousands of villages in an Afghan district wants to be able to determine its own rules and keep fighters from outside their community from entering the village to violently impose rules. But, they lack an agreement to defend one another or an external ally to defend them. So, they have little alternative to acquiescing. This outcome is neither what Afghan communities want, nor what the US and its allies want.

So, what could the US and its allies do to work with Wakils to solve this shared problem? The US and its allies could reach out to Wakils, village by village, to ask to work with them to repel Taliban, ISIS, Al Qaeda and other organizations that seek to base out of their areas and to violently impose rules on local residents. The US has special forces units that could perform these tasks, but has also trained Afghan special forces units that they could advise as they performed the tasks. Each Wakil could recruit and command a small security force[17] that would conduct patrols to identify personnel from the Taliban, ISIS, Al Qaeda and other organizations and

[17]This force would be unpaid and personnel would patrol and fight because they wanted to defend their homes and families and obey their Wakil. They would require only ammunition. Many security force assistance programs in Afghanistan have failed because the money associated with them

attack them. The special forces element could mediate between the Wakils of an area to get them to agree to come to one another's aid in case the Taliban massed to attack their village. The Wakils could hold a *jirga* that would govern their participation in this agreement.

Additionally, the *jirga* could govern a centralized security force that could serve as a quick reaction force to augment village forces if the Taliban massed to attack them, and that could put down any infighting between communities. The Wakil of each village in the *jirga* would select several young villagers to serve in the centralized force. They would provide an incentive for the centralized force to respond to attacks and would be a guarantee that the villages would come to one another's aid. The centralized force could be paid, trained and equipped by the US and its allies and would be able to request air and other support from them. This approach could not be implemented simultaneously throughout Afghanistan. It could be implemented in several areas initially, and then expanded out from each as soon as the system for identifying and attacking the Taliban, Al Qaeda and ISIS was consolidated (Jackson 2017).

The third challenge for great powers working with the Wakils is the most profound, but it is something that any great power wishing to compete in Afghanistan can change. It is the great power's instinct to create a mirror image, centralized government that purposefully or accidentally infringes on the order-generating, power aggregating function of the consensus based system of governance run by the Wakils. The instinct to impose a system to "fix" Afghan society is the impetus for its repeated rebellions against foreign powers. The centralized government organizations that great powers are prone to create have been ineffective in translating external resources into actions by their personnel, at least in part because they do not use the main Afghan value systems of patrilineal honor and Islam to motivate their personnel. The military organizations great powers have created similarly fail to leverage the patrilineal honor system and are ill-suited to the irregular warfare challenges that normally characterize great power competition in Afghanistan. This third challenge is something that any great power can change by choosing to work by, with and through the system of patrilineal consensus-building to achieve the shared aims of the great power and Afghan patrilines. The roots of great power failure in Afghanistan lie in the conceit that they can achieve their ends while ignoring this system.

References

J. Anderson, Khan and Khel: dialectics of Pashtun tribalism, in *Tribe and State in Iran and Afghanistan*. ed. by R. Tapper (Routledge, New York, 1983), pp. 124–125

F. Barth, Cultural wellsprings of resistance in Afghanistan, in *Afghanistan: The Great Game Revisited*. ed. by R. Klass (Freedom House, New York, 1987), pp. 188–227

---

attracted warlords who wanted the funds and the programs therefore failed to work with the people who are actually in charge in Afghan communities. Jackson (2017), pp. 718–747.

V. Brown, D. Rassler, *Fountainhead of Jihad: The Haqqani Nexus, 1973–2012* (Oxford University Press, Oxford, 2013)
L. Dupree, *Afghanistan*, 3rd edn. (Oxford University Press, Oxford, 1980)
D. Edwards, *Before Taliban: Genealogies of the Afghan Jihad* (University of California Press, Berkelely, 2002)
M. Goldman, Soviet military intervention in Afghanistan: roots and causes. Polity **16**(3), 394–403 (1984) (Spring)
A. Jackson, *When and Why Do Men Obey During a Civil War? A Study of Organizations Competing for Control in Afghan Communities*. Dissertation, Johns Hopkins University, Baltimore (2017)
A. Jalali, *A Military History of Afghanistan: From the Great Game to the Global War on Terror* (Kansas University Press, Lawrence, 2017)
B. Rubin, *The Fragmentation of Afghanistan: State Formation and Collapse in the International System*, 2nd edn. (Yale University Press, New Haven, 2002)
P. Tomsen, *The Wars of Afghanistan: Messianic Terrorism, Tribal Conflicts and the Failures of Great Powers* (Public Affairs, New York, 2011)

Afghanistan: Changing Dynamics of Peace and Conflict

Ali A. Jalali

Abstract As the world increasingly becomes embroiled in the struggle to counter the Coronavirus pandemic and face the resulting economic slowdown, Afghanistan enters yet another uncertain time. The country is on the verge of another transition that could lead either to a peaceful and stable future or descent into yet another cycle of violence. The challenges are great and evolving, the political context is changing, and uncertainties are compounding. Afghanistan's future depends on its internal political cohesiveness and positive and constructive interaction of its neighbors, other regional players, and the great powers whose direct and indirect engagement in Afghanistan has shaped the current situation in the country. The U.S.-Taliban Peace Agreement provides the basis for a process to bring peace in Afghanistan through direct talks among the Afghans. Its success, however, depends on its full implementation and close synchronization of different elements of the process including a comprehensive ceasefire, withdrawal of foreign forces from Afghanistan, the removal of the Taliban sanctuaries in Pakistani territory, progress in direct peace talks between the Taliban and the Afghan government, and long-term assistance to Afghanistan to ensure that the foreign troop withdrawal does not lead to collapse of the government.

Keywords Afghanistan · The United States · Taliban · U.S. interests in the region · Afghan peace process · Regional actors · Great powers · Intra-Afghan peace talks · Pakistan · Iran · Russia · China · Coronavirus pandemic

Introduction

As the world becomes increasingly embroiled in the struggle to counter the coronavirus pandemic and face the resulting economic slowdown, Afghanistan enters yet another uncertain time. The emerging global emergency will have a deep and extensive impact on a litany of pressing challenges faced by the country today: restoring internal political efficacy, fighting insurgency and terrorism, reducing poverty and

A. A. Jalali (✉)
3825 Shady Lane, Glenwood, MD 21738, USA
e-mail: aajalali@hotmail.com

© The Author(s), under exclusive license to Springer Nature Switzerland AG 2021
A. Farhadi et al. (eds.), *The Great Power Competition Volume 1*,
https://doi.org/10.1007/978-3-030-64473-4_7

economic decline, getting the peace process on track, and surviving the Coronavirus onslaught.

These challenges are complex, fluid and morphing. They are highly interdependent with multi-level chain effects, where the weakest links define progress and the nature of solution. Further, the involvement of a wide variety of actors at different levels impedes traditional paths to resolution. Optimistically, solutions may come through a coordinated and comprehensive approach at the national, regional, and global levels. However, the situation can take a downward spiral disrupting every aspect of life in a country that is not a stranger to hard times and enduring troubles.

Afghanistan's future depends on its internal political cohesiveness and positive and the constructive interaction of its neighbors, other regional players, and the great powers whose direct and indirect engagement in Afghanistan has shaped the current situation in the country. However, the multiplicity of regional players, some with competing demands, continues to complicate the issues of peace and conflict in the country. Fallouts of the great power competition and the third-party influence over bi-lateral relationships with Afghanistan confound the regional political dynamics.

The peace agreement signed by the United States and the Taliban in February of 2020, turns a new page in the history of the decades-long conflict in Afghanistan. The conditional peace deal sets the stage for the withdrawal of American military forces from Afghanistan ending the longest war in its history. Meanwhile, the ensuing shifts in the strategic dynamics of peace and conflict in and around Afghanistan opens new challenges and opportunities that will shape the political future of Afghanistan—a peaceful settlement or mutation of ongoing conflict.

The current complex situation in and around Afghanistan is rooted in the violent history of conflict in Afghanistan and the consequences of how the regional and global actors reacted to the war. This complexity came as a result of a changing political context caused by the stake holders shifting positions and policy choices. A review of the changing context is the starting point of understanding the situation, that have been directly and indirectly involved or influenced the course of events in the Central Asian country.

Changing Contexts

The War in Afghanistan, spanning the administration of four American presidents has become the U.S. longest foreign war surpassing the Vietnam war by eight years. It has been costly in blood and treasure for all the affected parties. In the intervening 19 years, the United States has suffered about 2,400 military fatalities in Afghanistan. Washington has reportedly spent up to one trillion dollars on the war including approximately $137 billion for reconstruction in Afghanistan (Congressional Research Service 2020). The casualty figures for Afghan civilians, militants and government forces are more difficult to quantify. Available estimates indicate that up to October 2018 nearly 60,000 Afghan soldiers and police have been killed,[1] close to 40,000 civilians have lost their lives in the violence and over 42,000 opposition forces have perished (Watson Institute at Brown University 2018).

For the U.S and its international partners, the main purpose for which they went to war was achieved in less than three months—the al-Qaeda network was disrupted, and the Taliban regime was overthrown. It was an accidental short war for the allies but became a prolonged process to their responsible exit—a process that would take another 18 years. The challenge has become how to prevent Afghanistan from again becoming a source of terrorist threats to the United States and its allies in the West. The notional response was to assist Afghanistan stabilize itself and to help other regional powers to stem the menace of trans-national terrorism and violent extremism. This has required a long-term strategy of stabilization. According to a Defense Department (2018) analysis, the United States greatly "overestimated its ability to build and reform Afghan government institutions" while its "stabilization strategy and the programs used to achieve it were not properly tailored to the Afghan context." The analysis further notes that the large sums of stabilization dollars the United States devoted to Afghanistan in search of "quick gains often exacerbated conflicts, enabled corruption, and bolstered support for insurgents" (SIGAR 2018).

Like the war in Vietnam, the war in Afghanistan has not been a conflict with a clearly defined strategic vision but inconsistent tactical reactions to a symmetrical escalation of violence with no coherent strategy. In fact, it evolved into several wars over time. Lt. Col. John Paul Vann, an American soldier in the Vietnam War famously said that the United States had not been fighting the war in Vietnam for 12 years, but for one year 12 times. The same can be said in Afghanistan where the international forces have fought many one-year wars inconsistently with each segment shaping the nature and complexity of the succeeding phase.

The fighting has been long and strategically inconsistent. From the outset, the reconstruction goals were too ambitious and the resources too limited. An under-resourced and inconsistent state building effort during a rising insurgency failed to match the real challenges and realities on the ground. Rather than pursuing a consistent reasonable effort building on cumulative progress, it was subject to constant changes due to shifting political factors and battlefield dynamics.

The military achievements in the first phase were consolidated in early 2002 with the dismantlement of the last pockets of al-Qaeda network in Afghanistan. In the following phase (2002–2005) the enemy force had disintegrated, public support for the policies of the Afghan government was overwhelming, and the international military forces had the hearts and minds of the population. However, valuable opportunities to stabilize the country and preempt the resurgence of the insurgency were squandered. From 2003 to 2005, the U.S. military executed a counterinsurgency (COIN) campaign in the east and south of Afghanistan with only two brigades. This could hardly prevent a Taliban resurgence.

---

[1] In January 2019, Afghan President Ashraf Ghani said more than 45,000 members of the country's security forces had been killed since he became leader in late 2014.

Insufficient deployment of troops and inadequate investment in institution building created a security vacuum that was filled by the rejuvenated insurgency. Meanwhile, unchecked movement of the Taliban and al Qaeda elements, harbored and regrouped by Pakistan intelligence across the border, led to the upsurge of insurgency in the following phase, which can be termed as the second Taliban war.[2]

Deterioration of security in Iraq kept U.S. attention focused on Iraq, at the expense of Afghanistan. The Chairman of the US Joint Chiefs of Staff, Admiral Michael Mullen, precisely summarized the challenges facing the international community in Afghanistan when he told the US Congress in 2007 that "in Iraq, we do what we must... In Afghanistan, we do what we can" (Gilmore 2007).

Donor fatigue was also becoming more and more visible as the war in Afghanistan dragged on. In the words of the ISAF Commander General David Richards (2014), "the international community as a whole, had not approached the reconstruction and development needs of Afghanistan with the sense of urgency that could meet the people's expectation. The Taliban had slowly exploited the resulting vacuum and this year (2006) it had reached a new pitch."

With a growing insurgency, increasing violence and the spread of militants' suicide attacks in 2008, Afghanistan was at a tipping point where its government's legitimacy (and that of its international backers) was being openly challenged by an array of antigovernment forces. Throughout 2009 and 2010, the U.S. government committed to an explicit stabilization strategy, surged more than 50,000 military forces to clear insurgents from the most dangerous and contested districts in the south and east, and deployed hundreds of civilians in stabilization programs to hold and build those areas so the Taliban would be unwelcome and unable to return. This was a great move, but too late and too short lived as the surge troops began to pullout after about one-year full deployment. However, a counterinsurgency campaign required patience and time to succeed. The full surge troops hardly stayed in Afghanistan for more than one year. It took them about six months to deploy and they began withdrawing in July 2011-just about one year after their full deployment.

Regardless of the announced intentions and the pace of the July 2011 drawdown, the deadlines became part of the strategic calculus for America's allies, the Afghan government and people, the Taliban, Afghanistan's neighbors, and other regional actors. In any rational committed strategic effort, force drawdown is determined by operational realities. Imposing time constraints on operations in an uncertain, dynamic environment sends messages of impending weakness. The perception of waning international resolve boosted insurgents' confidence that they could win and provided little incentive for them to support any peace talks. It also affected the morale of the population that did not support the Taliban's return to power but caused many to reconsider casting their lot with the eventual winner. This perception also drove

[2]In an interview with the Guardian, former Pakistan President Parvez Musharraf admitted that "Pakistani spies in the Inter-Services Intelligence directorate (ISI) cultivated the Taliban after 2001 because Karzai's government was dominated by non-Pashtuns, the country's largest ethnic group, and officials who were thought to favor India." See Jon Boone, *Musharraf: Pakistan and India's backing for 'proxies' in Afghanistan must stop*, The Guardian, February 13, 2015.

regional actors toward a hedging strategy impeding any hope of effective regional cooperation in stabilizing Afghanistan.

During the past four years, following the withdrawal of the bulk of the U.S. and NATO forces, the Afghan National Defense and Security Forces (ANDSF) faced an upsurge of insurgent attacks. To a large extent, it held its own, albeit with a higher casualty rate. Failing to seize or retain control of any provincial center, the Taliban made major inroads in the rural areas-seizing control of several districts and continuing to carry out suicide attacks in major population centers. All things considered, the military situation is one of a strategic stalemate. Estimates by the U.S. government show that as of 2019, 50 districts (12.3%), are under the Taliban control while 138 (33.9%) others are contested. This leaves 219 districts (53.8%) under government control (SIGAR 2019).

The rise of new threats of violent extremism in the region, including the local supporters of the Islamic State in Iraq and Syria, has added to instability and increased threat of terrorism in the region and beyond. Currently the United States Forces-Afghanistan (USFOR-A) execute two complementary missions. First, it conducts counter terrorism (CT) missions against al-Qaida, ISIS-K, and their associates in Afghanistan to prevent their resurgence. Second, in partnership with NATO allies and operational partner nations in the RS mission, U.S. forces train, advise and assist the ANDSF while providing enabling security and counterterrorism support in areas needed by the fledgling Afghan institutions.

Former President George Bush summarized the conduct of the Afghan war by saying "the history of military conflict in Afghanistan [has] been one of initial success, followed by long years of floundering and ultimate failure" (Bush 2019). The failure, however, does not strictly mean military defeat since the Taliban never beat the American military in major combat operations. Waging a war halfway around the world in difficult geography, while working with a difunctional and divided host nation and faced with a "frenemy" neighboring state (Pakistan) made a military resolution of the war unfeasible. Meanwhile the inconsistent strategy under which the war was persecuted made the political context for a political solution more and more complex and challenging over the years. Opportunities for better peace deals with the Taliban under favorable conditions were missed in 2001, 2003 and 2011, due to overconfidence in the ability of U.S. and allies to force the insurgents into a so-called "golden surrender."

U.S. Interest in Afghanistan

Afghanistan as a country and a state has never been in the domain of the U.S. vital interest. U.S intermittent engagement in the country has been mostly motivated by its transient regional and global security and strategic interests. Even in the context of the Cold War security and ideological alignments, a 1953 secret study by the Chiefs of Staff reads: "Afghanistan is of little or no strategic importance to the United States. It is situated on the perimeter of the USSR. It remains nominally independent of Soviet domination, but its geographic location, coupled with the realization by Afghan leaders of Soviet capabilities presages Soviet control of the country whenever the situation so dictates" (Poulada and Poulada 1995).

Afghanistan could not interest U.S. authors of the Containment Strategy at the advent of the Cold War to be considered a potential ally, even though the country was one of the three countries in West Asia that bordered the USSR. This was the time when the U.S Containment Policy aimed at countering real and perceived moves by the Soviet Union to expand communism abroad. Afghanistan's repeated requests for special relationship with America was ignored mainly due to Washington's higher priority in relations with Pakistan—a country that had a border dispute with Afghanistan (Dupree 1978). When Afghanistan made its final request for U.S. military aid in October 1954, the official response in December of the same year was extremely disappointing to Kabul. Secretary of State John Foster Dulles wrote, "After careful consideration, extending military aid to Afghanistan would create problems not offset by the strength it would generate. Instead of asking for arms, Afghanistan should settle the Pashtunistan dispute with Pakistan" (Poulada and Poulada 1995).

Consequently, the Afghan government was forced to reorient its policy toward its "great northern neighbor." A quarter of century later, Afghanistan was the only country in the region that fought a bloody war against Soviet communism and forced the Soviet Army out. It was also in Afghanistan that the United States waged the longest covert war against the USSR in the region. Less than a half century later, Afghanistan was the only country in the region from where the United States was attacked forcing the U.S. to invade Afghanistan.

However, when the Cold War was over, the Afghan conflict, in the words of a U.N. official, became a "forgotten war" and the Afghans became "a forgotten people" (Ermacora 1990). In the 1990s, the U.S. diplomacy in support of the U.N. efforts to end foreign intervention and find a peaceful solution to the Afghan civil war lacked consistency and perseverance. Washington's false hopes that the Taliban would be able to end the war militarily soon collapsed as the neo-fundamentalist militia proved to be the worst human rights abusers, record drug producers and hosts to the notorious terrorist network led by Osama bin Laden. Washington's 1998 pin prick cruise missile strikes against bin Laden's reputed terror training camps in Afghanistan, in the wake of the bombings of the U.S. embassies in Nairobi and Dar es-Salaam, only raised bin Laden's global standing among the anti-West militant groups. Further, the weak response helped create the perception that U.S. retaliation of future terrorist attacks would inflict minimal pain. The lack of international attention to the foreign-inspired conflict in Afghanistan haunted the world community in tragic ways.

The 2001 military invasion of Afghanistan by the United States and its allies was not initiated to fix the failed Afghan state through military action and stability operations. Had the 9/11 terrorist attacks in the United States not happened, a U.S. intervention into Afghanistan would have been unlikely. The United States military intervention and executing a long and costly war and stabilization effort served its vital interest to prevent another terrorist attack from an unstable region.

The peace agreement signed by the United States and the Taliban in Doha on February 29, 2020, turns a new page in the history of the decades-long conflict in

Afghanistan. The conditional peace deal sets the stage for the withdrawal of American military forces from Afghanistan ending its longest war in history. Meanwhile, the ensuing shifts in the strategic dynamics of peace and conflict in and around Afghanistan opens new challenges and opportunities that will shape the political future of Afghanistan—a peaceful settlement or mutation of ongoing conflict.

The Peace Process

The Doha agreement is predicated on guarantees by the Taliban to prevent the use of Afghan soil by any terrorist groups against the United States and its allies, and to enter into peace talks with the Afghan government, in exchange for an American commitment to withdraw its forces from Afghanistan within 14 months (Agreement for Bringing Peace to Afghanistan 2020). Basically, the US military pullout from Afghanistan is used as leverage to get the Taliban to enter into peace negotiations with the Afghan government aimed at bringing a sustainable peace in Afghanistan. For years, the Taliban conditioned their willingness to conduct peace talks on the withdrawal of the foreign forces from Afghanistan. It was almost 18 months ago (since October 2018) when the United States decided to hold direct talks with the Taliban in Doha, Qatar. Some observers see the deal as a U.S. exit strategy to extricate from an unwinnable war with a face-saving arrangement, others hope that the deal may provide an opportunity for the warring Afghan parties to reach a political settlement that may end one of the longest wars in recent history.

In justification of the U.S. peace agreement with the Taliban, in the absence of the Afghan government representation, the United States affirmed that negotiation with the insurgents was considered as a means to "pave the way for intra-Afghan negotiations on a political settlement and a permanent and comprehensive ceasefire" (Joint Declaration 2020). Still, the absence of the Afghan government gave legitimacy to the Taliban that can influence the intra-Afghan talks as it politically strengthened the Taliban status.

For as much as the Doha Peace Agreement looks transactional its implementation is more complex and subject to different interpretations. Several issues are of more importance:

- How the agreement can be monitored and verified?
- Are the Taliban ready to negotiate peace in good faith?
- The challenges facing the intra-Afghan Negotiations
- The dynamics of foreign troops withdrawal
- The role of global and regional actors—the wild card

How the Agreement Can Be Monitored and Verified?

The strategic transaction made in the Doha peace agreement puts the United States and the Taliban at different levels of control over implementation. The U.S. commitment to withdraw its forces from Afghanistan in 14 months, is a forthright pledge that has already begun. By contrast the Taliban's anti-terrorism guarantees are complex, indefinite, and hard to monitor or verify. Even in the event of verified Taliban cheating, the United States leverage to respond is expected to incrementally weaken as its military forces continue to withdraw. Although a pair of undisclosed annexes to the agreement, dealing with verification and enforcement issues, are mentioned by the U.S. representatives as a safeguard against lack of compliance, they are classified. In any case putting in place a reliable monitoring mechanism with international oversight is needed to verify the Taliban commitments.

Are the Taliban Ready to Negotiate Peace in Good Faith?

During my informal talks with Afghan political figures in July 2019, the Taliban dismissed a military solution to the conflict and made commitment to sharing power with all political groups. However, their commitment to specific issues such as human rights, democratic freedom and civil liberties, have been vague. During an interview with a senior Taliban associate in October of 2019, I was told that in order for the Taliban to maintain unity and cohesiveness among their ranks and files during the talks, they will be reluctant to make specific commitments. Privately, however, they indicate that they would hardly settle for less than maintaining certain ideological values of the "Emirate" including religious control and vetting of any social and political decisions in the government. With the legitimacy the insurgents gained by winning the withdrawal of "occupation forces" from Afghanistan, they favorably look at the intra-Afghan negotiation. Further, confident of representing the "Emirate" with one voice, the Taliban are willing to conduct talks as they see their negotiating position stronger than the assumingly divided team led by the government of the Islamic Republic of Afghanistan (IPA). If the Taliban see the movement in a position of maintaining battlefield momentum, they will prefer staying in negotiation to secure international approval and legitimacy.

The Taliban have made a commitment to start intra-Afghan negotiations with Afghan sides. This means that they consider the current government in Kabul as part of a team that represents a wider spectrum of Afghanistan political and societal segments. The initial political disarray in Kabul caused by a relentless election dispute between President Ghani and his opponent Dr. Abdullah provided an excuse for the Taliban to question the legitimacy of the Afghan government in Kabul which the insurgents call "the Kabul Administration." After the electoral dispute was resolved with the formation of an inclusive and unified government in Kabul, and more than 5000 Taliban prisoners were released by the Afghan government in exchange of

Taliban freeing 1000 Afghan government prisoners, the intra Afghan talks were kicked off in September following more than six months delay. Still it took another three months for the negotiating parties to agree on pressodurial matters. Meanwhile, the Taliban failed to reduce violence in Afghanistan even though the level of the U.S. troops in Afghanistan dropped to 4500. This has caused suspicions that the Taliban may intend to run the clock outwaiting the full withdrawal of the foreign troops by June 2021. Kabul-based observers have long sensed that the insurgents' vacillation to talk directly with the Afghan government is part of their strategic game plan. Their preference is to engage the political parties, members of the civil society and influential individuals in making separate peace deals. In interviews from 2019 to 2020 with several political observers, government officials and civil society members in Kabul I was told that such an approach would put them in control and in a position to accommodate others rather than being accommodated by a ruling power if they engage in talks directly with the incumbent government in Kabul. However, with increasing international pressure on the Taliban, the prospects for continuation of peace talks between the Taliban and the Afghan government negotiating teams are significantly improving.

The Challenges Facing the Intra-Afghan Negotiations

The intra-Afghan peace negotiation is expected to be the most difficult and lengthy phase where some very tough issues concerning the nature and the future of the Afghan state will have to be discussed. The sooner the talks get a momentum the better the chances are for conclusion before the completion of the scheduled withdrawal of American and Allied forces from Afghanistan. The intra-Afghan talks were envisaged to begin on March 10, 2020 but are long delayed and it is not certain when this can concludes. The fast spread of the coronavirus pandemic in the region can intensely restrict the pace of negotiations process. As the U.S.-Taliban peace agreement does not make withdrawal of U.S. troops contingent on progress in the Afghan peace process, the Taliban may see the delay to their advantage waiting for the pullout of the foreign troops.

The implementation of the deal ran into several glitches from the outset. The internal political impasse in Kabul due to the presidential election dispute caused major distractions. Meanwhile, President Ghani, dismayed by the Taliban failure to reduce violence as expected, disagreed to the release of 5000 Taliban prisoners called for in the U.S.-Taliban agreement as a prelude to the beginning of the intra-Afghan negotiation. The Taliban had agreed to reciprocate by freeing 1000 Afghan government security members in their custody. Claiming that the prisoner release was part of the negotiation that should be decided during the talks, Ghani refused to free the Taliban inmates in one group (BBC News 2020).

Later, in a reversal of his initial pushback, President Ghani agreed to a gradual release of the Taliban prisoners provided they pledge not to return to fighting.

The Taliban have repeatedly rejected Afghan government and the UN demand for ceasefire and stated that ceasefire without the complete implementation of peace agreement would be "illogical." Seeing violence as their only leverage in the process, the Taliban are unlikely to agree on ceasefire unless they come under pervasive pressure.

The uncertainty surrounding the intra-Afghan dialogue provides the Taliban with more choices to calibrate their commitment to peace talks with Kabul. As the opening of peace negotiation among the Afghans is delayed amidst continued violence across the country, emergence of more obstacles can hinder progress and may derail the whole process. There are serious concerns in Kabul that references to the Taliban "Emirate" in the U.S.-Taliban peace agreement hints to the insurgents' seriousness about being uncompromising about restoring their "Emirate" which means more than a name but represents a strict authoritarian ideological government system.

The process, will be prone to frequent interruptions and possible breakdown. The main question is whether the Taliban will negotiate in good faith with reasonable compromises on substantive issues including power sharing, preservation of democratic gains and constitutional rights especially for women. A major question is whether the Taliban have changed enough to be tolerant of different views? In an op-ed piece published in the New York Times ahead of the U.S.-Taliban peace agreement, Serajuddin Haqqani (2020), deputy leader to the Taliban movement, wrote "we are committed to working with other parties in a consultative manner of genuine respect to agree on a new, inclusive political system in which the voice of every Afghan is reflected and where no Afghan feels excluded." However, directly after the signing of the peace agreement the Taliban's public statements and messages in social media reflected a mood of victory over "foreign occupation" and continued struggle to establish an Islamic system in Afghanistan. This came as the longtime international support for the preservation of Afghanistan democratic, political, and social achievements are tacitly muffled in U.S.-Taliban peace agreement leaving such issues to be decided by the Afghans during their peace talks. The lessened U.S. leverage to preserve the major democratic gains of the past 18 years may help Taliban inflexibility on these issues.

One significant incentive for the Taliban to take a flexible position is the need to secure the much-needed foreign aid for running the post-conflict government. The aid is contingent on Taliban commitment to preserve certain democratic values and respecting human rights including human rights. The peace agreement confirms that "The United States will seek economic cooperation for reconstruction with the new post-settlement Afghan Islamic government as determined by the intra-Afghan dialogue" (Agreement for Bringing Peace to Afghanistan 2020).

The Dynamics Foreign Troops Withdrawal

Most Afghans accept that the foreign troops need to pull out eventually, but they see that the key to a responsible withdrawal is a reasonable timeline which is not front-loaded but benchmarked by progress in other elements of the peace process over the course of the withdrawal.

A uniliteral withdrawal of the American and allied forces before any peace settlement in Afghanistan will be a game changer. Such a situation can happen either by a sudden policy change in Washington, or as a result of a major fallout of the global COVID-19 pandemic and economic depression, or simply when the presence of international military would not improve prospects for political settlement in Afghanistan. Secretary of State Pompeo reportedly threatened Afghan leaders in March that if Afghan leaders fail to work together President Trump might decide on complete troop pullout and cut aids to Afghanistan (Lee et al. 2020).

As about 90% of Afghanistan operational and off-budget expenses are funded by foreign grants[3] the withdrawal could precipitate drastic cuts in financial aid to the Afghan government leading to possible political chaos and disintegration of the ANDSF. A situation some observers draw parallel with the 1992 when the breakup of the Soviet Union dried out foreign assistance to the post-Soviet Afghan government leading to its fast unraveling in the face of the Mujahedin pressure. The situation today, however, is a far cry from the one Afghanistan faced in 1992 when a fragmented Mujahedin faction were competing for power through making separate deals with elements of a divided Afghan polity. The country today has a constitutional order while the political and security context in and around Afghanistan is fundamentally different. But the dynamics of power competition can still plunge the country into the grips of a brutal civil strife with foreign interference, turning the country into the vortex of proxy war.

This does not mean that in case the peace negotiations fail, the Taliban will victoriously drive into the Afghan capital. Afghanistan today is more resourceful, resilient, and politically more ingenious and willing to fight. A recent Asia Foundation survey indicate that nearly nine in ten Afghans support efforts to negotiate peace with the Taliban. However, the price they are willing to pay for peace is measured at best. Protecting the current constitution, a strong central government and freedom of press are considered important achievements to preserve in any peace agreement. Most Afghans have no illusion about the nature of the Taliban real intention to return to their "Emirate" rule. The experience of the brutal rule of the Taliban in 1990s is still remembered by most Afghans. Despite the Taliban's recent charm offensive their "Emirate" identity has not changed. A recent survey indicates this year the public sympathy for the Taliban has further declined with the proportion of respondents who have a lot or a little sympathy with the Taliban to only 13.4%. Taliban are seen twice more of a threat in rural areas, where presence of the insurgents is more visible, than in urban centers (Asia Foundation 2019). Meanwhile, Afghan political and religious figures and foreign observers have continually questioned the Taliban's moral principles, given that their attacks harm mostly innocent Afghan civilians. The public support for the ANDSF continues to be high.

[3]The Afghan budget this year approved by the parliament amounts to $5.5 billion of which some $2.0 billion is funded by domestic revenue. Further the cost of ANDSF is another nearly $5.0 billion mostly financed by international donors.

Global and Regional Actors—The Wild Card

The principle stakeholders in the Afghanistan imbroglio include the warring sides in the conflict, the parties that are militarily backing the two sides, the countries impacted by the policies of foreign military backers and other interested parties.

The Great Powers and the regional actors generally share a common interest in the political resolution of the Afghan conflict. They see the return of peace and stability to Afghanistan as an essential element of regional stability and development. However, different countries view the challenge from the perspective of their own particular security and political interests—legitimate or otherwise.

The interests of certain actors are deliberate, others are opportunistic and have been shifting over the years as a result of changes in the great power relationship, the shifting inter-regional security dynamics, strategic uncertainties in and around Afghanistan, and the combat situation inside Afghanistan.

The United States and Pakistan are the two major actors who have substantial influence over the situation in Afghanistan. The U.S. is the main sponsor and backer of the post-Taliban state in Afghanistan and Pakistan provides infrastructural and material support to the leadership of the Taliban who wages their insurgency from sanctuaries in Pakistan.

The United States is interested in ending its long war in Afghanistan while leaving the country as a stable state that would not again become a source of terrorist attacks against its interests. The U.S. administration seek the attainment of this objective through the implementation of the recent peace agreement with the Taliban including a successful political settlement between the Taliban and the Afghan government.

Pakistan, a strong backer of the Taliban, has a major stake in conflict and peace in Afghanistan and is considered one of the most influential players in the Afghan conflict. Maintaining a strong influence in Afghanistan had long been the hallmark of Pakistan's Afghan policy. Islamabad's quest for a westward strategic depth to counter India's geopolitical edge had been one of the underlying principles of its strategy in the 1990s. Pakistan spared no effort to shape Afghan developments in accordance with its political ambitions. It helped keep the Afghan Mujahedin groups divided so they could be easily controlled—a move that fueled an intra-Mujahedin power struggle and civil strife following the collapse of the Communist regime in 1992. Although Pakistan may see the return of the Taliban to power in Afghanistan as neither possible nor desirable, it uses its control and influence over them as leverage to secure its regional geopolitical interests through its influence in Afghanistan.

Pakistan supports a peaceful resolution of the Afghan conflict in a way that would secure its strategic security and political interest in the region. The Pakistani military apparently applied selective pressure on Taliban leaders to join the U.S, sponsored peace process. Pakistan sees the U.S.-Taliban peace agreement as an opportunity that provides Islamabad with ample room to maneuver in securing its vital interests. If Pakistan does not see that its vital interests are secured, it will continue to keep the Taliban as a strategic ally and use it as a leverage in an unstable Afghanistan. Pakistan's main concerns are limiting the presence of its arch-rival

India in Afghanistan, the long-standing border dispute with Afghanistan, opposition to Afghanistan's hydroelectric dams on the Kabul basin waterways and Afghanistan's solidarity with the Pashtuns and Baluch activism. Further, the uncertain nature of the U.S. presence in Afghanistan and potential creation of a security vacuum by a premature departure of the American forces encourages Islamabad to invest heavily in the Taliban and to bear all the risks involved. As long as the Taliban strongly depend on their support infrastructure based in Pakistan, Islamabad will continue to exert significant influence over the insurgents and manipulate the implementation of the peace process in Afghanistan.

India and Iran are the two other main actors in the region whose interests clash with those of Pakistan and the United States. Seeing Pakistan as sponsor of violent extremists and using them as an instrument of policy against India, particularly in Kashmir, India staunchly opposes the Taliban, and remain a strong supporter of the Afghan government.

India's reaction to the U.S.-Taliban peace agreement was one of skepticism as it sees it only a first step toward U.S exit strategy leaving many uncertainties around the subsequent course of the peace process. India has concerns that the withdrawal of U.S. and other foreign troops from Afghanistan, before a political settlement, may enhance the influence of Pakistan. India's official statement shows the country's strong determination to remain a key stakeholder and partner in Afghanistan reconciliation and reconstruction. Calling India a "contiguous" neighbor of Afghanistan the statement makes references to India's sovereign claim on Pakistan-held Kashmir—hinting to the intention to maintain strategic depth in Afghanistan (Talukdar 2020). India also fears that U.S.-Taliban peace deal will provide new opportunities for a reinvigorated proxy war in Kashmir. One of India's main security goal is to minimize Pakistan influence in Afghanistan and block activities of anti-Indian militants from using Afghanistan territory against Indian interests.

India also has major economic and transit interests in Afghanistan. Barred by Pakistan from ground access to Central Asia and beyond through Afghanistan, India has joined Iran and Afghanistan to develop the Iranian port of Chahbahar on the Gulf of Oman linked by road to Afghanistan and on to a north–south corridor through Central Asia which is sponsored by several other nations including Russia and Japan.

Iran and the U.S. have certain shared interests in Afghanistan peace and stability, but their commonalities are overshadowed by growing tension between Washington and Tehran. Iran has not been a party to the U.S.-led peace process in Afghanistan but, in reaction to U.S.-Taliban peace agreement, Tehran welcomed "any initiative that helps to secure stability and peace in Afghanistan" (Reuters 2020). Meanwhile Iran strongly believes that peace in Afghanistan will be possible only through domestic Afghan talks and "consideration of the interests of Afghanistan neighbors" (Reuters 2020).

Iran, once a staunch anti-Taliban power has established working relations with the Taliban as the insurgents increased their influence in Afghanistan in recent years, and the bulk of foreign troops pulled out from the country. Iran and the Taliban share interests in the withdrawal of American troops from Afghanistan. Iranian relations with the Taliban are shaped more by their expediency for Iran's interests as well as a counterforce against Iran's ideological enemy, the Islamic State-Khorasan (IS-K) extremists also known as Da'ish. Tehran also eyes maintaining a sphere of influence in post-conflict Afghanistan. The Taliban ties with Iran, however, are mostly opportunistic and may not survive a change of the geopolitical situation.

In recent years, the Islamic Revolutionary Guard recruited thousands Afghan refugees in Iran and organized them into an all Afghan Shia militia to fight in Iran's proxy wars in the Middle East. Known as the Fatemiyoun Corps, the 10,000–20,000 strong fighters, took reportedly the brunt of the fighting in some of the tough battles in Syria (Jamal 2018). According to some sources, since 2013 as many as fifty thousand Afghans have fought in Syria as part of the Fatemiyoun (Jamal 2018). Most Afghan political and religious figures have aired their opposition to sending Afghan Shias to fight in Syria. The Government has also tacitly call for deactivation of the militia following the defeat of the ISIS. Insider reports indicate that although the militia has been downsized after the ISIS defeat, the ex-fighters could conveniently reactivated and deployed when needed (War on the Rocks 2018). Such a possibility has given air to some concern in Afghanistan regarding potential involvement of the Fatemiyoun in the Afghan politics with negative sectarian consequences.

Meanwhile, there is a long-standing dispute over water rights between Iran and Afghanistan that has been exacerbated in recent year due to prolonged drought and mismanagement of water sources. Despite the water sharing agreement signed by the two neighbors in 1973, Iranian President Hasan Rouhani warned in 2017 that his country "cannot remain indifferent to" construction of several dams in Afghanistan that impact the Iranian provinces of Khorasan and Baluchistan (Majidyar 2018). Further, the flow of narcotics from Afghanistan to Iran and continued presence of large number of Afghan refugees in Iran are other issues of potential tension in bilateral relation.

Among the great powers, Russia initially supported the U.S.-led coalition in removing the Taliban regime in Afghanistan. Russia shifted its position as the Taliban made a significant military comeback in the country, while new terrorist groups including the Da'ish emerged in Afghanistan with their potential threat to the stability of Central Asia.

Russia now supports a political settlement of the Afghan conflict and backed the U.S.-Taliban peace agreement. Like Iran, Russia has established working relations with the Taliban to influence the peace process and ensure leverage for Moscow in the post-conflict Afghanistan. Russia believes the political settlement in Afghanistan can open the way for Moscow to renew multi-faceted ties with Afghanistan and expand its influence in the country to counter U.S. and NATO presence in the region. During the months leading up to the Doha peace agreement, Russia hosted several meetings between Taliban delegations and Afghan representatives in Moscow.

China and the U.S. have a common strategic interest in stability in Afghanistan and Pakistan. The two countries share similar views on the parameters of a desirable political dispensation in Afghanistan. Late last year, Afghan and Taliban officials attended a conference in Beijing, and Chinese leaders supported the U.S.-Taliban agreement.

Unhindered implementation of the multi-billion-dollar China Pakistan Economic Corridor (CPEC) project depends on peace and stability in Afghanistan. Further, China faces growing inter-ethnic unrest, terrorism and separatist threats in its Uighur Xinjiang province which is considered the principal reasons for China's heightened interest in the stability and security of Afghanistan and Pakistan. However, Beijing's interests in Afghanistan are primarily economic, with a desire to tap into Afghanistan's vast natural resources. China supports its interest through working with a multi-lateral approach while focused on its long term economic and technological instruments of power, including its strategic Belt and Road Initiative (BRI) and intends to integrate Afghanistan into the BRI system.

Conclusion

Afghanistan is on the verge of another transition that could lead either to a peaceful and stable future or descent into yet another cycle of violence. The challenges are great and evolving, the political context is changing, and uncertainties are compounding. The ongoing political impasse in Kabul due to the presidential election dispute, and the spread of Coronavirus pandemic faced by the country's ill-equipped health system add new dimensions to the difficult situation. The growing economic slowdown due to the contracting economy further narrows options to deal with multiple challenges.

The U.S.-Taliban Peace Agreement, and its companion "Joint Declaration between the Islamic Republic of Afghanistan and the United States of America for Bringing Peace to Afghanistan" Provide the basis for a process to bring peace in Afghanistan through direct talks among the Afghans. Its success, however, depends on its full implementation and close synchronization of different elements of the process including a comprehensive ceasefire, withdrawal of foreign forces from Afghanistan, the removal of the Taliban sanctuaries in Pakistani territory, progress in direct peace talks between the Taliban and the Afghan government, and long-term assistance to Afghanistan to ensure that the foreign troop withdrawal does not lead to collapse of the government.

Such a solution, however, can hardly be sustained without the positive and constructive interaction of Afghanistan neighbors, other regional players, and the great powers whose direct and indirect engagement in Afghanistan has shaped the current situation in the country. Cooperation, not competition is the keyword to stave off the fallouts of the great power and regional actors' competition and to avoid third-party influence over bi-lateral relationships with Afghanistan. A domestic settlement in Afghanistan will require more than negotiations between the government and the Taliban. It will require Afghans of all ethnic groups and genders from across the political spectrum to agree on a unified national agenda for peace and the system of government.

Finally, the United States military intervention and its long and costly war and stabilization effort served its vital interest of preventing another terrorist attack from an unstable region. Now, would the U.S. peace agreement with the Taliban prevent the same dangers without the presence of U.S. troops in Afghanistan? Only complete implementation of the peace deal, and commitment of other involved regional and global actors to support the peace process will help put in place reliable and sustainable security arrangements that could prevent or counter future threatening situations.

By contrast, a uniliteral withdrawal of the American and allied forces before achieving any peace settlement in Afghanistan will be a game changer. Such a situation, however, does not mean the Taliban will victoriously drive into the Afghan capital. Afghanistan today is more resourceful, resilient, and politically more ingenious and willing to fight. But the dynamics of power competition can still plunge the country into the grips of a brutal civil strife. A war that with inevitable foreign interference, can turn the country into the vortex of another proxy war.

References

Agreement for Bringing Peace to Afghanistan between the Islamic Emirate of Afghanistan which is not recognized by the United States as a state and is known as the Taliban and the United States of America, February 29, 2020, which corresponds to Rajab 5, 1441 on the Hijri Lunar calendar (2020)

Asia Foundation, *A Survey of the Afghan People, Afghanistan in 2019* (2019)

BBC News Asia, *Afghan Conflict: President Ashraf Ghani Rejects Taliban Prisoner Release* (2020)

Congressional Research Service, *Afghanistan: Background and Policy in Brief* (2020)

L. Dupree, *Afghanistan* (Princeton University Press, Princeton, 1978)

F. Ermacora, U.N. Special Raporteur on Afghanistan, Situation of Human Rights in Afghanistan. Report to the U.N. General Assembly (1990)

G.J. Gilmore, American Forces Press. U.S. Department of Defense (2007)

S. Haqqani, What We, the Taliban, Want. The New York Times (2020)

A.S. Jamal, Mission Accomplished? What's Next for Iran's Afghan Fighters in Syria. War on the Rocks (2018)

A.S. Jamal, The Fatemiyoun Army: Reintegration into Afghan Society, Special Report 443. United States Information Agency (2019)

Joint Declaration between the Islamic Republic of Afghanistan and the United States of America for Bringing Peace to Afghanistan (2020)

C.E. Lee, C. Kube, A. Mitchell, D. De Luce, Pompeo to Afghan Leaders: Make a Deal with the Taliban or Risk Full U.S. Troop Pullout. NBC News (2020)

A. Majidyar, Iran and Afghanistan at Loggerhead over Water. The Middle East Institute (2018)

Pajhwok Afghan News, Taliban free no military personnel, Gen Tadin Khan (2020)

L.B. Poulada, L.D.J. Poulada, *The Kingdom of Afghanistan and the United States 1828–1973* (The Center of Afghanistan Studies at the University of Nebraska at Omaha and Dageforde Publishing, Lincoln, 1995)

Reuters Dubai, Iran dismisses U.S.-Taliban agreement over Afghanistan (2020)

D. Richards, *Taking Command* (Headline Publishing Group, London, 2014)

Special Inspector General for Afghanistan Reconstruction (SIGAR), *Stabilization in Afghanistan, Lessons from U.S. Experience in Afghanistan* (2018)

Special Inspector General for Afghanistan Reconstruction (SIGAR), *Quarterly Report to United States Congress* (2019)

S. Talukdar, US-Taliban peace deal: India remains deeply skeptical of 'agreement'; New Delhi has no options but to cross fingers and pray. Firstpost (2020)
The Watson Institute at Brown University, Human Costs of the Post-9/11 Wars: Lethality and the Need for Transparency (2018)
C. Whitlock, A war with the truth. The Washington Post (2019)

Humanitarian Aid in Conflict Regions: Refocus and Re-Imagine Creating an Architecture for Impact

Anthony J. Masys

Abstract Over the last decade conflicts have contributed to national, regional and global issues resulting in human suffering, displacement and humanitarian crisis. As reported in OXFAM (2019: 9)

> In 2018, 2 billion people were living in fragile and conflict-affected contexts. By 2035 it is predicted that this will be the case for 85% of the world's extremely poor people. Conflict is currently driving 80% of humanitarian need.

As described in the Global Humanitarian Overview (2020) 'Dozens of highly violent conflicts are causing widespread hunger, displacement, death and destruction around the world. Armed conflicts are killing and maiming a record number of children, forcing them to flee their homes and putting their lives on hold. Women and girls are at higher risk of sexual violence'. It is therefore incumbent on humanitarian and development actors, to ensure efficiency and effectiveness of their engagement strategies in protracted crises and conflict-affected regions (Mowjee and Garrasi 2015: 8). **Zürcher** (2019: **839) argues that 'Aid injected in highly insecure regions, where violence is a reality and insurgents retain some capacities, will increase, not dampen violence'**. The complex threat and risk landscape in conflict regions such as Afghanistan points to the requirement for the realization of the 'humanitarian-development-peace nexus' rooted in understanding and addressing the vulnerabilities of populations before, during and after crises. This chapter explores the complexity associated with humanitarian aid in conflict regions (in particular Afghanistan) and presents a framework, rooted in a systems thinking paradigm, to support the humanitarian-development-peace nexus in order to facilitate successful intervention with sustainable outcomes. Anticipatory policy/intervention design and deployment figure prominently.

Keywords Afghanistan · Humanitarian aid · Conflict · Implementation science · Developmental evaluation · Systems thinking

---

A. J. Masys (✉)
College of Public Health, University of South Florida, Tampa, FL 33620, USA
e-mail: anthony.masys@gmail.com

© The Author(s), under exclusive license to Springer Nature Switzerland AG 2021

A. Farhadi et al. (eds.), *The Great Power Competition Volume 1*,
https://doi.org/10.1007/978-3-030-64473-4_8

Introduction

Over the last decade, conflicts have contributed to national, regional and global issues resulting in human suffering, displacement and humanitarian crisis As described in the Global Humanitarian Overview (2020) 'Dozens of highly violent conflicts are causing widespread hunger, displacement, death and destruction around the world. Armed conflicts are killing and maiming a record number of children, forcing them to flee their homes and putting their lives on hold. Women and girls are at higher risk of sexual violence'. According to the World Bank, globally, conflict is currently driving 80% of humanitarian need. For perspective, consider that in 2018 two billion people were living in fragile and conflict-affected contexts. In support of these human security issues, development and humanitarian aid has been used to tackle the fundamental issues pertaining to poverty and violence reduction in these regions (Kaila et al. 2020: 1). This securitization of aid through humanitarian and development investment in conflict regions such as Iraq and Afghanistan has been viewed as a strategic tool to support peace building in the region. This '...reinforced the expectation that development assistance could be used as an effective means to stabilize war-torn countries. Multi-lateral and bi-lateral donors increasingly saw aid as an important instrument for addressing development and security issues simultaneously' (Zürcher 2017: 506).

A systematic review of the causal impact of development aid on violence in countries reported by Zürcher (2017: 506) argues that '...violence-dampening effect of aid appears to be conditional on a relatively secure environment for aid projects to be implemented. A violence-increasing effect occurs when aid is misappropriated by violent actors, or when violent actors sabotage aid projects in order to disrupt the cooperation between the local population and the government'. Development aid is widely recognized as a tool through which donor states attempt to deliver stability and prosperity to areas throughout the world (Boutton and Pascoe 2018: 1). With this in mind, over the last few decades there has been a growing demand to re-focus and reconfigure international assistance '...to improve coherence and effectiveness across the 'humanitarian-development divide'.

The difficulty in understanding the inherent complexity associated with humanitarian and development aid intervention strategies stems from foundational issues pertaining to mindset:

> Policymakers often have a linear view of the world, where pulling the right levers will get the economy and society back on track after shocks and crises. ... such an approach ignores how systems interact and how their systemic properties shape this interaction, leading to an over-emphasis on a limited set of characteristics, notably efficiency. (OECD 2020)

Working across the humanitarian, development and peace nexus in conflict regions is a complex and multifaceted problem. Bali et al. (2019) argues that in such complex problem spaces, '...not only do policymakers have to contend with the increasing 'wickedness' of these problems but must design and implement solutions in policy environments that are characterized by economic and political uncertainty, and technological disruptions. This is matched by an admitted decline in the abilities of governments to address these problems. How do we then design policies that are robust, durable and ultimately more effective?'

With this in mind, a new way of working (NWOW) is required that focuses on the Humanitarian, Development and Peace nexus in the conflict region. This chapter presents a new way of working rooted in an architectural framework and systems thinking that supports and develops synergies within the nexus.

Context

Afghanistan has been plagued with both natural disasters (drought) and man-made disasters (armed conflict) and in 2017 regarded as the second-least peaceful country in the world with the highest number of civilian deaths in conflict (over 10,000 in 2017). The amount of international aid disbursed since 2001—$57 billion against $90 billion pledged—is a fraction of what has been spent on the war effort. More importantly, it has largely failed to fulfil the international community's pledges to rebuild Afghanistan (International Crisis Group 2011).

Around 55% of the population lives below the poverty line, gender relations are extremely unequal and one in four people are displaced. In 2018, '…47% of the rural population was estimated to not have enough to eat, largely due to drought. The ongoing presence of armed actors in many areas means that agencies' ability to identify and understand the needs of communities, let alone secure long-term development gains, is severely restricted' (OFXAM 2019: 10). As reported in Zürcher (2019: 840) '…aid projects reduced violence in Iraq, but only in conjunction with increased numbers of troops, suggesting that the violence-dampening effect of aid is conditional on a relatively secure environment. … a community-driven development programme in Afghanistan reduced violence, but not in regions where initial levels of violence were already high. … aid projects in North Afghanistan were associated with more household security but only in an already relatively stable environment'.

One of the obstacles to achieving humanitarian and development outcomes stems from the observation that ***Humanitarian and development actors continue to conduct fragmented analyses that do not contribute to a shared vision and prioritised plan of action*** (Mowjee and Garrasi 2015: 29). Entrenched ways of working play a major role in discouraging shared analysis and planning. Similarly, obstacles stemming from a '…severe gap between policy and reality has had negative consequences for humanitarian and development programs, both tactically and in terms of physical implementation' (Blankenship 2014). As reported in Blankenship (2014: 7–8) 'There is a gap because the policies used by donors, implementers, and Afghan officials frequently fail to reflect or sync with the reality on the ground, both in terms of designing programs and reacting to accurate impact assessments. The gap has been widened by the reality of battling strategic frameworks: Viewing development as a "soft power" tool for stabilization has led to significant funding for aid activities, but it has also led to competition among counterterrorism, counterinsurgency, and development frameworks that continue to vie for dominance among policy makers and program teams'.

In short, investments in aid and development in the billions has been applied across highly insecure regions of conflict. Zürcher (2019: 839) argues that 'Aid injected in highly insecure regions, where violence is a reality and insurgents retain some capacities, will increase, and not dampen violence'. Our question then becomes, how do we design intervention strategies to support humanitarian, development and peace outcomes in this complex environment?

Human Security, Humanitarian Aid, Development and Peace

The human security concept (freedom from want, freedom from fear) certainly resonates with Afghanistan. The ongoing conflict and natural disasters that characterize the risk landscape of Afghanistan contribute to a cycle of vulnerability thereby making sustainable solutions not possible without peace. ODI (2016: 35) reports '…there is a mismatch between humanitarian mandates and coordination structures and the long-term strategies needed to respond to protracted crises'. The human security crisis characterized by complexity (Masys 2013) highlights that humanitarian relief, development programmes and peacebuilding are not sequential processes but are interconnected and interdependent. A new mindset is required: a mindset that is rooted in systems thinking.

To address these shortcomings, the concept of a 'humanitarian-development-peace nexus' has developed (OXFAM 2019: 6):

> The nexus approach stems in part from a recognition that emergency needs (and the identities of those most affected) are often symptoms of underlying issues that reflect broader inequalities and injustices. The nexus represents an opportunity to **engage with these root causes** and recognize that humanitarian crises can be caused and/or heightened by poor development policies and a lack of inclusive and appropriate development investment. Thus, meeting life-saving needs at the same time as ensuring longer-term investment addressing the systemic causes of conflict and vulnerability has a better chance of reducing the impact of cyclical or recurrent shocks and stresses, and supporting the peace that is essential for development to be sustainable.
>
> This is not about programmatic amendments but essentially to structural shifts across the aid system.

This derives from greater coherence across intervention strategies that meets immediate needs and '…at the same time as ensuring longer-term investment addressing the systemic causes of conflict and vulnerability – such as poverty, inequality and the lack of functioning accountability systems – has a better chance of reducing the impact of cyclical or recurrent shocks and stresses, and supporting the peace that is essential for development to be sustainable' (OXFAM 2019).

As noted in OXFAM (2019), the nexus has the potential to make aid more effective and efficient. It also provides a good opportunity to work with all stakeholders towards a common goal. The coherence approach associated with the Nexus, helps address systemic causes of conflict and vulnerability—such as poverty, inequality and the lack of functioning accountability systems. With this in mind, Mowjee and Garrasi (2015: 13) describes how 'with the formulation of the Sustainable Development Goals and the 2030 Agenda, the principle of "leaving no-one behind" will require both development and humanitarian actors to work together to address the needs of the most vulnerable and to create conditions for building resilient states and societies'.

This requires an organizational posture and framework that shapes organizational behavior to support '…consensus-building, brokering and building new partnerships; navigating and communicating complex ideas; using systems thinking; facilitating open dialogues; and co-creating ideas' (OXFAM 2019: 5). As will be discussed, an architectural framework rooted in DODAF can facilitate a consistent, high-level, cross-organizational conversation and space for co-creation recognizing the value of each sector. Bali et al. (2019: 7) argue regarding policy design that '…attributes such as adaptability, resilience, robustness, sustainability, explicitness and agility are argued to be central to 'good' designs'. These attributes are key requirements for the application of architecture frameworks.

> Architecture is the fundamental organization of a system embodied in its components, their relationships to each other, and to the environment, and the principles guiding its design and evolution. (Lankhorst et al. 2009: 2)

> Enterprise architecture: a coherent whole of principles, methods and models that are used in the design and realization of an enterprise's organizational structure, business processes, information systems, and infrastructure. (Lankhorst et al. 2009: 3)

Discussion

Afghanistan is overwhelmed with safety and security influences that shape the national 'human security' reality. As described in Masys (2017) 'Shocks (whether man-made or natural disasters) stress our 'security' ecosystem often resulting in failures at various scales thereby posing serious threats nationally, regionally and globally. We can define an ecosystem as an evolving and dynamic collection of actors which respond to its environment. This biological analogy emphasizes the interdependence of all actors in the environment who 'co-evolve their capabilities and roles'. Like natural **ecosystems**, the **security ecosystem** comprises a variety of diverse actors (human, physical and informational) as described in Masys (2017) that interact in complex and dynamic ways. To better manage black swan events that stress the security ecosystem, a fundamental redesign of our mental models and perspective is needed: essentially a paradigm shift in how we view and enable security. Woods (2006: 316) asks the question:

> How do people detect that problems are emerging or changing when information is subtle, fragmented, incomplete or distributed across different groups involved in production processes and in safety management. Many studies have shown how decision makers in evolving situations can get stuck in a single problem frame and miss or misinterpret new information that should force re-evaluation and revision of the situation assessment…

Understanding the human security landscape in Afghanistan is captured well by the comments of Cilliers (1998). Cilliers (1998: 2) remarks that 'a complex system is not constituted merely by the sum of its components, but also by the intricate relationships between these components. In cutting up a system, the analytical method destroys what it seeks to understand'. It has been shown in the literature across various domains of inquiry how small changes to a system can produce large effects. As applied to humanitarian intervention (Masys 2013), security (Masys 2018), thinking in terms of complexity provides a perspective that reveals emergent properties, nonlinearity and a 'dynamic system' of interactions and interrelations.

As discussed in Masys (2017), it is well articulated in the literature (Helbing 2013; Levine et al. 2011; Masys 2012a, b, c; Masys et al. 2014) that often policies and intervention strategies miss the mark with regards to humanitarian, development and peace. Linear views of the problem space confound planners such that interventions are thwarted. It arises from a 'narrow, reductionist world view' and a related 'mismatch between the complexity of the systems we have created and our ability to understand them' (Sterman 2006). Linear and 'siloed' thinking are too simplistic for 'security ecosystems' that are complex. Dekker et al. (2011: 941) argue that '…analytic reduction cannot tell how a number of different things and processes act together when exposed to a number of different influences at the same time. This is complexity, a characteristic of a system. Complex behavior arises because of the interaction between the components of a system. It asks us to focus not on individual components but on their relationships'. This is the realm of 'disaster forensics' (Masys 2016a, b): understanding the complex aetiology of intended and unintended consequences.

As described in Masys (2017), '…the complexity and dilemmas that characterize the security ecosystem (Fig. 1) requires a paradigm shift: a shift in the way we see; a shift in the way we conceptualize; a shift that recognizes the inherent interconnectivity and interdependencies. Such a shift challenges a reductionist approach and embraces a systems thinking paradigm. The paradigm of systems thinking permits a view of the world as a complex system in which as noted by Sterman (2000: 10) we come to the understanding that 'you can't do just one thing' and that 'everything is connected to everything else'. This is supported by Senge (1990: 73) who is of the opinion that the discipline of the systems approach lies in a shift of mind: in seeing interrelationships rather than linear cause-effect chains and seeing processes of change rather than snapshots'.

With regards to sustainability and development, a systems lens reveals striking insights and solutions by helping to frame the problem (Morgan 2005: 15–16):

- 'First, people see the part to which they are connected but are largely unaware of the bigger system(s) that surrounds it. They miss their impact on others and others on them.
- Second, people tend to lack a time dimension. They see the present but not the past. They are intent on figuring out where a particular system should be in the future. They have some interest in knowing where it is now. But they have little inclination to understand where it has been. They do not know the history of the

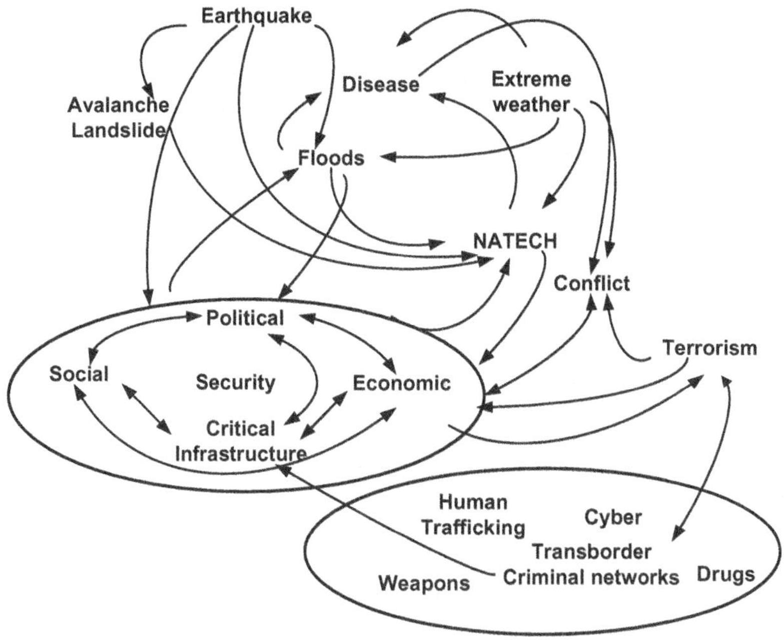

Fig. 1 Security ecosystem influence diagram (Masys 2017)

present. '…everything that was left unprepared becomes a complex problem, and every weakness comes rushing to the forefront' (Weick and Sutcliffe 2007)

- Third, participants miss—and in many cases mischaracterize—many of the key relationships that shape events.
- Fourth, people suffer from process blindness. They do not grasp the process dynamics, especially the deeper ones that are ongoing even within smaller systems. They suggest improvements which do not fit or even acknowledge the way the system actually works. They see individuals or events but not the processes of which they are a part'.

Described in Masys (2017), Levine et al. (2011: 7) argues that '…a system perspective can often reveal how behaviour that is competent from the standpoint of each individual actor does not contribute to achieving the overall goals which collectively all the actors in the 'system' say they are working towards, in different ways. System problems often result when different actors do not really share the objectives, or when they do not agree on which elements contribute to a single system'.

As shown in Fig. 1, the security ecosystem is integrated into the greater societal system showing how it influences and is influenced by disasters (both man-made and natural). Humanitarian, development and peace initiatives must consider this complex interconnectivity.

The mental models we have regarding humanitarian, development and peace within a security ecosystem (such as Afghanistan) incorporate ones biases, values, learning, experiences and beliefs about how the world works.

What this shows through the complexity lens is that the security ecosystem is shaped by factors seeded in advance. Recognizing this presents an opportunity for new strategic possibilities regarding Humanitarian, Development and Peace nexus. Reported in Masys (2017), Urry (2002: 59), in his discussion of complexity and systems, remarks that there exists a '...profound disproportionality of 'causes and effects'. Such systems possess a history that irreversibly evolves and in which past events are never 'forgotten'. Failure to recognize or understand the complex interdependencies associated with the security ecosystem can result in making incorrect assumptions regarding attribution and contribution of events and decisions. Local actions and decisions can have regional and global impacts. The application of systemic thinking across the security landscape is essential. As noted in Goldin and Mariathasan (2014: 66) 'systemic analysis must examine nodes, pathways, and the relationships between them, because catastrophic changes in the overall state of the system can ultimately derive from how it is organized- from feedback mechanisms within it and from linkages that are latent and often unrecognized'. This has been described in detail in Masys (2016a, b) leveraging Actor Network Theory to support disaster forensics.

Much of the challenges associated with managing shocks stems from the mind being '...blinded by optimism and confusion' or '...using out of date and unrealistic models of the world' (Ramo 2009: 6). Across the humanitarian, development and peace nexus '...The most common source of mistakes is not the failure to find the right answers. It is the failure to ask the right questions. Nothing is more dangerous in business (***or security***) than the right answers to the wrong questions' (Ramirez and Wilkinson 2016: 23).

The OECD (2020: 3) characterizes our current societal challenges stemming from the claim that 'We are not living in a linear, Newtonian world where actions cause predictable reactions. We are in fact part of a complex system of environmental, socio-political and economic systems that we are constantly reconfiguring and that are constantly affecting us. In such a world, a small change can be transmitted and amplified by the interconnectedness of the system to have enormous consequences, far beyond the time, place, and scale of the initial perturbation'.

The complex problem associated with humanitarian aid, development and peace operations requires a framework to support new ways of seeing to reveal the vulnerabilities and exploring the possibility/plausibility space in support of strategic decision making. Hynes et al. (2020: 2) argues, that

> Systems thinking allows us to identify the key drivers, interactions and dynamics of the economic, social and environmental nexus that policy seeks to shape and to select points of intervention in a selective, adaptive way. ...this allows us to emphasise the importance of system resilience to a variety of shocks and stresses, allowing systems to recover from lost functionality and adapt to new realities regarding international economics, societal needs, and human behavior.

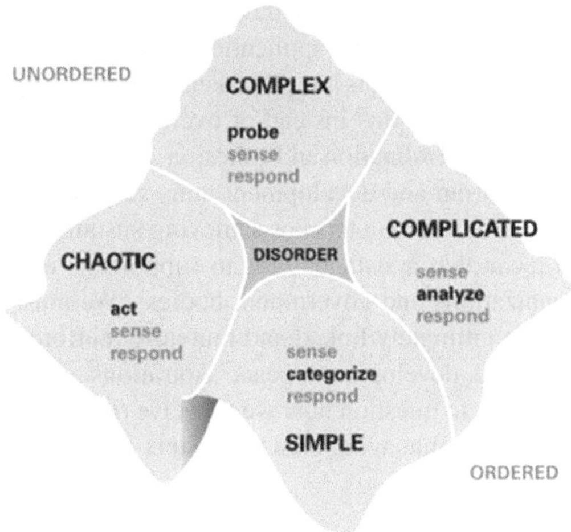

Fig. 2 Cynefin framework (Snowden and Boone 2007)

Cynefin Framework—a systems lens

Framing the dynamics associated with conflict and the humanitarian-development-peace nexus as a wicked problem replete with uncertainties, ambiguity and complexity lends itself to application of the Cynefin framework (Kurtz and Snowden 2003; Snowden 2003; Snowden and Boone 2007). Described in detail in Masys (2020), the Cynefin (kuh-ne-vin) is from the welsh word for 'habitat or rootedness' and was used by Snowden as a metaphor for a conceptual framework for 'time and space' to make decisions in complex situations.

The application of Cynefin helps to contextualize problems regarding causality across 4 + 1 domains: Simple/obvious; Complicated; Complex; Chaotic; and Disordered (Fig. 2).

The Cynefin Framework is apropos in contextualizing the humanitarian-development-peace nexus. The nonlinear behavior associated with interventions pertaining to the humanitarian-development-peace nexus is well documented. In matters pertaining to the impact and influence of humanitarian aid, development initiatives and peace initiatives in a conflict region, interdependencies, interconnectivity, indirect influences make mechanistic and linear approaches problematic. Understanding complex causality and influences requires a more holistic perspective. The Cynefin framework helps situate our conceptual understanding of humanitarian aid, development and peace initiatives thereby challenging linear event based mindset approaches to explore the uncertainty, ambiguity and complexity of the problem space. The research examining the Afghanistan humanitarian aid-development-peace nexus argues that competing visions, agendas and impacts created a condition whereby unintended consequences emerged. It is the inherent interdependencies within the nexus that places it conceptual identity within the complexity quadrant of the Cynefin. As part of the proposed solution space, the Cynefin framework informs impact analysis and implementation strategies through scenario development. This plays a key role in the application of the architecture framework.

Without a systems lens, Blankenship (2014) argues '…Programs have emphasized quantifiable outputs instead of overall outcomes. In addition, they have lacked the necessary coordination and cohesion, both internal and external, to achieve long-term humanitarian and development aims such as protecting advancements in women's rights over the long term or achieving sustainable livelihoods…problems delivering aid mean that assistance fails to support the effective capacity building of Afghan organizations and government bodies'. We must recognize that poverty, crisis and risk are intimately linked and mutually reinforcing. The gap between humanitarian assistance, development, peace aspirations and outcomes continues to grow. This is forcing us to question how we perceive risks and challenges our current mindset and approach to managing crisis, disasters and emergencies.

Complex Landscape of Humanitarian-Development-Peace Nexus

Intervention strategies that identify leverage points are critical. Identifying the actors (applying ANT) that characterize the problem space and their mapping in terms of influence becomes a key approach to understanding the problem space and the dynamics of interventions (Masys 2016a, b). Embracing a systems thinking paradigm such as the Cynefin, understanding the complexity associated with interventions in Humanitarian-development-peace nexus can be framed in terms of the social ecological model (SEM) (Fig. 3). As described in Hayden (2019),

> the basis of the SEM is recognition of the dynamic interplay among the various factors at different levels that affect behavior. Because of this interplay between and among the many levels of influence, changing one can have an impact on them all.

The implications for Humanitarian-development-peace interventions are manifest.

Across the Humanitarian-development-peace nexus are interactions that impact and influence across the SEM. The SEM provides a lens that supports the requirement for interdependency mapping to facilitate intervention and implementation strategy outcomes. It provides insights into unexpected outcomes that emerge from lack of understanding interdependency in a complex VUCA (volatility, uncertainty, complexity, ambiguity) environment. Applying an actor network theory approach (Masys 2014, 2016a, b), the elements of the interdependent problem space can be mapped across the intra and interpersonal through to societal. This is an important step in designing intervention strategies.

With the SEM in mind, we can better understand the root causes of insecurity and the challenges associated with humanitarian, development and peace initiatives. As described in the United Nations and World Bank (2018):

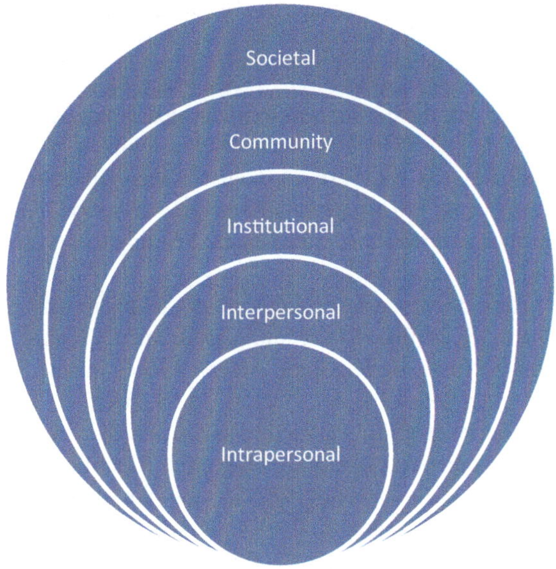

Fig. 3 Social ecological model

The best way to prevent societies from descending into crisis, including but not limited to conflict, is to ensure that they are **resilient through investment in inclusive and sustainable development**. For all countries, addressing inequalities and exclusion, making institutions more inclusive, and ensuring that development strategies are risk-informed are central to preventing the fraying of the social fabric that could erupt into crisis.

Growth and poverty alleviation are crucial but alone will not suffice to sustain peace. Preventing violence requires departing from traditional economic and social policies when risks are building up or are high. It also means seeking inclusive solutions through dialogue, adapted macroeconomic policies, institutional reform in core state functions, and redistributive policies. Inclusive decision making is fundamental to sustaining peace at all levels, as are long-term policies to address economic, social, and political aspirations. Fostering the participation of young people as well as of the organizations, movements, and networks that represent them is crucial. Women's meaningful participation in all aspects of peace and security is critical to effectiveness, including in peace processes, where it has been shown to have a direct impact on the sustainability of agreements reached.

Similarly, cross-country studies find evidence that high levels of **gender inequality and gender-based violence** in a society are associated with increased vulnerability to civil war and interstate war and the use of more severe forms of violence in conflict Changes in women's status or vulnerability, such as an increase in domestic violence or a reduction in girls' school attendance, often are viewed as early warnings of social and political insecurity. Prevention of violent conflict requires a strong focus on women's experiences and on measures to ensure their participation in political, social, and economic life.

Inequality and exclusion manifest most starkly in policy arenas related to access to political power and governance; land, water, and extractive resources; delivery of basic services; and justice and security. As the spaces where livelihoods and well-being are defined and defended, access to these arenas can become, quite literally, a matter of life or death. The arenas reflect the broader balance of power in society, and as such, they are highly contestable and often resistant to reform. **Competition for power** is an age-old source of conflict. Power balances and imbalances can put a society at risk of violence. Experience shows that more inclusive and representative power-sharing arrangements lower the risk of violent conflict. Decentralizing, devolving, or allowing autonomy of subnational regions or groups can help to accommodate diversity and lower the risk of violence at the national level.

The Humanitarian-development-peace nexus as applied here stimulates new thinking about the relationship of development, peace, and security—a relationship that takes concrete form in inclusive approaches to preventing conflict. A coherent strategy that can be sustained over time demands levels of **integrated planning and implementation** that are often challenging to development, security, humanitarian, and political actors.

How do we design a solution space? From intrapersonal to societal, the factors illustrated in Fig. 4 (stemming from an ANT analysis of the problem space: see Masys 2016a, b) permeate across all levels of the SEM model thereby illustrating the inherent complexity of the problem space.

The inner circle characterizes the major issues inherent in crisis regions such as Afghanistan. The surrounding circle identifies the nexus strategies to address the problem space. The outer circle denotes the framework to support the implementation of the nexus strategies. This framework is rooted in the Department of Defence Architecture Framework (DODAF).

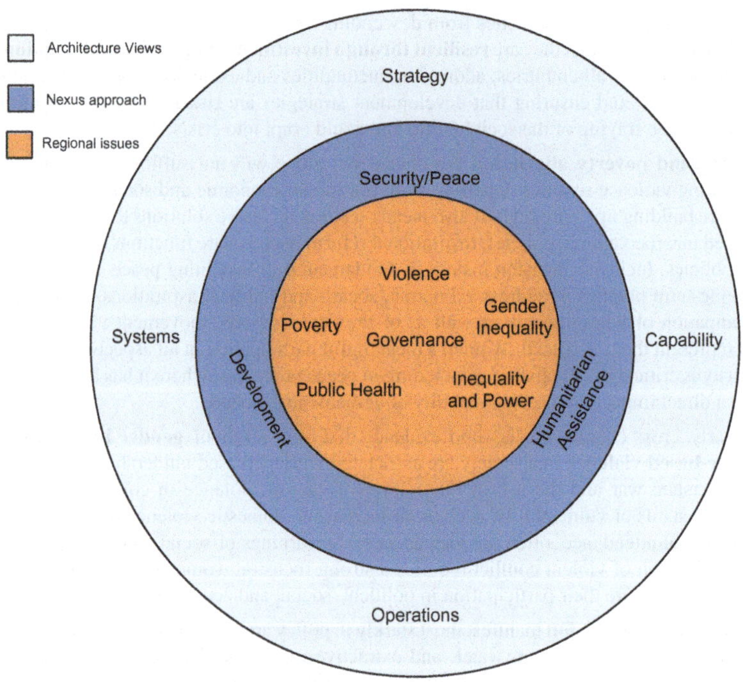

Fig. 4 Model of problem and solution space

Architecture Framework Methodology: DODAF

As described by Mowjee and Garrasi (2015: 22) '…navigating dilemmas in maintaining a principled stance, engagement with the government, and gaining access to populations in need of assistance is a classic challenge for humanitarian actors'. To help with navigating this complex problem space, a system architecture is presented.

An architecture can be simply viewed as a set of blueprints which model or represent a wide variety of relationships inherent to the overall capability being managed. The Institute of Electrical and Electronics Engineers (IEEE) defines architecture as "the structure of components, their interrelationships, and the principles and guidelines governing their design and evolution over time".[1] Architectures offer distinct advantages in structuring information and managing complexity, incremental development and implementation. They impose discipline and ensure use of a common language across diverse stakeholders.

Architecture frameworks enable one to capture the people, processes and technologies that exist (the "as is") or need to exist (the "to be") within a capability. The data elements captured in multiple views enable the analysis of interdependencies and interactions between people, processes and technologies within a system and across systems, thereby supporting a capability design and analysis. In so doing, the architecture approach promotes enhanced understanding regarding the humanitarian, development and peace operations and processes across organisational boundaries, including jurisdictional and security/safety boundaries, and fosters multi-agency integration and the adoption of common standards and protocols.

As described in Masys (2020), through the structured approach, applying an Architectural Framework for humanitarian, development peace nexus provide the means to describe roles and relationships, organizational, policy and governance drivers, to articulate operational concepts and to model critical 'business' processes. International defence organisations have promoted the use of architecture frameworks of which the US Department of Defense Architecture Framework (DoDAF)[2] is the most common. Through DODAF, the capability requirements are traced and supported from a top-down mapping of organizational guidance to capability development plans. Through this, opportunities and gaps can be described and addressed. From the top down approach emerges specific views: capability; operational; services and systems (Fig. 5).

This framework presents data as various views: Strategic View (StratV), Capability View (CapV), Operational Views (OV), and Systems Views (SV). These views are used to fully document the current state, or the proposed future state, of a system or capability. Appling the DODAF to the problem space assists with the analysis of enterprise-level systems (consisting of people, process and tools).

Framing and designing the humanitarian, development and peace nexus through DODAF operationalizes intervention strategies showing clear connectivity from

[1] Institute of Electrical and Electronics Engineers, Inc. (IEEE) Standard 610.12.

[2] http://dodcio.defense.gov/Library/DoD-Architecture-Framework/.

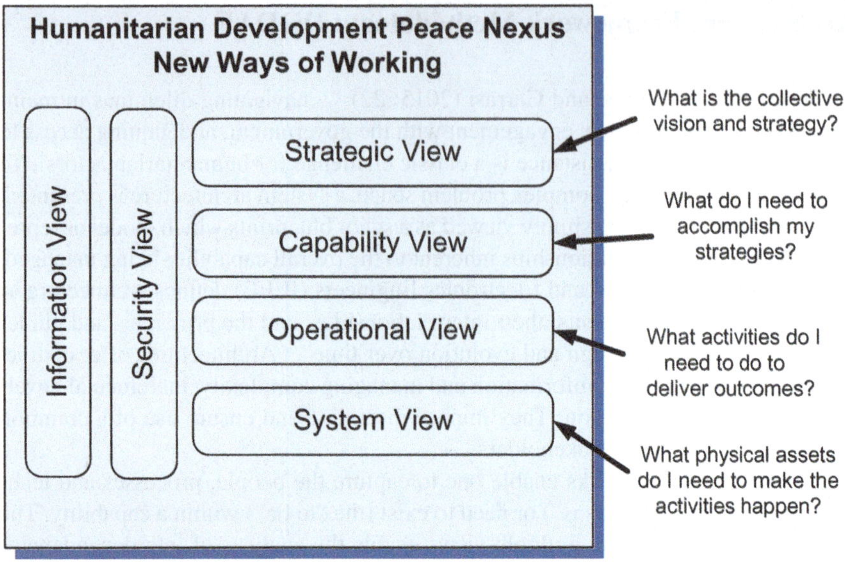

Fig. 5 Humanitarian, development peace nexus architecture framework

strategic to operational views. Insights into applying DODAF can be found in Masys (2016a, b, 2018), DODAF (2015).

Through adherence to this approach, the framework (Fig. 5) supports:

- Collaboratively developed priorities and strategies
- Systems understanding to support program development and implementation
- Agile and adaptive flexibility in strategic planning
- Evaluation strategies based on shared principles and standards
- Ensure monitoring and evaluation is continuous, realistic and outcome oriented
- Clear communication channels in planning and implementation
- Harmonized strategies
- Engagement with local actors
- Empower sustainability.

The landscape and problem space associated with the humanitarian, development and peace nexus is one characterized as a dynamic system with nonlinear behavior. This understanding has important ramifications in how we perceive and intervene within the problem space. As popularized by the Butterfly effect (Silver 2015: 119), '…many of our problems stem from inaccuracies in our data and in our assumptions. With nonlinear systems, inaccuracies are not forgiving'. As noted in OECD (2020: 2), '…in complex systems, tensions exist between efficiency and resilience, the ability to anticipate, absorb, recover, and adapt to unexpected threats. Resilience is a focus of specific parts of some systems, for instance military and health systems, but some systemic risks are the consequence of attempts to maximise efficiency in subsystems leading to suboptimal efficiency at higher levels'.

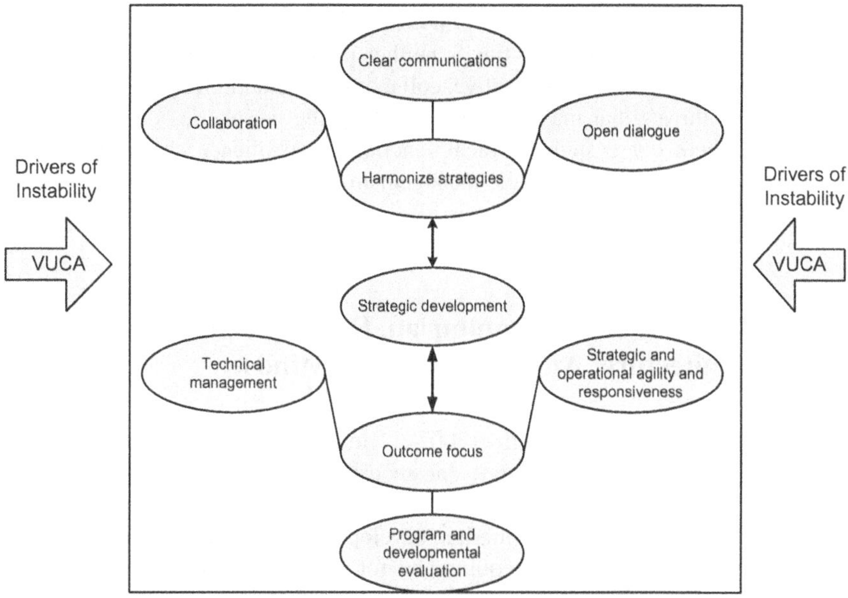

Fig. 6 Disciplined approach to humanitarian, development and peace nexus through the application of DODAF

To address the complexities associated with conflict area and the causes of risk, fragility and vulnerability and to design strategies for intervention without causing new risks and vulnerabilities requires a 'systems approach'. Hence the disciplined application of DODAF (Fig. 6). OECD (2020: 3), '...systems thinking allows us to identify the key drivers, interactions, and dynamics of the economic, social, and environmental nexus that policy seeks to shape, and to select points of intervention in a selective, adaptive way. Critically, this allows us to emphasise the importance of system resilience to a variety of shocks and stresses, allowing systems to recover from lost functionality and adapt to new realities'.

To support the development of the architecture framework, as described in Hynes et al. (2020) '...the IRGC highlights a multi-step procedure to identify, analyse, and govern systemic risks, as well as better prepare affected systems for such risks by mitigating possible threats and transitioning the system towards one of resiliency-by-design. As a cyclical process, the IRGC's process for the governance of systemic risk includes:

1. Explore the system, define its boundaries and dynamics.
2. Develop scenarios considering possible ongoing and future transitions.
3. Determine goals and the level of tolerability for risk and uncertainty.
4. Co-develop management strategies dealing with each scenario.
5. Address unanticipated barriers and sudden critical shifts.
6. Decide, test and implement strategies.
7. Monitor, learn from, review and adapt.

The purpose of IRGC's exercise is not to generate a deterministic model that applies to any and all systems—this is neither possible nor helpful. Instead, it is designed to produce more introspective, collaborative, and multi-system viewpoints regarding the threats that may be lingering along the peripheries of our systems, as well as where our system's critical functions or resilience challenges should be improved within future strategic management opportunities'. This is key in the development of the architecture framework.

Operationalizing the Humanitarian, Development and Peace Nexus Architecture: An Organizational Mindset

As noted in Weick and Sutcliffe (2007), 'Unexpected events often audit our resilience'. Disasters stress the safety and security of the region, impinging on the already fragile disaster response capabilities and affecting the most vulnerable. The challenges associated with humanitarian, development and peace initiatives regionally and globally highlight the requirement for a more responsive and impactful 'delivery' model of aid.

Creating the architecture views is only part of the solution space. One must operationalize them. Successful organizations that can manage complexity and avoid disasters though operating in hazardous and challenging conditions are called High Reliable Organizations (HRO). As described in Masys (2018), the roots of the HRO paradigm finds itself in the examination of aircraft carrier operations, air traffic control and nuclear power operations. They not only avoid dire consequences by catching problems early, but they have actually proven over time to have disproportionately fewer problems. These organizations do this by consistently noticing the unexpected, reporting it in an honest way, responding quickly and appropriately, learning from the things they did, and improving the process for the next time a challenge arises. This is the reality of Afghanistan.

Central to the success of HROs are the cultural dimensions. An HRO depends on:

> '…a higher-level system of knowledge and experience, interacting with and supporting an [organizational framework] to transform a high-risk system into a high-reliability system.
>
> The delegation is controlled culturally by a powerful system of selection, training, and mutual monitoring, criticism, and advice. The result is a pattern of extremely efficient communications, which gives the system the ability to absorb damage and surprises' (Spann 2006).

Weick and Sutcliffe (2007) suggest there are five common concepts that help organizations manage the threat of failure, absorb damage and surprises, and thereby become an HRO. The first three concepts fall under the category of "Anticipate the Unexpected" and the final two are listed under "Contain the Unexpected."

Masys (2018) describes in detail the key requirements of a High Reliability Security Organizations (HRSO), that being Mindset (Fig. 7); Security Culture and HRSO principles. Based upon Masys (2018), the HRSO is certainly applicable to support

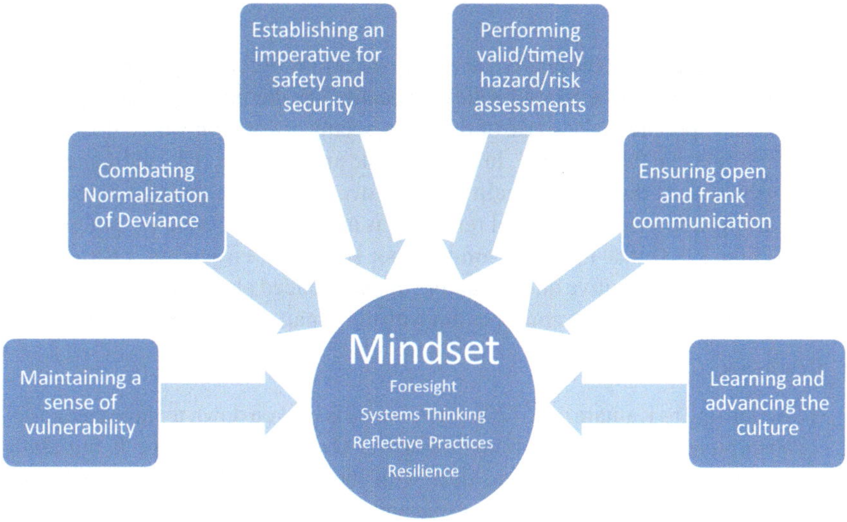

Fig. 7 Lessons learned and mindset (Masys 2018)

design and implementation of the humanitarian, development and peace nexus. At the core is mindset. How we view the world shapes what we see and then what we do. The complexity associated with humanitarian, development and peace nexus requires a key foundation (described in Masys 2018) in:

- Systems Thinking
  - As described in Masys (2015), in this complex problem landscape, systems thinking is both a worldview and a process in the sense that it informs ones understanding regarding a system and can be used as an approach in problem solving (Edson 2008). Senge (1990) describes systems thinking as 'a discipline for seeing wholes…a framework for seeing interrelationships rather than things, for seeing patterns of change rather than static snapshots'. The behaviour of the system is a result of the interaction and interrelationships that exists thereby acknowledging emergent behaviours and unintended consequences. The butterfly defect, as described in Goldin and Mariathasan (2014) draws attention to the new nature of systemic risk, which is such that small perturbations now have much greater effects and permeate all dimensions of society.
- Foresight
  - Foresight is the act of looking to and thinking about the future. Foresight enables you to get better at predicting, creating, and leading the future. This is a critical element of scenario planning.

- Reflective Practices
  - Sense making is at the heart of security operations. According to Weick (1995), sense making has a 'strong reflexive quality' to the process of interpretation because 'people make sense of things by seeing a world on which they already imposed what they believe'. In this dynamic and complex risk landscape reflective practices thereby can help to avoid overconfidence through a generative learning approach. Individual reflection is also being a 'reflective practitioner' (Schön 1983) that involves: 'on-the-spot surfacing, criticizing, restructuring, and testing of intuitive understanding of experienced phenomena; often it takes the form of a reflective conversation with the situation' (Schön 1983: 241–242).
- Resilience
  - According to Kamara et al. (2019) resilience is broken down to three categories: absorptive, adaptive, and transformative. Absorptive resilience is the ability to take protective action and deal with the stress a disaster brings as it is happening. Adaptive resilience is the ability to prepare and adjust according to a community's current or anticipated situation. Finally, transformative capacity is the ability for a community to identify, minimize, and/or correct risk hazards Kamara et al. (2019). All three areas must be addressed to minimize impact of disaster on a community.

To support the implementation of humanitarian, development and peace initiatives, a HRO embraces the 5 foundational principles +1 (as described in Masys 2018):

Principle 1: Preoccupation with Failure
HRO explore the hazard, threat and intelligence landscape to determine the space of possibilities. They encourage the reporting of errors and miscalculations with regards to crisis assessments, they elaborate experiences of a near miss for what can be learned, and they are wary of the potential liabilities of success, including complacency, the temptation to reduce margins of security.

Principle 2: Reluctance to Simplify
HROs take deliberate steps to create more complete and nuanced pictures of what they face and who they are as they face it. Knowing that the world they face is complex, stable, unknowable, and unpredictable. HROs position themselves to see as much as possible. They welcome diverse experience, skepticism toward received wisdom.

Principle 3: Sensitivity to Operations
They are attentive to the front line, where the real work gets done. …Anomalies are noticed while they are still tractable and can still be isolated. All of this is made possible because HROs are aware of the close ties between sensitivity to operations and sensitivity to relationships.

Principle 4: Commitment to Resilience
The essence of resilience is therefore the intrinsic ability of an organization (system) to maintain or regain a dynamically stable state, which allows it to continue operations after a major mishap and/or in the presence of a continuous stress. The hallmark of HRO is not that it is incident-free but that incidents don't disable it. Resilience demands a deep knowledge of the threat and space of possibilities (both in time and space).

Principle 5: Deference to Expertise
HROs cultivate diversity, not just because it helps them notice more in complex security environments, but also because it helps them do more with the complexities they do spot. Rigid hierarchies have their own special vulnerability to error and misjudgment.

Principle 6: Adaptability
HROs embrace a Safety II (Hollnagel 2012) approach that recognizes the need for dynamic and adaptive behavior that is derived from the first 5 principles. HROs understand that work as imagined often deviates from work as done. The volatility, uncertainty, complexity and ambiguity that characterize the humanitarian, development and peace landscape require a certain degree of adaptation to support the very fluid environment. Mindfulness (Weick and Sutcliffe 2007) and reflective practices (Masys et al. 2014) figure prominently.

The result is "collective outcomes". OCHA (2017: 4) describes this New Way of Working and '…frames the work of development and humanitarian actors, along with national and local counter-parts, in support of collective outcomes that reduce risk and vulnerability'.

Strategic View Coherence and Goals

Through the application of systems thinking, DODAF and HRO principles, humanitarian, development and peace initiatives can be designed and operationalized to execute the highest-priority objectives to support collective impact. The complex landscape associated with planning and decision making for the nexus requires attention to the dynamic complexity of the evolving scenario thereby making long-term forecasting problematic.

Systemic threats permeate the security landscape as shown in Fig. 1. Resilience thereby becomes a critical feature of the security ecosystem. Key to implementing a successful humanitarian, development and peace initiative, we look to Hynes et al. (2020: 2) and the NAEC report "Resilience Strategies and Policies to Contain Systemic Threats":

> …defines concepts related to systemic threats and reviews the analytical and governance approaches and strategies to manage these threats (including epidemics) and build resilience

to contain their impacts. This aims to help policymakers build safeguards, buffers and ultimately resilience to physical, economic, social and environmental shocks. Recovery and adaptation in the aftermath of disruptions is a requirement for interconnected 21st Century economic, industrial, social, and health-based systems, and resilience is an increasingly crucial part of strategies to avoid systemic collapse. Based on NAEC reports and the resilience literature, specific recommendations for building resilience to contain epidemics and other systemic threats include:

1. Design systems, including infrastructure, supply chains, economic, financial and public health systems, to be resilient, i.e. recoverable and adaptable.
2. Develop methods for quantifying resilience so that trade-offs between a system's efficiency and resilience can be made explicit and guide investments.
3. Control system complexity to minimize cascading failures resulting from unexpected disruption by decoupling unnecessary connections across infrastructure and make necessary connections controllable and visible.
4. Manage system topology by designing appropriate connection and communications across interconnected infrastructure.
5. Add resources and redundancies in system-crucial components to ensure functionality.
6. Develop real-time decision support tools integrating data and automating selection of management alternatives based on explicit policy trade-offs in real time. (Hynes et al. 2020: 2)

These resilience strategies certainly resonate with the systems thinking approach and the operational application of the Humanitarian, Development and Peace nexus through DODAF design.

Conclusion

As discussed in OXFAM (2019):

> The nexus approach stems in part from a recognition that emergency needs (and the identities of those most affected) are often symptoms of underlying issues that reflect broader inequalities and injustices. The nexus represents an opportunity to engage with these **root causes** and recognize that humanitarian crises can be caused and/or heightened by poor development policies and a lack of inclusive and appropriate development investment.

It is clear that, in general, conflict significantly affects the capacity to deliver aid effectively. As described in Jones et al. (2013: vi) '…development and humanitarian sectors continue to face criticism over their relative rigidity and short-termism with regards to project funding and delivery'. The World Bank (2018: xxvi) '…one of the objectives of Pathways for Peace is to stimulate new thinking about the relationship of development, peace, and security—a relationship that takes concrete form in inclusive approaches to preventing conflict. A coherent strategy that can be sustained over time demands levels of integrated planning and implementation that are often challenging to development, security, humanitarian, and political actors'.

In this chapter we outlined the challenges associated with humanitarian and development in conflict regions (Afghanistan). The complex causality that characterizes the conflict problem space points to the requirement for systems thinking to better understand the interdependencies and interconnectivity across the security ecosystem. With the systems perspective, architecture frameworks (DODAF) was introduced as an approach to support collective outcomes required in the humanitarian, development and peace nexus.

The approach supports consensus-building, brokering and building new partnerships; navigating and communicating complex ideas; using systems thinking; facilitating open dialogues; and co-creating ideas.

References

A.S. Bali, G. Capano, M. Ramesh, Anticipating and designing for policy effectiveness. Policy Soc. **38**(1), 1–13 (2019)

E. Blankenship, *Aid in Contested Spaces: Afghanistan.* Oxfam America Research Backgrounder Series (2014), http://www.oxfamamerica.org/publications/delivering-aid-in-contested-spaces-afghanistan

A. Boutton, H. Pascoe, Do foreign aid projects attract transnational terrorism? Peace Econ. Peace Sci. Publ. Policy **2018**, 20180027 (2018)

P. Cilliers, *Complexity and Postmodernism: Understanding Complex Systems* (Routledge Publishing, London, 1998)

S. Dekker, P. Cilliers, J-H. Hofmeyr, The complexity of failure: implications of complexity theory for safety investigations. Saf Sci **49**, 939–945 (2011)

DODAF, *Architectural Data and Models*, vol. II (2015), https://dodcio.defense.gov/Portals/0/Documents/DODAF2/DoDAF%20v2.02%20Chg%201%20Vol%20II%20Final%202015-01-19.pdf

R. Edson, *Systems Thinking. Applied: A Primer*. ASysT Institute (2008), http://www.anser.org/docs/systems_thinking_applied.pdf

Global Humanitarian Overview (2020), https://reliefweb.int/sites/reliefweb.int/files/resources/GHO-2020_v9.1.pdf

I. Goldin, M. Mariathasan, *The Butterfly Defect: How Globalization Creates Systemic Risks and What to Do About It* (Princeton University Press, Princeton, 2014)

J. Hayden, *Introduction to Health Behavior Theory*, 3rd edn. (Jones and Bartlett Learnings, Burlington, MA, 2019)

D. Helbing, Globally networked risks and how to respond. Nature **497**, 51–59 (2013)

E. Hollnagel, A tale of two safeties (2012), http://www.resilienthealthcare.net/A_tale_of_two_safeties.pdf

W. Hynes, I. Linkov, B. Trump, A systemic approach to dealing with COVID-19 and future shocks. New Approaches to Economic Challenges (NAEC). OECD (2020), http://www.oecd.org/naec/projects/resilience/NAEC_Resilience_and_Covid19.pdf

International Crisis Group, Aid and conflict in Afghanistan Asia Report No 210, 4 August 2011 (2011), http://www.operationspaix.net/DATA/DOCUMENT/3983~v~Aid_and_Conflict_in_Afghanistan.pdf

L. Jones, E. Ludi, P. Beautement, C. Broenner, C. Bachofen, New approaches to promoting flexible and forward-looking decision making: insights from complexity science, climate change adaptation and 'serious gaming' (2013). ODI February 2013

H. Kaila, S. Singhal, D. Tuteja, Development programs, security, and violence reduction: evidence from an insurgency in India. World Dev. **130**, 104911 (2020)

J.K. Kamara, K. Agho, A.M.N. Renzaho, Understanding disaster resilience in communities affected by recurrent drought in Lesotho and Swaziland—a qualitative study. PLOS One (2019), https://journals.plos.org/plosone/article?id=10.1371/journal.pone.0212994

C.F. Kurtz, D.J. Snowden, The new dynamics of strategy: sense-making in a complex and complicated world. IBM Syst. J. **42**(3), 462–483 (2003)

M. Lankhorst et al., *Enterprise Architecture at Work: Modelling, Communication and Analysis*, 2nd Edn. (Springer, 2009)

S. Levine, A. Crosskey, M. Abdinoor, System failure? Revisiting the problems of timely response to crises in the Horn of Africa. Network paper number 71. Humanitarian Policy Group. Overseas Development Institute (2011)

A.J. Masys, Human security: a view through the lens of complexity, in *Proceedings of the European Conference on Complex Systems 2012, Springer Proceedings in Complexity*, ed. by T. Gilbert, M. Kirkilionis, G. Nicolis, pp. 325–335 (2013)

A.J. Masys, Human security: a view from complexity, in *Proceedings of the European Conference on Complex Systems* (2012a), pp. 325–335

A.J. Masys, The emergent nature of risk as a product of 'heterogeneous engineering: a relational analysis of the oil and gas industry safety culture, in *Innovative Thinking in Risk, Crisis and Disaster Management*, ed. by S. Bennett (Gower Publishing, London, 2012b)

A.J. Masys, Black swans to Grey swans: revealing the uncertainty. Int. J. Disaster. Prev. Manag. **21**(3), 320–335 (2012c)

A.J. Masys (ed.), *Disaster Management: Enabling Resilience* (Springer, 2014)

A.J. Masys (ed.), *Applications of Systems Thinking and Soft Operations Research in Managing Complexity* (Springer, Cham, 2015)

A.J. Masys (ed.), *Disaster Forensics: Understanding Root Cause and Complex Causality* (Springer, 2016a)

A.J. Masys (ed.), *Exploring the Security Landscape: Non-traditional Security Challenges* (Springer, 2016b)

A.J. Masys, N. Ray-Bennett, H. Shiroshita, P. Jackson, High impact/low frequency extreme events: enabling reflection and resilience in a hyper-connected world, in *4th International Conference on Building Resilience*, 8–11 September 2014, Salford quays, United Kingdom (2014). Procedia Econ. Financ. **18**, 772–779 (2014)

A.J. Masys, Complexity and security: new ways of thinking and seeing, in *Asia/Pacific Security Challenges-Managing Black Swans and Persistent Threats*, ed. by A.J. Masys, L. Lin. (Springer Publishing, 2017)

A.J. Masys, Designing high reliability security organizations for the homeland security enterprise, in *Homeland Security Cultures: Enhancing Values While Fostering Resilience*, ed. by A. Siedschlag, A. Jerkovic (Rowman and Littlefield International, 2018)

A.J. Masys, Science informed major event security planning: from vulnerability analysis to security design, in *Science Informed Policing*, ed. by B. Fox, D. Reid, A.J. Masys (Springer Publishing, 2020)

P. Morgan, The idea and practice of systems thinking and their relevance for capacity development. European Centre for Development Policy Management, March 2005 (2005)

T. Mowjee, D. Garrasi, Coherence in conflict: bringing humanitarian and development aid streams together. Ministry of Foreign Affairs of Denmark, November 2015 (2015)

ODI, Time to let go Remaking humanitarian action for the modern era (2016), https://www.odi.org/sites/odi.org.uk/files/resource-documents/10422.pdf

OCHA (2017) New Ways of Working. https://www.unocha.org/sites/unocha/files/NWOW%20Booklet%20low%20res.002_0.pdf

OECD, A systemic resilience approach to dealing with Covid-19 and future shocks new approaches to economic challenges (NAEC), 28 April 2020 (2020)

OXFAM, The humanitarian-development peace nexus what does it mean for multi-mandated organizations? (2019), https://reliefweb.int/sites/reliefweb.int/files/resources/dp-humanitarian-development-peace-nexus-260619-en_0.pdf

J.C. Ramo, *The Age of the Unthinkable: Why the New World Disorder Constantly Surprises Us and What We Can Do About It* (Little, Brown and Company, New York, 2009)

R. Ramirez, A. Wilkinson, *Strategic Reframing: The Oxford Scenario Planning Approach* (Oxford University Press, Oxford, 2016)

N. Silver, *The Signal and the Noise* (Penguin Books, New York, 2015)

D. Schön, *The Reflective Practitioner: How Professionals Think in Action* (Temple Smith, London, 1983)

P. Senge, *The Fifth Discipline: The Art and Practice of the Learning Organization* (Doubleday Currency, New York, 1990)

D. Snowden, Complex knowledge, in *Building the Knowledge Economy: Issues, Applications, Case Studies*, p. 805 (2003)

D.L. Snowden, M.E. Boone, A leader's framework for decision making (cover story). Harv. Bus. Rev. **85**(11), 68–76 (2007)

D. Spann, Coping with the unexpected: integrating HRO and agile. Cutter IT J. **19** (2006)

J.D. Sterman, *Business Dynamics: Systems Thinking and Modeling for a Complex World* (McGraw-Hill Publishing, Boston, 2000)

J.D. Sterman, Learning from evidence in a complex world. Am. J. Public Health **96**, 505–514 (2006)

J. Urry, The global complexities of September 11th. Theory Culture Soc. **19**(4), 57–69 (2002)

K.E. Weick, Sensemaking in Organizations. Thousand Oaks, California, Sage Publications Inc. (1995)

K.E. Weick, K.M. Sutcliffe, *Managing the Unexpected: Resilient Performance in an Age of Uncertainty*, 2nd edn. (Wiley, San Francisco, 2007)

D.D. Woods, How to design a safety organization: test case for resilience engineering, in *Resilience Engineering: Concepts and Precepts*, ed. by E. Hollnagel, D.D. Woods, N. Leveson (Ashgate Publishing, Hampshire, 2006)

World Bank, *Pathways for Peace: Inclusive Approaches to Preventing Violent Conflict* (World Bank, Washington, DC, 2018), https://openknowledge.worldbank.org/handle/10986/28337

C. Zürcher, What do we (not) know about development aid and violence? A systematic review. World Dev. **98**, 506–522 (2017)

C. Zürcher, The folly of "aid for stabilisation". Third World Quart. **40**(5), 839–854 (2019)

Pre-revolutionary Iran and Great Power Rivalry

Mohsen Milani

Abstract In its long and fascinating history, Iran has experienced being a great global power, a formidable regional power, and the battle ground for great power rivalries. In ancient times, its empires were formidable global powers that were engaged in competition with other global powers. After the advent of the Industrial Revolution in the second half of the eighteenth century, however, a vulnerable Iran became the Great Prize in the Great Game by the Great Powers. In what follows, I will discuss the nature and dynamics as well as the dire consequences of different approaches prerevolutionary Iran utilized in the past few centuries in its tumultuous dealings with great power to protect its interests, territorial integrity, and independence. This historical analysis provides a valuable window through which we can better understand the evolution of the mind-set of Iranian leaders, including those of the Islamic Republic of Iran.

Keywords Iran · History · Revolution · Shah · US relations · Diplomacy · Great power competition · Great power rivalries

Introduction

Iran, or Persia as it has been known for millenia, is no stranger to global politics.[1] Iran belongs to a small group of countries that has experienced being a great global power, a formidable regional power, and the battle ground for great power rivalries. Iran was not, like the overwhelming majority of the Middle Eastern countries, created

[1] Most of the information in this section is from Milani 1988a, b and Milani 2013. In 1935, the Persian government requested from all foreign governments to use Iran, instead of Persia, in their official correspondence. Iran is a cognate of Aryans or the land of Aryans. Persia is derived from the Greek word Persis, which refers to the Pars province. Both names refer to the same country. I have used Persia throughout the chapter to discuss events prior to 1935. See Ehsan Yarshater, "Persia or Iran." *Iranian Studies* XXII, 1989.

M. Milani (✉)
University of South Florida, Tampa, United States
e-mail: milani@usf.edu

© The Author(s), under exclusive license to Springer Nature Switzerland AG 2021

A. Farhadi et al. (eds.), *The Great Power Competition Volume 1*,
https://doi.org/10.1007/978-3-030-64473-4_9

by the whims of Western powers after World War I. Iran, as a country and a civilization, is one of the oldest in the world—older than the West itself. During its long and turbulent history, Iran has experienced glory as well as defeat and humiliation. Three of its ancient empires were formidable global powers on their own right and were engaged in intense rivalry with other global powers. After the advent of the Industrial Revolution in the second half of the eighteenth century, however, feeble but strategically important, Iran became the Great Prize in the Great Game by the Great Powers, the United Kingdom and Russia. Still, Iran, unlike the vast majority of the developing countries, skillfully resisted being colonized by the West and retained its independence, even though it lost a vast swath of its territories.

In what follows, I will discuss the dynamics as well as dire consequences of different approaches that the prerevolutionary Iran utilized in its tumultuous dealings with foreign powers to protect its interests, territorial integrity, and independence.

Ancient Persia as a Global Power

Persia's debut as a global power took place during the Achaemenid dynasty (circa 546–334 B.C.) (Frye 1963 and Milani 2013). They created the "largest empire the world had seen" (MacGregor 2012). It stretched "from the coasts of Asia Minor and Libya to the deserts of Central Asia and the Indus River system" (Allen 1971). At its height, the empire covered a vast territory that today encompasses twenty-eight nations. To manage the empire, the Achaemenids created a centralized system of governance and communication. Their foreign policy was grounded on protecting and expanding the empire, carrying out military expeditions, and conducting diplomacy. The most powerful legacy of their governance philosophy, however, was the refusal to impose their religion of Zoroastrianism on the conquered people.[2] In the biblical books of Ezra and Nehemiah, Cyrus is praised as an "anointed" king who liberated the Jews from Babylonian captivity and allowed them to rebuild their temple in Jerusalem. In fact, the "Torah was established as the law of Israel" during the Persian rule (Milani 2013).

The Achaemenids were often in competitive conflict with the much smaller Greek city-states. Persia often relied on creating new divisions and accentuating the old cleavages among the Greek city-states to push its imperial ambitions.[3] Such intrigues, as well as the Greco-Persian Wars, became the raw materials from which the rather biased Greek historiography about Persia was developed for a Western audience.

The Achaemenid's rule ended when Alexander from Macedonia invaded Persia and plundered its huge treasure (circa 334 B.C.). His successors, the Seleucids,

[2]Zoroastrianism, one of the first monotheistic religions, started by Persian Zorastra or Zarathustra (circa 650 B.C.). It has had huge impact on all Abrahamic religions. Such notions as the messiah, hell, heaven, and evil are said to have come from Zoroastrianism.

[3]For a brief history of the Achaemenid Dynasty, see: https://www.iranicaonline.org/articles/achaemenid-dynasty.

ruled Persia for eight decades. They were overthrown by Parthians, who established another large Persian Empire (circa 247 BC to 224 AD). This time around, Persia faced a new Western power, the Roman Empire. Conflict and rivalries between the two empires continued and even intensified when Persia became a strategic bridge between the Roman Empire and China in the lucrative Silk Road.

Ardeshir ended the Parthians' long rule, and subsequently created the vast Sassanid Empire (circa 226–240) that "reflected a cultured and luxurious civilization" (circa 226–642 A.D) (Milani 2013). Unlike the enlightened Achaemenids who did not impose Zoroastrianism on the conquered people, Ardeshir established Zoroastrianism as the state religion and fused religion with politics. After the Roman Emperor Theodosius I made Christianity the Empire's official religion around 380 A.D., conflict between the two rival empires took an increasingly religious coloring. These religious conflicts increased after the Byzantine Empire (395–1453 A.D) split from Rome and moved its capital much closer to Persia.

The Persian and Byzantine empires were so distracted by their never-ending wars that they paid scant attention to the rise of the new religion of Islam in the Arabian Peninsula. They paid heavily for this historic neglect as both suffered humiliating defeats at the hands of the new Islamic army. In 637, after the Prophet Mohammad's death, the Islamic army defeated and occupied Persia, opening a new chapter in Persian history.

Persians converted to Islam slowly. However, when Abu Moslem-e Khorasani, a Persian warrior, helped the Abbasid dynasty ascend to power and run the Islamic Empire from Baghdad (*750–1256), their conversion accelerated as they became deeply involved in managing the Islamic empire (Bulliet 1979). As Persia became integrated into the new Islamic empire and civilization, various Persian and Persianate dynasties (non-Persians who adopted Persian language and culture) ruled over parts of Persia. However, an unprecedented calamity befell Persia when Genghis Khan invaded and destroyed much of Persia in 1219. After the invasion, Hulagu Khan, a grandson of Genghis, sacked Baghdad in 1258 and destroyed the Abbasid empire. From the ruin of the Abbasids empire, four empires eventually emerged: the Mughals in the Indian subcontinent, the Ottomans in Turkey, the Uzbek Confederacy of Central Asia, and the Safavids in Persia.

Despite all the calamities, the Persians miraculously protected their identity. Unlike the Egyptians who embraced Islam, adopted Arabic as a new language, and were Arabized, Persians were Islamized but not Arabized and succeeded in protecting their distinct identity, language and many of their traditions. No one contributed more to the survival of the Persian identify than Ferdowsi (940–1020) who wrote his masterpiece, *Shahnameh*, or "Book of Kings". In his 60,000-verse *magnum opus*, he combined history and myths to glorify pre-Islamic Persia, cleanse the Persian language of Arabic, and distinguish Persians from immigrants and invaders. He remains an icon of Persian nationalism.

The Achaemenid and Sassanid empires left a powerful legacy of glory among Persians that persisted until today. Their profound influence took roots in a vast swath of land beyond Iran's current territorial boundaries. For centuries, there has been nostalgia for Persia's past glory. Consider the rise of the Safavid in the early fourteenth century.

The Rise of Safavids as a Major Regional Power and Great Power Rivalry

At the turn of the sixteenth-century, Persia was in chaos and lacked a central authority.[4] For the previous few centuries when the country was invaded and plundered, Persians had resorted to soul searching, introspection, poetry, and mysticism to grasp and cope with their wretched conditions. It was from this chaos, passivity, and despair that the Safavid dynasty rose to power in 1501. They unified the country and, in short order, established an empire that was not as vast as the Achaemenids' or Sassanids', but was significantly larger than present-day Iran. At the time, Iranians were predominantly Sunni, the dominant sect in Islam. To distinguish the Persians from the Sunni Ottomans, who were Persia's main regional rival, the Safavids imposed on the population Twelver Shi'ism, a minority sect in Islam, and empowered its clerics to serve as the religious legitimizers of the new order. Under their rule, Persian art and architecture bloomed, handicraft industries flourished, and Shi'ism became an inseparable component of Persian nationalism. Additionally, Persia became a formidable regional power and exceedingly engaged in the power rivalries among new Western powers competing for global dominance (Amanat 2017).

At the beginning of their rule, the Safavids attempted to exploit great power rivalries to advance their ambitious agenda (Milani 1988a). Their goals were to undermine their Ottoman rivals and Portugal, the first Western power to establish a naval presence in the Persian Gulf in 1506. They sought cooperation from Spain and Hungary to weaken the Ottomans whose troops were threatening Europe with military advances. Having failed, they turned to the British who were also interested in a friendship with Persia. The serious interactions between the two countries started under Shah Abbas the Great (1588–1666). He enthusiastically received the two Shirely Brothers and their small delegation to his court and commissioned them with the urgent task of modernizing the Persian army. This wise move had positive repercussions for Persia. With critical support from the British, Persia finally expelled the Portuguese from the Persian Gulf in 1622 and reestablished its presence in that region (Savory 1980). Moreover, Persia scored a significant military victory in its lingering wars with the Ottomans (1603–18). Cooperation with Persia also paid generous dividends to the British For one thing, they became increasingly involved in Iranian politics and effectively used the country as a counterweight to restrain an ambitious Ottoman Empire which was posing a major threat to Europe. For another, Portugal, Britain's

[4]Some of the ideas in this section are or quoted from my "Iran's Ambivalent World Role," in *Comparative Foreign Policy*, edited by Steven W. Hook. New Jersey: Prentice Hall, 2002, pp. 219–244; and from *The Making of Iran's Islamic Revolution*. Boulder CO: Westview Press, 1988.

main naval competitor, was expelled from the Persian Gulf, and Britain established its first strategic footprint in that region—a development that expedited its rise as a dominant global power. Persia's passing alliance with Britain was arguably one of its most successful cases of using great power rivalry to achieve its own goals.

Persia was able to take advantage of power rivalries among great powers to advance its interests precisely because the power gap separating Persia and the West was not huge during the Safavid era.[5] That power gap, however, began to progressively widen at the beginning of the eighteenth century as Europe was transformed by the industrialization and democratization processes. These twin changes made Europe politically more vibrant and economically, militarily and technologically more powerful than the rest of the world. Exactly as Europe was taking off, circumstances in Persia were becoming unsetted. The country experienced three dynastic changes, incursions by the Ottomans, Russians and Afghans, and the futile foreign conquests by its ambitious shahs.

Shrinking of the Empire and Great Power Rivalry in Persia

By the time the Qajar dynasty (1796–1926) finally restored internal order, Persia found itself lagging far behind the West and Russia. Consequently, Persia became increasingly dependent on Britain and Russia and its capability to take advantage of great power rivalries was seriously diminished. Still the Qajars attempted, with limited access, to pit the great powers against each other by giving them major concessions and thus precluding either one to dominate Persia. In the process, Persia lost some of its fertile territories.

The two dominant foreign powers in Persia, Russia and Britain, had their distinct interests which sometimes converged and sometimes diverged. Both were interested in Persia's markets and its rich natural resources. Both sought to shrink the territories under the Persian rule, or "to cut down Persia to size." While Russia's strategic goal was to gain access to the Persian Gulf through Iran, which the British vehemently opposed, Britain's also aimed to make Iran a buffer zone to protect its most prized colony in India.

To restrain the growing power of these two foreign powers, Fath Ali Shah Qajar (1797–1834) relied on the "third force" strategy by forming a loose alliance with Napoleon I of France (1807–1809). In 1805, French diplomat, Amedee Jaubert, visited Persia with an offer of alliance from Napoleon. Persians, mesmerized by Napoleon's successes on the battle fields, signed the France-Persian Treaty of Finckenstein in 1807 (Amanat 2017). Persia lost this gamble and its loose alliance was quickly shattered even before Napoleon was decisively defeated at the Battle of Waterloo in 1815.

---

[5]For the most comprehensive analysis of Iranian foreign policy, see R.K. Ramazani, *The Foreign Policy of Iran: A Developing Nation in World Affairs, 1500–1941*. Charlottesville, University Press of Virginia, 1966.

Persia's powerlessness and the utter ignorance of its leaders about modern warfare became tragically exposed when Persia lost two major wars to Russia in 1803 and 1828. Humiliated, Persia was compelled to sign the devastating treaties of Golestan and Turkamanchay, pay significant war reparations, cede to Russia a large portion of its territories, and grant extraterritorial rights to the victor. Thereafter began the gradual shrinking of the Empire that the Safavids created. Not only the British did not raise an eyebrow about Persia's massive territorial losses, but they moved troops to southern Persia and practically controlled that region. More importantly, their troops defeated the Qajars and forced Persia to sign the 1857 Paris Peace Treaty in which Persia renounced its claims over Herat and other parts of Afghanistan (Amanat 2017). Not to be outmaneuvered by the Russians in Persia, the British successfully pressured the venal Qajar kings to grant them lucrative concessions as well. For the rest of their rule, the Qajars placed Persia on what seemed like an auction for the highest bidder, matching every concession given to Russia by a similar or a more lucrative concession granted to Britain (Keddie 1980).

Britain as the Dominant Foreign Power in Persia

By the turn of the twentieth century, Persia's economy was in shatters and its corrupt government bankrupt and controlled by Russia and Britain (Milani 1988b). Persia was a sovereign country, but only in name. To improve this pathetic state of affairs, merchants, intellectuals and clerics formed an alliance and launched the Constitutional Movement of 1905. Its key goals were to produce Persia's first written constitution, establish its first parliament, or Majles, and loosen the shackles of foreign domination of Persia. In that spirit, the Majles in 1911 employed the "third force" strategy to undermine Russian and British power. Consequently, the government hired an enlightened American, Morgan Shuster, to reform the country's archaic finances. Unknown to the constitutionalists was the signing of an infamous and secret Anglo-Russian Agreement in 1907 that practically divided Persia into British and Russian zones of influence and made a mockery of Persian independence. Predictably, Britain and Russia joined hands and compelled the inept Persian government to dismiss Shuster and shut down the parliament.[6] The government's utter helplessness was further exposed during World War I when Persia, despite its official declaration of neutrality, was invaded by the two warring sides.

Ironically, the October 1917 Bolshevik Revolution pushed Russia to the periphery of Persian politics and paved the way for the British to become the sole dominant power in Iran. Their sinister scheme for Persia was manifested in the 1919 Anglo-Persian Agreement. The agreement was never ratified by the Majles and after much public anger by nationalists, and even some Western powers, was declared null

[6]Upon returning to the United States, Shuster wrote a scathing criticism of the British and Russian intrigues in Persia. See W. Morgan Shuster, *The Strangling of Persia* (Mage Publishers, 1987). The book was originally published in 1912.

and void. Designed by Lord George Curzon, the contemporaneous Secretary of State for Foreign Affairs who had penned a two-volume tome about Persia, the agreement was a blatant attempt to turn Persia into a British protectorate by placing Britain in total charge of the county's political, economic and military affairs. For Curzon, Persia was the most pivotal country for stopping the spread of communism and preventing the Bolsheviks from threatening the British-controlled Mesopotamia and the British colony in India and elsewhere.

The Bolsheviks were fully aware of the British machinations and tried in vain to prevent it from dominating Persia. The revolutionary government in Moscow unliterally terminated all previous "imperialistic" agreements with Persia and signed a new Treaty of Friendship with Persia in 1921. Persia welcomed Moscow's goodwill and concessions, but its Anglophile politicians were uninterested in opposing the British. Even though the 1919 Angelo-Persian Agreement was rejected, Britain nevertheless achieved some of its goals through other means.

It was in these tumultuous times that Reza Khan, the commander of the British-controlled Cossack Brigade, staged a coup in 1921 and quickly reestablished internal order (Ghani 2000). In five years, the Majles elected Reza Khan as the new Shah, and established the Pahlavi dynasty (1926–79). The British were irrefutably behind the coup because they saw in Reza Khan an authoritarian leader who would support most of the strategic objectives enunciated by Lord Curzon. It is most revealing that shortly before the coup, General Edmund Ironside, Commander of British forces in Iran, had prompted Reza Khan to lead the Cossack Bridge, a force that was created in 1879 and was under Russian control until 1920.

The British calculations about Reza Shah were only partially prescient. Reza Shah centralized authority, restored stability, modernized the economy and the armed forces, protected Persia from communist temptation, and lowered Soviet influence. In foreign policy, Reza Shah hardly challenged Britain: When the Iraqi Shias staged a massive uprising against the British rule in Mesopotamia, he did not raise a finger to help the Shi'a brethren.

However, Reza Shah, much to London's chagrin, had a strong nationalistic proclivity, albeit in its most authoritarian genre. He became less subservient to the British as he consolidated power. He was a visionary who laid the foundation of a modern state and launched an impressive campaign of industrialization and modernization that improved the standard of living of its population. At the same time, he explored every opportunity to curtail British influence. In 1928, he ended the humiliating "Capitulation Laws" that gave legal immunity to Europeans and Russians living in Persia. He nationalized the country's customs, took away the British right to print the Iranian currency, and, instead, granted it to a newly-established bank. He also renegotiated a new oil contract with Britain that slightly increased Iran's share of profits from the Anglo-Persian Oil Company, which controlled Iran's expanding oil industry.

Reza Shah, like many shahs before him, used a mild form of "third force" strategy to weaken the British and further marginalize the Soviet Union in Persia. At first, he enticed the United States to become involved in Persia. He invited Arthur Millspaugh, a U.S. State Department advisor, to reform the country's finances. Iran even signed an oil agreement with a small American oil company. However, the U.S. had no appetite for any engagement in Persia and was determined not to alienate Britain, its main ally and the world's preeminent power.

Reza Shah also flirted with Nazi Germany to undermine Britain and the Soviets, a perilous move that became his Achilles heel. In 1935, as Hitler was spreading propaganda about the superiority of the Aryan race, Reza Shah officially requested foreign governments to use Iran, or the land of the Ayrans, instead of Persia in their correspondence. This move was interpreted by the British as an indication of Reza Shah's sympathy with German fascism. There were other reasons for British anxiety, such as the strengthening of Iran's economic and political ties with Germany. Thousands of German advisors and civilians helped built the 850-mile Trans-Iranian Railroad, perhaps Reza Shah's most consequential developmental project. Britain regarded these advisors as intelligence agents and demanded their immediate departure. At first, the Soviets, who had signed the disastrous German-Soviet Non-aggression Pact of 1939, tolerated the German presence in Iran. However, once the futility of their misguided pact became abundantly clear, the Soviets, too, demanded the departure of German advisors from Iran. Although Reza Shah agreed to this demand, the Allies still accused him of pro-German sentiments. Therefore, he was forced to abdicate in 1941 and was subsequently exiled. His young son, Mohammad Reza, inherited the Peacock Throne at a time when Iran was occupied by the British, Soviet, and American armies.

A New Foreign Policy Paradigm and End of x British Domination of Persia

From the beginning of his ascension to power, Mohammad Reza Shah sought to fundamentally change the direction of Iranian foreign policy. Having witnessed first-hand how the British and, to a lesser degree the Soviets had humiliated his father, the Shah hoped to entice the United States to become engaged in Iran. Applying the "third force" strategy, he viewed the United States as a powerful partner and a counterweight against the British and Soviet's nefarious aspirations to dominate Iran. What irrevocably persuaded him to follow this strategy was the positive role the United States played in helping Iran resolve its first major internal crisis after WWII. Not only did the occupying Red Army refuse to leave Iran, as it was obligated to, but it had also shamelessly assisted separatists to establish two puppet governments in the Iranian provinces of Azerbaijan and Kurdistan. Thanks to the generous support of the United States and the adroit diplomacy by the Shah and Prime Minster Ahmad Qavam, the Red Army was pressured to leave the country and the Iranian military crushed the two secessionist governments. Henceforth, the Shah became determined to form a strategic alliance with the United States to consolidate power at home, expand influence abroad, and diminish Soviet and British power in Iran. However, at that time, he was too timid and too inexperienced to achieve his goals. Ironically, he achieved them after defeating his most formidable foe in Dr. Mohammad Mossadegh, who served as prime minister for 27 eventful months.[7]

Mossadegh championed a new foreign policy paradigm in the early phase of the Cold War based on liberal nationalism, democracy, and friendly relations both with the West and the East to safeguard the country's independence. This paradigm eventually became the ideological foundation of the Non-Aligned Movement. A French educated Qajar aristocrat, he saw a symbiotic relationship between domestic despotism and the capability of great powers to undermine Iran's independence. On one hand, he called for democracy and restoration of constitutionalism at home. On the other hand, he made a compelling argument that without the nationalization of the oil industry, which the British controlled, Iran could not enjoy genuine independence or democracy. To some degree, he achieved both goals during his short tenure.

As a deputy in the 14th Majles from 1944–1946, Mossadegh championed the policy of "Negative Equilibrium." Intoxicated with their World War II victory, an assertive Britain and Soviet Union were demanding major oil concessions. Their boisterous supporters sought to establish "positive equilibrium" by demanding that the Majles should grant similar concessions to both powers. Mossadegh, on the other hand, sponsored a bill that disallowed all concessions to foreign powers unless approved by the Majles. Emboldened by the massive popular support he had received, he introduced the bill to nationalize the oil industry. He was subsequently nominated by the Majles to become prime minister and implement the nationalization project. A reluctant Shah approved the Majles decision in 1951 (Cottam 1964).

Mossadegh skillfully linked control over the oil industry to the Iranian sense of national pride. This message resonated with a large swath of the population and their recollection of their recent history. Oil was first discovered in Iran in 1907 by a British subject who had received a concession from the Qajars. When the Royal Navy changed the source of its fuel from coal to oil in 1915, the British government became a major stockholder in the Anglo-Persian Oil Company. Thereafter, the British did not relinquish their unequivocal jurisdiction over the company, whose name Reza Shah had requested to be changed to the Anglo-Iranian Oil Company (AIOC). Gradually, the company became Britain's largest and most profitable foreign investment. Mossadegh considered it an abject national humiliation that Iran had no right to scrutinize the AIOC's accounting books and was receiving less revenues than the taxes the British government was collecting from the company.

Mossadegh's bold nationalization project was a devastating blow to the foundation of the British power in Iran. With its prized Indian colony gaining independence in 1947, Britain relentlessly applied every conceivable method to undermine Mosaddeq. They were terrified that a successful nationalization project in Iran could inspire others to emulate Mossadegh.

Mossadegh was well-aware of the British intrigues. Recognizing that he was fighting against the world's major global power, he expanded his popular base of support at home, and counted on the United States, with its legacy of anti-colonialism,

[7]For the biography and legacy of Mossadegh, see Christopher de Bellaigue. *Patriot of Persia: Muhammad Mossadeqh and a Tragic Anglo-American Coup*. HarperCollins, 2012.

as a potential ally against Britain. He traveled to Washington, met President Harry Truman, and made a futile request for a loan. Truman was sympathetic toward Iran and proposed "nationalization with compensation," which supported nationalization but demanded that Iran pay compensation to the British. Mossadegh rejected the proposal, maintaining that it was the British who must compensate Iran.

A decisive moment came when General Dwight Eisenhower became president on January 20, 1953. Unlike Truman who did not view Iranian nationalism as a threat, Eisenhower looked at that movement entirely from the lens of the Cold War and dismissed it as a menace. His perspectives were somewhat identical to that of Prime Minister Winston Churchill, whose animosity toward Mossadegh was well-known. He believed Mossadegh was too soft on communists and feared that the Soviet Union could ultimately emerge as the real victor in Iran. He conveniently ignored that Mossadegh was neither anti-West nor anti-Soviet Union and endeavored to establish good relations with both. What Washington and London opposed was Mossadegh's refusal to entangle Iran deeply in the evolving Cold War. The exaggerated fear of "Iran going communist," the insatiable appetite of the American oil companies to enter into the lucrative Iranian oil industry, and Mossadegh's refusal to compromise with the West untied Washington and London to overthrow Mossadegh. Thus, through a joint coup operation by the Central Intelligence Agency and British Military Intelligence, called operation Ajax, Mossadegh, whose popular base of support had shrunk, was overthrown.[8]

Pax-Americana in Iran

Immediately after the coup, Mohammad Reza Shah, who had fled the country a few days earlier, returned home and gradually became Iran's undisputed autocrat.[9] In retrospect, the coup was a pyrrhic victory, for it sowed the seeds of the 1979 revolution that ended the Pahlavi rule as well as U.S. relations with Iran. The coup had negative consequences: The Shah lost political legitimacy and was perceived by many Iranians as a U.S. puppet, and the United States lost credibility for toppling a democratic and popular prime minister. Both were so mesmerized by the success of the coup that they overlooked these adverse consequences. While the Shah admired the United States as a global power capable of protecting his Peacock Throne, the United States judged the Shah to be a valuable asset in its Cold War against the Soviet Union. That Iran shared more than 1,700 miles of common borders with the Soviet Union further augmented its strategic appeal for the United States.

---

[8]For the 1953 coup, see Mark Gasiorowski, and Byrne, M. eds. *Mohammad Mossadegh and the 1953 Coup in Iran*. Syracuse: Syracuse University Press, 2004; and Ervand Abrahamian. *The Coup: 1953, the CIA and the Roots of the Modern U.S.-Iranian Relations*. New York: The New Press, 2013.

[9]For the best biography of Mohammad Reza Shah, see Abbas Milani. *The Shah*. New York: Palgrave 2011.

The coup also marked the end of Britain's preeminent position in Iran and the emergence of the United States as a new hegemonic foreign power. The post-coup government, fully supported by the United States, jailed Mosaddeq and kept him under house arrest until he died in 1967, killed and jailed many of his supporters, banned unfriendly parties, and suppressed the press. Simultaneously, the government received $45 million in financial assistance from Washington to improve the country's anemic economy. With U.S. support, Iran's security forces were centralized and modernized: The national police was reorganized in 1954 and the infamous secret police, SAVAK, was created three years later. The Iranian oil industry was practically "denationalized" and a new oil consortium was formed, and the Anglo Iranian Oil Company was renamed British Petroleum. The new consortium comprised mostly of British and, for the first time, American oil companies. Many people saw the involvement of American oil companies as the price the Shah's government had to pay for America's critical role in the coup.

The Shah, the sole architect of the country's foreign policy, formed a strategic alliance with the United States to consolidate his rule at home, radically change the orientation of the country's foreign policy, and expand power abroad.[10] Iran quickly became an integral part of the U.S. Perimeter Defense Strategy, which established alliances with countries bordering the Sino-Soviet bloc (Milani 1988a). Iran, for example, joined the anti-Soviet Baghdad Pact in 1955. Most significantly, Iran and the United States signed the Joint Defense Agreement in 1959, an agreement that terrified the Soviet Union. The historic agreement substantially increased U.S. military presence in Iran, gave Washington the leading role in modernizing the country's armed and security forces, and guaranteed U.S. support if Iran were to be attacked.

The nascent strategic alliance was not free of tensions. In the early 1960s, President John Kennedy pressured the Shah to launch political and land reforms to immunize Iran to a Chinese-style peasant revolution. The Shah not only championed a progressive land reform, but he also initiated modest political reforms, such as granting suffrage to women. As he was initiating reforms, the Shah was challenged by an obscure religious leader named Ayatollah Ruhollah Khomeini. In October of 1964, the Majles approved a controversial legislation that granted American military personnel and their dependents legal immunity (the Status of Forces Agreement), known in Iran by its pejorative label of Capitulation Laws. The new law touched a raw nerve in the consciousness of Iranians and reminded them of the past British and Russian manipulations. It was mystifying that the Shah supported this legislation, even though his father, Reza Shah, had terminated the "Capitulation Laws" in Iran in 1928. Once the new bill was passed, Iran received a $200 million loan from the United States to buy American military hardware. Opponents of the Shah saw the loan as a bribe for the passage of the "Capitulation" legislation.

Although these twin actions by the Majles were vigorously condemned by nationalists and leftists, it was Khomeini who stole the show and emerged as the most outspoken and intrepid opponent of the Shah and his alliance with the United States.

[10]For a favorable analysis of the Pahlavi era, see George Lenczowski, ed. 1978. Iran under the Pahlavis. Stanford, CA: Hoover Institution Press.

He bitterly denounced the Shah for turning Iran into a "U.S. colony" and lambasted him for developing close relations with Israel. In significant ways, the roots of the Islamic Republic of Iran's anti-Americanism can be traced back to Khomeini's rise as the uncompromising enemy of the Shah and the United States. When Khomeini's devotees took to the streets in opposition to the Shah's policies, the government mercilessly crushed them. The event is remembered as the June Uprising of 1963 and was a dress rehearsal for the Islamic Revolution of 1979. The Shah, backed by the United States, was in no mood to tolerate the cantankerous Ayatollah and exiled him to Turkey in 1965. Fourteen years later, in 1979, Khomeini triumphantly returned to Iran and exiled the Shah.

Having temporarily eliminated the threat posed by Khomeini, the Shah became even more resolute in deepening his multifaceted relations with the U.S. The problem was that the United States looked upon him more as a subservient client who, in the words of Gary Sick, "had always been kept on a short leash" (Sick 1985). The Shah had probably sensed this disparaging sentiment toward him but hoped to change it. His wish came true as three major events in the late 1960s and the early 1970s turned him into Washington's valuable junior partner, rather than its subservient client: The withdrawal of the British forces from the Persian Gulf; the United States' decision not to become directly involved in the Persian Gulf; and the astronomical increase in the price of crude oil in the early 1970s. As a consequence of these changes, the Shah found more maneuverability space and independence to pursue his ambitions goals.

In 1967, Prime Minister Harold Wilson announced that the UK would withdraw its forces from the Persian Gulf by the end of 1971. The Shah confided in his powerful advisor, Alam (1993), and asked him to inform the Americans that he strongly opposed the military presence of any Western power in the Persian Gulf and intended to fill the power vacuum that would inevitably be created by the British withdrawal. At that time, the United States was intensely engaged in the Vietnam War and the American people had no appetite for another military intervention. Consequently, the Nixon Administration designed its new Twin Pillar Policy, formulated in the classified National Security Study Memorandum (NSSM) 66 in 1969. Accordingly, the U.S. "would not assume the former British role of protector in the Gulf area, but the primary responsibility for peace and stability should henceforth fall on the states of the regions" (Clare 1984). With its new policy, the U.S. increased its military sales to both Iran and Saudi Arabia so that they could play the new role assigned to them and deter Soviet advances. However, Iran, by the virtue of its size, ideal strategic location, much larger population and economy, and much bigger armed forces, played a much bigger role in this strategy than Saudi Arabia did (Alvandi 2014). The Shah, who had a close personal relationship with Nixon, emerged as the "policeman" or the "gendarme" of the Persian Gulf. He cherished this new status as a requisite stepping-stone to become a player on the global stage.

Convinced that without a powerful military, Iran could not expand its power, the Shah used the petrodollar to undertake an unprecedented militarization program, arming the country to the teeth. By the mid-1970s, Iran became the world's number one buyer of American weapons. Kissinger recalls that after 1972, "we adopted a policy which provides, in effect, that we will accede to any of the Shah's requests for arms purchases from us" (Sick 1985). In fact, Iran had a carte blanche to buy any weapons, except nuclear weapons, without going through the customary review process for arms sales conducted by the Defense Department. Between 1970 and 1978, according to Michael Clare, Iran was "the leading overseas customer for U.S. arms and accounted for 25% of all FMS [Foreign Military Sales] orders received during this period" (Clare 1984). In the words of one American lawmaker, Representative Gerry Studds, the military build-up by the Shah was "the most rapid build-up of military power under peacetime conditions of any nation in the history of the world" (Clare 1984).

The massive militarization emboldened the Shah to expand Iran's regional power and its global influence. Dr. Henry Kissinger, former U.S. Secretary of State, recalls that "On all major international issues, the policies of the United States and Iran have been parallel and therefore mutually reinforcing" (Bill 1988). With U.S. and Israeli support, he provided financial and military support to Iraqi Kurds to weaken Iraq, which had signed a friendship treaty with the Soviet Union in 1971. Despite the mild opposition by the United states and Israel, the Shah signed the 1975 Algiers Accord with Iraq's Saddam Hossein and received major concessions. For the first time in decades, he also sent Iranian troops to Oman to defeat the Moscow-supported rebels.

Iran's economic relations with the United States changed as well. Thanks to the sudden increase in oil prices, Iran, which had previously been the recipient of U.S. financial assistance and loans, became independent of U.S. aid. Moreover, in the 1970s, Iran experienced impressive economic development, unparalleled in its recent history. The Shah even began to invest in the U.S. and other Western countries. Iran became an irresistible magnet for American and Western entrepreneurs. In 1975, Iran and the United States signed an economic agreement that obligated Iran to buy $15 billion in goods and services from the U.S. in the next five years. This was the largest deal any two counties had ever signed (Bill 1988). Hundreds of American companies, small and large, rushed to Tehran to sign new contracts. Distressed Western companies went to the Shah for being bailed out. American companies were granted contracts to build eight nuclear power plants in Iran. On the eve of the 1979 revolution, there were more than 50,000 Americans living in Iran, perhaps more than any other Middle Eastern country except Israel. At the same time, there were more than 55,000 Iranian students studying in the U.S.

While the Shah was developing a strategic alliance with the United States during the Cold War, he also pursued a complex and successful policy toward the Soviet Union. He was exceptionally vigilant not to overtly antagonize Moscow. Throughout his rule, he feared the possibility of a Soviet incursion into Iran or a communist takeover of the country, or a coup, spearheaded by the Moscow-controlled Tudeh party and other leftists. In fact, the Moscow-controlled Tudeh was the largest and most organized party in the country and the principal conveyer of leftist ideologies during most of the 1940s and early 1950s. What increased the Shah's deep suspicions of the Soviet plans was the 1953 discovery of a relatively large secret networks of the Tudeh members and sympathizers within the armed forces. That network was, of course, dismantled by the government.

To appease Moscow, the Shah improved relations with the Soviet Union and occasionally used the Soviet card to receive concessions from the United States, particularly in buying America's most advanced weapon systems. This balancing act was not always easy. When Iran and the United States signed the Joint Defense Agreement in 1959, which gave the U.S. the lending role in modernizing Iran's armed forces, Soviet hysteria reached its climax. The agreement shattered any illusions the Soviets might have had about possible Iranian neutrality during the Cold War and seemed to have convinced them that the Shah's regime must be undermined. Therefore, they pressured and threatened the Shah to reconsider its close alliance with Washington. The single most consequential concession the Shah gave Moscow during his entire reign was the 1962 Non-Aggression Treaty between Iran and the Soviet Union. The agreement, which the U.S. reluctantly tolerated, practically prohibited Iran from ever allowing the United States to use Iranian territory as a base to launch missiles and nuclear weapons against the Soviet Union. It also laid the basis for a much less contentious relationship between the two countries. The agreement, however, did not prevent the United States from placing its most super secretive electronic listening devises in Bishahr and Kapkan in northern Iran to monitor Soviet activities (Bill 1988).

Soon after the Non-Aggression Treaty, Moscow provided a $286 million loan to Iran in 1966 and built its first steel mill, which the West had refused to build. The Soviet Union also completed the Trans-Iranian Gas Pipeline to carry gas from Iran to the Soviet Union. As Iran's oil revenues increased in the 1970s, the Shah purchased limited quantities of weapons from the Soviet Union. In short, the Shah's policies toward Moscow paid off as there is no evidence that the Soviet Union seriously attempted to undermine his regime in the last decade of his long reign.

End of Pax-Americana in Iran and the Islamic Revolution

In the last two years of the Pahlavi rule (1977–79), Iran-U.S. relations appeared to be based on solid foundations. However, friction surfaced when Jimmy Carter was inaugurated as the president on January 20, 1977. A highly moralistic and religious man, Carter pressured the Shah to liberalize and open up the repressive political system, which the Shah reluctantly did. Carter's inability to find a practical balance between his advocacy of human rights and his responsibility to protect America's enormous strategic and economic interests in Iran became tragically visible when he visited the Shah in Tehran on December 31, 1978. Oblivious to the fact that political unrests were slowly brewing in Iran, Carter toasted His Imperial Majesty's leadership and declared that "Iran, because of the great leadership of the Shah is an island of stability in one of the more troubled areas of the world. This is a tribute to you, your Majesty, and to your leadership and to the respect and the admiration and the love that your people give to you." The Shah listened approvingly as President Carter, the world's most powerful man, was profusely glorifying him, recognizing his "popularity," and making no mention of human rightists. He could not be more pleased.

Carter did not know that in less than 57 weeks after his misdiagnosis of the situation in Iran, a popular revolution would demolish the 2500-year old monarchy and transform Iran and U.S. into bitter enemies. Carter recalls that, according to a CIA assessment in the summer of 1978, only six months before the 1979 revolution, Iran was "not in a revolutionary or even a prerevolutionary situation" (Carter 1983). Of course, revolutions are intrinsically unpredictable phenomena. But there was no justification in concluding that Iran was not even in a prerevolutionary situation. Part of the problem was that the United States was so obsessed with the Cold War and the role Iran was playing in that conflict that it deliberately decided not to pay much attention to Iran's internal situation. The United States knew much about the elites, but knew precious little about the people and the suppressed forces that opposed the Shah. No wonder the revolution shook Washington. The Shah, too, was shaken to his core, for he learned the hard way that internal security cannot be guaranteed or be bought by forming alliances with external forces and that security is ultimately a function of popular support and political legitimacy.

For the first nine months after the 1979 Revolution, the Provisional Revolutionary government and the United States tried hard to continue their relations, albeit with the recognition of the new realities of a revolutionary country. The government did not seek to end Iran's military relationship with the United States, but it did seek to reduce the level of cooperation with the United States and become considerably less involved in the Cold War. However, when the exiled Shah was granted a visa to come to the United States for medical treatment of his terminal cancer, a group of young students stormed the American embassy in Tehran and took all its personnel hostage. In reaction to this blatant violation of international law and diplomatic traditions, Carter severed diplomatic relations with Iran (Milani 2012).

The hostage crisis ended Pax-Americana in Iran. It also marked the start of anti-Americanism as the defining feature of the foreign policy of the Islamic Republic of Iran.

References

A. Alam, *Gofteguhaye Man ba Shah* (My Conversations with the Shah). Translated into Persian by Abdolreza Hoshang Mahdavi. Tehran: Tarh-e Nau

L. Allen, *The Persian Empire* (University of Chicago Press, Chicago, 1971).

R. Alvandi, *Nixon, Kissinger, and the Shah* (Oxford University Press, New York, 2014).

A. Amanat, *Iran: A Modern History* (Yale University Press, New Haven, 2017).

J. Bill, *The Eagle and the Lion: The Tragedy of American-Iranian Relations* (Yale University Press, New Haven, 1988).

R.W. Bulliet, *Conversion to Islam in the Medieval Period* (Harvard University Press, Cambridge, 1979).

J. Carter, *Keeping Faith* (Bantam Books, New York, 1983).

M.T. Clare, *American Arms Superpower* (University of Texas Press, Austin, 1984).

R.W. Cottam, *Nationalism in Iran* (University of Pittsburgh Press, Pittsburgh, 1964).

R. Frye, *The Heritage of Persia* (The World Publishing Company, New York, 1963).
C. Ghani, *Iran and the Rise of Reza Shah: From Qajar Collapse to Pahlavi Power* (I.B. Tauris, New York, 2000).
N. Keddie, *The Roots of Revolution* (Yale University Press, New Haven, 1980).
N. MacGregor, 2600 Years of History in one Object. WordPress (2012). https://vialogue.wordpress.com/2012/02/29/ted-neil-macgregor-2600-years-of-history-in-one-object
M. Milani, Iran's ambivalent world role, in *Comparative Foreign Policy*, ed. by S.W. Hook (Prentice Hall, New Jersey, 1988a)
M. Milani, *The Making of Iran's Islamic Revolution* (Westview Press, Boulder, 1988b).
M. Milani, Hostage crisis, in *Encyclopedia Iranica*, vol. VII (2012). https://www.iranicaonline.org/articles/hostage-crisis
M. Milani, The political system of the Islamic republic of Iran, in *Comparative Governance*, ed. by P. Kirzen (McGraw-Hill, New York, 2013)
A. Millspaugh, *Americans in Persia* (Da Capo Press, Boston, 1976).
R. Savory, *Iran Under the Safavids* (Cambridge University Press, Cambridge, 1980).
G. Sick, *All Fall Down: America's Tragic Encounter with Iran* (Random House, New York, 1985).

Afghanistan—Development Options in the Context of US-Taliban Agreement and Great Power Rivalry

Ashok Nigam

Abstract The history of Afghanistan and great power conflict in the region suggests that many challenges lie ahead for peace and security despite the US-Taliban Agreement of February 2020. On the strong assumption that Afghanistan is able to achieve peace and security, this paper looks at the possible role of the UN and development scenario in the context of great power rivalry and competition given its geographical location. Agriculture including poppy cultivation, is mainstay of the people. But the mineral resource potential of Afghanistan provides possibly the only source of greater economic autonomy. The article argues that UN's role in the next phase would primarily be as facilitator and independent monitoring of the implementation of the peace agreements and reporting on progress in peace and security, beyond the humanitarian and development. Afghanistan has been an aid-dependent country since 2002. Although foreign aid will be needed for many years in the post intra-Afghan agreement phase, the country will need to rely increasingly on its domestic resources as many of the traditional donors retrench not only on military support but also from development assistance. Though Afghanistan will need aid for some years, the record of its ineffectiveness in Afghanistan raises the question of its criticality in the next phase. The mineral sector provides the most promising area for growth, revenue generation and economic independence. To avoid the resource curse, a transparent Public Investment Fund should be set up for all mineral resource income. Mineral processing provides another avenue for employment generation. Agreements on sharing the mineral income should be part of the intra-Afghan agreement. Such a peace dividend could make the chances of sustainable peace and security stronger.

Keywords Afghanistan · Minerals · Rare earth elements · Lithium · Development · Economy · Peace

A. Nigam (✉)
New York, NY 10044, USA
e-mail: akn55ajs@gmail.com

© The Author(s), under exclusive license to Springer Nature Switzerland AG 2021

A. Farhadi et al. (eds.), *The Great Power Competition Volume 1*,
https://doi.org/10.1007/978-3-030-64473-4_10

Introduction

With the signing of the peace agreement between the US and the Taliban, the prospects of a 'form of peace' may emerge in 2020 after 19 years of conflict (US Department of State 2020). The nature of the peace and security remains to be seen with the next phase of intra-Afghan talks on which Afghanistan's future development prospects will rest. Significant uncertainties remain, not least those introduced by the consequences and the priority of Covid-19. This paper analyses the development options on the assumption that peace and security will prevail or at the very least that the situation is better than it has been for the last 18 years. Under this strong assumption the paper examines (i) the possible role of the UN; and (ii) the development options of the nation in the context of progress since 2001 and the great power rivalry that prevails in the region.

The post 2001 period was characterized by hope that with the removal of the Taliban, Afghanistan could develop, as measured by the global indicators of progress. While there has been progress, reliable and disaggregated data on Afghanistan is still very difficult to obtain. This was true back in 2001 and the data situation is probably only marginally better today given the capacity and security constraints in the collection of data. Learning from Afghanistan's history and the uncertainties in peace and security, it would be too ambitious to even attempt an articulation of development options. However, on the strong assumption that the US. Taliban peace agreement holds and the intra-Afghan talks are successful, this paper looks at the possible role of the UN in the political peace process and the development options in the medium-term.

The Political Scenarios and the Role of the UN

The key elements of the Agreement for Bringing Peace to Afghanistan between the United States and the Taliban are fourfold:

1. Guarantees and enforcement mechanisms that will prevent the use of the soil of Afghanistan by any group or individual against the security of the United States and its allies.
2. Guarantees, enforcement mechanisms, and announcement of a timeline for the withdrawal of all foreign forces from Afghanistan.
3. Complete withdrawal of foreign forces and timeline in the presence of international witnesses, and guarantees and the announcement in the presence of international witnesses that Afghan soil will not be used against the security of the United States and its allies, the Taliban will start intra-Afghan negotiations with Afghan sides.
4. A permanent and comprehensive ceasefire will be an item on the agenda of the intra-Afghan dialogue and negotiations. The participants of intra-Afghan negotiations will discuss the date and modalities of a permanent and comprehensive

ceasefire, including joint implementation mechanisms, which will be announced along with the completion and agreement over the future political roadmap of Afghanistan.

While the first three elements will unfold over the next 14 months or so, though these elements may also not unfold smoothly and without violence, it is the fourth element that is perhaps the most uncertain and problematic. Given the history of Afghanistan and the great power rivalry, the odds of peace in the post US-Taliban agreement are not promising. Past political analyses have explored options but these have not materialized.[1] Political analyses will no doubt speculate on the possible outcome and options under alternate political scenarios following the US-Taliban agreement. This paper does not attempt to do so except to note that many challenges lie ahead to peace and security in Afghanistan. On the strong assumption that the US-Taliban peace deal holds, and the intra-Afghan talks are successful, though one hastens to add that it is unclear what would qualify as 'success', a key question is what role can the UN play in this fourth element.

The future of Afghanistan has to be determined by Afghans themselves. The US-Taliban agreement is primarily an exit agreement for the US, not a framework for peace and security in Afghanistan amongst the different warring factions and interests. It is as yet unclear if the outcome will be successful or will end up in further conflict and other regional or global interests supplanting the US and NATO. Many difficult issues remain. Should the intra-Afghan talks be successful, there is a long-standing Afghan process of a Loya Jirga for arriving at decisions amongst the many ethnic groups of the country. It is not an easy process, but it is the traditional respected process. Eventually, any agreement would undoubtedly go through the Loya Jirga.

The Taliban views seems to be that: (i) the current system of administration would continue for the day-to-day operations of the country until such time that the intra-Afghan talks yield an agreement and determination of the next steps; and (ii) the Taliban and representatives of the current Afghan administration/government would draft an entirely new Constitution, although the view of President Ashraf Ghani appears to be that the Taliban should suggest changes to the current constitution. It is yet to be determined as to what would be the final agreement on the approach to be adopted. The victors would probably have much of their way.

Until an agreement is reached, there could be a logjam in the negotiations. This logjam could be (a) with or (b) without conflict. If there is conflict, then it is uncertain what the outcome would be and it may be too early to speculate on next steps. If it is without conflict, or low-grade conflict, then it is worth considering what the next steps could be to address the two questions posed in this paper.

Since 2002, the United Nations has had substantial presence in Afghanistan through the Afghanistan Assistance Mission (UNAMA) and a number of UN agencies. Multiple analyses and reports provide an assessment of the role and effectiveness

[1]See for example "Stefan Olsson, Erika Holmquist, Samuel Bergenwall, Helene Lackenbauer, "Afghanistan After 2014 Five Scenarios" FOI Swedish Defence Research Agency," April 2012.

of the UN presence on the political, development and humanitarian front.[2] The UN has not succeeded as a mediator for peace in Afghanistan[3] and it is unlikely that it would succeed in this particular case largely because of a lack of trust from both sides. So what role can the UN play in the next phase?

The primary role that the UN could play is that of facilitator in the intra-Afghan process—facilitator of logistics and information sharing. It could also support Afghan capacities on both sides in the negotiating process. In the negotiations itself, the two sides would need to find an independent Afghan negotiator. One option would be for the two parties to secure the facilitator services of Zalmay Khalilzad who has been the instrumental negotiator of the peace agreement and would be best placed in a personal capacity. Alternately, other post-conflict situation have relied on individuals or countries such as Norway to facilitate the process. This option is envisaged with the US continuing to play a facilitator role, but the former could well be a fall back, if needed.

Given the complexities of Afghanistan with it many ethnic groups, finding an appropriately knowledgeable and acceptable individual may be challenging. Regardless, the role of the UN would be limited in the intra-Afghan peace negotiations to that of an observer, logistic support, and monitoring. The UN has played a role in a number of countries in the drafting of a constitution, though it would be preferable if an indigenous national process is followed in this case. The primary reason is that a constitution is not a piece of paper that is drawn from the experiences of other countries. It is and must be inherently national and culture specific with understanding of the internal cultural, social and political dynamics of the country. So it would be advisable if a national team is put together to develop the next constitution of Afghanistan to which the UN could provide advisory support with lessons from other countries. The constitution writing process, if it has to sustain, is not a quick process—it will need considerable deliberation and consensus gathering, prepared by Afghans who are respected by the people of the country.

With respect to monitoring of the implementation of the intra-Afghan peace agreement when reached, this too should be an Afghan-led process. Monitoring teams can be set up for each of the provinces comprising of one from the current Afghan administration and one from Taliban supported by a UN official. The role of the UN would be to act as facilitator and independent reporting of the on-the-ground situation supported by technical experts as required. The UN could be the secretariat for logistical issues at the provincial level.

The UN would, of course, continue to be a trusted partner in humanitarian and development support. The UN has a history of working on both these fronts regardless of the government/authorities in power and there is no reason to think it would not

[2]Periodic reports of the Special Representative of the UN Secretary General to the UN Security Council provide information on the role and assessment of the work of the UN in Afghanistan, in particular of UNAMA.

[3]This is evidenced by the fact that despite a political mission in Afghanistan since 2002, the talks with the Taliban were ultimately direct talks with the US with little role for the UN.

continue to do so in the next phase though its scale and spread would depend on funding received from member states.

In the past, the UN has helped to develop strategic directions for the country such as the Afghanistan Compact 2006; Afghanistan National Development Strategy etc. The lesson is that these ended up as paper exercises which could not be implemented. Effective strategic planning requires the full ownership and participation of the relevant Afghan leaders and authorities. Post the intra-Afghan agreements, the UN and World Bank could continue to propose development options for the country for consideration by the relevant Afghan authorities. The second part of this paper proposes approaches to the development option in the context of great power competition.

Political scenario remains uncertain

Given the history and uncertainties in Afghanistan, it is worth considering the current indications of the political scenario that may emerge and its implications which are relevant for the development options as well. Following the intra-Afghan deal, the Taliban would come out of the process in a leading role to dictate the terms and conditions of the peace. While it is perhaps not productive to argue who won and who lost, clearly in the future the Taliban would be in a strong position to dictate the terms of the peace.

On the assumption that 'to the winner goes the spoils', the Taliban may well succeed in dictating a new constitution. Whether there is consensus and a new constitution can be drafted successfully and is accepted by all the ethnic groups is a major question. The UN has played in role in other countries in the drafting of a constitution but a truly national process is required. With the withdrawal of the US troops, the Afghan government will not be in a position to put up any armed resistance to the Taliban. Furthermore, the allegiances of the Afghan Army may well be questionable if the power dynamics in Kabul changes. The Taliban have already proposed the C5 route for much of the current Afghan government officials viz. that all those Afghans who have come from outside the country after 2001, should be put on a C5 aircraft and be airlifted out of the country.[4] This would include all the progressive groups that have been instrumental in improving the economic and social conditions of the country since 2001. It is unlikely that this would happen since the Taliban would be dependent on the C5 group to enable the governance of the country to continue to avoid chaos.

The other factor that is often pointed out is the role that the youth, particularly women, of Afghanistan would play. The last 18 years have seen progress in their education and rights and in particular the rights of girls and women. The youth, it is professed, would not be agreeable to return to the Taliban ideology, even if it is modified from what existed in the Taliban world of 2001. The youth make up about 63% of the country so they should have a sizeable say.[5] While the role of youth and

---

[4]Source: personal interviews with individuals that were party to the talks.

[5]Afghanistan has one of the youngest and fastest growing populations in the world—with approximately 63% of the population (**27.5 million** Afghans) below 25 years of age and 46%

women is very important, their acceptability in determining the nature of the peace agreement remains to be seen. It is not clear as to what kind of Islamic democracy will prevail in the post intra-Afghan world even if elections are held. Further, it is uncertain as to what kind of an Islamic Republic would Afghanistan be and what would be the role of Sharia?

Youth and women have made much progress in their civil and political rights and in education in the non-Taliban held areas since 2002.[6] But the role of women and youth in the post intra-Afghan agreement cannot be underplayed. Youth numbers have grown both in the Afghan Government and Taliban ruled areas. How the role of youth will play out is uncertain, particularly if the Taliban will control all their armed youth. Much depends on how the tribal loyalty plays out which is difficult to predict.

If the C5 group is marginalized and no longer influential in the policy decision making and administration of the country, then the ideology that existed prior to 2001, with whatever changes that may have taken place during this 18-year period that are currently not well known, would be in-charge of policy and administration. Indeed, the governance capacities of the Taliban and of provincial administration have probably not been much enhanced. During this period 2001–2019, the Afghan government has had a heavy reliance on ex-pat Afghans to run the country. Development assistance and also security relied heavily on Afghans who had returned from the West since the international forces and donors felt they could communicate and understand them and trust them more. If this group is marginalized with the ruling ideology dictated by the Taliban, the direction of change and governance capacity is difficult to predict. This injects more uncertainty in the outcome for governance in the post-intra Afghan talks.

Given the uncertainties, the possibility of civil conflict and years of continued security concerns, should there not be an amicable and peaceful solution, cannot be ruled out. Political analysts could speculate on the various possible pressures, scenarios and outcomes, including an Afghanistan with different degrees of autonomy in different regions.[7] Alternately, the predominant ideology and authority over much of the country could be as determined by the Taliban, resulting in what some have argued as a Vietnam situation after the withdrawal of the US.

(11.7 million children) under 15 years of age according to the National Statistics and Information Authority (NSIA). (Source: https://afghanistan.unfpa.org/en/news/young-people-make-their-voices-heard-through-afghan-youth-parliament).

[6]See for example, Cheshmak Farhoumand-Sims, International Journal, "Unfulfilled promises Women and peace in post-Taliban Afghanistan," Summer 2007; and A conversation with Aziz Royesh and Orzala Ashraf Nemat, "Investing in Women's Education in Afghanistan," Yale Journal of International Affairs, Winter 2011.

[7]For a recent analysis see Maryam Jami, "Intra-Afghan Peace Talks: A Channel to Peace," Glocality, 3(1): 1, 1–6. https://doi.org/10.5334/glo.18 and Ashok Behuria, Yaqoob Ul Hassan & Sanya Saroha (2019) US-Taliban Talks for Afghan Peace: Complexities Galore, Strategic Analysis, 43:2, 126–137, https://doi.org/10.1080/09700161.2019.1595483 To link to this article: https://doi.org/10.1080/09700161.2019.1595483.

Added to the internal political dynamics are consideration of great power rivalry. The objective of this paper is not to delve into the external political dynamics of great power rivalry, of which there are numerous analyses, except to outline the possible influence and implications for Afghanistan's eastern neighbor, India. A number of analyses have suggested that a chief loser in the US-Taliban agreement was India with Pakistan's closeness to the Taliban and India's limited influence on them. Furthermore, it is felt that with the closeness of Pakistan to the Taliban, India would be particularly vulnerable to increased terrorist attacks in its state of Jammu and Kashmir (J&K) with support from the Taliban as a quid-pro-quo for Pakistan's support to it. This expectation of increased terrorist activities in J&K has likely been partly mitigated by India's amendment of Article 370 of its constitution under which J&K is now a union territory under direct administration by the central government, though this does not mean that attacks from across the border will cease; the changes in the residency rules in J&K which allows people from other parts of India to get residency status; and the division of the state into two union territories. These actions by the Indian government under the Indian constitution have de-facto ended the pressure for separation of the state from India—an objective supported by Pakistan. India's chief interest with regard to Afghanistan would, therefore, be twofold: (i) ensuring that the Taliban does not join terrorist groups in Pakistan to wage such acts in J&K or other parts of India; and (ii) economic and trade relationship with Afghanistan for mutual benefit, in particular in agriculture and minerals.

Development Outcomes Have Improved Substantially Since 2001[8]

Afghanistan has an estimated population of 32.2 million in 2019. According to the World Bank GDP per capita has increased from $120 to $580. Life expectancy has increased from 44 to 61 years. Under-five mortality rate has nearly halved since 2000. The percentage of people using at least basic sanitation services increased from 23 to 40% from 2000 to 2015 and the percentage of people with access to safe drinking water increased from 27% in 2007–2008 to 65% in 2013–2014. Virtually no one in Afghanistan had access to electricity in 2000, and by 2016, nearly 85% of the population had electricity (CSIS Briefs 2019).

School enrollment has increased from 0.8 million to over 8 million. In 2017, 3.5 million girls were enrolled in school. More than four-fifths of primary-aged girls are enrolled in school. One-third of students enrolled in universities are women (CSIS Briefs 2019). Maternal mortality has decreased from 1600 to 327 per 100,000 births.

---

[8]Much of this section is extracted from the World Bank Group (2016): "Afghanistan to 2030—Priorities for Economic Development under Fragility", World Bank Group, (2016). It draws from this publication for lack of authoritative national sources. In could easily synthesize and present data and analysis from UN-DESA, ESCAP, UNDP and other sources but the objective of this section is not do the analysis of the veracity of the data but to examine potential options for development based on the limited data that is available.

Gender equality has gone from 0 to 27 seats in Parliament. In infrastructure, mobile phone subscriptions have gone from 0 to over 18 million. Phones are being used for mobile payments, among other functions. But national poverty increased from 38.3% in 2011–12 to 54.5% in 2016–17. Over 18 million Afghans now live in poverty. Poverty is concentrated in rural areas but the difference is becoming less pronounced over time due to rural–urban migration. Unemployment and underemployment have increased. The number of Afghan males aged 25–50 years in full-time employment as a share of the population has declined from 76.3% in 2011/12 to 53.1% in 2013/14 to 67.5% in 2016/17. The Human Development Index ranking stands at 169 out of 188. Corruption Perceptions Index Ranking stands at 177 out of 180. Agriculture share of total employment stands at 56%. Afghanistan is heavily reliant on agriculture with poppy cultivation being a source of income for millions, The economy is remarkably undiversified with the only other major contributors to GDP being wholesale and retail sector and transportation and Information Communication Technology (ICT).

Afghanistan Economy Remains Undiversified and Reliant on Agriculture (World Bank 2016)

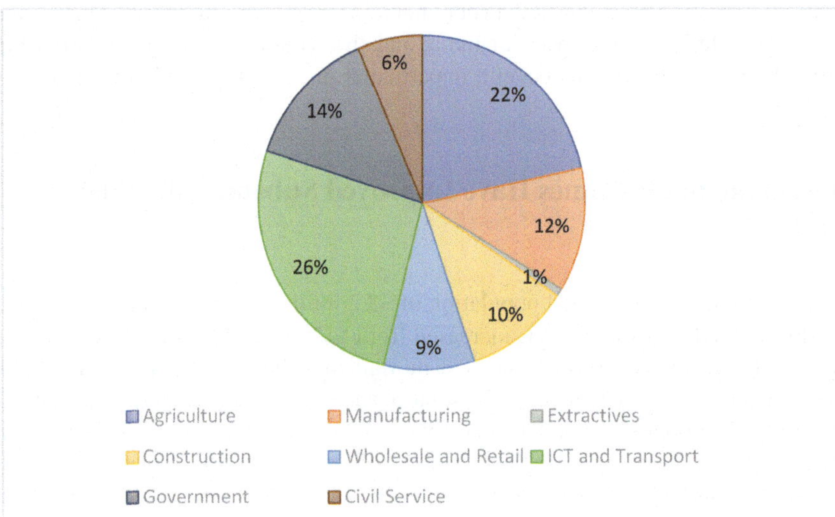

Government revenue has gone from 3.3% to 11.9% of GDP. The Afghan government collected $10 billion in government revenue in 2002. By 2016, this figure exploded to $2.4 billion—equal to 12.5% of GDP. The total is expected to surpass $4 billion by 2023 (CSIS Briefs 2019).

Civilian casualties increased to a post-2011 high of 11,418 in 2016, while an increasing proportion of Afghanistan's land area under the control of the anti-government elements.

Opium cultivation remains a key sources of income for many.[9]

During 2017, Afghan farmers gave more land to opium cultivation than at any other time since UNODC monitoring began. A year later, the area under cultivation had plummeted by 20% to 263,000 hectares. As a result, the amount of opium produced also dropped sharply, from roughly 9,000 tons in 2017 to an estimated 6,400 tons a year later—a 29% reduction. Despite these decreases, the area under poppy cultivation was still at its second highest level since the beginning of monitoring.

Impoverished farmers experience bleak years of declining income. While cultivation remained historically high, the amount of money farmers made from opium in 2018 tumbled by 56% from 2017 levels—from US$1.4 billion to US$604 million. The income estimated for 2018 was the second lowest since 2010. In areas where the drought had its strongest impact, many farmers faced a complete loss of income from opium poppy. The rapid decline in opium poppy income was largely prompted by the reduction in area under cultivation and a fall in the average opium yield—down 11% down on the previous year.

Income levels were further dented as opium prices fell and supply remained relatively high. Farmers were getting an average of US$94 per kilo of dry opium at harvest time in 2018, one of the lowest prices ever recorded. Afghanistan gets only a small share of overall profits from opium production.

Along with the fall in cultivation, production and income for farmers, the overall opiate economy in Afghanistan—which includes heroin production and trafficking to the border—contracted sharply between 2017 and 2018. The gross value of the Afghan opiate economy fell by two-thirds, from between US$4.1 and US$6.6 billion in 2017 to between US$1.2 and US$2.2 billion in 2018. The opiate economy was still worth between 6 and 11% of Afghanistan's GDP and it exceeded the value of the country's officially recorded licit exports of goods and services. For individual Afghans, the poppy has become a crucial element of their livelihoods. The farmers, however, are not the people who benefit most from the opiate economy. By far the largest share of income is generated by opiate transformation and exports to neighboring countries. Based on seizure and use data of opium and heroin in Afghanistan and neighboring countries, an estimated two-thirds of the opium available for export was converted into heroin or morphine within Afghanistan and the remainder was exported unprocessed. Although the amount of money generated by the opiate economy is large relative to the size of Afghanistan's economy, this income is just a tiny share of the profits generated by the trade. In 2015, UNODC estimated the trafficking of opium and heroin on the Balkan route alone was worth US$28 billion. Trafficking from Afghanistan's borders to consumer markets appears to be organized by nationals of other countries, meaning the proceeds generated in the international trade hardly feed into the domestic economy.

In 2018, farmers employed the equivalent of roughly 190,700 full-time workers to help them weed and harvest opium poppy. The actual number of people engaged

[9]Much of this section is extracted from UNODC and Government of Islamic Republic of Afghanistan, "Afghanistan opium survey 2018 Challenges to sustainable development, peace and security," 2019.

in the process is likely to be higher as the figure does not include family members engaged in such activities. In 2018, the combined wages for opium poppy labor amounted to US$270 million, or 44% of the farmers' income from opium over the year. The workers who cut the seedpods of mature poppies and collect the gum that oozes out are among the most vital workers in this process. Yet the role of these workers, known as lancers, has often been overlooked.

On average, lancers reported working for 15 days and harvesting opium for two farmers over the course of the season. They reported an average daily wage of US$12 in 2018, equivalent to US$170 per season. Farmers gave a lower estimate of the pay they offered lancers, at US$7.70 per day, which did not include payments in opium. However, even this lower estimate is almost twice as much as wages for other farming related jobs, and substantially more than construction workers, who can expect to be paid US$ 4.80. Roughly 16% of farmers reported that they also worked as lancers to earn extra money, another indication of the allure of the opium trade. Reported incomes for lancers were highest in the Southern region and lowest in Eastern and Western regions, which reflects wider survey data on the availability of workers.

Afghanistan mining is the key source of domestic resources for development[10]

Afghanistan has abundant mineral resources with an estimated value of $1 to $3 trillion. In 2005—2006 United States International Development (USAID) scientists from the US Geological Survey (USGS) and the Afghanistan Ministry of Mines made an assessment of the mineral resources of Afghanistan (Table 1). The US geologists found charts and data at the Afghan Geological Survey in Kabul collected by Soviet mining experts during the Soviet occupation of Afghanistan in the 1980s.

Over 43 days and 23 flights, the USGS flew nearly 23,000 miles to collect data. It covered about 70% of the country.

The USGS concluded that Afghanistan has vast mineral resources. It includes vast reserves of gold, platinum, silver, copper, iron, chromite, tantalum, lithium, uranium and aluminum. In the north and south, there are several hydrocarbon sites. The country also has massive deposits of dimension stone. The deposits of copper around 60 million metric tons and iron of around 2,200 million metric tons are some of the largest in the world. The Hajigak iron deposit could generate an average of $900 million per year until 2031.

It is estimated that oil and natural gas reserves could be worth more than $220 billion. Afghanistan may hold 1.4 million tons of rare earth elements such as lanthanum, cerium and neodymium, and lodes of aluminum, gold, silver, zinc, mercury and lithium. The Khanneshin carbonite deposit in Afghanistan's Helmand province is valued at $89 billion—in other words Afghanistan could become the Saudi Arabia of lithium. Rare earth elements and uranium are present in Helmand province. The biggest deposits discovered are iron and copper, which are estimated

[10]Unless otherwise referenced, this section is drawn from Ahmad Shah Katawazai, Foreign Policy, "Afghanistan's Mineral Resources Fueling War and Insurgency," Foreign Policy, Asia Pacific Economy Essay, May 29, 2018.

Table 1 Summary of known resources and estimated undiscovered resources for selected commodities in Afghanistan identified by the US. Geological Survey—Afghanistan Ministry of Mines Joint Mineral Resource Assessment Team. [Values in metric tons unless otherwise indicated; kg, kilogram; m^3, cubic meter; wt.%, weight percent; approx., approximately]

Commodity	Province	Deposit type	Known resource estimates from Abdullah and others (1977)	USGS-AGS assessment of undiscovered deposits (mean expected values)
Metals				
Aluminum	Zabul, Baghlan	Bauxite	4,535,000 (bauxite at 50.5 wt. % alumina and 12 wt. % silica)	Further study recommended
Copper	Kabul, Logaar	Sediment-hosted copper	12,340,600	16,880,000 (copper); 600,000 (cobalt)
	Kandahar, Zabul, Herat	Igneous-related copper	68,500	28,469,200 (copper);724,010 (molybdenum); 682 (gold); 9,067 (silver)
Gold	Takhar, Ghazni	Placer gold	918 kg	Further study recommended
	Badakhshan, Ghazni, Zabul	Lode gold	Approx. 1,780 kg	Further study recommended
Iron	Bamyan, Baghlan	Sediment-hosted iron	2.26 billion (>62 wt. % iron)	Further study recommended
	Badkhashan, Kandahar	Igneous-related iron	178,000,000 (at between 47 and 68 wt. % iron)	Further study recommended
Lead and zinc	Kandahar, Herat, Paktia	Igneous-related lead and zinc	90,000 (combined lead and zinc)	Further study recommended
	Ghor	Sediment-hosted lead and zinc	153,900 (combined lead and zinc)	Further study recommended
Mercury	Farah, Ghor	Hot-spring mercury	May contain gold and silver	32,000
Tin and tungsten	Herat, Farah, Oruzgan	Tin veins, tin and tungsten skarns and greisen	No previous estimate	Further study recommended
Industrial minerals				
Barite	Parvan, Herat	Bedded and vein barite	151,500,000	Further study recommended
Brick clay	Kabul	Clay	2,200,000 m^3	Further study recommended
Celestite	Baghlan, Kunduz	Celestite	>1,000,000 (at 75 wt.%)	Further study recommended
Chromite	Logar, Paktia	Chromium oxide	Approx.. 200,000 (at about 43 wt. %)	980,000 (chromium oxide)
Fluorite	Oruzgan	Fluorspar	8,791,000 (ore averaging 46.69 wt. %)	Further study recommended
Graphite	Badakhshan	Disseminated flake graphite	5,000	1,050,000 (flake graphite)
Halite	North Afghanistan	Evaporite	No previous estimate	Further study recommended
Kaolin	Baghlan Baghlan	Residual kaolin Sedimentary kaolin	100,000 to 150,000 (clay) 385,000 (clay)	Further study recommended Further study recommended
Lazurite	Badkhshan	Skarn lazurite	1,300	Further study recommended
Potash	Balkh, Samangan, Kunduz	Evaporite	No previous estimate	27,514,000
Rare-earth elements	Helmand	Carbonatite	No previous estimate	1,405,000 (REE), 3,480,000 (niobium and phosphorous, uranium and thorium)
Sulfur	Balkh, Badakhshan	Bedded and fumarolic	450,000	6,000,000
Talc, asbestos, and magnesite	Nangarhar	Metasomatic/metamorphic replacement magnesite	1,250,000 (talc); 31,200 (magnesite)	Further study recommended
	Nangarhar	Ultramafic-hosted talc magnesite	50,000 (mined previously)	13,400,000 (asbestos)
Building materials				
Aragonite	Helmand	Dimension stone	770,000	Further study recommended
Dolomite	Bamyan	Building stone	1,040,000	Further study recommended
Glass sand	Balkh	Sand	110,000 (siliceous sand); 10,900,000 (sandstone)	Further study recommended
Limestone	Bamyan	Building stone	3,500,000	Further study recommended
	Badakhshan, Herat, and Baghlan	Cement and flux	>500,000,000	Further study recommended
Marble	Various	Building stone	1.3 billion (coarsely crystalline marble)	Further study recommended
Sand and gravel	Badakhshan	Aggregate	136,000,000 m^3	Further study recommended

(continued)

Table 1 (continued)

Commodity	Province	Deposit type	Known resource estimates from Abdullah and others (1977)	USGS-AGS assessment of undiscovered deposits (mean expected values)
Sandstone	Bamyan	Building stone	650,000 (siliceous sandstone)	Further study recommended

Source United States Geological Survey and US Agency for International Development. (2007) "Preliminary assessment of non-fuel mineral resources of Afghanistan."

to be large enough and the Hajigak iron deposit could generate an average of $900 million per year until 2031.

Using a geology-based assessment methodology, the US Gas Resource Assessment Team estimated mean volumes of undiscovered petroleum in northern Afghanistan. They estimated that Afghanistan may have 1.6 billon barrels (0.2 billion metric tons) of crude oil, 16 trillion cubic feet (0.4 trillion cubic meters) of natural gas, and 0.5 billion barrels (0.8 billion metric tons) of natural gas liquids. Most of the undiscovered natural gas is in the Amu Darya Basin (US Geological Survey 2006). The US Department of Defense's Task Force for Business and Stability Operations estimated that oil and natural gas reserves could be worth more than $220 billion (Katawazai 2018).

Natural resources, particularly precious stones, have funded war and insurgency in Afghanistan for decades. In the 1990s, the United Islamic Front allegedly earned between $60 and $200 million per year from illegal extraction and trafficking in northern Afghanistan. Afghan coffers see very little income from the mining sector, and there are an estimated 1,400 illegal mines in the country. In 2014, the two mining areas of Deodarra and Kuran Munjan in Badakshan province alone provided around $20 million to armed groups. Mineral resources are thought to be the Taliban's second-largest source of revenue, while contributing less than one percent to state income in 2013. Illegal mining is rampant throughout Afghanistan, with more than 2,000 such sites raising money for warlords and the insurgents. A Special Inspector General for Afghanistan Reconstruction (SIGAR) report found that illegal mining has been costing the state up to $300 million annually since the Taliban collapse in 2001.

1. Afghanistan—development options

Considerable analysis has been conducted on what Afghanistan needs to do for social, economic and political development. One major attempt at this was in 2006 with the endorsement of the Afghanistan Compact at the London Conference in 2006 (The London Conference on Afghanistan 2006). That Compact encompassed all the sectors of the country from security, human rights, economic, social and political development. It was a most comprehensive framework developed in partnership with all the stakeholders except the Taliban. Subsequently, the Afghanistan National Development Strategy (ANDS) and the Medium-Term plans were developed. These frameworks could not be implemented because two critical assumptions failed—that there would be peace and security in the country and that donors would be willing to fund the huge financial requirements even though the frameworks did not put a precise dollar value to it but it would have been significant were it to be implemented. The other constraints of course was lack of Afghan capacity and governance to implement even if the first two assumptions held and the ineffectiveness of aid.

How critical will foreign aid be in the post intra-Afghan agreement phase?

Afghanistan is highly aid dependent. Grants as a per cent of GDP stands at 37% (World Bank 2016). Security and civilian aid amounted to an estimated $112 billion between 2002–2015, peaking in 2011 at approximately $15.7 billion, equivalent to 97% of GDP (World Bank Group 2016). Official development assistance (ODA) as a percentage of central government expenditure decreased from 206% in 2006 to 59% in 2015. In 2016, net ODA received was $4 billion—a sharp decrease from a peak of nearly $6.9 billion in 2011 (CSIS Briefs 2019).

In their analysis Thomas Ruttig and Jelena Bjelica (2018) quoting from a joint report by Oxfam and the Swedish Committee for Afghanistan entitled "Aid Effectiveness in Afghanistan" note that "Around 66% of Afghanistan's budget in the financial year of 1396 (March 2017-February 2018) was funded through international donor support, according to the Oxfam/SCA report. Only 33% came from domestic revenues, even though revenue has tripled over the last ten years, from around USD 750 million in 2008/09 to USD 2.5 billion in 2017/18. These figures reflect a continuing high level of aid dependency. Additionally, the report said (with reference to a 2016 Afghan government update through the Self-Reliance Through Mutual Accountability Framework (SMAF)) that 59% of development aid provided by the international community had gone through the government's core budget that year. But this percentage fluctuates on a yearly basis: in 2015, for example, of 3.73 billion USD disbursed, around 40% was provided through the budget."

"The funds Afghanistan receives are mainly channeled through donor-run projects as well as trust funds, which are usually designed as multi-year endeavours and managed by donors. For example, for a major aid investment such as the Afghan Reconstruction Trust Fund (ARTF), although it is a fully 'on-budget' programme, programmatic decisions are still made by the World Bank. This is strongly influenced by donors' investment choices and restricts government ownership of the ARTF, as the Swedish government's development agency SIDA reported in a 2015 evaluation. This approach to budgeting, however, has resulted in the 'over-aiding' of the country by allocating more funds than needed for the multi-year cycles. The Oxfam/SCA report stated that each donor maintains its own agenda based on separate aid agreements with the government and highlighted that donors mainly choose to fund areas that appeal to their constituents back home."

According to the Oxfam/SCA report, the area that has received most financial support is social infrastructure and services, with over 14 billion USD from 2011 to 2015. This is followed by economic infrastructure and services (4 billion USD), humanitarian aid (2 billion USD), and production sector support (1.6 billion USD). It further offers a detailed overview and effectiveness assessment of the projects funded in each of the above-mentioned sectors.

It also found that in 2014 the majority (58.7%) of donor financed 'off-budget' projects were below one million USD, and nearly a third between one and ten million USD. "By contrast," the report stated, "more than a quarter of 'on-budget' projects in

2014 were in the 10–50 million USD category and only 21.7% of on-budget projects were less than 1 million USD." This, according to the report, means that donor support is more fragmented when provided off-budget.

The implementation of a large number of small projects, involving a large number of implementing agencies, despite existing coordination mechanisms, can lead to increased transaction costs for both donor agencies and the Government. Donors may decide to deliver aid this way to reduce their reputational risk, however it can actually increase their fiduciary risk with more resources needed to keep an eye on multiple projects, and eventually increase the long-term development risk as government ownership of this type of development approach remains limited.

Given its aid dependence, it is generally recognized that Afghanistan will continue to need foreign aid for many years into the future even after the intra-Afghan agreement. But it has also generally been accepted that foreign aid to Afghanistan has generally been ineffective. A 2008 Oxfam report on aid effectiveness in Afghanistan concluded that there was continuing fragmentation of aid leading to ineffective aid. Thomas Ruttig and Jelena Bjelica (2018) further note quoting from a joint report by Oxfam and the Swedish Committee for Afghanistan entitled "Aid Effectiveness in Afghanistan" "(1) assessed efforts by both donors and the Afghan government to align with criteria provided by the 2005 Paris Declaration on aid effectiveness and builds on a 2008 Oxfam report on aid effectiveness. (2) The report—which was informed by a comprehensive review of the ministries' and other government institutions' primary data as well as interviews with key-informants among government, donors, development partners and civil society—comes to the conclusion that the continuing "fragmentation" of aid provided in many sectors still "leads to ineffective aid." The report noted: "*There are over 30 different international donors disbursing aid in Afghanistan, each with their own agenda and aid agreement with the government, and effective donor coordination and harmonisation is not a practice adopted universally [...] Yet there are still major issues of fragmentation, with donors bypassing government systems in multiple areas of the development sector, and it is this fragmentation that leads to ineffective aid.*" The report revealed that during the period 2001–2008, over two-thirds of the aid—then around 20 billion USD in total—bypassed the Afghan government. Donors justified this practice by pointing to the endemically corrupt Afghan system, arguing that channeling money through government and non-governmental organizations guaranteed better financial accountability. This argument was bolstered by the findings in the report that the Afghan government did not know how the remaining third of the aid, which was around five billion USD, had been spent, due to a lack of internal communication and coordination." The 2008 report also found that "over half of aid is tied, requiring the procurement of donor-country goods and services." Furthermore, it pointed out that large sums of development money was siphoned off, particularly through reconstruction contracts "for international and Afghan contractor companies," where profit margins were "often 20% and can be as high as 50%." It also criticized the exaggerated salaries of "most full time, expatriate consultants, working in private consulting companies, [who] cost 250,000–500,000 USD a year." It estimated that 40% of aid since 2001—around six billion USD—had returned to donor countries in corporate profits and consultant salaries, effectively turning aid into donor countries' export promotion. A World Bank report published in the same year stated that spending "on" Afghanistan did not equal spending "in" Afghanistan, and that ***"only 38 cents of every aid dollar spent in Afghanistan actually reaches the economy through direct salary payments, household transfers, or purchase of local goods and services."***

2. Development options

The World Bank Group (2016) has outlined what would be needed for Afghanistan to "stand on its own feet" though this focuses primarily on agriculture, including alternative livelihoods to poppy cultivation, manufacturing and services sector, less so on the mineral sector. In the mineral resources sector, with its huge potential, Katawazai (2018) concludes "Mining in a country like Afghanistan won't be easy without a strong commitment from the Afghan government and robust support from the international community. Afghanistan's rich mineral resources, if exploited effectively, could prove to be the best substitutes for foreign aid and could decrease the country's dependence on donor countries and foreign support. These resources, if properly managed, provide an opportunity for Afghanistan to write its own story of economic success. Robust policies and strong institutional arrangements together with clear policy direction will help attract both domestic and foreign investors. Better management of mineral resources could result in sustainable economic growth paving the way for a lasting peace."

The experience of many resource rich extractive industry countries has not been positive for the masses of their population. Afghanistan with its history and complexities cannot promise to do better—if anything it would be much more challenging for the vested groups to give up their power, influence and control of resources. Couple that with rampant corruption, poor governance and lack of capacity and we have to ask the question as to what makes us so optimistic that the resource riches of Afghanistan can be the engine of development?

It is in this context that great power rivalry comes into the equation. Any analysis of the economic and development prospects of the country cannot be looked at in isolation of the political economy. Geographically, Afghanistan is at the crossroads of many historical and current dynamics. The old silk road, the new silk road, the Belt and Road initiative, the proximity to Iran—an avowed security threat noted by a super power, spheres of influence and power—Russia, US and China; centers for terrorism for the US and other countries in the region and next to another unstable country, Pakistan, which has its own interests including that of its rivalry with India and desire to prevent Indian influence in Afghanistan.

Given the great power rivalry what are the prospects for development even if there is peace and stability through an agreement with the Taliban and US and between the intra-Afghan groups? This question does not need much economic analysis but rather an analysis of the political economy of the country, its ethnic composition and the related tensions and rivalries, the mineral wealth of the countries and great power politics.

Given the degree of dependence of Afghanistan on external assistance it is inconceivable that the country could manage to run the machinery of government without significant external assistance for a considerable time into the future. Disengagement from aid-giving will need to be gradual if only to prevent Afghanistan from going back into conflict. Such assistance would of course be conditional in the post intra-Afghan agreement period. Already it is seen that the US has been using its leverage of foreign aid to bring the Ashraf Ghani and Abdullah Abdullah factions to cooperate and undertake talks with the Taliban with the recent announcement by the U.S. Secretary of State, Mike Pompeo, of the suspension of $1 billion in assistance should the two parties not dialogue with the Taliban. Reportedly Saudi Arabia, U.A.E. and Qatar have committed development assistance. Afghanistan in the post intra-Afghan talks environment will continue to need extensive foreign aid. This dependence provides an important lever of influence that the donors will be able to maintain well into the future. But it is also worth asking that given the ineffectiveness of foreign aid, how critical will it be in the next phase especially if a Taliban inclusive government is not endeared to agree to all the conditionality that comes with foreign aid?

Moreover, given the lack of current capacity, which will likely be amplified if and when the Taliban comes into governance, there will be a steep learning curve before any impactful actions can be taken. This is quite apart from the challenges of the merging of different cultures of the Afghan factions finding common ground to be able to function and govern. Already, there were challenges of government officials from different ethnic groups working together in government. Once the Taliban is in government the challenges will only grow. Even on the strong assumption that workable relations between the Taliban and the current government can be found, it will take considerable time for working relations to be built—certainly more than 5 or even 10 years.

Afghanistan has not been able to overcome its capacity constraints in the 18 years since 2001. Capacity is particularly weak at the provincial levels. In 2005–2006, the challenge was to build capacity at the provincial level given the security challenges under which those with capacity at the Kabul level were unwilling to go to the provinces. It is still unknown as to what policies the Taliban will adopt with respect to the economy; health and education; rule of law; governance, including provincial governance; federal-provincial relations; women participation and democracy in general. Understanding and sorting all this out could well take up another 5–15 years going by the experience of the last 18 years. Any development planning needs to take account of the evidence and lessons from experience. It is possible to repeat with modifications the development and strategic planning initiatives since 2001 but the question that still remains is what is the guarantee of greater success this time round?

With these challenges it is worth exploring the 'realistic' development options for Afghanistan. The major value-adding sector as noted earlier is the mining sector. This is the sector that can provide immense financial resources for Afghanistan *if* there is peace and security. At the same time, given that the Taliban has been out of power for so long and still needs to learn the ropes of governance, and there needs to be a cultural merging/ understanding in the governance of the country—huge challenges—the mining sector should not be looked at as the 'cash cow'. If the mining sector is exploited in a haphazard manner for short-term gain, then much of the benefits will flow out of the country and into the hands of the corrupt few. Absence of the rule of law will ensure that Afghanistan remains an aid-dependent country and likely aid-deprived for years to come, or is forgotten and left to a destiny based on whatever domestic resources it can mobilize with its ensuing implications for peace and security.

Maryam Jami (2020) rightly concludes that, "It has been proved that the Afghan war cannot be won by relying on violence. Instead, diplomacy and reconciliation can be better options to end this war." While there is a common interest of the great powers to see peace in Afghanistan, this does not obviate continued competition among them—this time for the mineral resources of the country. It not too far-fetched to think that the reasons for the invasion by the Soviet Union into Afghanistan and subsequently the ISAF-led forces was in part to gain access to the mineral resources of Afghanistan. Indeed, the peace agreement with the Taliban is being sought not so much because the great powers want to extricate themselves from the country but to allow for the exploitation of the mineral resources of Afghanistan under the Taliban with the help and support of the great powers. It has been widely reported in the media that Afghanistan's mineral resources are the key motivation for the US-Taliban agreement. Since the resources could not be extracted by US companies through an armed intervention in the country with the continued conflict with the Taliban, the next route would be through bringing out peace agreement with the Taliban to allow exploitation of the mineral resources. The US is reported to have already spent $1 trillion in Afghanistan, though much of it flowed back to the US with payments made to US companies and armed forces. The US would be interested in getting returns from the mineral wealth of Afghanistan. However, if this is a motivation then the experience from Iraq suggests that it may be too optimistic an expectation. Further, the extraction of mineral resources will require foreign companies and expertise along with whom will come security mechanisms for them to operate—so perhaps foreign forces will not be able to fully extricate themselves from Afghanistan and a different security dynamic and tensions could emerge.

Great power rivalry will turn into great power competition for the rapid extraction of the resources, besides sucking in ISIS to also want to take advantage of the weak security situation with the departure of any meaningful US-NATO forces. This is the history of countries with rich mineral resources in Africa and elsewhere in the world. Resource rich countries find it difficult to live in peace. At the same time, it would be unrealistic to expect that the Taliban inclusive government could resist the pressures for resource extraction that would undoubtedly come from the aid giving countries also who would like to reduce their commitments as well as other great powers of the region interested in the mineral resources.

Key elements of a development strategy

The dilemma for Afghanistan is what to do about foreign aid, resource extraction, the new silk road and the Belt and Road Initiative which would be primarily for its mineral resource sector, in the next phase of its development?

While uncertainties abound, this paper suggests that under the strong assumption that peace and security can prevail, the key elements of the development strategy of Afghanistan in the post-Taliban-US-Intra Afghan agreement and in a region, which has and will remain a fulcrum of great power rivalry could be the following:

- Ensuring peace and stability through whatever political agreements maybe necessary, *including decentralization and allowing administration and resource sharing with other regions, including through agreements to set up semi-autonomous regions given the ethnic diversity of the country*. This is an imperative and a sine-qua-non for the development of the country. Return to conflict should not be an option.
- There are limited possibilities for Afghanistan to generate significant revenues from agriculture, manufacturing, and service sector as seen from the data and conclusions from World Bank analysis cited earlier—at least in the short to medium-term. Significant additional revenue generation would be possible from the mineral resources of the country, as seen from the studies cited above, depending on the manner of their extraction and sharing of revenues in any public–private partnership agreements, corruption and governance. So Afghanistan should prioritize additional revenue generation from the mineral resources of the country but in a phased manner with the following options:
  - Limited extraction until such time as the governance system in the country is strong enough to ensure that the bulk of the mineral wealth will be passed on to the people of Afghanistan and not be repatriated beyond an agreed rate of return for the mineral extraction companies and/or lost to corruption. The rate of extraction of each of the mineral resource to be determined by the efficacy of the receipt and expenditure of the mineral income for national development.
  - Extraction as partnerships are agreed upon and the capacity and security situation allow but all revenues from mineral extraction to be deposited in a *Public Investment Fund* from which disbursements are made for capital and operating expenditure of the government based on efficiency, effectiveness and absorptive capacity of the country.
  - Processing of much of the mining extraction within the country, thereby creating manufacturing and service sector jobs for Afghans. The rate of the extraction of the minerals can be set to be in congruence with the rate at which national processing and manufacturing capacity can be set up. Indeed, foreign partners in mineral extraction should be compelled to process the minerals in Afghanistan rather than simply take the raw minerals out of the country.

- Promotion of the potential that exists in non-poppy agriculture through investments from the Public Investment Fund. There is no certainty that non-poppy agriculture will be able to compete successfully with poppy cultivation in the short to medium-term but it is imperative that non-poppy agriculture is incentivized.
- Promotion of tourism in the service sector. Undoubtedly not a priority action given the huge fear and security concerns but success in this sector can also change the perceptions of the country and allow for its quicker integration in the region and beyond. However, it is not clear how willing the Taliban would be to open up the country.

Conclusion

A major concern would be the ideology of the Taliban and the next government vis-à-vis its citizens and the outside world. This may emerge only over a period of time. Afghanistan's development will remain conditional on the ruling ideology in the post peace agreement world, beyond of course the necessary condition of peace and security.

In the context of the prospects for the huge mineral resources of the country and the great power rivalry, *this paper concludes by postulating that it may well be worthwhile for all the Afghan parties to prioritize the huge gains from Afghanistan's mineral resources and agreement on how these resources can be shared as part of the intra-Afghan peace agreement. A resource sharing agreement as part of an economic agreement alongside a political agreement maybe the way to go*—not that this may be easier but at least there would be a strong motivation for arriving at a peace agreement and seeing the *benefits of the mineral resources of Afghanistan shared across the different income and ethnic groups. Afghanistan's mineral resources are spread across the country so it may be possible to come to mutual agreement on their exploitation and formulae for resource sharing.* Admittedly a tall order with its own complexities and complications and with almost certain influence-peddling by non-Afghan parties in a region fraught with great power rivalries, but also an incentive for peace! *Depending on the resource sharing agreement, the people of Afghanistan may well see the economic and social benefits to them of peace, thereby, giving peace a chance.*

References

A. Behuria, Y. Ul Hassan, S. Saroha, US-Taliban talks for Afghan peace: complexities galore. Strat. Analy. **43**(2), 126–137 (2019). https://doi.org/10.1080/09700161.2019.1595483

CSIS Briefs, Finishing strong: seeking a proper exit from Afghanistan. (2019) https://www.csis.org/analysis/finishing-strong-seeking-proper-exit-afghanistan

C. Farhoumand-Sims, Unfulfilled promises women and peace in post-Taliban Afghanistan. Int. J. **62**(3), 643–663 (2007)

M. Jami, Intra-Afghan peace talks: a channel to peace. Glocality, **3**(1), 1, 1–6 (2020). DOI: https://doi.org/https://doi.org/10.5334/glo.18

A.S. Katawazai, Afghanistan's mineral resources fueling war and insurgency. Foreign Policy (2018). https://www.foreignpolicyjournal.com/2018/05/29/afghanistans-mineral-resources-fueling-war-and-insurgency/

The London Conference on Afghanistan, The Afghanistan compact—building on success (2006). https://gsdrc.org/document-library/building-on-success-the-afghanistan-compact/

National Statistics and Information Authority (NSIA), Young people make their voices heard through the Afghan Youth Parliament. UNFPA Afghanistan. (2018) https://afghanistan.unfpa.org/en/news/young-people-make-their-voices-heard-through-afghan-youth-parliament

S. Olsson et al., Afghanistan after 2014 five scenarios. FOI Swedish Defense Research Agency. (2012) https://www.studentredaksjonen.com/uploads/6/1/0/7/61076221/foir_3424.pdf

A. Royesh, O.A. Nemat, Investing in women's education in Afghanistan. Yale J. Int. Affairs, Winter, 21–24 (2011)

T. Ruttig, J. Bjelica, The state of aid and poverty in 2018: a new look at aid effectiveness in Afghanistan. J. Res. Econ. **3**(2), 110–138 (2018)

United States Geological Survey Open-File Report, Assessment of undiscovered technically recoverable conventional petroleum resources of Northern Afghanistan (2006). https://pubs.er.usgs.gov/publication/ofr20061253

United States Geological Survey and U.S. Agency for International Development, Preliminary Assessment of Non-fuel Mineral Resources of Afghanistan (2007). https://pubs.usgs.gov/of/2007/1214/

UNODC and Government of Islamic Republic of Afghanistan. (2019). Afghanistan opium survey 2018 Challenges to sustainable development, peace and security. https://www.unodc.org/documents/crop-monitoring/Afghanistan/Afghanistan_opium_survey_2018_socioeconomic_report.pdf

US Department of State, Agreement for bringing peace to Afghanistan between the Islamic Emirate of Afghanistan which is not recognized by the United States as a state and is known as the Taliban and the United States of America (2020). https://www.state.gov/wp-content/uploads/2020/02/Agreement-For-Bringing-Peace-to-Afghanistan-02.29.20.pdf

World Bank Group, Afghanistan to 2030—priorities for economic development under fragility (2016). https://www.worldbank.org/en/country/afghanistan/publication/afghanistan-to-2030-priorities-for-economic-development-under-fragility

Building Human Security and an Entrepreneurial Middle Class with Natural Resource Partnerships: Propinquity, Behavioral Economics, and Blockchain Game Theory Strategy

Michael H. Peters

Abstract A 2500-year history of invasion, conquest, and resistance has left Afghanistan ethnically diverse and culturally rich. And yet, this storied history has bequeathed to the nation's people a legacy of political fragmentation, economic impoverishment, and in recent times, even some international isolation. The last 40 years of relentless conflict, encompassing two superpower invasions and decades of a smoldering civil war, have placed entire generations of Afghans at risk for post-traumatic stress disorder and have pushed the nation into a cycle of poverty, reduced educational opportunity, and violence. It is the contention of this paper that several academic concepts–the neuroscience principle of propinquity, the scholarly disciplines of behavioral economics and game theory, and the blockchain technology promoted by "Satoshi Nakamoto"—can be woven together into a practical application to create business partnerships, jobs, and an entrepreneurial middle class for Afghanistan, thereby decreasing poverty and the fallout of crime and violence significantly. Based on these principles, a genuinely non-exploitative partnership structure can be built between United States and Afghan interests. The central role of military can be replaced by private sector players, to achieve through human security what could not be achieved through kinetic force. In such a diverse culture, and amidst the difficulties of Afghanistan's geographical and topographical scenario, military dominance in Afghanistan has always failed, for thousands of years. Yet, that never would have been the true solution anyway. It seldom is. Afghanistan may utilize its vast mineral wealth as a starting place for trade, human security and national security, assisted but unexploited. And done in partnerships rather than as suppliers, they may then begin to extract the wealth of its human resources and find its future as a contributing and secure world citizen. The discussion below will sketch a broad outline of a possible path forward, employing twenty-first century innovative business structure to solve for the underlying long-standing impediments to security and prosperity. Like gravity, applications of these principles are true whenever humans compete for value or security. Anywhere self-interests have vetoed the common good,

M. H. Peters (✉)
Expert In Residence, Harvard University Graduate School of Design, 611 Pennsylvania Ave SE, Suite 404, Washington, DC 20003, USA
e-mail: mpeters@trustedblockchain.com

© The Author(s), under exclusive license to Springer Nature Switzerland AG 2021
A. Farhadi et al. (eds.), *The Great Power Competition Volume 1*,
https://doi.org/10.1007/978-3-030-64473-4_11

to the ultimate demise of both, these principles and guidelines that follow will prove useful for building the hope of a nation.

Introduction

The United States has spent $2 trillion in the war in Afghanistan. This comes to 32 times more than in the war's predecessor, the Gulf War. Add to that cost lost lives, lost capacity, US and Afghan internal political strife, regional foreign policy mayhem, and incalculable damage to an entire generation of children, and the cost of war can be seen as incalculable. There are consequences to the decades of war, poverty, uncertainty, surge and withdrawal, aid and no aid, missing limbs, widowed heads of household, and lost educational opportunity. What can we do to promote peace and human security that we have not already tried?

In the centuries of interaction between the West and East, more narrowly focused in this document on the United States and the Silk Road of Central Asia and the Middle East, the working model has been vendor and purchaser, supplier and buyer. While this seems reasonable, of course, there are scenarios where national security, human security, and economic development must be valued more heavily than simply getting a good price for raw material.

The true cost of exploitative transactions are seldom considered for natural resources. Whether short-term exploitation of trees in Malawi or the rain forests, or of minerals and gemstones in Afghanistan, or of cheap labor markets, the true long term costs of shortcuts are generally not pondered or weighed. The short-term gain and stockholder appeasement model of capitalistic enterprises is naturally first priority, even if unsustainable in a timeframe as short as ten years. The billions of dollars spent in aid and trillions in war are proof enough that the effort to control resources and advantageous labor markets, while neglecting the people themselves of resource-rich nations, will eventually create a hot spot of strife, if not induce moral decay and corruption.

In what follows, I will put forth the structural revolution of building innovative private sector partnerships to harvest human potential, and to capture the compounded value of the supply chain for Afghanistan and developing nations for security and international prosperity. This can be accomplished by combining already-recognized elements of Behavioral Economics, Game Theory, intentional propinquity, a twenty-first century project-based entrepreneurial education model, and mobile technology to create the multiplying power of an entrepreneurial middle class for small business job creation and human security. The undisputed engine of the most powerful economy in history, twentieth and twenty-first century United States, has been its entrepreneurship, and the "small business" job creation that naturally precipitates. Small businesses are also the businesses that become large businesses, and continue to create jobs. The development of human creativity, the ability to have and follow dreams in security fuels growth in knowledge and skills, health and culture, while reducing crime. Private sector seeding, via partnerships, can accomplish what the military is not designed to accomplish. We will explore all of this together, and how the principles may be intentionally employed, so to speak.

Aid is helpful, but cannot accomplish what private sector non-exploitative partnerships can create amongst the ambitious youth and the next generation. This chapter will discuss this strategic approach to building a nation's future for its children and children's children in the pages that follow. Blockchain game theory for business partnerships, mobile technology previously unavailable, "scalable project-based entrepreneurial education," together with trillions in Afghanistan's own natural resources, will jumpstart their future of creativity and an entrepreneurial generation.

The Bonus: This model will simultaneously develop international human potential to tackle issues such as water and food security, emerging diseases, climate change, and other potentially existential issues in the days to come. Every community has genius that is tapped because of opportunity or remains untapped for a lifetime. Developing human potential while reducing the number of emotionally damaged youth holds a promise that kinetic military action coupled with even the best aid programs can never accomplish. Meaningful opportunities and inspirational "education" absolutely reduces crime and violence, and creates a future for the children of the world.

> Far and away, the best prize that life has to offer is the chance to work hard, at work worth doing. (Roosevelt 1903).

A Dilemma of Historic Magnitude

Referred to as "the Central Asia roundabout" due to its passageway of Silk Road trade from the Mediterranean, Middle East, India, and China, Afghanistan has acquired resultant settlers for millennia (Starr 2011). Yet there is even more to Afghanistan's ethnic and cultural diversity: the region has suffered a relentless series of invasions and accompanying seismic cultural shifts throughout all of recorded history. From Darius I (516 BCE) and Alexander the Great (330 BCE) to Genghis Khan (1219–1221), from the British (three times from 1838–1919) to the Soviets (three times from 1929–1979), each invader has left its deposit of culture and values. This interplay of cultural influences is dizzyingly complex. Darius I managed to extend Persian control into parts of modern Afghanistan before the arrival of Alexander. The Macedonian's bloody foray into Bactria resulted not only in a Hellenistic-Persian-Buddhist hybrid culture that persisted for centuries, but also in an enduring Western fixation with the strategic importance of the region, a view that has persisted to this day (Holt 2012). Persianizing influences continued with the emergence of the Sassanian Empire as a regional power (224–651 (Payne 2014)). A seismic shift was set in motion by Arab conquest of the region in the seventh and eighth centuries, although full Islamization required nearly 500 years (Kennedy 2007). During a respite coinciding with Europe's Middle Ages, Central Asia—including much of Afghanistan—experienced what has been termed its Golden Age (Starr 2013). In the ensuing centuries, waves of Turkic invasion and peaceful settlement left their own mark on Afghan ethnicity and culture. In the thirteenth and fourteenth centuries, the invasions of Genghis and Timur swept through the region, causing massive depopulation (Saunders 2001). After the rise of European colonialism, Afghanistan found itself caught up as a buffer state between Russia and British India in the notorious Great Game contest for Central Asian influence, suffering invasion and attempted regime change in the nineteenth century—and arguably the twentieth as well—in consequence (Murray 2016). The twenty-first century has yet to see a truly new paradigm emerge.

This historical sketch, while all too brief, serves to underscore an inevitable conclusion: the legacy of this cycle of invasion and resistance has been incalculably costly. This region has fiercely retained its independence at the price of internal fragmentation, economic underdevelopment, political isolation, and frequent radicalization. In the apt words of military historian Tanner (2009), "In between enduring or resisting invasions from every point of the compass (and most recently from the air), the Afghans have honed their martial skills by fighting among themselves, in terrain that facilitates divisions of power and resists the concept of centralized control."

The dream that a unified nation state could be created by pencil strokes on a map guided the decisions of Western political elites working hurriedly and secretively in the postwar world to respond to the collapse of British colonialism and the emergence of new geopolitical realities. In truth, their artificially created national boundaries for Afghanistan stitched together a patchwork of people groups. The Constitution of Afghanistan officially lists 14 distinct ethnic groups before adding "and others," an admission that even this extended list is far from complete (The Constitution of the Islamic Republic of Afghanistan 2004). Indeed, modern Western academics have tallied no fewer than 55 Afghan ethnicities (Orywal 1986). Pashtuns (40%) and Tajiks (36%) comprise the largest ethnic groups, followed by Hazara (10%), Uzbek (8%), and Turkmeni (2%). Each of the other recognized groups accounts for 1% or less of Afghanistan's population (Akseer et al. 2019). Besides the official languages of Dari and Pashto, eight other tongues, including English, are spoken by at least 1% of the population (Central Intelligence Agency 2020). Ethnicities remain regionally concentrated. "The effort of European surveyors has borne fruit mainly in the form of constant civil wars among the Afghan people. The Uzbeks, Tajiks, and Turkmen north of the Hindu Kush, as well as the Hazaras among the mountains, have constantly resisted rule by the Pashtuns of the south; and vice versa" (Tanner 2009). As Nigel J. R. Allen has commented, "Much of the problem in Afghanistan today stems from the Eurocentric view that Afghanistan is a sovereign state internally consistent with an undifferentiated population" (Allan 2001).

With each group come anthropologically diverse folkways and mores and dramatically different embedded convictions about forms of governance, law, and leadership. For a nation state to bring unity to a heterogenous mixture of peoples with such a high degree of diversity is a staggering challenge, one with few parallels on the planet. Western concepts of artificial national boundaries, forms of government, contract law and property rights, partnerships (transactional versus relational), and definitions of honesty and fairness, cannot be assumed. In Afghanistan, the corresponding concepts are as diverse as the people groups and as varied as the radical and isolating regional topography of the nation.

To illustrate an essential priority in Afghanistan, little understood by presumptuous or non-studious outsiders, by and large the country prefers *jirga* decisions—respected village or regional leaders sitting in an egalitarian circle so that none are viewed as superior—over the Western model of contract or criminal law, argued by a lawyer in front of a judge and jury (Taizi 2006; Reichel 1998). That changes quite a lot in one's approach to the country and business and unwelcome intervention, if one were to understand and respect this fact. In a 2019 nationwide survey of nearly 16,000 Afghans, local *jirgas* were rated much higher than state courts for being "fair and trusted" (81% vs. 66%), "following local norms and values" (74% vs. 57%), "effective at administering justice" (74% vs. 53%), and "resolving cases promptly" (73% vs. 47%) (Akseer et al. 2019). When asked how much trust they have in any certain groups to "do their jobs," 67% of these Afghans expressed "some" or "a lot" of confidence in local *jirgas*, while only 53% expressed confidence in their local member of parliament, 47% in international NGOs, and 44% in government ministers (Akseer et al. 2019). This example from the justice system offers a warning and a lesson. To ignore the value system and governance convictions of a people would be to diminish them—and to welcome the consequences of having attempted to superimpose one's own will and values over theirs.[1] How different would this be from trying to impose one's religion or non-religion over an unwilling participant? Such a misstep would be unsustainable and futile. Instead, the way forward for a viable future for Afghanistan must resonate with local values, as diverse and nuanced as they may be. Relationships and decisions must be embraced from within rather than imposed from without.

Even guided by this understanding, there are formidable obstacles to helping Afghanistan resolve the internal pain points and press outward. It is a landlocked nation, with no rivers that can be navigated to an open sea. Roughly the size of Texas but with a topography more analogous to Switzerland's, Afghanistan has only 12% arable land, with 50% of its land mass at 6550 feet elevation (2000 m) or above, and 80% of its inhabitants living in rural areas with little infrastructure or communication (Library of Congress 2008). As a result, education for most has been informal at best. In 2014, 59% of a national sample of Afghans reported that they had never attended a school of any kind, while only 10% had received a high school diploma and a scant 1% had completed a university education (Warren et al 2014). The national poverty rate has soared in recent years, with 55% of all Afghans forced to subsist on less than one dollar per day in 2016–2017 (Akseer and Rieger 2018). In urban areas, 29% live in crushing poverty, while in rural areas the problem is even more severe (Ministry of Economy 2010). There are demonstrable correlations between poverty of this magnitude with its subsequent lack of hope and the occurrence of violent crime stemming from convenience and need for means of subsistence (Kelly 2000). In the United States, low-income youths with few opportunities for education or meaningful employment are much more likely to commit violent crimes than are their higher income peers (Kearney & Harris 2014). When asked in 2019, Afghans

[1] See, for example, the carefully reasoned analysis of Wardak, A. (2004) Building a post-war justice system in Afghanistan. *Crime, Law and Social Change 41*, 319–341.

troubled about the direction of their nation most often named violence and crime (75%) and a weak economy (42%) as the cause of their pessimism (Akseet et al. 2019). When asked how often they fear for their personal safety, three-fourths of all Afghans responded "always," "often," or "sometimes," while fewer than half did so as recently as 2012 (Akseer and Rieger 2018). One in six Afghans reported that they or someone in their family had experienced crime during the previous year, most often physical attack and beating (mentioned by 36% of those who reported crimes) but also including extortion, burglary, suicide attacks, and murder (Akseet et al. 2019). And the vicious cycle relating diminished educational opportunities and increased violence continues full circle; in 2018, more than a thousand schools remained closed due to security concerns, while 192 schools suffered violent attacks (Akseer et al. 2019). There is no reason to believe that hopelessness in Afghanistan has not affected the region and the world at the cost of tens of thousands of dismembered or lost lives and trillions of dollars, as well as a staggering opportunity cost for the Afghan people and the world community.

Firsthand experience brings the point home. During the author's visit to Kabul in January 2020, virtually every building of significance was surrounded by one of thousands of high concrete walls protected by some of the miles of razor wire in view, and descriptions of real estate that are unimaginable to the western world (Fig. 1). Tens of thousands of blast walls line streets and divert traffic from main roads to side streets. To get some idea, a reader in the West could visualize the main streets of any familiar city and imagine that they were all blocked indefinitely with concrete barricades and armed security checkpoints. Traffic moving at 2 km/h through key parts of the city and diesel smoke hovering in the air are amongst the maddening consequences of the security concerns that continue unabated even after so many years.

Fig. 1 Kabul homes for rent, January 2020. Photograph by author

High Level Solution

Over the last twenty years, the United States, the UNDP, USAID, NGOs and many additional players have sacrificed deeply in attempts to establish their closest approximation of best practices for governance, economic development, and human security in Afghanistan. According to the Watson Institute for International Affairs at Brown University, the total cost to the US federal government alone of the so-called "War on Terror" in Afghanistan, Iraq, and Pakistan has been \$6.4 trillion, financed primarily by borrowing at interest (Watson Institute at Brown University 2020). Human costs on the Afghan side are harder to quantify but no less staggering. To gain some intuition, we can consider the emotional cost.

Posttraumatic Stress Disorder (PTSD) is the subject of voluminous research (Shalev et al. 2000). In response to a threatening or horrifying experience, victims who develop PTSD exhibit a range of symptoms, including recurrent, involuntary memories ("flashbacks") and dreams; intense mental and physiological reactions to cues that trigger these memories: avoidance of external reminders, such as people, places, and topics of conversation; persistent negative beliefs about oneself, others, or the world; persistent fear, horror, anger, guilt, or shame; detachment from others; and low interest in participation in important activities (Bisson et al. 2015). PTSD is also associated with a whole range of negative effects on overall health, taking a toll on the immune, cardiovascular, and digestive systems and leading to unexplained dizziness, blurred vision, sleep disorders, and chronic pain (Gupta 2013). Although PTSD has been especially well studied in combat veterans, it has a significantly high prevalence in civilian populations and may become widespread in a community experiencing traumatic events. At the beginning of 2020, for example, one in three Hong Kong adults were exhibiting symptoms consistent with PTSD after experiencing nine months of sporadic protests and isolated riots (Ni et al. 2020). A similar proportion of teens in impoverished Los Angeles and Chicago neighborhoods have been found to exhibit mental and behavioral traits consistent with PTSD (Gudiño 2013). One in three direct victims of terrorist attacks likewise experience long-term PTSD symptoms (Paz Garcia-Vera et al. 2016). In this light, what have the 40 years of gunshots, brawls, hatred, civil war, invasions, landmines, poverty, and joblessness done to two generations of Afghan adults and youth?

At this point, a bit of risk for a new and creative approach may be welcome for preserving lives, livelihoods, human security, and national/regional/world security. What is there to lose? There is widespread realization that a change in the current paradigm is clearly needed (Cornell and Starr 2018). If we were to look for solutions to protect the future, taking into account the high cost of PTSD on their youth, while recognizing the diverse priorities and convictions of Afghanistan's ethnicity and culture, we're faced with a choice: We could give up (to their peril and our own, as history has shown), or we could begin here.

Behavioral Economics and Anthropological Levers

If there are some common themes of passion and priority amongst all languages and cultures and ethnicities, even in a land so diverse as Afghanistan, what would they be? If there is a common language, non-spoken and beyond ethnic boundaries, how might that play a role in solving security and crime problems, knowing how crime and violence are directly correlated to economic hope and self-determination?

What drives and motivates the human species, and how might this information serve to develop a war-torn nation without kinetic force, and without eternal dependence on aid-related approaches? Here is a list of prime movers most influential to individuals and governments, in varying degrees:

- Greed
- Fear
- Pride
- Survival instinct
- Family: protection, provision, love
- Power hunger
- Drugs, creating irrational or unintended behavior
- Exceptional individuals (the notion that nothing terribly good or terribly bad happens without an extraordinary person who has escaped the limitations and mediocrity of his time, for better or worse, the "great man theory" of history)

Any and all of these listed factors can alter or define the path of a man or woman or child, or of a nation. Religion and nationalism, per se, are not listed, since humans are prone to use these to manipulate others and to contort ideas to serve their own ends, using deity or national pride and raw emotion to justify their actions. The list above encompasses most religious zeal or nationalistic fervor in such cases. We will leave the rest outside of the scope of this conversation.

Let's return to the search for a Rosetta Stone solution to anguish and chaos, when the tools of hope, quality of life, and intellectual stimulation are generally unavailable to most. Such a search must begin with this list above, topics that remain relevant—regardless of language and culture. How can this list then be woven into a cross-cultural, multi-lingual, education-and-wealth-agnostic approach to building relationships, business, an economy, and human security? It should be apparent that the solutions to these problems, or circumvention thereof, must be the goals. Regardless of language or customs, preferences for a type of governmental construct or lifestyle, religion or geographical constraints, humans are much the same in what they fear and what they desire. Is there a way to get to the heart of this, using internal means, rather than simply laws or military, or external constraints fraught with the perils of loopholes or rebellion? How might we develop human potential in a region or nation or neighborhood filled with bad memories and a primary need for self-preservation? Is "aid" the solution? Do natural resources alone provide a reason for hope? I submit the answer has always been a firm "no," generation after generation.

As a case study, while much of the world's natural resources are found in what are still the poorest and most violent countries on earth, South Korea became an economic powerhouse from zero, via human development and a focus on human capital (Kim 1991). In the spirit of Adam Smith, we can also agree that Great Britain developed human resources and technology—and soon owned the gold and wealth collected by Spain in that same era. Money does not solve generational problems, directly. Afghanistan currently ranks 170 in the Human Development Index (HDI) and relies on foreign aid for as much as 90% of the national budget (UNDP 2019). The literacy rate is only 31% of the adult population—17% among females nationwide, with only 1.6% in two southern provinces (UNESCO 2020). As one of the world's youngest nations, with nearly two-thirds of the population under 25 years of age, the upside potential is staggering (UNFPA 2020). So is the downside potential: if the opportunity today in Afghanistan is lost, the consequences for future world security cannot be ignored.

Afghanistan, with massive yet largely untapped mineral wealth, can be likened to a family sleeping on the driveway of their $100 million mansion, because they don't have the keys to their own home (Risen 2010). Successful transformation of mineral mining from a hobby economy to modern industry (Fig. 2) is a path that leads to both human development and resource development, in contrast to merely selling the material to others, who become wealthy at the developing country's expense. It is the thesis of this article that the same path of partnerships, human development, and internal resource development (starting with the less capital expenditure-intensive and more scalable minerals) will also lead to human security, breaking the cycle of poverty, crime, and violent ideology—and all of this with much lower risk to human lives or national economies.

Partnerships

With the above list of prime movers in human culture always in mind, let us now explore how to define a cross-ethnic strategy, based on Behavioral Economics, Blockchain Game Theory, intentional propinquity, a twenty-first century project-based entrepreneurial education model, and mobile technology. To what end? To foster the creation of an entrepreneurial middle class, unleashing its massive power for small business job creation, in the process decreasing crime and violent conflict.

I suggest the seeding of a generationally secure solution for Afghanistan is not primarily a military endeavor at this point, nor merely an "investment" of CapEx industry or infrastructure by outside agencies or nations alone, but rather lies in creating human development and micro-disposable income to create an entrepreneurial creative middle class. The seed of the future is to bring the compounded value of the supply chain for Afghan minerals "home" to them via business ***partnerships*** between Afghans and Americans (in particular). Contrast a partnership concept with a conventional supplier/purchaser model for raw materials. Historically, the latter ultimately nearly guarantees failure for the nation supplying the

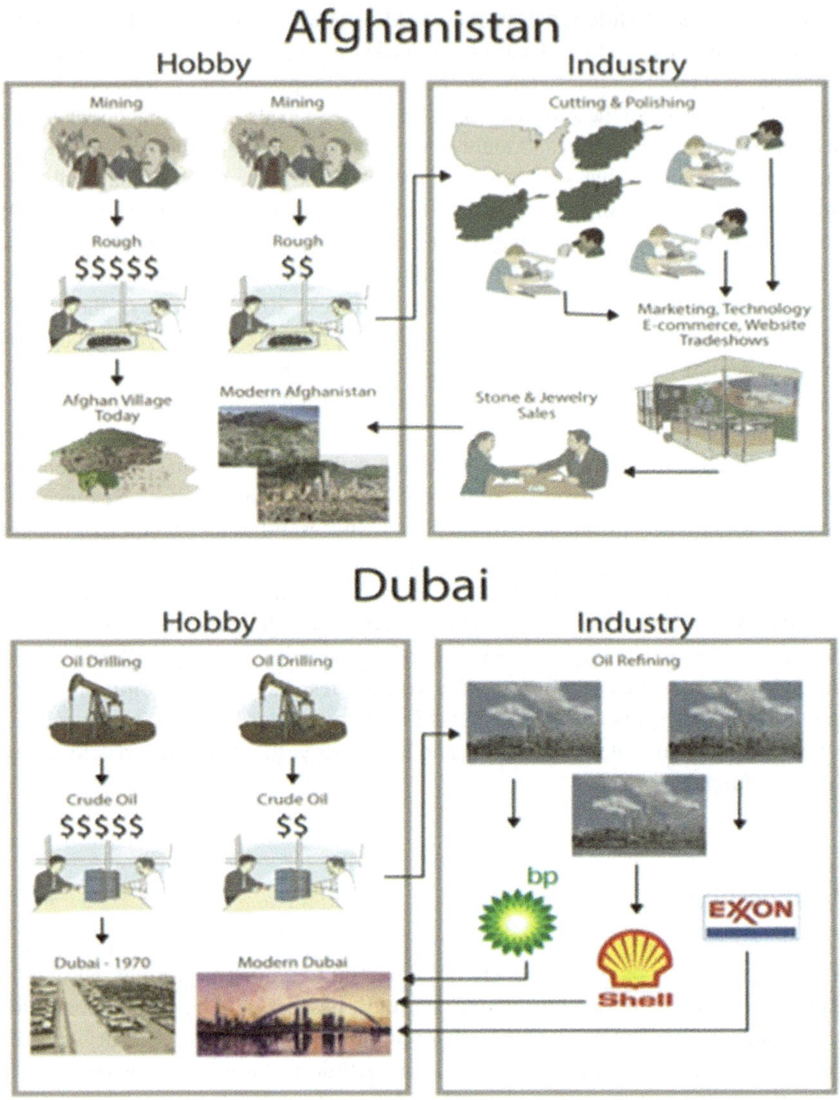

Fig. 2 Transformation of mineral resources from "hobby" economy to modern industrial economy

raw material. Exploitative removal and sale of lithium, copper, rare-earth minerals, precious and semi-precious gemstones and other raw materials from the Hindu Kush and other regions does not provide a future for Afghanistan, as similar extraction models have not for other mineral rich nations.

The alternative concept reflects the reality that a simple and low purchase price for raw materials is universal in the world of manufactured and marketed goods.

As an example, in 1996 it was estimated that the raw material of rubber used to manufacture an automobile tire costs $8.87 (equivalent to $14.62 in 2020 dollars); a typical manufactured tire retails today for $200 (Ferrer 1997). Another classic example is "bottled water" that costs between 300 and 2000 times as much as tap water (Colas 2013). And yet, selling bottled water at market prices does not yield massive profits. Why? Because the cost of pallets, truckers' wages, fuel, plastic bottles, insurance, warehousing, marketing and the rest has increased the actual manufactured cost of a third of a penny of water into seventy-five or eighty cents of actual cost to the manufacturer. That's how it is in the world of supply chain management and wholesale/retail. The rubber farmer makes very little; everyone along the chain makes at least as much. The actual profit at the point of sale may be small, but far exceeds the rubber farmer's portion of the product's sales price, after he or she has supplied the very material to create the product.

What if the raw material owner had the skills and access to control half or more of the supply chain, processing, and marketing? What if the raw material supplier and the rest of the supply chain were partners, rather than vendor/buyer? What if all parties could avoid the implied struggle over price cuts, profit distribution, and gamesmanship? Instead of the golden rule of the twenty-first century, "He who has the gold, rules," we might harken back to the original quotation and see what the results may be.

Let us consider a very different model of creating wealth, capacity, and security using a business partnership structure that leverages behavioral economics, blockchain game theory, the neuroscience concept of propinquity, and technology for a synergistic future. True private sector partnerships in Afghanistan with United States partners would share the value-add multiplier that typically lands largely in the pocket of the refiners, marketers, developers, manufacturers, branders, retailers, and other governments. In this model, in return for providing the natural resource at no cost to their US partner (thus solving innumerable issues), Afghans would receive training and human resource development across the full spectrum of the required steps to market. Afghans would share in the full revenue model as partners—rather than remaining in the dead-end role of being a supplier only. This model creates an entirely different and powerful dynamic, reminiscent of the "Satoshi Nakamoto" masterpiece, the 2008 Blockchain Game Theory strategy whitepaper, which we will address soon (Nakamoto 2008). There is a way to build security into a business relationship that simultaneously changes the life trajectory of everyone involved.

Defining Propinquity

The Merriam-Webster Dictionary (2020) says that "Propinquity and its cousin 'proximity' are related through the Latin root *prope*, which means 'near.' That root gave rise to 'proximus' (the parent of 'proximity') and 'propinquus' (an ancestor of 'propinquity')."

The field of social psychology has identified ***propinquity*** as a principal cause for interpersonal attraction (Small and Adler 2019). It is nearness, rather than absence, that tends to make the heart grow fonder—a nearness in physical proximity that in turn fosters the increased frequency of positive experiences (Alphonso 2016). And in any cosmopolitan or multi-ethnic society, this transfer of spatial closeness to relational closeness is the key to building a civil society. The concept of "neighbor" must translate into "neighborliness," the point of the story of the Good Samaritan (Chan 2019). Without the trust that comes from relationships that are personal, the list of motivators in the human condition listed above, such as greed, fear, pride, power and the like will succeed in making business and cultures mistrusting and always searching for ways to win, rather than partner for the greater good. Humans are not "wired" to trust, and the self-fulfilling prophecies of bad experiences only reinforce the self-preservation motives that impede progress and human security. And yet, "propinquity" breaks down seemingly unbreakable barriers. Whether "accidental" or intentional, prejudices and fears are often removed by propinquity, close proximity, particularly in the context of shared goals or experiences.

This "propinquity effect" has been well documented for nearly a century. A classic study traced the development of friendships in a small, two-floor New York City apartment building in the 1940s. Unsurprisingly, near neighbors tended to be the closest friends, while those living on separate floors were unlikely to form close relationships (Festinger et al. 1950). Another frequently cited study followed the relationships formed in the Maryland State Police Training Academy in the 1970s. Researchers discovered that most friendships formed between trainees whose last names were close together in an alphabetized list. Why? Because this list was used to assign the officers to dormitory rooms and classroom seating (Segal 1974). Frequent contact allows the discovery of commonalities and the experience of mutually enjoyed interaction, leading to greater relationship intimacy.

That is the point. Propinquity is a neuroscience building block. Studies have shown that proximity helps overcome racism, ethnic bias, clothing style differences, language barriers, income inequality, education diversity, and nearly any other barrier to relationship (Korol et al. 2018). A near-desk coworker, a next-door apartment neighbor, mailbox neighbor, a roommate or bunkmate, or even an alphabetically sequential classmate is up to six times more likely to be a friend than someone further away or seen less frequently, regardless of race or religion or unfamiliar dress, or age or education or marital status. In fact, to make the point on the negative side, but just as revealing of the premise, the frequency of marriages or trysts between coworkers who share space and meetings and deadlines is a staggering 62%, statistically. Even in a world where business interactions are increasingly becoming more "virtual," they can still be infused with a sense of "social presence" if built with care—a significant factor in harnessing the benefits of propinquity in an environment marked by geographic isolation (Oh et al. 2018).

Propinquity is an international language and neuroscience magnet. Propinquity applied to people intentionally empowers the development of relationships between diverse people groups, religions, ethnicities, languages, age, and more. Breakthroughs can create a ripple effect throughout society: if my friend makes a friend outside our "group," this person will likely become my friend as well, a dynamic frequently observed in children but possible for adults (Jugert and Feddes 2017). Previous enemies for many years, suspicious and hostile due to rumor or history or physical or cultural differences, often become friends when propinquity is introduced, even against advice and contrary to their own prejudices and intentions. It's like gravity—it's very hard to fight.

The power of propinquity has historically been unrecognized as a tool, and only studied as an accidental effect. Propinquity relationship, security, and business value have been especially difficult for Afghanistan to harness, in no small part because of geographic and cultural barriers. What if now, propinquity could be fostered through business structure innovation, and priming the economic and security pump by developing Afghanistan's mineral wealth? Mutually beneficial, positive shared security and prosperity among people of different ethnic and tribal loyalties, while simultaneously better integrating the nation into the global economy is surely a legitimate objective for any nation on the world stage. Might we utilize intentional propinquity in our structure of partnerships amongst rival tribes or international partners previously untrusted, to build both relationships and wealth, security and quality of life, all of which diminish crime and violence?

Intentional propinquity has been built into security structure for centuries, such as European royalty and native Americans who intentionally shared grandchildren across previously hostile borders and tribes to build loyalties and shared time and mutual interests. The propinquity neuroscience effect has proven to overcome ethnic and racial boundaries, economic, educational, language or social barriers, and of being suspicious of the unfamiliar. The latter is often referred to today as xenophobia or islamophobia, as well as terms yet uncoined involving distrust of American "cowboys" by UK business, or white carpetbaggers distrusted due to legitimately bad experiences in many eastern countries. Intentional propinquity is to execute strategically the principles of the many studied cases of happenstance propinquity in a neighborhood or workplace.

How might this be done? Employed intentionally in business partnership structures, bring together (with a stake in the outcome of a business enterprise) the children of leaders and youth from previously warring or competitive or untrusted families or tribes, or foreign companies or countries, into mutually dependent business relationships. The neuroscience effect of propinquity is to build trust where there was only suspicion or even hostility. This, combined with behavioral economics, twenty-first century project-based entrepreneurial scalable education, and Tragedy of the Commons game theory to make every transaction or partnership "everyone must win or everyone will lose," can allow business relationships structure to be utilized for human security, job creation, and improved standards of living, rather than merely the historic model of "extreme wealth to the most shrewd and most powerful," and the resulting hot embers of income inequality.

Since lack of challenging education and meaningful work are at the root of 60–70% of criminal activity in the US, as revealed by a quick search of "education employment and crime," it stands to reason that international partnerships done well, for human development rather than exploitation, will help break a cycle of poverty, crime, and violence. This will significantly reduce the need for law enforcement and military intervention, with their financial and social costs, as well as assist in replacing illicit and dangerous agricultural crops often used for survival of local villagers.

Blockchain Theory

The brilliance of Satoshi Nakamoto (likely a pseudonym) in his 2008 whitepaper outlining a new method of a decentralized and secure cryptocurrency (which morphed into blockchain in 2013) lies not in the cryptographic methods (Nakamoto 2008). That is a technology with a long and continuous history of development. The power of Satoshi's big idea is actually, in my opinion, the ***game theory*** at the root of blockchain (a word that is not used in his whitepaper, actually). What challenge was he facing, as he hoped to create a currency that could not be controlled by a central bank, could reflect true market value internationally, and could not be stolen by bank robbers or embezzlers?

Satoshi baked "Tragedy of the Commons" game theory into blockchain, for the purpose of de-incentivizing what would have been inevitable continuous "hacks" to steal the billions in US dollars deposited in a decentralized international ungoverned "blockchain" as cryptocurrency. Tragedy of the Commons, like "chicken" and "prisoner's dilemma" and "Nash bargaining" and "Pirate," are studies in human propensities and statistical probabilities in decision making. Tragedy of the Commons, in particular, involves the war between personal self-interest versus the common good. If an action (say, the exploitation of a resource) creates both a benefit and a cost, and the benefit is experienced by the individual while the cost is shared among the entire community, an ethical dilemma is born (Hardin 1968). Originally applied to human impact on the environment, Tragedy of the Commons theory has been shown to apply to any number of human interactions (Hardin 1998). Antibiotic overuse, for example, may benefit an individual yet promote the spread of antibiotic resistant superbugs to the entire community (Porco et al 2012). Or, consider debris in near-earth orbit: deployment of a satellite, with no ability to safely deorbit it after failure, may benefit the nation or organization that owns the craft, while simultaneously adding to the risk experienced by others who wish to use near-earth orbital space, or not to experience the dangers of loss of orbit by something from which they never benefitted (Salter 2015). The applications of Tragedy of the Commons theory to economics are legion.

That is the Tragedy of the Commons game theory strategy, in short. Who benefits? Who gains? What are the motivations for making a deal, or violating a deal, or stealing from that deal? How might the deal be constructed so that the common good, the good of others, coincides rather than competes with individual personal benefit? Satoshi built "the blockchain" (although this word was not present in his initial whitepaper, and supply chain use of this idea and "blockchain" was not introduced until 2013) in such a way that a person seeking to steal or destroy would lose more (mining, "proof of work," computing and electrical costs, peer reputational destruction, and immediate devaluation of stolen assets) by the theft than would be gained by that same player spending his energy protecting the blockchain from others.

One can make the case that the deep mathematical encryption within the blockchain structure is always conquerable to the clever, gifted, and determined. And that encryption, no matter how powerful, is far less powerful in mitigating corruption than his introduction of Tragedy of the Commons Game Theory where the dishonest have their own motives used against them to preserve the value of the deal. Satoshi turned the insidious hackers of the world into defenders and protectors of the blockchain, they themselves closing doors as quickly as anyone else could open them. The enemies became allies, because self-interest and the common good have become identical.

The technical details of the technology are not our discussion point here, perhaps to the readers' relief. The underlying principles are remarkably simple. In a business structure "use case" of blockchain theory the risk of a tarnished reputation is so visible that a violator may be forever ending future business relationships. In addition, there is the financial cost of dishonesty or broken trust exceeding the value of disingenuous, greedy, or corrupt behavior. The loss of financial and business value to themselves due to the nature of the "blockchain partnership structure" of shared responsibility is a powerful disincentive to engage in behavior that benefits "self" at others' expense. In short, selfish interests resolved in the common good and "reputation helps solve the 'tragedy of the commons,'" even when dealings are with individuals outside one's own social group, such as from another country but built into their "blockchain" business partnership (Millinski et al. 2002). Blockchain forges a link between ethical social behavior and individual reputation by creating a transparent, verifiable, traceable, and permanent record of the deal and business transactions, distributed openly to the entire community (Iansiti and Lakhani 2017). Unethical behavior is immediately daylighted, and the consequences are likewise immediate.

As we build partnerships and foster propinquity in Afghanistan, we are proposing to utilize the power of blockchain philosophy social and economic "technology" as a structural component for conducting business in an environment where security and trust concerns play a major role—as, in fact, they always do. If partnerships are structured to incorporate "we all win or we all lose," trust is built based on self-interest and the structure itself.

Our team has developed and patented supply chain blockchain software and hardware for Afghanistan's mineral industry to demonstrate and prove anti-crime ethical sourcing, and fair trade, from the mine to customer (ethicallysourced.com, trustedblockchain.com) using human-free oracles for geotagging, time stamps, barometric pressure and AI to chart topography to market, RFID control points, fair trade biometrics, material science scans, photos of the miners, and all data unalterably on the blockchain and IPFS. This blockchain application is likewise available for Afghanistan voting, licenses, property rights, records, medical use, and other areas of need. But, more importantly in some ways, we have also quite successfully implemented this "tragedy of the commons blockchain business structure" in Afghanistan, to the benefit of all. Win–win.

Behavioral Economics

A recent Nobel Prize in this field and the publication of many books and papers on the topic[2] have made unnecessary a deep discussion here. Instead, we will consider a quick listing of the Behavioral Economics (BE) principles that should be considered as business structure is framed to overcome language, culture, bias, distance, and security issues.

Defaulting. Keep it simple. The simplest path is usually the path most chosen, especially in complex scenarios with a proliferation of options (Huh et al. 2014).

Social Proof. Among participants in a study, 44% more people will make a choice because others are known to be doing it, rather than basing their choice on the factual options themselves (Cialdini 2001). In the marketplace, customer ratings and reviews have emerged as a powerful motivator—or demotivator—for sales (Amblee & Bui 2011). Word of mouth and visible participation builds community trust.

Loss Aversion. Humans are twice as likely to make the exact same decision if the question is framed in terms of protection against losing something they have or might have had, versus promoting what they might gain from a decision (Kahneman & Tversky 1979). Loss aversion is a deeply human trait, apparently hardwired in certain dopamine-regulated regions of the human brain (Tom et al. 2007).

Greed, Fear, Pride, Survival, Family, Power, Drugs, Great Man Theory... I submit that, just as the list above reflects some of the reasons that "rational" economics don't always "work"—this list has more power over decisions than normal logical behavior might have predicted. Flight-or-fight adrenaline and norepinephrine hormones, while they are helpful for exercising power, are not useful for promoting business collaboration, whereas dopamine, serotonin, and oxytocin are relationship and deal builders (House and Singh 1987; Shamay-Tsoory and Abu-Akel 2016). Everyone who has lived with a marriage partner, raised children, or negotiated a difficult business deal already has experience in initiating hormone dumps of either category. An awareness that these same principles affect human interactions at all levels will be required to forge a way forward in Afghanistan.

In short, building mutually beneficial, propinquity-promoting business relationships in Afghanistan will not come by defining rational behavior in a policy meeting and enforcing it by a show of power. Experience teaches us otherwise. Instead, it will rely, among other things, on the principles of Behavioral Economics, such as Nobel Laureate Richard Thaler's "Nudge": decision makers "choosing actions that are intended to make the affected parties better off *as defined by themselves*," and

[2]See, for example: Camerer, C. F., Loewenstein, G., & Rabin, M. (Eds.). (2004). *Advances in Behavioral Economics*. Princeton, NJ: Princeton University Press; Camerer, Colin F. and Loewenstein, George (2003) Behavioral Economics: Past, Present, Future. In: *Advances in Behavioral Economics. Roundtable Series in Behavioral Economics*. Princeton, NJ: Princeton University Press., pp. 1–61; Wilkinson, N., & Klaes, M. (2017). *An Introduction to Behavioral Economics*. Macmillan International Higher Education; Camerer, C. (1999). Behavioral economics: Reunifying psychology and economics. *Proceedings of the National Academy of Sciences, 96*(19), 10,575–10,577.

creating a "choice architecture" that will make entering into collaboration "easier, but no one is forced to do anything" (Thaler).

Business and Partnership Structure to Rebuild a Nation

Imagine a group of people trapped in an auditorium by a criminal gang, bent on extorting a ransom—or they will murder all of the hostages. We can't know what would cause the gang leader to gas to death all of the hostages. We do know, however, what would guarantee that he would *not* gas the auditorium: if he were told his grandchildren were now inside the auditorium. The siege will turn 180 degrees and be lifted in a heartbeat. Regardless of culture or religion or form of governance preference, regardless of greed, fear, pride, power, or other motivators, the "deal" can be constructed to turn self-interest into the protection of the common good. Partnerships created this way, the way of propinquity/game theory/behavioral economics, project based scalable education, and the macroeconomic concept of creating a multiplying economy via a seeded entrepreneurial middle class, rather than "investments" and "aid" and "supplier–buyer" supply chain management. As one renowned expert on this part of the world said recently, "This can work. It has the right stuff" (Shivers 2020).

We have established that trust is difficult enough to come by, and guarantees are nonexistent on this planet. We have further established that PTSD of a nation weighed down by 50 years of unspeakable stress does not help foster trust or stability. We have mentioned the lack of technology, the isolating topography, the multitude of ethnic groups and cultures in Afghanistan effecting philosophies of governance and relationships and religious nuance, and the Silk Road and invasion influences.

But what is common among the diverse groups? They all, we all, have children, and we all are someone's child. And regardless of race or religion or age or education or financial status, everyone desires their children have hope and provision and a future. There are motives that exceed greed and fear, when opportunities are presented and executed fairly, and the alternatives are understood to be "more of the same" that no one desires.

So, using the "technologies" mentioned above, how do we create an entrepreneurial middle class? In the US, 60% of jobs are "small business" jobs, originally created by an "average" person with a great idea, and a little bit of money to hire two of his friends in a garage (Erlanger and Govela 2019). If the reader requires convincing, he or she need only consult the histories of now immense companies such as Amazon, Apple, Disney, Google, Harley Davidson, Hewlett Packard, and Mattel, to name but a few examples. Even the photos of their original garages are available to one searching for them. The "design" of the powerful US economy is the freedom of a little time, connected relationships, and a little bit of extra money to create and explore new ideas. How do we open this door for developing nations? As mentioned, "investments" alone are not the secret sauce, nor are aid programs or intervention by NGO's or governments. That is merely feeding and teaching a nation to fish, and walking away leaving them without bait or fishing poles or nets, or a place to market their goods. These well-meaning measures have never worked to build human security or to develop human resources and a sustainable, secure economy or life for a nation, long term. In truth, no one denies the fact that "aid" can create the exact opposite effect of the self-determined future everyone would prefer. We need different measures, different structures built on the above principles where exploitation is nearly impossible and partnerships are win–win for all of the private sector partners and governments, in international and local relations.

A Current Example of These Principles Implemented in One Industry

The classic business structures of contract law, hierarchy, debt and equity, supplier and buyer are deeply inadequate in an environment as diverse and difficult as Afghanistan. Let's reimagine the paradigm. The following simple but powerful model is already proving successful in its early days in Afghanistan, with colored precious gemstones as a test case. What if we were to structure our business deals that might surround the $3 Trillion or so in natural resources in Afghanistan in a way that:

- Afghans and US are equals in the business, 50–50. Equals. No vendor/buyer environment—where a struggle for better prices may result in dishonesty, or lack of safety for workers, or other forms of loss for one party or another. This is inevitable in old-school competitive structures.
- Afghans add the raw materials to the supply chain, incentivized to keep costs low and safety high because they share in the final profit margin of the entire compounded value of the supply chain, rather than single transactions. 1.2 times 1.2 times 1.2 times 1.2 equals a doubling of cost with four steps of profit taking in the supply chain after the raw material stage. If villagers are partners rather than suppliers, they can easily double their income with exactly the same amount of work.
- All of the other elements of Intellectual Property, and the manufacturing and marketing and distribution, are provided by the US partner with all of the upline profits being split with the Afghans.
- In return for raw material access, the children of the leaders from all areas of the country, as well as villagers with previously little hope for upward mobility, are trained in every aspect of the chain, from lapidary skills, to inventory control and software, to app development and ecommerce, to blockchain coding, to graphic design, and laboratory skills.

As with Blockchain game theory, it is in no one's best interest to damage the business or steal from it or misrepresent it to others. To the contrary, everyone who may have been inclined to use their power to damage the business or steal its assets now has every interest in protecting that business, for their children and for their nation. If one wins, everyone wins. If anyone loses, everyone loses. If bias or prejudice or culture get in the way, everyone loses. Propinquity, game theory, and behavioral economics are harnessed to create a scalable entrepreneurial middle class with its generational multiplier effect. The opportunities for creative application of these principles are endless.

As an academic, a scientist, and an entrepreneur it is my observation that we, in the west, have been trapped for many years into our perspectives and structures borrowed from Athens, Rome and the British Empire. In so doing, we have been at a loss of how to overcome barriers, build trust, and do business by using universal communicators rather than contract law and western (often alienating) assumptions.

While the various means of applying the above principles are endless, it is my contention that these are the principles required to overcome the tragedies of poverty, war, crime, and alienation, as well as for building strong families and communities and domestic businesses.

Days are early, but the author can report the following observations:

- During a January 2020 visit to Kabul, from the highest level of Afghan officials, "Contact me anytime day or night," and "Expect my full support, and always at no charge," with "no corruption."
- During the autumn of 2019, previous tribal leaders, some referred to by others as "warlords" (though perhaps more aptly sometimes "those who were trying to protect their families in a nation at war") and businessmen who had been at odds for two or three generations, now warmly shake hands and share meals together and with us.
- During the spring of 2019, Afghan young people whose lives were at risk from bad influences, drugs, and crime, express excitement that they are learning a lucrative life-skill.
- In 2018, in the executive lunch hall on Capitol Hill, a 30-year US Congressman who had "opposed Afghanistan involvement" his entire career, changed his mind during lunch after discussing these principles.

The above evidence is anecdotal, of course, and many challenges must be overcome. However, I put forth the proposition that it is possible to create an entrepreneurial middle class that multiplies jobs, rather than merely adding jobs. Henry Ford created unknowingly a happy accident at the turn of the last century Ford paid assembly line workers double daily pay to slow turnover rates, resulting in many workers with a little extra money starting a new business with a friend. Since then, well over half of all jobs in the United States, the powerhouse economy of the world for 100 years, are "small business" jobs, created by an entrepreneurial middle class. The driver of the US economy is still a middle class with a tiny amount of extra cash, and a good idea. Investment in large capital expenditure projects and aid programs is incapable of accomplishing the multiplier effect of an entrepreneurial middle class. In fact, at some point such expensive projects may damage self-determination and human growth and security, more than they assist, as anti-trust advocates have argued at times. In contrast, there is reason for optimism that the principles outlined in this paper can prove successful in creating a long-term platform for peace and economic growth in Afghanistan and potentially in other developing nations.

Consequences for Great Power Competition

The time has come to reimagine the role of Afghanistan in the world. No longer should it be considered a "buffer state," a shock absorber to keep superpower combatants from direct conflict. It is possible for US businesses and public–private partnerships to build a truly collaborative relationship with Afghanistan interests. Both parties would gain from the partnership, and the increase in regional stability and security would truly benefit the entire planet. The consequences of missing an opportunity to build deep relationships in "the roundabout of Central Asia" is reflected in this statistic from the International Monetary Fund regarding the Great Power competition: In the year 2000, 80% of all nations traded with the United States as their top trading partner. By 2018, that number was less than 30%. China now holds the top trading position with 128 out of 190 nations (Leng and Rajah 2020). While most trading partners and allies still regard the US as the world's leading economic power, China is moving rapidly toward that position in the eyes of many (Wike et al. 2017). China is utilizing both scale (the engine of profit and growth) and international dependent relationships to become the great power of the future. As they control relationships internationally, resources for electronics such as rare earth minerals, as well as transportation and energy hubs and pathways, the case for the United States playing a stronger strategic role via partnerships is strong. Instead of the inevitability of losing tender bidding wars to the cash-heavy Chinese (government controlled and subsidized) operators, or Russia or Iran, we can simultaneously solve national security, regional security, human security, and business opportunities, reduce military costs, and build lasting relationships based on trust and friendships of mutual benefit, across cultural and religious barriers.

How? A Marshall Plan of Sorts.

Help create an entrepreneurial middle class anywhere a footprint is useful or help is needed for security, via scalable entrepreneurial project-based education (this does not require dedicated structures and the exceptional teachers required by a reused-but-now-deficient nineteenth century "lecture" and repeat "simon-says" model of education). Add to this business partnerships in any industry that incorporate the structure previously discussed of game theory, propinquity, and behavioral economics. Sharing of the supply chain can replace the vendor/buyer relationships that persist in extracting wealth from those who need it most. As has been proven repeatedly, the resultant loss of societal equilibrium, loss of security and prosperity, and the crescendo of anger and hopelessness—is a gap soon filled by nefarious characters or by internal turmoil for neighborhoods, nations, and the world. With some creativity, a little courage, and an innovative strategy we can take back what has been taken from us in the last 18 years we've been busy, engaged in other matters.

References

T. Akseer, J. Rieger (eds.), *Afghanistan in 2018: A Survey of the Afghan People* (The Asia Foundation, San Francisco, 2018)

T. Akseer, K. Hayat, E.C. Keats, S.R. Kazimi, C. Maxwell-Jones, M.S. Shiwan, D. Swift, M. Yadgari, F.A. Yousufzai, *A Survey of the Afghan People: Afghanistan in 2019* (The Asia Foundation, San Francisco, California, 2019).

N.J.R. Allan, Defining place and people in Afghanistan. Post-Sov. Geogr. Econ. **42**(8), 545–560 (2001)

C. Alphonso, Propinquity, in *Encyclopedia of Family Studies* (2016), pp. 1–3

N. Amblee, T. Bui, Harnessing the influence of social proof in online shopping: the effect of electronic word of mouth on sales of digital microproducts. Int. J. Electron. Commerc. **16**(2), 91–114 (2011)

J.I. Bisson, S. Cosgrove, C. Lewis, N.P. Robert, Post-Traumatic Stress Disorder. BMJ (Clinical Research Ed.) **351**, h6161 (2015)

Central Intelligence Agency, The World Factbook 2019–20. Central Intelligence Agency (2020), https://www.cia.gov/library/publications/the-world-factbook/geos/xx.html.

J.K.H. Chan, Propinquity, in *Urban Ethics in the Anthropocene* (Palgrave Macmillan, Singapore, 2019)

R.B. Cialdini, Harnessing the science of persuasion. Harv. Busin. Rev. **79**(9), 72–79 (2001)

N. Colas, As quoted in Rosenberg A. Message in a Bottle: What Water Sales Tell Us About the Economy. CNBC, July 12, 2013 (2013), https://www.cnbc.com/id/100883042. Accessed 22 March 2020

The Constitution of the Islamic Republic of Afghanistan (Ratified) (2004), https://www.embassyofafghanistan.org/page/constitution/. Accessed 22 March 2020

S.E. Cornell, F.S. Starr, Modernization and Regional Cooperation in Central Asia: A New Spring. Central Asia-Caucasus Institute & Silk Road Studies Program, Silk Road Paper, November (2018)

O. Erlanger, L.O. Govela, *Garage* (MIT Press, Cambridge, 2019)

G. Ferrer, The economics of tire remanufacturing. Resources, conservation and recycling. Sci. Direct**19**(4), 221–255 (1997)

L. Festinger, S. Schachter, K.W. Back, *Social Pressures in Informal Groups: A Study of Human Factors in Housing* (Harper, New York, 1950)

O.G. Gudiño, Behaviorial inhibition and risk for posttraumatic stress symptoms in Latino children exposed to violence. J. Abnorm. Child Psychol. **41**(6), 983–992 (2013)

M.A. Gupta, Review of somatic symptoms in post-traumatic stress disorder. Int. Rev. Psychiatr. **25**(1), 86–99 (2013)

G. Hardin, Extensions of the tragedy of the commons. Science **280**(5364), 682–683 (1998)

G. Hardin, The tragedy of the commons. Science **162**(3859), 1243–1248 (1968)

F.L. Holt, *Into the Land of Bones: Alexander the Great in Afghanistan*, vol. 47 (University of California Press, Oakland, 2012)

R.J. House, J.V. Singh, Organizational behavior: some new directions for I/O psychology. Annu. Rev. Psychol. **38**(1), 669–718 (1987)

Y.E. Huh, J. Vosgerau, C.K. Morewedge, Social defaults: observed choices become choice defaults. J. Consum. Res. **41**(3), 746–760 (2014); S.S. Iyengar, E. Kamenica, Choice proliferation, simplicity seeking, and asset allocation. J. Pub. Econ. **94**(7-8), 530-539 (2010)

M. Iansiti, K.R. Lakhani, The truth about blockchain. Harv. Busin. Rev. January–February, 118–127 (2017)

Islamic Republic of Afghanistan Ministry of Economy and the World Bank Economic Policy and Poverty Sector, Poverty Status in Afghanistan. PowerPoint Presentation (2010)

https://siteresources.worldbank.org/AFGHANISTANEXTN/Resources/305984-1326909014678/8376871-1334700522455/PovertyStatusLaunchPresentation.pdf. Accessed 22 March 22 2020

P. Jugert, A.R. Feddes, Children's and adolescents' cross-ethnic friendships, in *The Wiley-Blackwell Handbook of Group Processes in Children and Adolescents* (2017), pp. 373–392

D. Kahneman, A. Tversky, Prospect theory: an analysis of decision under risk. Econometrica **47**(4), 263–291 (1979)

M. Kearney, B.H. Harris, *Ten Economic Facts about Crime and Incarceration in the United States* (The Hamilton Project, Brooking Institute, Washington, D.C., 2014)

M. Kelly, Inequality and crime. Rev. Econ. Stat. **82**(4), 530–539 (2000)

H. Kennedy, *The Great Arab Conquests: How the Spread of Islam Changed the World We Live In* (Da Capo Press, Philadelphia, 2007)

K.S. Kim, The Korean miracle (1962-1980) Revisited: Myths and Realities in Strategy and Development (Helen Kellogg Institute for International Studies, University of Notre Dame, 1991); J. Minns, Of miracles and models: the rise and decline of the developmental state in South Korea. Third World Q. **22**(6), 1025-1043 (2001)

L. Korol, A.W. Fietzer, J.G. Ponterotto, The relationship between multicultural personality, intergroup contact, and positive outgroup attitudes toward Asian Americans. Asian Am. J. Psychol. **9**(3), 200 (2018)

A. Leng, R. Rajah, Chart of the week: global trade through a US-China lens (2020), https://www.lowyinstitute.org/the-interpreter/chart-week-global-trade-through-us-china-lens. Accessed 25 March 2020

Library of Congress, *Country Profile: Afghanistan* (Library of Congress—Federal Research Division, Washington, D.C., 2008)

Merriam-Webster (2020), https://www.merriam-webster.com/dictionary/propinquity#note-1. Accessed 25 March 2020

M. Milinski, D. Semmann, H.J. Krambeck, Reputation helps solve the 'tragedy of the commons'. Nature **415**(6870), 424-426 (2002); P. Barclay, Trustworthiness and competitive altruism can also solve the "tragedy of the commons". Evolut. Human Behav. **25**(4), 209-220 (2004)

C. Murray, *Sikunder Burnes: Master of the Great Game* (Birlinn, Edinburgh, 2016)

S. Nakamoto, Bitcoin: A Peer-to-Peer Electronic Cash System (2008), https://bitcoin.org/bitcoin.pdf

M.Y. Ni, X.I. Yao, K.S. Leung, C. Yau, C.M. Leung, P. Lun, G.M. Leung, Depression and post-traumatic stress during major social unrest in Hong Kong: a 10-year prospective cohort study. The Lancet **395**(10220), 273–284 (2020)

C.S. Oh, J.N. Bailenson, G.F. Welch, A systematic review of social presence: definition, antecedents, and implications. Front. Robot. AI **5**, 114 (2018). https://brandongaille.com/19-unbelievable-workplace-romance-statistics/

E. Orywal, *Die Ethnische Gruppen Afghanistans: Fallstudien Zu Gruppenidentitaet Und Intergruppenbeziehungen (The Ethnic Groups of Afghanistan: Case Studies of Group Identity and Intergroup Relations)* (Ludwig Reichert, Wiesbaden, Germany, 1986)

R. Payne, 16: The reinvention of Iran: The Sasanian Empire and the huns, in *The Cambridge Companion to the Age of Attila* (2014), p. 282

M. Paz García-Vera, J. Sanz, S. Gutiérrez, A systematic review of the literature on posttraumatic stress disorder in victims of terrorist attacks. Psychol. Rep. **119**(1), 328–359 (2016)

T.C. Porco, D. Gao, J.C. Scott, E. Shim, W.T. Enanoria, A.P. Galvani, T.M. Lietman, When does overuse of antibiotics become a tragedy of the commons? PLoS One **7**(12) (2012)

P. Reichel, *Comparative Criminal Justice Systems* (Prentice-Hall, New York, 1998)

J. Risen, US identifies vast mineral riches in Afghanistan. *The New York Times* (2010), https://www.nytimes.com/2010/06/14/world/asia/14minerals.html

T. Roosevelt, *A Square Deal* (Speech given to New York State Agricultural Association in Rochester, NY, 1903)

A.W. Salter, Space debris: a law and economics analysis of the orbital commons. Stan. Tech. L. Rev. **19**, 221 (2015)

J.J. Saunders, *The History of the Mongol Conquests* (University of Pennsylvania Press, Philadelphia, 2001)

M.W. Segal, Propinquity, alphabet and attraction: an unobtrusive measure of the effect of propinquity in a field setting. J. Pers. Soc. Psychol. **30**, 654–657 (1974)

A.Y. Shalev, R. Yehuda, A.C. McFarlane, *International Handbook of Human Response to Trauma* (Kluwer Academic/Plenum Press, New York, New York, 2000)

S.G. Shamay-Tsoory, A. Abu-Akel, The social salience hypothesis of oxytocin. Biolog. Psychiatr. **79**(3), 194–202 (2016); M.C. Soares, R. Bshary, L. Fusani, W. Goymann, M. Hau, K. Hirschenhauser, R.F. Oliveira, Hormonal mechanisms of cooperative behaviour. Philosop. Trans. R. Soc. B Biolog. Sci. **365**(1553), 2737–2750 (2010)

M. Shivers, Great Power Competition Conference (Tampa, Florida, USF, January, 2020)

M.L. Small, L. Adler, The role of space in the formation of social ties. Ann. Rev. Sociol. **45**, 111–132 (2019)

F.S. Starr, *Afghanistan Beyond the Fog of Nation Building: Giving Economic Strategy a Chance* (Central Asia-Caucasus Institute & Silk Road Studies Program, Washington, D.C., 2011)

F.S. Starr, *The Lost Enlightenment. Central Asia's Golden Age from the Arab Conquest to Tamerlane* (Princeton University Press, Princeton, 2013)

S. Taizi, *Jirga System in Tribal Life* (Tribal Analysis Center, Williamsburg, VA, 2006)

S. Tanner, *Afghanistan: A Military History from Alexander the Great to the War Against the Taliban* (Da Capo Press, Philadelphia, 2009), p. 1

S.M. Tom, C.R. Fox, C. Trepel, R.A. Poldrack, The neural basis of loss aversion in decision-making under risk. Science **315**(5811), 515–518 (2007)

UNESCO Office in Kabul, Enhancement of Literacy in Afghanistan (ELA) Programme (2020), https://www.unesco.org/new/en/kabul/education/youth-and-adult-education/enhancement-of-literacy-in-afghanistan-iii/

UNFPA Afghanistan, Young People (2020), https://afghanistan.unfpa.org/en/node/15227

UNICEF spokesperson, as quoted in "Attacks on Afghan Schools Nearly Triple, Warns UNICEF," Deutsche Welle, May 28, 2019 (2020), https://p.dw.com/p/3JGJ9. Accessed 24 March 2020

Z. Warren, S.A. Kabuli, S. Sarabi, S.M. Sadat, C. Satkowski, A. Steppe, *Afghanistan in 2014: A Survey of the Afghan People* (Asia Foundation, 2014)

The Watson Institute for International and Public Affairs, Brown University. "Cost of War." (2020), https://watson.brown.edu/costsofwar/costs/economic. Accessed 24 March 2020

R. Wike, J. Poushter, L. Silver, C. Bishop, Globally, more name US than China as world's leading economic power. Pew Res. Center **13** (2017)

[illegible]

Defining Great Power Competition

Dexx Pierce

Abstract Great power competition is a complex definition combining competition and great powers, yet few are able to define the term as a whole in context of the current reality. In this chapter, we provide a definition of great power competition which reflects the reality of competition while taking into account the realities and nuances of today's global environment. Beginning with games played by children or recreationally within societies and progressing to the great powers competing at a global scale, competition is freshly defined using familiar and understandable concepts and language. Competition at any level consists of two elements at its most fundamental form: offense and defense. Those two competitive elements are each comprised of a balance of subordinate elements: sustainment, strategy, tactics, adaptability, consistency, competence, and misdirection/deception. Each of these concepts play an essential role in effective competition but are often considered in isolation rather than in context of how the element fits into the greater schema of the competitive effort. We address the definitions, function, value, and utility of each element to communicate and set a baseline for the concept of employment as part of the larger tapestry of competition, from low-stakes games to nation-state struggles to gain control over their desired spheres of influence, both at home and across the globe. Several common distractors (power, advantage, leverage, and influence) are more symptoms, measures, or expressions of competition rather than the elements themselves, in that they misdirect understandings of competition are also examined, to bring to light pervasive misunderstandings of the nature of competition which are often discussed through misrepresentative reductionist discussions of great power competition. This chapter redefines what a great power competitor is, establishing a new description of great power competition separate from the cold war era definitions which have dominated conversation since the 1950s—with the hope of beginning a new, impactful conversation on the nature of this important topic.

D. Pierce (✉)
Department of Defense, 7406 Germer St., Tampa, FL 33616, USA
e-mail: bdplayer@gmail.com

© The Author(s), under exclusive license to Springer Nature Switzerland AG 2021
A. Farhadi et al. (eds.), *The Great Power Competition Volume 1*,
https://doi.org/10.1007/978-3-030-64473-4_12

Introduction

Great power competition is a common political buzzword—yet attempts to define the term often result in descriptions or discussions which never provide a definition or examine only excerpts of the concept. Endeavors to define great power competition have taken the shape of describing what great power competition looks like rather than defining the term a usable, sustained form. In this chapter we will introduce a definition of great power competition[1] with the hope that it will be refined and improved on by others as time goes on (Carse 1986). To describe great power competition, we will first frame and define what competition is at a foundational level—as well as describing its goals, elements, context, and nuance. We will then define what a great power is and conclude by defining great power competition. The intent of this chapter is to establish a starting point for conversation over how to define great power competition and to provide a definition to be built upon and refined.

Why define great power competition? What does this mean for those concerned with the interplay of nation states in today's world? Let us begin at the beginning. Shared language is what establishes and enables communication between people. Speaking the same language can be full of misunderstandings and wasted communication leading to significant breakdowns in exchanges of ideas, if people use different meanings for the same words. The resultant consequences can be enormous, especially when decisions of international significance are made based on fundamental understandings of the words used in communication. The concept of great power competition is an anchor to much national and foreign policy, not only in the United States but in many other nations around the world. Great power competition is carried out between nation states capable of competing across the spectrum we will discuss, but requires the participation of all elements of society, from the individual citizens of a country to the engines and mechanisms of industry to the intricate bureaucratic workings of governments. Furthermore, the most effective competition at any level is carried out by each member of the team or entity understanding not only their own role, but the overall goals and intents of the competitive effort. Understanding the totality of the elements at play in guiding a nation or organization through highly nuanced and complex competition is essential both for leaders and those supporting leader decision-making through well-informed academic research or thought and well considered policy recommendations. Omitting or neglecting one or more elements of competition in crafting and orchestrating highly complex competition strategies with uncounted moving parts carries the latent potential for failure or defeat by an opponent. This is true especially with an opposition which pays attention to all the essential elements and shapes their competitive efforts to take advantage of their adversary's weaknesses.

Unlike the Cold War, where only two great powers competed against each other and were defined primarily by military capability, today's great powers are not defined primarily by military competency or capability. Instead, today's great powers

[1] There are two primary identified types of competition, finite and infinite—we will focus on finite completion for this analysis.

are characterized and defined through a multivariate implementation of competitive efforts across the information, economic, military, political, and legal spheres—name just a few at play. Even without military capability being the primary defining characteristic of great power competition, the nation states competing amongst each other are no less acrimonious than any other time in history, no matter how carefully or politely veiled their policies may be. Many theorists and researchers believe the consequence of all-out war is the primary mitigating in remaining in the sphere of competition rather than escalating to committed and declared conflict, especially in the era of nuclear armaments and other potentially uncontrollable means of conducting warfare. Exploring these second and third order effects are outside of the scope of this chapter, yet are dependent upon an understanding of the foundational elements of competition we explore and define here.

This chapter will explore in detail the concepts of competition and the logical extension of great power competition. To provide context for this discussion we will provide the definitions at the beginning. These definitions will be developed and explored through the course of this chapter.

Competition: The interaction among actors as a means of securing interests and advantages while denying other actors either the ability to dictate or shape the rules of the game or the ability to thoroughly pursue interests within given limits or constraints.

Great Power Competition: The interaction among notably powerful nation state actors capable of operating globally and on any dimension as a means of securing national interests and operational space while denying other actors of relative power either the ability to dictate or shape the rules of the game (geopolitical, economic, military, etc.) or to deny the cohesive pursuit of national interests, below the threshold of war.

A Description of Competition

Let us begin with exploring a fundamental understanding of competition. One of the most universal exhibitions of competition is gameplay. In every society, children play games—with nearly universal characteristics. The games girls and boys play are different, where young girls naturally focus on games of cooperation and subtlety, while boys play games of competition and conflict from the earliest ages.

These games are training for later in life, where the rules and principles followed on the playground shape the behaviors of people running governments and enterprise.

Observing competition in a laboratory-like conditions of a sports field can yield valuable insights to the nature of competition.

Competition, at its most basic principles consists of one goal with two elements: (goal) achieving an objective before the opponent while (element 1) doing the things it takes to achieve that objective, e.g., *offense*, and (element 2) preventing your competitor from achieving the goal before you, e.g., *defense*. While other subordinate goals may emerge such as winning a series or showing off athletic prowess, those secondary goals or mechanisms are dependent on winning. Take a basket ball game—the primary goal is to score more baskets than the opposing team. In order to score more baskets, the team must work together to move the ball down the court, maneuver around the other team's players, and throw (or dunk) the ball into the hoop. When the other team has the ball, the entire focus of the team pivots to preventing the ball from being placed in the protecting team's hoop through blocking the opposing team's movements, attempting to the take the ball from the opposing team, or trying to demoralize or confuse the other team in any way (permittedly) possible. The entire spectrum of competition is present within these games—*tactics*, *strategy*, *adaptability* and even misdirection (faking a pass before attempting a goal).

The same can be true for board games. The quintessential Western board game, chess, seeks to eliminate the other team's pieces, while defending the king piece from capture. The quintessential Eastern board game, Go (or Wei Qi), seeks to control more territory at the end of the game than the opponent, while preventing the opponent from doing so (Lai 2004).[2] Many, if not most, board games have elements of offense and defense, whether capturing the opponent's pieces, gaining more tokens, or simply obstructing your opponent's movements. Board games *appear* to require more strategy than games played physically, but still require the same elements of *tactics*, *sustainment*, and *consistency* even though they may manifest in different ways.

Keeping these principles in mind, we can test these observations and expand our testing hypothesis of goals, offense, and defense to the larger, more complex world of business. When Toyota entered the American market in the late 1950s, very few expected them to compete with Ford, Chrysler, and GM. At the time, Toyota was viewed as a newcomer who would quickly disappear. And disappear they did, for all practical purposes, until the early 1980s following the oil crisis. Taking advantage of decades of developing competency in producing a dependable and inexpensive vehicle, Toyota quickly emerged as a dominant competitor in the American automotive market by doing more than producing a dependable vehicle that was affordable to maintain and operate. Toyota persuaded consumers to purchase their cars and trucks rather than their opposition's familiar and iconic vehicles which epitomized the American way of life. Using simultaneous offensive and defensive tactics of locating manufacturing plants in America and employing Americans to build, publicize, and sell from local showrooms at low prices, Toyota established themselves as a part of American society in only a couple of decades. They opened the door to other Asian companies of all types and challenging the American automotive industry writ large.

In a more recent example of business, Amazon challenged the entire physical goods retail industry by developing online purchasing to a level never before seen, thanks to the recent emergence of the World Wide Web three years previously in 1991. While other companies had pioneered online sales in the 1980s with select

[2]David Lai's 'Learning From the Stones is an excellent primer on the game of Go and how it applies to understanding the Eastern approach to strategy and competition. The author of this chapter teaches Go to any newcomer seeking how to understand competition in the Asian sphere.

corporate customers, Amazon opened the market to anybody with a credit card and a computer. Beginning with book sales, Amazon refined its business processes and quickly expanded to just about every other purchasable retail good in just a few short years. Today, Amazon has expanded from exclusively online sales to in-person sales through the purchase of Whole Foods and experimentation with brick-and-mortar Amazon stores. In response, traditional retailers moved to defend their sales by offering online or hybrid shopping to avoid losing their place in the market, if not going out of business completely. In addition to the elemental principles of *offense* and *defense* found in board games, competition in a complex business environment requires *adaptability* and *consistency* for a company to successfully compete. Adapting and remaining trustworthy requires *functional competency* to meet expectations and deliver what is promised to customers able and willing to switch loyalty if their needs and desires are met better by someone else.

Board games are elective—one can choose to play or not to play. The stakes for playing, winning, or losing a board game are low. The stakes for competing in business are much higher. Not only is the lead businessmen's job is on the line, but all of his or her employees and stakeholder interests. Customers are also affected, as they depend upon a business to supply their wants or needs, sometimes exclusively sourcing their goods through the business. As we ascend the ladder of complexity, the consequences of competition tend to rise as well.

The highest magnitude competition of all is between nation states. In a board game, the objectives and environment are simple. Follow a limited set of rules, a player must accomplish a defined goal while preventing the opponent from placing or capturing more pieces. In business, the complexity of competition scales with the number of individuals involved and the breadth of environment. Rules of behavior become less cut and dry as size and complexity increase; the rules become more flexible and adaptive, more open to debate. The objectives of competing in business are less clearly defined. The objectives often change as the capabilities of the primary company and its competition change—think Amazon moving from bookselling to selling everything. Competition between nation states can be understood as scaling on a similar level. The complexities involved in competition between nation states involve every element of society: the multitude of industries present within its borders, geopolitics, the rules and laws—of not only the two countries at play but international laws and norms—and then, for good measure, the entirety of every other country (industry, societal, legal, etc.) which has involvement in either or both of the two competing primary countries. In today's global environment, it is almost easier to list the countries not involved than the ones which are Nation state competition has existed for so long many people take it for granted.

Some of the earliest recorded histories of mankind describe commerce with other states and wars between powers, characteristics of nation state competition (even if not properly defined as a nation, as in the case of the Vikings or Mongols). With the settling of the Americas and exploration of Asia, the entirety of the known world was increasingly defined and described with boundaries. As the Treaty of Westphalia settled into place, those who controlled a given area within certain boundaries were viewed as the rulers of that nation. Any attempts to expand the boundaries of the nation required diminishing the boundaries of an adjacent nation. But territorial conquest or annexation was not the only way to expand a nation's borders. The Dutch, British, Portuguese, and many others used colonization to compete in the race to gain control of the known world and control the flow of resources across the globe. Rather than erasing the name India and replacing it with Great Britain, Great Britain 'allowed' the Indians to retain their country (defined as not removing them from their land and replacing them with British Islanders), while controlling the administration, trade, laws, religions, etc., as did the Dutch, Portuguese, French, Spanish, and other players competing for expansion of the Western world.

At a less global scale, European history has been a nearly unending saga of competition at every level. Taking into account just the last quarter millennia, Europe has seen internal strife within countries of different factions competing between themselves to control a country or countries attempting to compete through military actions, influence, finance, religion, culture, and even marriage. Some countries, such as Poland, have ceased to exist for decades, only to emerge after the nation state which subsumed them was no longer able to maintain monopoly. Today, with the European Union, countries compete with economics and politics no less competitive than when military conflict settled disputes.

Indeed, at the nation state level, and perhaps seen at the lower scales of competition, the ultimate goal is the complete monopoly of the competitive environment. Once a complete monopoly exists, there is no competition. Thus, a competitor must be able to continue competing without succumbing to internal or external forces and to sustain competition beyond the initial weeks or months of the competition. Many cases in history have demonstrated that a winner in competition or conflict is the competitor who is able to endure the longest—not necessarily to the strongest, most technologically advanced, or smartest.

Adding further complexity to nation state competition are three main pillars within the makeup of a given country. These three pillars are *population*, *military*, and *government* and often operate to separate ends—sometimes aligning, sometimes conflicting (Clausewitz et al. 1984). Many times the population or military have a different goal or desire than the government, resulting in different behavior not in line with the government. At times two of the elements can be in conflict with the third, and *rarely* are all three seen operating in concert—except against existential threats. This internal conflict can add extremely complex variables which increase challenges for those trying to make sense of a single nation state in context of larger competition, especially when examining competition between multiple nation states. Each has their own internal conflicts swirling amidst vying equities between the different levels of the nation state. Eastern approaches to the separation between the sectors of society are difficult to determine, as many have not been translated into English.

The *population* level consists of the general masses of a nation who are not in government or military. In some ways, population is both the easiest and hardest to define as they are defined by exclusion from the other two levels. Individuals from the population make up both the government and the military of a nation. Individuals are not monolithic in nature and embody the complexities of the culture (the multitude of cultures) within the nation, business, and other catchalls not defined by the labels of government and military. Some observers would comment that business or industry should be a tier of identified components of a nation state. For the purposes of our definitions, business and industry will not be included as a key element of the population, as they are not directly responsible for policy and law in majority of cases examined when forming our typology defining the categories. The population is the base of ALL members of society—the government, the military, business and commerce, medical, media, etc. All sectors of society are comprised of members of the population, who bring their culture(s) and interests to any competition. Those interests are inextricably entwined with (represented by) the government's interests, and often rely on engagement with entities outside of the nation state to function. This of course, complicates matters. Engagement—or interaction with these outside entities—does not require competition outside the nation state, yet competition at the nation-state level often occurs when fields of enterprise encounter similar activities across the global field.

The *military* is a level which is not always applicable in every nation, but is present and relevant in most nation states large enough to maintain a military. By and large, even small countries maintain some form of military force which can be used by the government to protect national borders, defend national interests, or—in extreme cases—control the population to the benefit of the government. It is true there are cases which involve the military changing the government, but by and large those are the exceptions. Those cases are highly visible outliers, we do not consider them consistent enough to build a definition applicable to the majority of situations. The military possesses considerable power in any country, as a rule. Far exceeding the financial investment of the nation state complex structure, the military has the ability to shape and direct national policy and decision-making because of the capability to exert force and power both internally and externally. In the vast majority of situations, the military's power is harnessed and directed by governmental leadership. While systems of checks and balances exist to direct and govern military behavior and actions, potential still exists for the military to exert power and influence decision making and actions among both the population and the government. This leads to complexity in understanding internal dynamics of competition even though militaries are the least active level in the nation state to engage in competition, being reserved for escalation outside of competition to war or defensive actions against attempts to escalate into conflict. Rarely do we see a nation's military engaging in conflict or competition in absence of governmental direction. Militaries are designed for conflict rather than competition, a topic we will discuss later in the chapter.

Government is the smallest proportion among the levels of the nation defined by Clausewitz (1984), and can be likened to the bit in a horse's mouth or a rudder on a ship—small, but shaping the direction of the entire nation state. The government comes from the population and is charged (ideally) with care and maintenance of the nation state and all elements therein. Governments are theoretically able to orchestrate all instruments of national power to compete with other nation states, or to compete in narrow bands such as industry, medical, etc. Because of this, governments are—for the most part—able to set the tempo and conditions for competition, at least within their own borders or sphere of influence. Nation states which seek to secure interests through non-cooperative means can do so through either competition or through conflict. Common knowledge among the field of international relations describes one of the goals of competition as pursuit of interests, but the defining national interests becomes increasingly contentious as one increases the fidelity of detail within a definition or analysis of interests. Defining interests is not the purpose of this chapter, yet national interests *are* a key element of great power competition, as nearly every actor in competition is pursuing some sort of advantage or interest. The difficult part of defining or describing national interests arises when the different levels of population, military, and government begin to cross and blend. The people who make up the government are difficult to separate from the population (in some cases, the military as well) and often bring their own personal interests which may be conflated with national interests, making analysis of competition difficult at best (Bueno de Mequita and Smith 2011).

Many definitions of competition describe competition in terms of *advantage*, or seeking to gain advantage. A foundational assumption rests on whether or not advantage is an essential element of competition. Said differently, is advantage even an element of competition or merely a by-product? Well, in order to win a competition, a competitor must gain advantage. One cannot exceed the other's ability, skills, or accomplishments without advantage—otherwise the competition will end in loss or a tie. However, much attention is placed on advantage without examining the other side of the coin—denial of advantage, or simply *denial*. A competitor can pursue advantage endlessly, only to find themselves shut down repeatedly by the opposition or adversary if they do not guard themselves, their goals, and their interests. Thus, advantage is an essential aspect but cannot exist without the other elements of competition identified throughout this section—*offense*, *defense*, *sustainment*, *adaptability*, *strategy*, *consistency*, *competence*, and *misdirection and deception*. In the following section, we will examine and define each of these factors briefly.

Components of Competition

In this section, we will look at the primary elements of competition, and examine the function, value, and utility of each element. In the first section we looked at the key constituent parts which comprise competition at its most fundamental form, as applicable to any application. Defining and describing the different components of competition help us make sense of and analyze the complexities of competition. Those key components identified are offense, defense, sustainment, strategy, tactics, adaptability, consistency, competence, and misdirection.

Offense is the most visible and essential element of competition because it gains ground and sets the initiative. Offense is the first of the two prime elements of competition. Without offense, a player or actor cannot fully compete. Inherent to the nature of competition is the idea of movement. The function of offense is the movement of the player or actor from which other elements are nested. Offense is characterized by initiative, purposeful movement towards a goal, and pushing boundaries as far as the rules of competition (or referees—observers in the competition) will allow. Offense does not have to be effective to be present or exist. There simply must be effort and intent, even if feeble or inept. For offense to be effective however, a host of other elements must be present in comparatively high levels—at minimum more effective than the opposing competitor. An effective offense emerges from the totality of its parts being more effective than the opposition's efforts. Many times, for an offense to be effectual, it must be different than the competitor's offense, since no two competitors are ever equal. Each brings different strengths to bear on the situation—often shifting and adjusting emphasis or changing internal relationships as the situation shifts and adjusts as a means of moving faster than the opposition. To be dependable and of utility, offense must be consistent, forceful, and of a unified—or at least coordinated—purpose of execution (Twain 2014).[3]

Defense is the less visible of the primary elements of competition. It protects goals and equities against oppositional offense and is often ignored when overshadowed by the more flashy offense; it is the essential second of the two prime elements of competition. With an exceptional, unrelenting offense against a less-than-equal opponent, a defense is not essential in competition but against an opponent with either relative parity or against a stronger opponent, defense is paramount to success. The function of defense is to create space and time for the offense to achieve effect, which can be done by denial of action or desired action, disruption of the opposition's offense, misdirection or deception, and a host of other elements designed to halt the initiative or create disruptions, diversions, or delays within the opposition's decision-making cycle synchronizing and administering the overall competition effort (Boyd 2007). Effective defense can effectively neutralize a more powerful offense and bring capability disparities into equilibrium—if not completely upending the dis-equilibrium. Often, the disruptions a defense brings are more psychological than 'real' because they cause the opposition to invest energy into efforts which are ultimately distractions or unfruitful, and hopefully to waste effort preparing for a different game than is ultimately played.

Sustainment is the first of the subordinate elements essential to competition.

Sustainment is the ability to not only continue to the end, but to ensure all parts of the competition are maintained. In the game of chess, the game pieces could theoretically be reduced to two pieces (kings) chasing each other around the board and go on for days until one player gives up. In baseball, a game can go on indefinitely until one team is so tired they cannot maintain a defense and yields a run to the opposing team. In business, two competitors can go back and forth, as Ford and Chevrolet have since their competition began in 1911, each striving for a monopoly on the market yet never succeeding. Today we see nation states competing at small levels for regional prestige and at global levels to—still—unclear ends. The first player to yield completely loses the competition. Yielding should not to be confused with a retreat

[3]The role of luck should not be diminished, but it should never be planned upon. Mark Twain's famous short story entitled 'Luck', is a wonderful illustration of the effect of luck with geopolitical effect.

from a position of disadvantage. A retreat or pause may provide later strategic gains, as in the case of soccer or hockey where moving away from the goal provides the ability to breath, assess the situation, and reposition players. Ensuring the members of the team are rested, nourished, and hydrated are essential sustainment roles for athletics; ensuring supply chains can operate uninhibited are essential sustainment roles for business and nation states. While many may view sustainment as from a logistics perspective, sustainment also relies on morale, direction, or vision to sustain an organization or entity, such as coaching, training, or management to equip those at the worker level with an understanding that their efforts are not in vain.

Strategy is the next subordinate element essential to competition. Most dictionaries define strategy as some variation on 'carefully developed plans to achieve a goal.' Strategy can be considered as the big picture or long-term framework to operate within. To borrow an example from music, strategy is the structure of musical work—in classical music the structure of a Symphony consists of four parts each with defined parameters while the specifics of each are left up to the composer. In jazz or rock, a basic chord progression within a certain structure provides the elemental building blocks from which to diverge and create innumerable variations. In each of these musical examples, the 'strategy' or basic shape is immediately understood by all participants while the specifics while for unique and innovative implementation are left to individual execution. In games or business, strategy involves describing and defining goals to achieve while setting basic parameters for what people should and should not do. For a nation state, strategy describes a clearly defined coordinated actions to arrive at an articulated end state. The benefits that a strategy brings are that it—when well-articulated—provides mission and vision for the individuals and groups are taking the actions at a daily basis and forms a baseline from which to develop tactics which are both beneficial and NOT counter-productive for effective competition. The utility of a strategy is that it provides boundaries for all members of the team, rules of the game internal to one side (which may or may not be visible to the adversary or competitor), and helps to eliminate an overwhelming assortment of choices which often lead to 'paralysis by analysis.' Put another way, strategy is a way of framing a problem to focus on a positive competitive outcome rather than obstacles or challenges to success.

Tactics are the compliment of strategy. In our example of musical frameworks, tactics are the specific melodic and rhythmic implementations which bring to life the vision of the composer (in the case of classical music) or the soloists and rhythm sections (in the case of jazz or rock). Tactics are the minute, highly detailed specific implementation of a vision tailored to the culture and strengths of the players on the field and the environment. Tactics can change at the drop of a hat. Tactics are not set in stone, but are malleable and fluid to adapt to a situation, yet still rehearsed and practiced by all members of the team. An example would be a football play which set up one team member to pass the ball to another, with defense team members channeling the opposing team away from the offensive effort. The value of tactics is they are understood by all members of the team, yet can be changed at a moment's notice to another rehearsed tactic while still moving the team towards the goal in within strategic parameters. A team or organization with an abundance of tactics tailored to specific environments are able to maneuver within the opponents OODA loop[4] (Boyd 1986, 2007) much faster than the opponent can react. This freedom to maneuver creates space and time to develop additional strategies or tactics to provide advantage and/or additional opportunities which would not have been had otherwise.

Adaptation can be viewed as the nerve system of competition. While one team engages in strategy and tactics, the other team engages in them as well. The ability to adapt and adjust one's strategy and tactics as part of sustaining the ability to conduct both effective offense and defense is essential for successful competition in a complex environment. Adaptation identifies changes in the environment and sends signals to the decision-making elements (coach, board of governors, governmental leaders, etc.) that the environment has changed and requires changes of behavior to avoid future pain or setback. In complex, rapidly changing environments (the more complex the environments, the more essential adaptation is), the ability to adapt to changes within the environment and to opponent actions becomes an essential interplay between strategy and sustainment. A competitor must be able to adapt within the framework of a strategy in order to sustain an effective offense and defense. Often, this involves changes to tactics although minor adjustments to strategy may be essential to achieve competitive effectiveness.

Consistency is perhaps one of the most overlooked elements of effective competition. Consistency provides all members of the competitive team, organization, or nation with the ability to sustain a sense of identity and integrity (wholeness). Consistency can be wholly compatible with adaptation, if the identity includes adaptive characteristics. Individual members of an organization are human beings, with all the characteristics of humanity, meaning they are inherently resistant to change and possess the need for identity and belonging. When too much change—especially in the wrong areas—is enacted too fast, those individuals or component units begin to lose their identity and their connection to the larger identity. This loss of consistency removes their ability to contribute to the benefit of the overall end state. Individuals without this continuity begin to operate independently for the benefit of the individual or unit rather than towards a greater purpose. By maintaining consistency of effort, identity, and strategy, an entity made of many subordinate elements is able to maintain its identity as a single entity rather than a conglomeration of individuals.

Competence is another fundamental assumption underpinning all of the elements of competition from offense/defense to consistency. Competence is the ability to carry out given tasks at a satisfactory level which consistently meets goals and needs. Competence is the skill which distinguishes idealists from competitors. An effective competitor is able to carry out the tasks and skills of every level of competition, both in developing and communicating strategy to implementation of tactics and sustainment. The function of competence is to focus on a discrete skill and carry it out to the highest possible level (or at a minimum to an acceptable standard).

---

[4]Observe-Orient-Decide-Act (OODA) is a decision-making framework for taking the most effective action in a high-paced, complex environment, developed by legendary strategist Colonel John Boyd.

Competency is the foundation of trust—quite possibly the most core element of building an effective competitive team—and the prerequisite for the ability to carry out effective offense and defense in competition.

Finally, *misdirection and deception* are most often defensive actions used to cause an opposing competitor invest resources in an unproductive area, reducing the opposition's ability to defend against eventual offensive actions or actions used to conduct an effective offense with less resistance. Misdirection and deception are present at every level of competition, from board games where a tempting pawn is moved into an easy-to-capture position which draws out a knight or bishop, fake-pumps on the basketball court to draw the opposition off balance, or in great power competition as was seen by Russian disinformation following the downing of Flight 17 over Ukraine in 2014. Misdirection and deception play an important role in any type of competition but are often viewed as playing dirty and are the subject of complaint by those it is used against. Likely because it is so effective.

Common Distractors in Understanding Competition

One thing we must at least be aware of is the potential for conflating subordinate elements of competition with the primary elements—if we are to clearly understand the holistic nature of competition. If we are going to understand competition in its entirety, we must remain cognizant of which elements are part of the essential make up of competition and what are subordinate (or complementary) elements. Often, subordinate elements are used as 'red herrings' or distractors to epitomize competition using a single aspect rather than competition as a whole. Common distractors or mischaracterizations of competition writ-large are the terms *Power*, *Advantage*, *Leverage*, and *Influence*. Each of these concepts play a role, but can be taken as the entirety of competition at the neglect of other potentially more important elements, or at risk of loss of context and orientation to the bigger picture. While not part of our main definitions, these aspects of the concept are crucial to our conversation defining great power competition. This is not to minimize the utility of these concepts. Perhaps they may be best understood as measures of effectiveness (outcome evaluation) rather than fundamental components of competition.

Power is a necessary element of both offense and defense and is the ability to carry out plans with little regard for opposition. Power, while possessing a quality of its own, is dependent on the skill and application of both force and mass. Power can be analogized to driving a bulldozer—under the guidance of a skilled operator, excellent work can be done in much less time than an entire team of construction workers with hand tools. But it can also deeply screw things up, even set back a competitor severely if misapplied to the point of destroying the environment (not limited to physical ecosystems) or playing field. Power can arguably be considered as competence—although many would debate that power is actually a subset of its own, focused primarily on force, with little to do with proficiency or capability. If viewed from a brute-force perspective, this relevant, technical competency enables a competitor to place greater force in a given place and time with less effort than a less skilled competitor. An example of this would be a boxers punches—a trained, competent featherweight boxer is typically able to deliver more power in a well-trained blow than a strong barroom fighter.

Advantage provides opportunities for offense, while reducing the demands on defense. We define advantage as the opportunities and space to maneuver freely, to employ one's abilities and/or assets, or to shape the environment with acceptable push-back. Continued advantage is not required however, if a successful strategy is employed which lands a player with the final win—however small the margin (Greene 2006). Many of the memorable televised games in history (and many Go or Chess games for that matter, although less exciting to the average viewer) are of teams winning at the last moment, coming from behind after possessing little to no advantage for the majority of the game. In many competition discussions, writers tend to spend an exceptional amount of time on advantage—especially on gaining and maintaining advantage—without considering that constant advantage is not the key factor in winning a competition. Viewing advantage as the primary element of competition may mislead a player into a false sense of security and cause them to trade strategy (or any of the other necessary elements) for immediate advantage.

Leverage is a subset of advantage which maximizes existing advantage to greater effect or utility. We define leverage as the ability to exert one's advantage or assets to create a larger advantage or space to maneuver more freely or to be able to create a larger advantage or maneuver space where one may not have existed previously (Nye 1990). In a business context, leverage may be a lowered manufacturing cost (advantage) which one competitor offers an ally to maximize overall lowered costs to undercut the opposition. In nation state competition, leverage may be policies or laws which control the import and export of competitor's goods, providing an advantage.

Influence is subset of both power and advantage, indirectly expressing and contributing to the two prime factors. We define influence as the ability to convince or manipulate other actors—often allies—to behave in ways conducive to the primary (influencing) actor's benefit (Nye 1990; Wu 2018).[5] Influence is one of the squishiest elements of competition because it is not easily measured. Influence effects are often not directly attributable to a single driver or initiating action, as many other conflating variables may have contributed to the *influenced* actor's decisions and behaviors.

Spectrum of Engagement

Competition, as we have defined it so far, is relatively straightforward—albeit with many components. By examination of the definition itself, this concept is relatively simple. So why is competition considered to be so complex? It is because competition

[5]Influence is sometimes referred to as soft power, a term coined by Joseph Nye in his prescient article following the collapse of the Soviet Union.

is only a component of a larger spectrum of engagement, which we shall explore as we attempt to define competition by looking at its *adjacent concepts*.

Clausewitz (1984) was one of the first to identify that the spectrum of interaction begins at cooperation and moves through competition to an extremity state of conflict/war. International relations theorists have built upon these spectrums, providing useful categories of *harmony* as the foundational, least contentious element of interaction and moving to *cooperation*, *competition*, and finally *conflict*/war (Keohane 1984).[6] Let us examine these four aspects briefly. Harmony indicates everyone is getting along, operating with no dissension in the mutual reinforcement of each other's beliefs, values, and behaviors. This is, of course, a rare occurrence. Cooperation means individuals are working together within a given context, even if they do not align on all elements of overarching purpose. Competition means a further divergence of commonality, although it does not categorically mean a difference in mutual interests. Competition is often characterized by a common mutual interest in obtaining a goal—often a monopoly in the area of focus—but with agreed-upon constraints in how the competing parties interact. For instance, one opponent does not pull out a revolver and shoot the other in a chess game. Neither do chess players kick each other in the shins or knock each other flat on their back, as you would in a game of football (European and American, respectively). Direct conflict/war is the furthest, most extreme end of this spectrum, where mutual interests either take a backseat or are completely disregarded: disabling or destroying the opponents become the foremost goal of a competitor in war. We describe conflict as non-persistent armed or violent encounters between powers or nations, while war is defined as a persistent commitment to violent action. Great power competition *almost* inevitably involves conflict, although usually carried out through proxies to avoid direct conflict or war. War is much more difficult to reverse as resolution almost always requires one side to concede to another, with resulting loss of ability to compete, even if for a time.

For the purposes of this chapter, we are focusing primarily on understanding the true nature of competition within the context of a larger environment, with minimal description of that environment. The actions and behaviors of an actor engaged in competition may stray outside arbitrary limits within the neat categories described in the harmony-cooperation-competition-conflict/war spectrum. While engaging in the primary mode of competition, an actor may sometimes emphasize cooperation or sometimes emphasize conflict. Actions which blur across categorical lines are inevitable and will occur. War, however is a point of embarkation in which there is little easy return. Thus, competition may be a dangerous realm in which to engage for some nation states because of the narrow gap between competition and war.

Humanness, with all its complexity, makes separation between the different elements of the spectrum of engagement additionally difficult to define and maintain. People, contrary to conventional wisdom, are often more emotional than logical.

[6]Robert Keohane's seminal work 'After Hegemony' describes a three-part typology of Harmony, Cooperation, and Discord. Much International Relations work uses Keohane as a foundation. Keohane includes conflict as part of cooperation and even assumes competition cannot exist without conflict.

Decisions are very often made based on emotion, with retroactive rationality applied to decisions to justify choices. Once actions and behavior are implemented, rational 'justification' for those choices and actions become even more important to reduce cognitive dissonance from throughout the entire entity, from leadership to rank and file. *Cultures*, which are based on beliefs and values shaped by one's environment, create different understandings of what it means to be human and who those 'deserving of humanity' are. Cultures with clashing beliefs and values increase the friction and misunderstanding which instigates conflict and lack of cooperation. Too often, dehumanizing the opposing entity is a tool used by competitors to mitigate human sympathies and unwillingness to engage in brutal tactics or strategies. When an opponent is viewed as 'other', the human element which regards fellow humans as equals is desensitized to viewing the opponent as 'not as human', and therefore worthy of treatment or consideration which removes humanness.

It is with these considerations that we can understand how the Eastern and Western *cultures* view each other in competition or escalation to war. Shaping western thinking for the last 200 or more years, Clausewitz's description of war and conflict, combined with Westphalian understanding of what comprises a nation state, has formed the basis for much of the western cultural understanding of competition. Westerners understand competition from a paradigm of parity, with equivalent rules and norms for all actors who operate from a shared paradigm, often framed in military terms (Haynes 2020). In contrast to this, Eastern cultural understanding has been shaped by Sun Tzu, Taoism, and Confucian thought for the last nearly 2,000 years, which articulate there are few rules to be adhered to (Sun Tsu and Sawyer 1994). Rather, to those immersed in the eastern ways of thinking such as Sun Tzu, the single overarching rule is success and adaptability rather than a structure of rules and processes adhered to in the Western mindset. This means different approaches to competition by different actors, depending on their subscription to Western versus Eastern thinking. A way to view this disparity is imagining an American football team taking the field against a European football team, each thinking they are playing the same game simply because it is called by the same name. The rules, goals, norms, and cultures informing each team's understanding of the game are different, yet with the same goal of scoring more goals than the other. This clash of cultures often leads to overly complicated analysis or misdiagnosed relevant factors when examining great power competition without adequately defined elements of what competition is.

The Nuance of Competition

Competition, in context of well-defined parameters, is fairly straightforward. Yet, with in context of the larger spectrum of engagement, competition is full of nuance. As with most definitions which are useful, this definition for this complex concept merely begins the conversation of understanding to be built upon by further work. In this section we will discuss some of the nuances of competition as a frame of reference for future work.

Competition is rarely a solo effort. In its most rarefied and 'pure' forms, such as a board game or tennis match, a single individual is able to compete against another individual. But rarely if ever do we see in larger arenas a single competitor. Competition is often characterized by *alliances* among competitors who find it in their best interest to partner with another competitor—or at least, 'an enemy of my enemy'—to avoid defeat, create space and time to gain or make up advantage, or to adjust defense (Wishnick 2011).[7] Alliances can also be formed by competitors seeking a similar goal, as was seen in the World Wars or by businesses facing a major shift in the environment, such as the rise of Amazon.com in the 1990s. Rarely do allies have completely mutual interests and goals. Consequently, the actions and behaviors of each actor may be contrary to an allies' goals, strategy, or tactics. This is an important element when conducting an analysis of competition with multiple actors, as a behavior which seems self-defeating or contrary to successful competition may actually be a completely logical behavior when considered in light of another goal.

A subset of alliances is *dependencies*. Dependencies are formed when a major competitor draws weaker or 'lesser' layers into their sphere of influence in making the lesser players dependent on the greater power and thus behold and to acting on the larger powers behalf at times. Sometimes called proxies, dependencies often do so to avoid conflict with the larger power and the focus of competition themselves. When a competitor is attempting to conduct an effective offense and defense, they often have *simultaneous streams of effort* in play. The use of simultaneous streams of effort provide contingency plans to implement strategies if the environment or playing field were to change. Multiple ongoing efforts may also a defensive tactic, such as exhibiting the capability to do multiple things at once, which increases the difficulty for a defender to know how to distribute resources. Simultaneous streams of effort may also be used to blunt an offense by requiring diversion and dispersion of resources, reducing the competition's ability to focus effort.

Mutually conflicting efforts are typically unplanned, although some competitors skilled in the art of misdirection can use them to great effect to divert competitor's offensive or defensive resources, making them less effective. Mutually conflicting efforts are often considered to be a symptom of lack of unity, consistency, or even competence. Yet, the consequence of ignoring the significance of intentional mutually conflicting efforts (resulting in a successful misdirection or deception operation) is high enough to pay attention to the impact or importance of the seemingly contradictory behaviors and actions (Fig. 1).

Disparate goals and values are inherently contradictory yet ubiquitous in larger and more complex competition environments, making it more difficult to make sense of the truly foundational goals of the actors, much less assess the effectiveness of strategy or tactics in play to achieve those goals. Analyses which only identify a small handful of goals for a large organization or nation state may often result in

[7]Elizabeth Wishnick's (2011) analysis of the Sino-Russian partnership alliances and separation during the Georgian crisis nicely encapsulates the complexity of alliances and competing interests. The cited article is also available as a full length monograph from SSI.

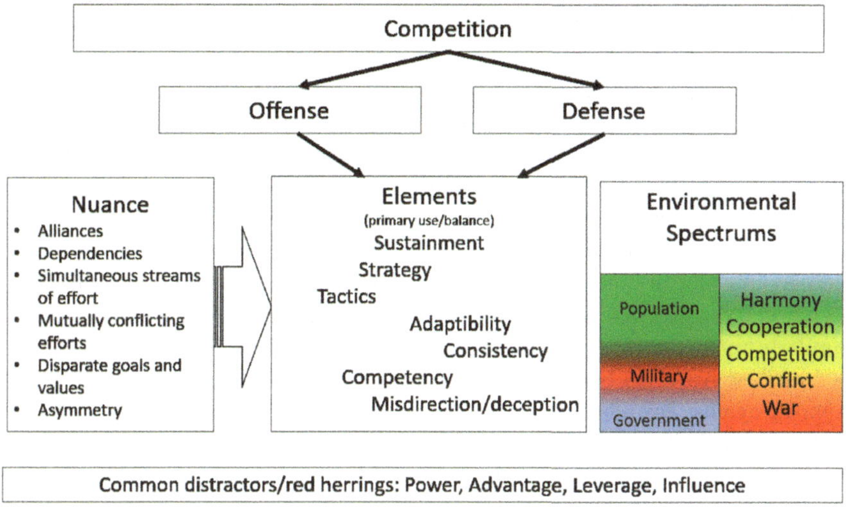

Fig. 1 Great power competition defined at a glance

harmful decisions and planning assumptions. Multiple goals or values which are not always aligned within an organization should not automatically be viewed as a negative, but rather it be initially viewed as a strength for an organization because of the diversity which differing perspectives bring. While this complicates developing an understanding of an opposing competitor, it can also be used as a tool to create disunity amongst the opposition, at the opposition is not always adept at ensuring internal consistency.

The final nuance of competition which we shall describe is *asymmetry*. Asymmetry is an imbalance of resources or competencies when compared to the opposition. In team sports, one team may have a strength of passing the ball and using teamwork to achieve success, while the other team may have a strength of a single talented player who is able to monopolize possession of the ball to great effect. In the business world, customer service and relationship may be the great strength of one competitor, while their opposition's product quality tends to be the primary source of advantage. Among nation states seeking to maintain equilibrium, one nation may have excellent political relationships, while another nation of similar size possesses a military able to repel any invasion. Asymmetry between competitors should be a given; attempts to overcome asymmetry and develop symmetry between competitors without maintaining asymmetric advantage is usually an indicator of eventual loss of the competition.

The Great Powers Today

To this point, we have talked about nation states in broad, general terms. As of early 2020, there are 195 countries in the world, many of which are small and unable to compete with larger countries in direct competition (examples would be Monaco, Liechtenstein, or Tuvalu). Many countries are able to compete at a regional level, but not at a global level (examples would be El Salvador, Hungary, or Thailand). Most countries compete effectively in industries such as manufacturing or finance globally, but not in other areas such as political influence or military power and presence. Examples of this would be Germany and Ireland, or Indonesia and Taiwan. Similarly, many countries compete regionally in multiple spheres of influence, such as Brazil or Australia but not at a global level. Only a few countries are able to compete against any other country in the world, at any given time. These countries are what we call 'The Great Powers', because of their ability to compete on nearly any dimension—information, economic, military,[8] political, and legal (meaning ability to set or influence international policies and rules), against any other country in the world and on any continent (Overfield and Tallis 2020). Countries which may be considered great powers in today's world are those capable of countering—and opposing—other nations across most of these dimensions without fearing debilitating retribution or potentially catastrophic consequence.

At the time of writing, there are only three countries in the world capable of fitting this definition of a Great Power: United States of America, People's Republic of China, and the Russian Federation. Russia only barely fits the definition, because of their lack of capability to compete at parity with the United States or China economically at a macro scale. However, they are capable of competing militarily due to their nuclear arsenal of intercontinental ballistic missiles, and narrowly through economics with their ability and willingness to arm other countries anywhere around the globe.

Defining Great Power Competition

Based on our definitions of competition and our exploration of the spectrum of engagement, we can finally define what great power competition is with confidence. Our definition for Great Power Competition is recapped as:

> The interaction among notably powerful nation state actors capable of operating globally and on any dimension as a means of securing national interests and operational space while denying other actors of relative power either the ability to dictate or shape the rules of the

[8]Most definitions of Great Power Definition have historically centered around military power, as the only two great powers, the US and USSR, were primarily defined by ability to conduct military offensives against each other. Today's environment has changed how we understand the term to a much more encompassing definition. Overfield & Tallis provide one of the better overviews for understanding the evolution of previous definitions.

game (geopolitical, economic, military, etc.) or to deny the cohesive pursuit of national interests, below the threshold of war.

Conclusion

We have methodically walked through what competition is, what the essential elements of competition are, how competition fits in a spectrum of engagement from harmonious to war, and have addressed some nuances of competition which enables us to make better sense of the complexities of competition between great powers. Our analysis of the concept of great power competition has enabled us to define what great power competition is. The lack of communication using similarly defined language and terms often prevents the ability to reach a common understanding and agreement on what a problem actually is, much less how to address the problem. Our hope with presenting this definition of competition, especially in context of the great powers, is not to establish the quintessential definition but to begin a deeper discussion leading to a more holistic and common dialectic which frames the discussion and understanding and of the forces at play in our world today.

Understanding the complexities, principles, and elements of competition from the basic to the global level permits us to more fully understand competition as it plays out on the global stage.

References

J.R. Boyd, *Patterns of Conflict*. Unpublished briefing (1986). https://www.ausairpower.net/APA-Boyd-Papers.html

J.R. Boyd, *Patterns of Conflict* (live). Original video archived at the Marine Corps University Research Archives, Quantico, VA (2007). https://www.youtube.com/user/Jasonmbro/videos

B. Bueno de Mequita, A. Smith, *The Dictator's Handbook: Why Bad Behavior Is Almost Always Good Politics* (PublicAffairs, New York, 2011).

J.P. Carse, *Finite and Infinite Games: A Vision of Life as Play and Possibility* (The Free Press, New York, 1986).

C.V. Clausewitz, M. Howard, P. Paret, B. Brodie, *On War* (Princeton University Press, Princeton, 1984).

R. Greene, *The 33 Strategies of War*, Ch. 12 (Viking Press, New York, 2006)

P. Haynes, What U.S. Navy strategists and defense planners should think about in the era of maritime great power competition. Def. Secur. Anal. **36**(1), 101–108 2020. https://doi.org/10.1080/14751798.2020.1712027

R.O. Keohane, *After Hegemony: Cooperation and Discord in the World Political Economy* (Princeton University Press, Princeton, 1984).

D. Lai, *Learning from the Stones: A Go Approach to Mastering China's Strategic Concept, Shi* [Mongraph] (Strategic Studies Institute, US Army War College, Carlisle, 2004).

J.S. Nye, Soft power. Foreign Policy **80**, 153–171 (1990)

C. Overfield, J. Tallis, *Great Power Competition: What Makes Powers Great and Why Do They Compete*? [Monograph] (CNA, 2020)

M. Twain, *The Complete Works of Mark Twain: The Novels, short stories, essays and satires, travel writing, non-fiction, the complete letters, the complete speeches, and the autobiography of Mark Twain* (Kindle edition) (Amazon, 2014)

R.D. Sun-Tzu and R.D. Sawyer, *The Art of War*. Translated by R. D. Sawyer (Westview Press Books, Cambridge, MA, 1994)

E. Wishnick, Russia, China, and the United States in Central Asia: prospects for great power competition and cooperation in the shadow of the Georgian crisis. Curr. Polit. Eco. Russia, Eastern Cent. Eur. **26**(6), 763–808 (2011)

I.S. Wu, *Soft Power Amidst Great Power Competition* (Wilson Center, 2018). https://www.wilsoncenter.org/sites/default/files/media/documents/publication/2018-05-soft_power_-_wu.pdf

Seize the Opportunity: Craft a Smart Power Strategy for the Central Region

Ambassador (ret.) Michael E. Ranneberger

Abstract The Central Region is strategically important to the interests of the United States as a global power. The drawdown of U.S. forces from Afghanistan provides the most important opportunity since the intervention in 2001 to craft a "smart power" strategy for the region in order to advance American interests over the medium and longer-term. An integrated approach combining the hard power of a limited military presence with the arsenal of American soft power will enable the United States to balance and compete with growing Chinese and Russian influence, and will be cost effective. Dynamic economic and political changes in the region, and the attitudes of local populations create a conducive framework for such a strategy. Given the region's proximity to Russia and China, American efforts will have greater success if they are not couched in terms of a zero sum approach to competition, which risks stoking regional concerns. Robust American engagement in the Central Region should take into account possibilities for limited cooperation with Russia and China on issues such as countering terrorism and narcotics trafficking. For purposes of this paper, the Central Region encompasses Afghanistan, Pakistan, Kazakhstan, Tajikistan, Uzbekistan, Turkmenistan, and Kyrgyzstan. I approach these issues through the perspective of my involvement in the Central Region as the State Department's Senior Political Adviser to the Commanders of the U.S. Central Command (CENTCOM) from 2011 to 2016, and through my more than 30 years of diplomatic experience.

Keywords Great power competition · Smart power · Soft power · Strategy · Central region

Engagement and Competition

The Great Power Competition between the United States, Russia and China has been proposed as the geopolitical framework in which American policy should now operate in the Central Region (and globally according to the current National Security

Ambassador (ret.) M. E. Ranneberger (✉)
Tampa, USA
e-mail: merambfs@yahoo.com

© The Author(s), under exclusive license to Springer Nature Switzerland AG 2021

A. Farhadi et al. (eds.), *The Great Power Competition Volume 1*,
https://doi.org/10.1007/978-3-030-64473-4_13

Strategy). A conference on "The Great Power Competition" at the University of South Florida in January of this year considered this question and the issues involved. While there were many take-aways from the conference, the broad agreement that the U.S. lacks a coherent and coordinated regional strategy was striking. This paper builds on the presentation I made as part of a panel on "regional diplomacy," in which I urged a more concerted use of soft power as part of an integrated smart power strategy (Ranneberger 2020; Colby and Mitchell 2020).

For purposes of this paper, the Central Region encompasses Afghanistan, Pakistan, Kazakhstan, Tajikistan, Uzbekistan, Turkmenistan, and Kyrgyzstan. I approach these issues through the perspective of my involvement in the Central Region as the State Department's Senior Political Adviser to the Commanders of the U.S. Central Command (CENTCOM) from 2011 to 2016, and through the lens of my more than 30 years of diplomatic experience.[1]

The Central Region is strategically important to the interests of the United States as a global power. The drawdown of forces from Afghanistan provides the most important opportunity since the intervention in 2001 to fashion an integrated strategy for the region in order to advance American interests over the mid and long-term. A smart power approach combining the hard power of a limited military presence with the arsenal of American soft power will enable the United States to balance and compete with growing Chinese and Russian influence, and will be cost effective. The changing economic and political situation in the region and the attitudes of local populations create a conducive framework for expansion of soft power.

Central Asia, in some respects, remains what it was for centuries: a crossroads between East and West, as evidenced by its significance in the Silk Road, and as a field for empires. "Here," as Frederick Starr has written, "on the vast territory between real or imagined modern empires, lies a dynamic region with historical and cultural connections in all directions and with deep ties with all the major powers…." Islamic armies, the Mongols, Russians, Chinese, and Persians traversed and conquered much of the region (Starr and Wimbush 2019). The "Great Game," played between Great Britain and Russia during the 19th century, was a competition for spheres of influence in the region, with Afghanistan seen by the British as a strategic buffer to protect India (Global Security 2020).

Russia and then the Soviet Union absorbed Central Asia. The collapse of the Soviet Union made the region a strategically significant theater of competition. The United States moved quickly to develop positive relationships with the newly independent states (U.S. Department of State 2020a, b).

Based on geographical proximity, security concerns, commercial linkages, shared history, and cultural connections, Russia and China have special connections to the region, and American policy must realistically take these into account. From one perspective, it could be argued that these Russian and Chinese interests make Central

---

[1] I served, at the U.S. Central Command headquarters in Tampa, Florida, as the State Department's Senior Policy Adviser for General James Mattis from September 2011 to March 2013, for General Lloyd Austin from March 2013 to March 2016, and for General Joe Votel from March 2016 until June 2016. I accompanied them on all overseas trips and participated in virtually all their meetings with foreign officials.

Asia part of their "spheres of influence." The advent of the unipolar world at the end of the Cold War led to the euphoric American declaration of a "New World Order" in which consideration of spheres of influence and great power competition became obsolete (Allison 2020). In fact, the United States was never in a position to fully exploit that unipolar world; the interventions in Afghanistan and Iraq highlighted the limitations of American power. The limits of American power, however, should not lead to the default position of accepting tacit spheres of influence. In the Central Region, as Frederick Starr and Enders Wimbush have observed, everything is in motion, including a seminal process of generational change. "Only the most wooden strategist would still try to characterize this vast region in terms of traditional balances of power or spheres of influence," they conclude (Starr and Wimbush 2019).

The United States is embroiled in a multi-faceted global competition with China and Russia, which involves economic, political, and security interests—but also values. The competition is too often viewed solely through the prism of military and economic power and, more recently, as competition for or within tacit spheres of influence. The shifting dynamics in the Central Region coupled with the American drawdown from Afghanistan offer an historic opportunity to re-examine our approach to the region and ramp up the "soft power" America possesses through its global appeal based on culture, language, and values. The Great Power Competition underway in the region will not be carried out in a conventional way; the competition is fundamentally asymmetrical, as Dr, Jackson said at the conference—and that plays to the comparative advantages of the United States (Jackson 2020).

Central Asia, some long-time observers have concluded, "is in the midst of a major geopolitical shift…which will elevate China's influence and importance to Central Asian states" (Rumer et al. 2016). China's Belt and Road Initiative—which echoes the abortive New Silk Road initiative of the United States—is building interconnections which give it influence in the region. However, this need not diminish the region's ties to the West (Rumer et al. 2016). In that regard, Henry Kissinger's observation is relevant: "the absence of both an overriding ideological or strategic threat (following the end of the Cold War) frees nations to pursue foreign policies based increasingly on their immediate national interest." The resulting world order, he goes on to say, "will have to emerge much as it did in past centuries from a reconciliation and balancing of competing national interests" (Kissinger 1994).

Polling data indicates that the people of the region are focused more on national interests than foreign entanglements (Stewart and Quillian 2019). The fact that the countries of the Central Region are pursuing their national interests and do not want to be anyone's pawns establishes a constructive context for American engagement, but only if the United States avoids the pitfall of framing policy in terms of a zero sum competition between the United States, and China and Russia. A policy of confrontation with China risks America's isolation in the Central Region. None of those countries would want to be—or could afford to be—supportive of America in any political conflict with China which it considered to be the result of misguided United States policy, Kissinger has observed (Kissinger 1994). "And we would really not like to feel on ourselves unfavorable political consequences in relation to some competition in our region between large powers," the Uzbek foreign minister has said (Wong 2020).

Equally important, casting policy in the Central Region primarily in terms of competition with Russia and China will serve as additional motivation for growing cooperation between those two countries. While Russia and China do compete in Central Asia and globally, they are also accommodating each other to some degree (Khan 2019). This is driven in part by their shared competition with the United States. The caution that "the United States must not find itself alone in a world where Russia and China collaborate to achieve common ends, and undermine U.S. interests" should be on the minds of those who are shaping policy toward the region (Khan 2019).

At the same time, "Central Asians themselves increasingly support some kind of U.S. presence as a means of balancing Russia and China," the authors of an article published by the Central Asia-Caucasus Institute maintain (Starr and Wimbush 2019). This appreciation of the importance of the United States may be reinforced by the potential threat posed to the region by the growing relationship between Russia and China. The U.S. can best play this "balancing" role through the application of smart power.

Central Asia is in the midst of an "amorphous but general realignment, the likely outcome of which will be new and unprecedented alliances, relationships, and transactional trade-offs. Shaping the geopolitics of this region into landscapes that affirm long-term American strengths," Frederick Starr continues, "will require thinking and actions that transcend our conventional strategic paradigms" (Starr and Wimbush 2019). An innovative approach requires the exercise of smart power; maximizing America's comparative advantage in soft power should be a major component.

The Current Central Asian Strategy

The Central Asia Strategy published by the U.S. Government in February is framed in constructive rhetoric about broad engagement with the region, but the approach casts American policy mainly in terms of security interests and competition with Russia and China, and that will not be welcomed in the Central Region. The strategy opens with this unassailable statement: "The United States' primary strategic interest in this region is to build a more stable and prosperous Central Asia that is free to pursue political, economic, and security interests with a variety of partners on its own terms; is connected to global markets and open to international investment; and has strong, democratic institutions, rule of law, and respect for human rights." The policy objectives of the strategy start with: "support and strengthen the sovereignty and independence of the Central Asian States, individually and as a region" (U.S. Department of State 2020a, b). The intent of the strategy—and of this coded language—was revealed, before it was issued, by Secretary of State Pompeo during his January visit to Kazakhstan and Uzbekistan, and in a speech he delivered in London. The overriding American interest in constraining China was presented in stark terms. During his visit to the region, Pompeo urged leaders to be wary of Chinese influence. In his

London speech, Pompeo declared that "The Chinese Communist Party presents the central threat of our times" (Lee 2020; Eurasianet 2020).

Secretary Pompeo met with Central Asian leaders under the C5+1 framework, which was established by the Obama administration as a means of gaining regional cooperation on Afghanistan. The Central Asia Strategy lists 6 objectives for the C5+1, the first three of which focus on security issues, rather than on development or fostering broader ties with the United States (U.S. Department of State 2020a, b).

The Central Asia Strategy seeks to reassure the region that the U.S. is committed to expand and maintain support for stability in Afghanistan. "The United States recognizes that a secure and stable Afghanistan is a top priority for the Central Asian governments, and each has an important role to play in supporting a peace process that will end the conflict," the strategy states. "The United States will encourage the Central Asian states to develop economic and trade links with Afghanistan and to model stable governance of multi-ethnic, Muslim-majority countries. The Central Asian nations will become stable, secure, and engaged partners of the United States and continue support for international efforts to stabilize Afghanistan" (U.S. Department of State 2020a, b). At another point, the strategy lists the objective to "Encourage connectivity between Central Asia and Afghanistan. The Central Asian states will develop closer ties with Afghanistan across energy, economic, cultural, trade, and security lines that directly contribute to regional stability" (U.S. Department of State 2020a, b). Juxtaposed to this language, Pompeo announced only a small amount of funding for regional efforts on Afghanistan (Eurasianet 2020).

The strategy mentions nothing about collaboration with countries outside the region. The European Union launched a new Central Asia Strategy in June 2019, which focuses heavily on support for development and other aspects of soft power (Council of the European Union 2019). There are likely multiple areas on which the United States and European countries can work together to advance shared interests. Such collaboration is particularly relevant to balance and compete with Russia and China.

Whenever CENTCOM commanders discussed Afghanistan with Central Asian leaders, the leaders' skepticism about American efforts to promote stability and development there was palpable. They did not believe that our almost two decades of involvement had brought about the fundamental change needed to position Afghanistan to become a politically and economically stable partner. I knew that reactions to periodic briefings by State Department officials were the same. There is no reason to believe that regional skepticism about America's commitment to remain engaged in Afghanistan has changed since then. Any strategy for Central Asia will fail unless we can persuade the region that, first, we want to expand engagement with the region in its own right; and, second, that the United States will remain robustly engaged in Afghanistan over the mid and longer-term following the withdrawal of our forces.

Continuing to place policy primarily within a security framework is not credible, because the region understands that security interests will become less of an American focus following the completed withdrawal from Afghanistan. The region is already at a disadvantage in terms of the level of American interest, given its remoteness and small diaspora. Unlike most other regions, "no Central Asia lobby exerts any considerable influence within the U.S. political system" (Sullivan 2019). And, the administration's preoccupation with China infers a focus on East Asia, echoing the "pivot to Asia" heralded by the Obama administration.

Crafting a Broader "Smart Power" Policy

Because the United States has dealt with the Central Region mainly through the prism of security interests, we have not sufficiently employed the tools of soft power to complement hard military power. Joseph Nye, who coined the terms "soft power" and "smart power," has said that "Hard power is the use of coercion and payment. Soft power is the ability to obtain preferred outcomes through attraction… Thus the need for smart strategies that combine the tools of both hard and soft power" (Nye 2009). The lack of such a holistic smart power approach has been particularly glaring in Afghanistan, but is evident throughout the region.

A number of factors highlight the importance and timeliness of a smart power strategy.

- The region is a significant field for competition among major powers; to be a global power the United States must be engaged.
- The soft power deployed by Russia and China have not gained them a decisive advantage.
- At the same time, it may be possible to forge some limited cooperation with China and Russia focused on the region, particularly with respect to security issues, but perhaps broadly to foster regional stability.
- Challenges in Afghanistan and Pakistan can only be addressed through a regional approach.
- The region's energy and mineral reserves—and private sector investment—are economically important to the West, including the United States and Europe.
- The region has a key role to play in countering extremism, terrorism, and narcotics trafficking.
- We have an interest in promoting fundamental freedoms and positive relationships.
- Fostering regional stability is implicit in and essential to all of the above.

Afghanistan and Pakistan

During almost two decades of involvement in Afghanistan, the United States has missed or failed to seize opportunities to promote its long-term interests both there and broadly in the region (Malkasian 2020). The inevitable winding down of American military presence in Afghanistan, however, provides perhaps the most crucial opportunity—since our intervention to overthrow the Taliban—to forge a more constructive strategy to advance the interests of the United States in the Central Region.

The intervention against al Qaeda and its Taliban hosts was not delimited as a surgical operation, but there was no strategy for targeted mid and longer-term goals that would effectively address our security and political interests. We did not complement military action with the policies and support necessary to make the Afghan government credible to the people. For the first 12 years of its intervention in Afghanistan, the United States faced the classic dilemma of how far to push President Hamid Karzai to address non-military issues, which were vitally important to defeating the Taliban. The issue which loomed from the outset of American involvement was the glaring need to address, or at least mitigate, the related problems of staggering corruption, poor governance, and the failure to deliver services to the Afghan people. Assessments of the war effort by the Obama administration concluded that the "incompetence and venality of the Afghan government were the principal reasons people were siding with the Taliban" (Chandrasekaran 2012).

We were, however, inherently unable to tackle these issues effectively, because we were never able—or willing—to exert decisive leverage through the threat of disengagement. Half-hearted, piecemeal efforts in Afghanistan turned out to be a proving ground for similar approaches in Iraq. It seems almost as if the mistakes made in each country were mutually reinforcing. *Quid pro quos* on policy issues are a staple diplomatic tool, but we are too often unable and unwilling to use them. A series of highly qualified State Department envoys, ambassadors, and military commanders were flummoxed by this problem. While Ashraf Ghani came into office with an internationally respected reputation, there has nevertheless been no major progress on the key non-military issues essential to the fight against the Taliban and now, as the U.S. starts drawing down its forces, the country is politically divided.

In 2014 I attended a briefing for CENTCOM Commander General Lloyd Austin given by a senior Afghan general in a hard-hit conflict zone. The Afghan commander, who had a good professional reputation, presented an impressive power-point presentation which delineated the military situation and attendant challenges. The commander said he was making progress militarily. I noticed a footnote at the bottom of one of the slides which read: "However, we must remember that winning hearts and minds is 70% of the battle." When I was discussing the briefing later with General Austin, I ventured the opinion that the briefing was at best sobering, at worst depressing. If winning hearts and minds was 70% of the war effort, then efforts to defeat the Taliban were doomed, since the Afghan government lacked credibility in the eyes of the people. On visit after visit to Afghan forces throughout the country over all the years I was at CENTCOM, the troops appeared to be well-trained and sometimes even well-motivated. Yet, the lack of any meaningful "hearts and minds" effort was a chasm in their, and our, efforts. "The easy part will be finding and killing the Taliban," General Nicholson told his forces. "The hard part is everything else" (Chandrasekaran 2012). Ambassador Eikenberry wrote to Secretary Clinton that "a credible partner in Kabul, basic governance and services" were essential in order to defeat the Taliban (Chandrasekaran 2012).

A “hearts and minds” effort was not impossible in Afghanistan and did not require the herculean process characterized as “nation building,” but it did require a comprehensive smart power strategy and steps to address the concerns of ordinary Afghans. President Obama’s national security team understood this, agreeing on the need to “build a stable, reasonably functional Afghan state” (Chandrasekaran 2012). In trying to achieve this, we necessarily dealt with and through the elites, but much more could have been done to employ soft power by engaging the grassroots and fostering civil society.

Afghans did not see a more effective central government delivering services, nor was there a vitally needed effort to facilitate political, social, and economic devolution of power. As Rajiv Chandrasekaran points out, the U.S. failed to empower the Afghan government, instead funneling assistance through American contractors and non-governmental organizations (Chandrasekaran 2012). At the same time, American programs were generally focused on short-term goals, which were often not sustainable, and 70% or more of the value of most contracts went to overhead (Chandrasekaran 2012). Ineffective counter-narcotics and alternative livelihood programs were part and parcel of this (Chandrasekaran 2012). In the process, we inadvertently reinforced Afghans’ perception that their government was, in effect, a creature of the Americans and outside forces.

The security prism, and the attendant failure to develop a smart policy approach, have also characterized relationships with Pakistan and with the rest of the Central Region. The United States has faced an acute policy dilemma with Pakistan, which has blunted efforts to achieve peace and stability in Afghanistan. Because of its possession of nuclear weapons and its strategic location, Pakistan is a country that we cannot allow to fail and, therefore, on which we cannot exert decisive influence. Time and again CENTCOM commanders have had blisteringly frank—and profoundly frustrating—exchanges with the Chief of Army Staff, the ultimate authority in the country, about the need to act against the Haqqani network. The Pakistani strategic calculus remained consistent: that the use of proxies was crucial to project strategically important influence in Afghanistan (Chandrasekaran 2012).

The U.S. Government at times conditioned some assistance, but the Pakistani leadership always knew that we would not, and could not, act decisively. The Pakistani Chief of Staff of the Armed Forces once summed it up very well in a conversation with one of the CENTCOM commanders. “We’ve always had these sorts of issues,” he said with a smile, apparently oblivious to how profoundly troubling his statement was, “but we have always found a way around them, and we always will.”

The Necessity of Hard Power

Discussion of the application of hard power in the Central Region is beyond the scope of this paper, but it is important to recognize it is a necessary component of a smart power policy. A credible residual American military presence in Afghanistan will be essential to reassure the countries of Central Asia. Law enforcement and intelligence collaboration are needed both to track former fighters for the Islamic State who have been repatriated and to assist efforts to counter extremism and radicalization (Helf 2019). There may be some basis for the United States to collaborate with Russia and China to forestall terrorism in the region and to suppress narcotics trafficking. Even very modest cooperation will send a signal to the countries of the region that the United States is not pursuing a complete zero sum approach (Shivers 2020).

There is no clear dividing line between hard and soft power. Exploiting opportunities to support professionalization of armed forces throughout the region, and training and assisting deployment of peacekeepers will exercise soft power, since polling data shows strong regional support for the UN (Inglehart et al. 2014). Helping regional countries resist the threat of Islamic extremism will "remove one of the main concerns that both Russia and China cite in defense of their meddling in Central Asian affairs," experts believe (Starr and Wimbush 2019). In addition to law enforcement and intelligence collaboration, America's public diplomacy support for programs to counter violent extremism should resonate in Central Asia given its history of religious moderation dating back to the tenth century (Starr and Wimbush 2019).

Soft Power Approaches

Security has also been the paramount issue for the United States in the rest of the Central Region (that is, the 5 "stans" of Central Asia) as the Northern Distribution Network was forged in order to establish alternatives to Pakistan for the supply and movement of our forces into and out of Afghanistan. In addition, security issues relating to terrorism, extremism, and narcotics trafficking remain valid concerns. These security interests, however, exert too much influence on American policy. Coalescing regional cooperation on Afghanistan is a top priority for the U.S.; it can best be achieved in terms of a broader smart power policy.

Weighed against the enormous financial costs of the military intervention in Afghanistan and the cost savings of withdrawal, a modest investment in soft power will yield disproportionate benefits. Expanding soft power in the region should take advantage of America's comparative advantages with respect to culture (which includes the favorable perception of our open society), education, health, digital engagement, civil society, and business, among other areas.

Attitudes and developments in the Central Region suggest a framework conducive to the application of soft power.

- On governance, according to polling data, more than 60% of the people of the region believe that democracy is important (Inglehart et al. 2014).
- Despite ethnic and linguistic links, people of the region, most notably in Kazakhstan, do not have an exclusive pro-Russian focus (Marlene and Royce 2019).
- Despite its attempts to wield soft power in the region, the approval rating for the Chinese leadership in much of the region is at or below 50%, China is not admired

by many people, and China worries about "Sinophobia" in the region (Shapiro 2019).

- People identify themselves overwhelmingly in nationalist, not ethnic or religious terms (The Central Asia Barometer 2017).
- There is a positive perception of American culture (Laruelle and McGlinchey 2017).
- There is broad appreciation of the need to be engaged globally, including through trade and contributions to international peacekeeping. Two-thirds of people in the region have some or great confidence in the UN (Inglehart et al. 2014).
- The generational change which is underway makes it timely to expand engagement with youth (Stronski and Zanca 2019).
- Economic and governance reforms—most notably in Uzbekistan and Kazakhstan, but percolating more broadly in the region, offer opportunities for enhanced interaction on a wide range of issues (Sullivan 2019).

The accelerating soft power engagement of major powers in the region should motivate the United States to expand its efforts. China and Russia have steadily expanded their soft power. Russia is pursuing what commentators have called either a "hybrid imperialism" or a "neocolonial" strategy, in part by using soft power to exploit advantages due to geographical proximity, its history with the region, linguistic and ethnic affinities, the Russian diaspora, and Central Asians working in Russia (Rumer et al. 2016; Khan 2019; Starr and Wimbush 2019). The Collective Security Treaty Organization and the Eurasian Economic Union advance Russian influence. China is making deep economic gains through the Belt and Road Initiative (though the countries of the Central Region are increasingly aware of the debt trap implications). India is expanding its economic relations with the region, partly to offset Pakistan and China. Iran exerts significant influence through proxy ties to Shiites and other groups, and commercial interests. Turkey and the Gulf States are less involved, but have interests (sometimes competing) to cultivate ties based on religious identity and economic interests. The European Union wants to project influence economically and politically.

While each country in the Central Region has distinct characteristics, there are building blocks for soft power which are relevant to all. First, the populations of the region will welcome expanded linkages with the American people. Second, there will be receptivity to expanded outreach between American officials and the people of these countries, including the burgeoning civil society. Third, U.S. Government assistance will be welcomed in education, health, and other non-political sectors. Fourth, there are abundant private sector opportunities not just with respect to oil and minerals, but related to infrastructure, expanding digital networks, and other areas.

In examining the potential to substantially expand use of "soft power," it must be acknowledged that the United States has not been adept at this. In order to fashion a smart power policy, and incorporate the advantages the United States has in terms of soft power, a "whole of government" approach is required, yet that has proved to be a chimera more often than not. When I was at the U.S. Central Command, the Obama administration launched an initiative to develop coordinated "whole of government" strategies for every country in the world. I soon realized that most Ambassadors had tasked junior officers to draft them, and paid little attention. The process petered out into an unceremonious end. Another example demonstrates the obstacles to operational and tactical cooperation. I proposed in 2014 that the Command send a team to assist the Near East Asia Bureau with planning on Syria. General Austin, the Commander, welcomed the idea as did Assistant Secretary Anne Patterson, yet it never happened due to bureaucratic obstacles. Such whole of government planning is, however, possible and is crucial for American engagement in the Central Region as we approach a watershed moment for American interests. Whole of government approaches have been fashioned on issues such as Sudan, but doing so is only possible if there is sustained senior-level oversight.[2]

In considering the range of soft power, a relevant caveat is that, although the United States has a major interest in promoting good governance and democracy, it is counter-productive to be perceived as trying to impose our approaches and models. Smart power requires balancing our sometimes competing national interests and values. We should, though, not be reticent in advocating the values of freedom and openness. Demands for reform and fundamental freedoms should not be imposed, but the U.S. reform agenda should not be solely focused on "improving social and economic conditions rather than on democracy promotion," as has been suggested (Rumer et al. 2016). Polling data, it is worth reiterating, indicates that more than 60% of people in the region consider democracy important, and people still look to the United States as the world's leading democracy. Abdicating that responsibility, which reflects American values, inadvertently complements the strategies of China and Russia, rather than competing more effectively based on our comparative advantages. Support for reform and fundamental freedoms can sometimes be explicit, but it is usefully implicit in all American soft power.

Soft power embraces a myriad of sectors and activities. I have selected several which are especially relevant for the Central Region: engagement with civil society; support for the free flow of information; public diplomacy; expanded people-to-people exchanges; and support for health, education, and research.

Civil Society: Following the end of the Cold War, civil society expanded rapidly in the former Soviet republics of Kazakhstan, Uzbekistan, Tajikistan, and Kyrgyzstan. While there are wide variations, civil society is robust in all those countries. Even Tajikistan, under authoritarian President Rahmon, has several thousand civil society organizations, while Uzbekistan has more than 5,000, Kyrgyzstan as many as 12,000,

[2]Despite the fact that Sudan was a highly charged domestic political issue, a whole of government approach was developed and implemented during 2001–5. This was due primarily to the fact that President George W. Bush was personally interested in the issue and drove the process. Secretaries of State Powell and then Rice carefully oversaw the policy process. Rice enlisted her deputy Robert Zoellick to handle the issue. Zoellick chaired daily inter-agency meetings. The whole of government approach fell apart following Zoellick's departure from State in 2006. My experience with this—and other issues—convinces me that a whole of government approach can be achieved, but only if the top leadership takes direct personal interest and runs the process.

and Kazakhstan around 40,000. The number of non-governmental groups in these countries has expanded particularly during the last two decades.

Turkmenistan, which imposes draconian restrictions on non-governmental activities, is the exception. Civil society in Afghanistan started to emerge following the ouster of the Taliban. While civil society is still nascent, there are thousands of non-governmental groups. Pakistan has a well-developed civil society comprising tens of thousands of groups (Asian Development Bank 2020).

There are several common denominators among these civil societies. First, many of these organizations are focused on education (both expanding literacy and civic awareness). Second, a substantial number of these groups are involved directly or indirectly in fostering human rights. Third, women's issues are a focus of many groups. Fourth, most of these groups struggle to sustain themselves financially. Thus, civil society throughout the Central Region merits American engagement (Asian Development Bank 2020).

While interaction with civil society in these countries is a very important element of soft power, it must be deftly handled in order to forestall potential criticism that groups are being manipulated by the United States. This means ensuring that demands for change and reform are "locally driven," as the Carnegie Endowment has pointed out (Rumer et al. 2016). While supporting reform, American assistance can focus on a wide range of non-political areas that are not the most politically sensitive. Productive engagement requires a combination of some official assistance to civil society, and encouraging expanded involvement by American non-governmental groups (including providing assistance to American civil society groups to empower such engagement).

President Obama launched the *Stand with Civil Society* initiative to support and defend civil society. This was both a recognition of the importance of civil society globally and a response to increasing restrictions being placed on non-governmental organizations in many countries. USAID's Civil Society Support Program in Central Asia is providing $18 million over five years, through the Eurasia Foundation, to support civil society in Kazakhstan, Tajikistan, Turkmenistan and Uzbekistan. Notably, one of the goals of the program is to build a new generation of forward-looking civil society leaders (U.S. Agency for International Development 2020). This is critically important given the youth bulge in the Central Region as reflected in the fact that the median age is under 27 (Stronski and Zanca 2019). The current senior leadership in Washington, however, has not highlighted the need to support civil society and empower youth.

The funding level is far too low. Increasing funding for soft power activities will be key to developing a smart power strategy, but this will be a large challenge. Assistance to promote democracy, human rights, and civic engagement has consistently been the smallest portion (about 10% or less) of overall U.S. foreign aid, which in itself is only around $50 billion globally, most of which goes to just a five countries (McBride 2018).

Information Flow: One indicator of the extent of American soft power in Central Asia is the fact that the United States remains "culturally attractive." American films and music, for example, are popular (Laruelle and McGlinchey 2017). More important than that, however, is that the United States is admired for its openness and freedoms. Radio Free Europe and Radio Liberty play a crucial role in advancing that perspective by providing professional news programming into the Central Region. Their reach, which was already in the millions through more than 1,000 hours of programming a week, has been expanded through the creation of a Russian language television station (Radio Free Europe 2020). Further expanding similar programming would be enormously helpful to an overall smart power approach. Studies have shown that such media broadcasting does not greatly influence popular attitudes per se, but people want access to news not encapsulated in propaganda—and that implicitly underscores the openness and values of the American system (Chapman and Gerber 2019).

The United States has a huge comparative advantage through its English language, which has become international, and is now crucial for business, science, and access to information. Funding for these programs was cut back years ago, but should be greatly expanded. Related to this, American cultural centers (now dubbed American Spaces) should be established in countries where they do not exist, and expanded where they do exist. According to the State Department, 1.4 million Central Asians have visited American Spaces to learn English and to learn about American culture (U.S. Department of State 2020a, b). The goal should be to at least quadruple this over the next 5 years.

Public Diplomacy: Public diplomacy is *the* most important dimension of smart power. The purpose of public diplomacy is to tell our story, which includes discussing the nature of the United States, advancing appreciation of the value of fundamental freedoms and open societies, and promoting understanding of American foreign policy. Whether or not people "like" or "respect" the United States, they are at a minimum interested in the country due its status as a superpower, and because of the reach of its culture and openness. The United States has greatly under-exploited this advantage in the Central Region, and globally.

Expanding contact with ordinary people is a vital component of public diplomacy, yet few American diplomats do this on a sustained basis. American diplomats engage with elites far more than they engage with ordinary citizens of foreign countries. Some of this is the result of the "observe and report" culture that is fostered by the State Department which, as Chandrasekaran observed in Afghanistan, is "predisposed to dwell on the limits of American power instead of its transformative potential" (Chandrasekaran 2012). But the hidden truth is that our diplomats too often self-censure themselves in order to avoid making waves with governing elites. Ambassadors and senior officials in the Central Region need to reach out to the people of the countries in which they serve. This can be done through town hall meetings, web chats, media appearances, project launches, and speeches. Involving the indigenous media in these contacts shows the general public that we are engaging in a frank and open way.

While security concerns are a legitimate issue with respect to outreach, this can be overcome. Establishing Facebook pages and Twitter accounts, as Ambassadors have been mandated to do, are no substitute for going outside of embassies in order to engage people directly. While elites may well feel threatened by such engagement, people welcome it, as I can attest from my diplomatic experience. Many times while I was serving as an Ambassador, people in my host country told me that I was the first person who had brought them together. This involved sponsoring dinners, receptions, and events at which people from different cultural spheres, political viewpoints, and social strata mixed. The catalytic role that American officials can play, though intangible, exerts influence and is a significant dimension of soft power.

During one of my visits to Pakistan with General Mattis, I was seated next to a Pakistani general at dinner. I deliberately asked him a provocative question regarding why the United States had such a dismal approval rating (about 15% at the time) in his country. I had my views regarding why this was so, but his answer startled me because it echoed what I was thinking. The reason, the general replied, is that our people never hear from your officials, and don't know what you are doing in the country. I knew that the Pakistani government was partly responsible for the negative perception of the U.S., but his response resonated.

Ambassadors only infrequently deliver detailed policy speeches, in part because American leaders, including the President and Secretaries of State and Defense, do not publicly articulate a coherent foreign policy. These pronouncements, to be credible, should be accompanied by discussions about America itself, both the strengths and weaknesses of our democratic system.

"I don't know what our integrated strategy is, or specifically what it is for my region," the Commander of the U.S. Central Command, General James Mattis, said when the issue arose at a briefing for the Defense Policy Board (Mattis 2019). The sad fact is that there is still no "integrated policy" for the Central Region. In the Digital Information Age, American presidents and senior officials have multiple ways of engaging international audiences; not doing so cedes our comparative advantages and inadvertently enables our competitors.

Various elements of the American government (including the U.S. Central Command) are involved in communicating with populations in the region but, as CENTCOM's Chief of Information Operations Andrew Whiskeyman pointed out at The Great Power Competition conference, efforts within our government are not synchronized (Whiskeyman 2020). The Central Asian strategy and the jarring rhetoric coming from Secretary Pompeo and other officials do not facilitate an effective public diplomacy effort. The administration's rhetoric against globalization and free trade will only make the United States less attractive to the region. As Sean Ryan noted at the conference, we are at a "strategic disadvantage" vis-à-vis Russia and China with respect to information operations in the Central Region (Ryan 2020). This is due both to lack of coordination and insufficient resources. The budget for public diplomacy activities is tiny, and it is small for information operations. Incorporating the United States Information Agency into the State Department has made public diplomacy a neglected bureaucratic step-child, rather than empowering our diplomats as it was supposed to do.

People-to-People Exchanges: The chance to study and live in the United States is a "dream shared by many" in the Central Region (Laruelle and McGlinchey 2017). Many people throughout the world make an artificial distinction between the American people and the U.S. Government. That distinction only underlines the importance of exchanges and highlights why they are essential to cultivate influence. A few years ago, when I was Ambassador to Kenya, I was speaking to a group of Muslim students who were about to graduate from secondary school. When I invited questions, a young man in the front row excoriated the United States for killing innocent men, women, and children in Iraq and Afghanistan. I responded, took other questions, and then invited one last question. The young man raised his hand and asked how he could get a "green card," i.e. become a permanent resident. That vignette illustrates this perceived dichotomy within people who excoriate the U.S. Government, while clearly perceiving America as attractive due to openness and opportunity.

The United States allocates only a miniscule portion of its budget for exchanges. Since the independence of the Central Asian countries two decades ago, 40,000 people from the region have been funded by the U.S. Government to visit the United States (U.S. Department of State 2020a, b). While this sounds impressive, the reaction should be: *only* 40,000. Officially funded exchange programs help America develop positive relationships with people who influence their societies. In the Central Region there is vast scope to expand programs to sponsor international visitors and encourage youth to study at American universities. Criteria to issue visas for study at American universities are too restrictive. Some who study in the United States will never return to their home countries, but many others will, having been influenced in a way that will likely make them seek transparency and reform in their societies. When I was talking in Saudi Arabia with a director in their intelligence service, the topic of the King Abdullah scholarships arose. The King had initiated a program to send tens of thousands of Saudis to study in the U.S. each year. The director said his son was studying at UCLA but, he added, "I hate the program." Why, I wondered. He responded that when his son returned on visits to the Kingdom he came with all sorts of "reformist" ideas! "However," he added, "my wife loves the program."

Foreigners who study at American universities will often play important roles in their countries. There are *only* about 1 million foreign students in the United States. *Only* about 4–5,000 students from the entire Central Region come to study in the U.S. each year (U.S. Department of State 2020a, b). American universities welcome and compete for foreign students, who tend to excel and who are a significant source of revenue. The low number is due to visa restrictions, inability of foreign students to finance their education, and insufficient sources for counseling and information to facilitate entry into American universities. The United States can increase the number of students from the region by addressing these issues. It can also forge collaborative programs with regional governments to incentivize study in the U.S. and the return of those students back home. In my experience, it is rare to encounter returned students who do not have, on balance, a positive view of life in America.

Health, Education, Research: Among the many other potential dimensions of soft power, support for development of the rapidly expanding health care, education (including literacy, civics, and modernization of curricula), and research sectors in the region are important (Khan 2019). Expanding engagement in these sectors will complement programs to support civil society and people-to-people exchanges. As this is done, we should brand programs in every area of official assistance as coming from the people and government of the United States. In some instances, due to security and political constraints, it may not be possible to do so, but in most cases there is no excuse for not doing so. As an Ambassador, I learned that people at the grassroots were too often not aware that a particular project (such as a health clinic) had come from the American people and government. Not infrequently, non-governmental groups which receive official funding take sole credit for projects.

Expanding American private sector activity in the Central Region—beyond the energy sector—will add critical weight to soft power. The Build Act is a good start, and can be used to cultivate useful public-private partnerships, but the funding level will have to be increased if the United States wants to compete effectively, especially with China.

In formulating approaches to soft power, we should weigh the fact that the countries of the Central Region are part of the Muslim world. In 2009, President Obama gave a much anticipated speech at Cairo University which proposed a new, innovative framework for relations between the United States and the Muslim world (The New York Times 2009). Unfortunately, there was little follow up to implement that framework—which included proposals, for example, to expand exchanges and research ties, which are highly relevant to many Muslim countries in the Central Region. The Obama speech was a well-intentioned initiative to deploy soft power to engage Muslim societies. It ended as a lost opportunity because there was no comprehensive strategy to implement it. We do not now need such a high profile initiative, but the concept—and its disappointing aftermath—are worth keeping in mind as we consider how best to engage Muslims in the Central Region.

Cautionary Notes

When considering soft power, a few cautionary notes are warranted. While the overall climate in the Central Region is conducive to utilize and expand soft power, this will not be an easy process given the complex attitudes of the populations of the region, and the reality of growing Chinese and Russian influence. American approval ratings in Kazakhstan hover at about 50%, 30% in Kyrgyzstan. Russia, on the other hand, enjoys a 90% favorability rating in both countries, while Chinese favorability hovers in the 60–80% range (Laruelle and McGlinchey 2017).

A commentary by Europe-Central Asia Monitoring attributes this to a "disconnect between cultural and political soft power." American and other Western assistance is received with mixed feelings due to the perception—or misperception—that a "Western political reform agenda" is being imposed (Laruelle and McGlinchey 2017). In addition, many people of the region, in part driven by the influence of Russian media, are sensitive to alleged efforts to promote certain values, such as respect for LGBTI rights. Russian media spreads narratives of "decadent Western values and the need for a so-called Eurasian civilization to safeguard the region's traditional values and morality" (Laruelle and McGlinchey 2017). While the United States has provided assistance for cultural preservation projects in the Central Asia, expanding those efforts would help offset such criticisms.

An "America First" approach to foreign policy—sometimes perceived as xenophobic or neo-isolationist—should logically prioritize expansion of soft power, because doing so will directly benefit the American people. The use of soft power brings wide-ranging benefits—tangible economic ones, less tangible but important contacts and interactions—for American non-governmental organizations, universities, businesses, the tourist sector, and others. Foreign students alone account for billions of dollars in revenue for universities (Loudenback 2016).

Moving Forward Toward a Smart Power Strategy

At the heart of any discussion of a smart power policy for the Central Region is the future of Afghanistan. We went into Afghanistan without a comprehensive strategy; we must exit with one which will enable the United States to remain constructively engaged. "Any American policy that seeks to lessen or withdraw U.S. support from Afghanistan is bound to impact negatively all the other states of Central Asia," Frederick Starr of the Central Asia-Caucasus Institute has concluded (Starr and Wimbush 2019). Without an approach that is credible to the region, no smart power strategy will succeed.

Crafting a smart power approach toward the region will prove challenging—mainly due to the dysfunctional nature of policymaking within our government—but not impossible. "Many official instruments of soft power—public diplomacy, broadcasting, exchange programs, development assistance, disaster relief, military-to-military contacts—are scattered across the U.S. government.," Joseph Nye has written. "There is no overarching policy that even tries to integrate them with hard power into a comprehensive national security strategy" (Nye 2009).

A serious impediment to developing smart power approaches is the disparity between the resources of the Defense and State departments. "The United States spends about 500 times as much on the military as it does on broadcasting and exchange programs. Is this the right proportion?" Nye has asked (Nye 2009). Given the disproportionality of resources in personnel and funds, State remains wary of being absorbed into closer coordination and planning with Defense. While understandable, seeking more synergy between the policies and activities of Defense and State is crucial to crafting smart power policy. Late one Friday I was alone in my office at CENTCOM when someone pounded on the door. It turned out to be a colonel who was frustrated because he needed a State Department clearance on his 30-page Yemen strategy and action plan, and he could not reach the "Yemen unit" at State. Colonel, I told him, you will probably never get a response; State has one desk officer for Yemen, who is overwhelmed. He was literally speechless.

CENTCOM and other Defense Department elements had dozens of people working on Yemen. Such circumstances underscore the difficulty of inter-agency coordination and collaboration, but also the need for it.

The Strategic Multi-Layer Assessment program at the Pentagon, which was discussed at the opening of the Great Power Competition conference, draws upon expertise within the U.S. Government, including State, and outside experts (Strategic Multi-Layer Assessment Program Presentation 2020). This mechanism could be expanded, tailored, and given more weight in the inter-agency process in order to facilitate smart power planning. Both State and Defense understand the importance of soft power; the hurdle is how to map this out and incorporate it into integrated smart power policy. "It is increasingly clear that the DOD needs to complement "control" with an explicit focus upon "influence" factors and forces that produce desired behavioral outcomes across complex and intermeshed human and technical systems," a senior general has commented (Strategic Multi-Layer Assessment Periodic Publication 2017). "By complementing its military and economic might with greater investments in its soft power, the United States can rebuild the framework it needs to tackle tough global challenges. That would be true smart power," Nye concludes (Nye 2009). Commenting on the unprecedented flux in world order, Kissinger argues that America should not abandon its essential idealism. Doing so, he warns, will undermine our influence (Kissinger 1994). In order to advance its interests in the Central Region, which involves balancing and competing with Russia and China, the United States needs a smart power policy, with soft power—and its values—at the core.

References

G. Allison, The new spheres of influence, in *Foreign Affairs* (2020)

Asian Development Bank, Civil Society (2020), https://www.adb.org/site/ngos/publications.html. Accessed 12 March 2020

The Central Asia Barometer, Public Opinion Poll in Central Asia (2017), www.centralasiabarometer.org. Accessed 10 Feb 2020

R. Chandrasekaran, *Little America* (Vintage Books, New York, 2012)

H. Chapman, T. Gerber, Opinion formation and issue framing effects of Russian News in Kyrgyzstan. Inte. Stud. Q. **63**(3), Abstract (2019). https://doi.org/10.1093/isq/sqz046.html. Accessed 21 Feb 2020

E. Colby, A. Mitchell, The age of great-power competition, in *Foreign Affairs* (2020)

Council of the European Union, Central Asia: Council Adopts a New EU Strategy for the Region (2019), https://www.consilium.europa.eu/en/press/press-releases/2019.html. Accessed 18 March 2020

Eurasianet, U.S. faces headwinds pushing Central Asian Integration. Eurasianet (2020), https://eurasianet.org/us-faces-headwinds-pushing-central-asian-integration.html. Accessed 20 Feb 2020

Global Security, The great game—Russo-British Rivalry (2020), https://www.globalsecurity.org/military/world/war/great-game.htm. Accessed 14 Feb 2020

G. Helf, Blog. United States Institute of Peace (2019), https://USIP.org/blog/2019.html. Accessed 10 March 2020

R. Inglehart, C. Haerpfer, A. Moreno, C. Welzel, K. Kizilova, J. Diez-Medrano, M. Lagos, P. Norris, et al. (eds.), *World Values Survey: Round Six—Country-Pooled Datafile Version* (2014). http://www.worldvaluesurvey.org. Accessed 27 Feb 2020
A. Jackson, Panel on Regional Security. The Great Power Conference (University of South Florida, Tampa, 2020)
M. Khan, Where is Central Asia in the Current U.S. Grand Strategy? Center for Global Policy (2019), https://cgpolicy.org/author/mkhan.html. Accessed 13 March 2020
H. Kissinger, *Diplomacy* (Simon and Schuster, New York, 1994), p. 805
M. Laruelle, E. McGlinchey, Renewing EU and US Soft Power in Central Asia. EUCAM Commentary, number 28, October (2017), https://eucentralasia.eu/2017/10/renewing-eu-and-us-soft-power-in-central-asia-3.html. Accessed 17 March 2020
M. Lee, Pompeo Message in Europe, Central Asia Trip: Beware of China. AP News (2020), https://apnews.com. Accessed 20 Feb 2020
T. Loudenback, International students are now 'Subsidizing' public American universities to the tune of $9 billion a year. *Business Insider* (2016), https://www.businessinsider.com. Accessed 5 March 2020
J. McBride, How does the U.S. spend its foreign aid?, Foreign Affairs (2018), https://www.cfr.org/backgrounder/how-does-the-us-spend-its-foreign-aid.html. Accessed 10 March 2020
C. Malkasian, How the good war went bad, in *Foreign Affairs* (2020)
L. Marlene, D. Royce, Kazakhstani public opinion of the United States and Russia: testing variables of (Un)favourability. Cent. Asian Surv. **38**(2), 197–216 (2019). https://www.tandfonline.com.html. Accessed 19 March 2020
J. Mattis, *Call Sign Chaos* (Random House, New York, 2019)
J. Nye Jr. Combining hard and soft power, Foreign Affairs (2009), https://www.foreignaffairs.com/articles/2009-07-01/get-smart.html. Accessed 1 March 2020
Radio Free Europe/Radio Liberty, Central Asia (2020), www.rferl.org. Accessed 12 March 2020
M. Ranneberger, Panel on Regional Diplomacy. The Great Power Conference (University of South Florida, Tampa, 2020)
E. Rumer, R. Sokolsky, P. Stronski, U.S. policy toward Central Asia 3.0. carnegieendowment.org (2016), https://carnegieendowment.org/2016/01/25/u.s.-policy-toward-central-asia-3.0.html. Accessed 23 Feb 2020
S. Ryan, Panel on Information Environment. The Great Power Conference (University of South Florida,Tampa, 2020)
D. Shapiro, China and Russia in Central Asia: A Tricky Balance. Harvard Kennedy School, Belfer Center (2019), https://www.russiamatters.org/blog/china-and-russia-central-asia-tricky-balance.html. Accessed 18 Feb 2020
M. Shivers, Panel on Regional Economics. The Great Power Conference (University of South Florida, Tampa , 2020)
F. Starr, S. Wimbush, U.S. Strategy Towards Afghanistan and (the Rest of) Central Asia. The American Interest (2019), https://www.the-american-interest.com/2019/01/24/u-s-strategy-towards-afghanistan-and-the-rest-of-central-asia.html. Accessed 3 Feb 2020
G. Stewart, S. Quillian, Sitting on the Sidelines? U.S. Decline in Central Asia. Pacific Council on International Policy (2019), https://www.pacificcouncil.org/about/network/profile/sumaya-quillian.html. Accessed 16 March 2020
Strategic Multi-Layer Assessment Program presentation, The Great Power Conference (University of South Florida, Tampa, 2020)
Strategic Multilayer Assessment (SMA) Periodic Publication, White Paper on Influence in an Age of Rising Connectedness (2017), https://info.publicintelligence.net/SMA-InfluenceConnectedness.pdf. Accessed 12 March 2020
P. Stronski, R. Zanca, Societal Change Afoot in Central Asia (2019), https://carnegieendowment.org/2019/10/18/societal-change-afoot-in-central-asia.html. Accessed 1 March 2020
C. Sullivan, The superpower and the "Stans": Why Central Asia is Not "Central" to the United States. SAIS Review of International Affairs (2019), https://www.saisreview.org/2019/03/27/the-superp

ower-and-the-stans-why-central-asia-is-not-central-to-the-united-states.html. Accessed 2 March 2020
Text: Obama's Speech in Cairo, The New York Times (2009), https://www.nytimes.com/2009/06/04/us/politics/04obama.text.html. Accessed 23 Feb 2020
A. Whiskeyman, Panel on Information Environment. The Great Power Conference (University of South Florida, Tampa (2020)
E. Wong, U.S. faces tough 'Great Game' against China in Central Asia and Beyond. New York Time (2020), https://www.nytimes.com/2020/02/13/world/asia/china-great-game-central-asia-trump.html. Accessed 24 Feb 2020
U.S. Agency for International Development, Civil Society Support Program for Central Asia (2020), https://www.usaid.gov/central-asia-regional/fact-sheets/civil-society-support-program-central-asia.html. Accessed 2 March 2020
U.S. Department of State, United States Strategy for Central Asia (2020a), pp. 1–2, https://www.state.gov/united-states-strategy-for-central-asia-2019.html. Accessed 3 March 2020
U.S. Department of State, United States Strategy for Central Asia (2020b), pp. 1–2, https://www.state.gov/united-states-strategy-for-central-asia-2019.html. Accessed 3 March 2020
U.S. Department of State, Report of the Visa Office 2019 (2020), https://travel.state.gov/content/travel/en/legal/visa-law0/visa-statistics/annual-reports/report-of-the-visa-office-2019.html. Accessed 28 Feb 2020

The Downside of the U.S. Marginalizing Economics in Great Power Competition

Leif Rosenberger

Abstract A strategic shift has occurred in great power competition. The Pentagon says this strategic shift calls for an emphasis on U.S. grand strategy (linking economics and security) and geopolitics. Unfortunately, the U.S. keeps failing in its great power competition in the Greater Middle East. A militarized U.S. foreign policy marginalizes economics against Russia in Syria, against Iran in Iraq and against China in Afghanistan. Peaceful foreign policy opportunities are squandered. Therefore, the current U.S. administration's economic nationalism is problematic. A better approach would be to turn enemies into friends by using shared prosperity in intra-state relations to reduce the demand for violence in the Syrian and Afghan civil wars. The French and Germans did much the same thing in inter-state relations after World War II.

Keywords Great Power Competition · Syria · Iraq · Afghanistan · Economic coercion · Marginalizing economics · China · Economic white flag · Root causes · Lost opportunity · Mismanagement of natural resources · Reconciliation · Post-conflict risk · Trade diversification · New silk road · Warnings · Russia · China · Iran

Introduction

A militarized U.S. foreign policy marginalizing economics didn't happen so much in the United States 150 years ago. Then, U.S. foreign policy was deeply infused with economic logic. But U.S. economic statecraft has changed over the years. The U.S. foreign policy became increasingly militarized. Robert (Zoellick 2012), former President of the World Bank, says that the U.S. started to marginalize economics after the start of the Cold War The U.S. too often reached for the gun instead of

L. Rosenberger (✉)
ACERTAS, Los Angeles, USA
e-mail: leifrosy@yahoo.com

Modern War Institute, 26 Regiment Drive, Gettysburg, PA 17325, USA

United States Military Academy, West Point, NY 101996, USA

© The Author(s), under exclusive license to Springer Nature Switzerland AG 2021

A. Farhadi et al. (eds.), *The Great Power Competition Volume 1*,
https://doi.org/10.1007/978-3-030-64473-4_14

the purse. Not surprisingly, the use of economic instruments to achieve geopolitical objectives became a lost art (Blackwill and Harris 2016).

The attacks on the World Trade Center in New York and the Pentagon on 11 September 2001 triggered another global war, this time against terrorism. Would the U.S. keep marginalizing economics? To answer this question, one needs to separate thoughtful discussion from foreign policy action.

In terms of thoughtful discussion, the answer is clear. Three years after the 2001 attacks, the 9/11 Commission issued a report which underscored the need to address the negative underlying economic conditions that fostered violent extremism.

In fact, of the 27 recommendations of the report, only one is advocating the use of military force. One problem is that U.S. foreign policy has been exporting anger and fear rather than hope and opportunity. The U.S. needs to foster hope and opportunity if the war on terrorism is to achieve anything but tactical success.

The 9/11 Commission's Report said that when people lose hope, when societies break down, when countries fragment, breeding grounds for terrorism are created. Backward economic policies and repressive political regimes slip into societies that are without hope, where ambition and passions have no constructive outlet.

In other words, spreading prosperity, while not a silver bullet, does help in combating violent extremism. Conversely, poverty and illiteracy are easy prey for violent extremists to exploit.

But terrorism was not the only cause for concern. Full blown wars were also breaking out, most of them civil. In academe, speculative political debates broke out regarding the root causes of these civil wars. Thankfully, a definitive study by Paul Collier and his cohorts (Collier 2003) at the World Bank found that civil wars were heavily concentrated in the poorest countries.

Thus, the World Bank and the 9/11 Commission were mutually supportive of one another. The World Bank research showed that the root cause of civil wars was the failure of economic development. The 9/11 Commission Report recommended addressing the economic conditions that fostered violent extremism. But this was rhetoric, not action. American foreign policy was still militarized and out of balance. Douglas Lovelace, former Director of the Strategic Studies Institute at the US Army War College, notes that the administration of President George W. Bush was frequently accused of favoring the use of unilateral military power over multilateral diplomacy and development as the primary tool of American national security" (Deni 2015).

When Donald Trump became president, former General Martin Dempsey and former Admiral Michael Mullen, two former chairmen of the Joint Chiefs of Staff, voiced their concern that having too many generals and admirals in the highest US national security positions could lead to the over-militarization of US national security policy (Rosenberger 2016).

President Obama took office in 2009 with a different mindset than that of President Bush. He said he wanted to rebalance American foreign policy, away from what he felt was an over-reliance on the military toward greater reliance on diplomacy and development. If one looks at his rhetoric and National Security Strategy documents from 2010 and 2015, it's clear that the Obama administration favored diplomatic solutions over military ones.

While President Obama deserves some credit for his rhetoric about the need to rebalance American foreign policy, US Army War College Professor John (Deni 2015) says the tragedy of the Obama administration is that Obama failed to rebalance American foreign policy. Deni says there is plenty of evidence showing that US foreign policy remained militarized. The most obvious indicator was federal spending. Deni's research indicates that the international affairs expenditure (which includes money for diplomacy and development) remained a small fraction of what the Pentagon was spending throughout the eight years of the Obama administration.

Militarizing U.S. Foreign Policy in Syria

Currently, the U.S. military likes to use the term DIME to refer to diplomacy, information, military and economic instruments of national power. But the M in the DIME keeps overpowering the rest of the DIME, especially marginalizing the E and the I.

In the past, the Syrian regime has successfully used brute force to suppress periods of Sunni protests and demonstrations in Homs and Hama in 1964–1965 and Hama again in 1982. President Assad tried the same approach with brute force in March 2011, but this time, the attempted crackdown did not work so well. For one, radical Sunni Islamists (ISIS) had hijacked the revolution. For another, the Arab Spring inspired a higher level of violence from the opposition. Feelings of Sunni revanchism, which had been suppressed for almost 30 years, burst out into the open. To make matters worse, radical Islamists highjacked the Syrian revolution. They saw Arab Spring as an opportunity to spread the rule of Islam.

President Obama was politically correct at home. He gave moralistic speeches about how President Assad was illegitimate and had to relinquish power. He also gave the opposition to Assad every reason to believe that the US military would intervene in Syria with enough military force to shove him out of power. But in the eyes of this opposition, it was a big buildup for a big letdown.

Before long, the Syrian civil war morphed into a war by proxy. Assad was receiving overwhelming military support from Russia, Iran and Hezbollah. In contrast, the Syrian opposition was receiving much smaller amounts of military assistance from the US, Turkey, Saudi Arabia and other the Gulf Cooperation Council (GCC) states, Qatar, France and the UK. Obama only gave the opposition enough military assistance to intensify and prolong the fighting, but never enough to help the opposition to decisively win.

As Richard (Haas 2017) puts it, the West had limited will and limited means, so, it had to set limited goals for limited amount of good. The opposition was a victim of false expectations and ultimately felt betrayed by American's ambivalence and dithering. America may have been a superpower elsewhere, but the U.S. was a paper tiger in Syria.

Obama was fixated on the idea that the conflict could only be resolved if Assad was removed from power. The problem was that Obama grossly underestimated the strength of the Assad regime. The only way Western intervention would work would be if it put lots of American boots on the ground. Obama was not going to do that. He knew what he did not want—Assad to remain in power—but he had no idea who or what he wanted instead of Assad. Obama wanted to see a moderate, democratic, secular, pluralistic government replace Assad. The only problem was that such a possibility was not a realistic prospect (Van Dam 2017).

In contrast, Russia, Iran and Hezbollah knew exactly what they wanted: for Assad to remain in power. This was a core interest for them. In contrast, Syria was never a core interest for Obama. Therefore, there was never any US political will to arm the political opposition to such an extent that they would have a chance to win battles against the Assad regime. There was never any clear US strategy in Syria, except that of defeating ISIS. Obama had no geopolitical clue in Syria other than counterterrorism tactics against ISIS.

What's interesting is that Robert Ford, US Ambassador to Syria, was reportedly opposed to calling for Assad's departure; Ford argued that America would not be able to bring this about (Van Dam 2017). In contrast, President Obama delivered stirring rhetoric calling for Assad's departure, but his demand carried no intent to enforce that wish. The US may have been a superpower within its own borders, but it was a paper tiger in Syria.

Worst of all, Obama created false hopes among the opposition to Assad that decisive Western support was forthcoming. In the end, US support was not nearly as powerful as Obama's rhetoric. His moral high ground prevailed over pragmatism. Sadly, it took the political opposition to Assad a long time to accept the fact that they had been victims of false expectations (Van Dam 2017).

Militarizing U.S. foreign policy in Syria, therefore, didn't work out so well. Since the U.S. military opted to retreat from Syria rather than compete for power and influence, that raises the question: Did the American I in information and the American E in economics show up for duty in Syria? Perhaps we should combine economics and information as "economic information." If so, how did American "economic information" do?

Lost Opportunity: Assad's Economic Reforms in 2009

Henry Kissinger once said, "Opportunities cannot be hoarded; once past they are usually irretrievable (Starr 2015). In Syria, a good opportunity for shared prosperity instead of war came in 2009. Before then, the international community had been reacting negatively to one wave of domestic repression after another in Syria. The first wave started in 2001, and the second started in 2006. After each wave of repression, the international community treated Syria like a pariah, and when Syria feels threatened by a hostile external environment, it almost always becomes intolerant of domestic dissent and cracks down with a vengeance (Wieland 2015).

To appreciate just how unique this 2009 window of opportunity was for shared prosperity in Syria, it is important to understand just how deep-rooted Syrian socialism had become since its roots in the 1960s. When the Syrian Ba'ath party came to power in 1963, nobody in the Syrian government was thinking about free market reforms. In fact, the wind was blowing in the opposite direction. The economy was essentially closed. One of the first things that the Ba'ath party did when it seized power in 1963 was to nationalize most of the economy. An inefficient state controlled all the banks and many of the major industries after nationalization (England 2008).

At this point in time, Syria frankly wanted nothing to do with America, so, there was no realistic opportunity for shared prosperity.

In the twenty-first century, it is clear that Syria's economic growth was relatively buoyant from 2004 until late 2008 because of high oil prices. But then things took a turn for the worse. While the production of oil still dominated the economy, Syria's oil output fell from 600,000 barrels a day (B/D) in the late 1990s to just 380,000 B/D in 2008. In fact, it was a net-oil importer in 2008 for the first time in decades. Syria's oil revenues fell in 2008, despite high oil prices in the first half of the year.

Facing adversity, the Syrian government belatedly grasped the reality that it needed to adapt to survive. And so, in 2005, the government opted for a "social market economy." The new policy was a de-facto acknowledgement that the old economic model was stagnant, and it was time to shake up the economy. The International Monetary Fund (IMF) helped Damascus draw up a new economic blueprint, whose purpose was to shift Syria away from a command economy and closer to a free market economy (Gelvin 2018).

2008 saw the growth of economic diversification and stronger non-oil sectors. Non-oil exports rose 23% in 2008 alone. Projects were underway to double Syria's natural gas output over next three years.

Frankly, Syria's economic transformation was impossible to miss. In 2003, its statist economy was closed to markets. It only had six state banks and only serviced the public sector. The private sector had to leave Syria to do its banking. In addition, Syrians could only dream of borrowing to buy a house, and ATMs were an alien concept. But between 2004 and 2008, Syrians started to use credit cards and apply for loans, not just to buy homes but to also buy new cars and computers. Syria boasted nine private banks, including two Sharia-compliant entities. Restrictions on foreign currency transactions were relaxed. As new sectors opened, and the business climate improved, private-sector contribution to growth of non-oil GDP rose 80%.

That said, Syria's movement toward a free market still had a long way to go. To make matters worse, the global financial crisis hit Syria in October 2008. Luckily, its tight regulations and weak global links to international banking partially protected it. Although Gulf construction firms were still buying land in Syria, Damascus worried that the financial crisis might cause these countries to cut investments.

An additional concern in 2009 was a reduction in remittances from abroad if foreign firms were forced to cut staff. This was already a big threat in Central Asia, but most Syrians working abroad were working in skilled sectors, and so, such a shift could mark a reverse of the brain drain. The problem, of course, would still be the lack of opportunity which drove them out in the first place.

Still, the global financial crisis hurt Syria's overall economic picture by causing a fall in Syrian export markets and restricting government spending due to lower oil revenue. That translated into a slowdown in the Gross Domestic Product (GDP) growth, from 5% in 2008 to 3% in 2009. As export growth slowed, the current account deficit was on track to worsen from 1% of GDP in 2008 to 5% of GDP in 2009.

Syria's economic reformers argued that economic reform was the only way to attack the massive unemployment rate, rampant inflation and dwindling oil reserves. But those calling for free market reforms—convinced that this was the only path out of chronic economic malaise—faced an uphill struggle.

The global financial crisis and an economic slowdown also had an impact on support for Syrian economic reforms evolving from pessimism to ambivalence to optimism. How did this roll out? Immediately following the collapse of the big US banks, it seemed that Syria's nascent reform program was doomed to fail. Among Syrian economic reformers, there was concern that the return of government bailouts to save the Western banking system would play into the hands of those hard-liners in Syria arguing against free market reforms. President Bashar al-Assad personified those in Damascus who struggled to decide which way to go. At times, Assad seemed to back reform, but he also put obstacles in their way. These same conflicting signals hurt Syria on the international stage, provoking even more US economic sanctions and frustrating accord with EU countries.

But then Syrian economic reformers had a second chance. French President Sarkozy travelled to Damascus in September, opening the door to rehabilitation. In addition, the Syrian foreign minister visited London. Although there was a noticeable slowdown in the pace at which Syria was implementing economic reforms, there was reason to be optimistic about their continuing in 2009. Syria continued its gradual shift to diversification. Inherent in this approach were more free market reforms. A case in point was the creation of a Syrian stock exchange. Just a few months before, the stock market appeared to be shelved, but then economic reformers were winning out. The Syrian stock market opening was back on track and slated for 23 February 2009.

Re-establishment of a stock market in Syria was arguably a crucial step to take the economic reform process forward. The stock market was the next piece of the puzzle. There were between 26 and 45 companies ready to list. Companies would require a minimum market capitalization of $5 million and three years of audited financial statements. Most importantly, the stock market would also be open to foreign investors.

Opening the Syrian stock exchange also sent an important signal to the Syrian bureaucracy. Once a stock market was up and running, it would be a symbolic success for those promoting free market reforms in Syria—an issue that became more pressing as the country's limited oil reserves declined. Opening a stock market was a positive signal meant to attract foreign investors. It's having a stock market showed that Syria was on its way to becoming an emerging market economy and was obviously an important indication that it was open for American trade and investment.

Thus, the Syrian domestic situation had changed. President Bashar al-Assad was pursuing economic reforms at home and abroad, and Syrian free market reforms created a strategic opportunity for America to reward Damascus with economic interdependence and shared prosperity. If the Syrian people were happy with job creation and rising incomes as a result of Western trade and investment in Syria, hope and opportunity would replace the anger and fear of Arab Spring. Therefore, there would be no incentive for President Assad to crack down on muted political dissent at home. He would think twice before throwing away shared prosperity with America and the West.

US Secretary of Defense James Mattis recently said, "You don't want to miss an opportunity because you were not alert to the opportunity. So, you need to have that door open" (Lamothe 2018). Unfortunately, neither President Bush nor the Obama administration were alert to the opportunity or had a door open to support Syria's movement toward a free market economy. Why did they squander this strategic opportunity? President Bush militarized American foreign policy after 9/11. Syrian President Assad would not support President Bush's Iraq War (2003–2011), which was based on the fiction that there were weapons of mass destruction (WMD) in Iraq.

President Obama promised to stop militarizing American foreign policy, but as Dr. John R. Deni, an analyst at the Strategic Studies Institute at the Army War College points out, Obama failed to implement his rhetoric about rebalancing American foreign policy with a greater emphasis on diplomacy and development (Deni 2015). Syria was a case in point.

If the US and Syria found common ground in shared prosperity, it is possible that they could build on this progress to find common ground in shared security. Or the US could follow the George Clemenceau legacy of economic coercion. After World War I, this economic coercion led to German resentment, the rise of Hitler and World War II. In the case of Syria, a US policy of economic coercion and no viable alternative option could well lead to Syrian resentment and another wave of Syrian repression like those in 2001 and 2006. And as cited earlier, when Syria felt threatened by a hostile external environment, it almost always became intolerant of domestic dissent and cracked down forcefully for regime survival. America chose the George Clemenceau approach of economic coercion and another cycle of forceful Syrian repression at home.

Earlier, this chapter cited Henry Kissinger, who said, "opportunities cannot be hoarded; once past they are usually irretrievable." Not surprisingly, America's decision to squander that creative economic opportunity with Syria led directly to Syrian resentment and civil war. The point is that President Assad would have arguably been more relaxed toward dissent at home if a more peaceful version of Arab Spring happened in Syria and had America pursued serious economic interdependence and shared prosperity with Syria.

A Decade of Lost Chances

That said, Ken Pollack, a long time Middle Eastern political-military affairs expert and a resident scholar at the American Enterprise Institute (AEI), likes to say that "the conventional wisdom in the Middle East is almost always wrong." If this is true, there may not be a perfect time for the US to engage Syria. In fact, Carsten Wieland, a diplomat with the German Foreign Office, has written a book with the descriptive title Syria: A Decade of Lost Chances (Wieland et al. 2013).

Wieland documents numerous opportunities for Western engagement with President Bashar Al Assad when he was President in the lost decade from 2000 to 2010. For one reason or another, these opportunities were squandered by the West or by Assad himself.

Syrian Mismanagement of Natural Resources

In thinking about the Syrian civil war and whether it could have been avoided, many scholars tend to look to ideological, religious or political factors as the major reasons for the war.

However, what is often overlooked is the Syrian government's gross mismanagement of its natural resources as an underlying root cause of the uprising in March 2011.[1]

First, Syrian agricultural planners grew wheat and cotton, which used way too much water. Second, instead of using sprinklers or drip irrigation, they used flood irrigation, which wasted water in a few areas while using way too little in other areas. Half of all Syrian irrigation came from groundwater systems that were over-pumped. That, in turn, led to groundwater levels dropping (Salman and Mualla 2003). In fact, estimates said that 78% of all groundwater withdrawals in Syria were unsustainable (Wada et al. 2012).

Water Depletion

Syria was thus rapidly depleting its non-renewable water resources. By 2007, it was withdrawing 19.2 billion cubic meters of water against renewable resources of 15.6 billion cubic meters of water. In two badly affected areas—Mhardeh in Hama governorate and Khan Shaykhun in Idleb governorate—the groundwater table fell by up to 100 m from 1950 to 2000. In the aquifers around Damascus, the water table is plummeting at a rate of 6 m a year or more—and springs have dried up in many areas. Since 1999, the Khabur River has had no perennial flow (Ward and Ruckstuhl 2017).

One of the largest karst springs in the world, the Ras al Ain Springs on the Syrian-Turkish border, has disappeared completely since 2001, following extensive

[1]For instance, the Syrian construction of the Tishrin Dam on the Euphrates River. While the dam brought benefits to a privileged few, families living in the area were driven off their land. They were forced to migrate into tents outside Damascus.

over-extraction in the spring catchment area over the last 50 years. The area north of Damascus, which used to be renowned for its vines and wheat fields, has turned to desert following extensive over-exploitation of groundwater (de Chatel 2014).

This mismanagement of Syrian water resources made it increasingly vulnerable to drought. The extreme drought came in 2006 and lasted until the Syrian uprising broke out in 2011. Scientists are divided as to whether this drought was the worst in 500 or 900 years. Gary Nabhan, a respected agricultural ecologist, says this dry period was the "worst long-term drought and most severe set of crop failures since agricultural civilizations began in the Fertile Crescent many millennia ago" (Nabhan 2013).

The drought was especially intense and devastating to many people in Syria. The rainfall was 60% less than usual, and some regions received no rain at all. The consequences for Syrian agriculture were devastating. The wheat harvest came in at 2.1 million tons in contrast to an average of 4.7 million tons before the draught. This forced Syria to import wheat for the first time in 15 years (Ward and Ruckstuhl 2017).

That said, it's important to understand that drought is not unusual in Syria. In fact, Syria has experienced a drought in half the time in the last 50 years. Neighboring countries, like Iraq, Jordan, Lebanon and Israel, have also experienced high levels of drought, starting in 2006. But these four countries had all built robust resiliency programs before the droughts to help them soften the impact and recover afterwards. Syria, however, never bothered to create a resiliency program. So, it was predictable that it was there that a grave humanitarian crisis occurred.

Instead, the negative dominoes are falling. The longstanding deterioration of Syria's natural resources was especially severe in its impact on the country's northeast. This was a region that has historically been poor and neglected by the government. In the two decades leading up to the Syrian uprising in 2011, the people in the northeast were hopeful things would improve. Twice, there was a big build, and twice, there was a big let-down.

First was the government's discovery of oil. This brought considerable benefit to government revenues and to the economy, but it did little to relieve the poverty in the region. Second was the rapid development of the northeast water resources (de Chatel 2014).

The paradox for this natural resource disaster and its grave humanitarian and ultimately political and military disaster occurred in this same a region where the government invested heavily in water resources. Over 35 years (from 1985 to 2010), Syria doubled its irrigated area, from 651,000 hectares in 1985 to 1.35 million hectares in 2010 (de Chatel 2014).

But the water was diverted once again for a privileged few. The UN estimated that between 2008 and 2011, 1.3 million people were affected by the drought, with 800,000 people severely affected (Solh 2010).

During this period, yields of wheat and barley fell 47% and 67%, respectively, and livestock populations also plummeted (ACSAD 2011).

Food Insecurity

The Syrian people became less and less able to cope with the disaster. With no crops for two consecutive years, farmers no longer had seeds, while herders were forced to sell or slaughter their flocks. The incidence of nutrition-related diseases soared. By 2010, the UN estimated that 3.7 million people or 17% of the Syrian population were food insecure (UN Food and Agriculture Organization 2012).

Stephen Starr, Founder and Editor in Chief of Near East Quarterly, says that the drought and the food shortages were the single most important factors setting off the revolt in 2011.

Migration

300,000 people migrated due to the drought, leaving more than two-thirds of the villages in two governorates (Hassakeh and Deir ez-Zor) deserted. 65,000 families migrated from the northeast to the tent camps that lie around Damascus and Aleppo. By 2012, the United Nations Food and Agriculture Organization 2012 assessment said that 3 million Syrians were in urgent need of food aid, and agricultural water use was unsustainable (UN Food and Agriculture Organization 2012).

The combination of severe drought, persistent multiyear crop failures and the related economic deterioration led to a very significant dislocation and migration of rural communities to cities. These factors further contributed to urban unemployment, economic dislocations and social unrest (Femia and Werrell 2016).

Political Unrest

All these factors added to growing economic and political uncertainty. Early warnings were prescient: Some of the earliest political unrest began around the town of Daraa, the historic breadbasket of Syria, where a particularly large influx of farmers was displaced off their lands by crop failures. Political unrest was also visible at Deirez-Zour and Hama. Deirez-Zour, one of the most dangerously dry areas, was full of deep-seated dissent. Hama was a major destination for drought-displaced farmers, despite its own water scarcity problems (Saleeby 2012).

It was not the drought, the decline in food production or the rise in food insecurity that ultimately contributed to the disaffection of the Syrian people from the government. What fueled the Syrian uprising was the indifference of the country's government to the social, economic and humanitarian consequences of its policies on water and its failure to protect vulnerable populations from the effects of climactic disaster using adequate social safety nets (Ward and Ruckstuhl 2017).

Suzanne Saleeby—writing in the February 2012 issue of Jadaliyya, a magazine from the Arab Studies Institute—provides an excellent analysis of the links between economic and environmental conditions, and the political unrest (Saleeby 2012).

She says that the uprising was triggered by the lamentable failure of the Assad government to respond with adequate humanitarian assistance or help farmers to ride out the drought and restore their productive capability.

US Embassy and Syrian Government Warnings

P.H. Gleick notes that as early as 2008, the US embassy was warning the White House that the drought, food insecurity and social unrest could trigger an uprising. UN FAO Syrian Representative Abdullah bin Yehia warned that the impact of the drought, combined with other economic and social pressure, could undermine stability in Syria. In July 2008, the Syrian Minister of Agriculture said that the economic and social fallout from the drought was "beyond our capacity" to fix as a country (Gleick 2013).

Despite the clear and present humanitarian crisis developing year after year, Francisco Femia notes that "the day before the revolt in Syria, many international security analysts were predicting that Syria was stable and immune to the Arab Spring. They concluded it was generally a stable country" (Plumer 2013).

What these analysts missed was an obvious connection between economics and conflict. The contrast between indications and warning of a direct military threat and economic warnings could not be more different. If a Syrian ballistic missile was launched somewhere in the Middle East, shared early warning systems would alert affected nations while the missile was airborne. But if a country like Syria grossly mismanages its natural resources and year after year makes itself increasingly vulnerable to a massive drought, migration and social unrest, the White House, regardless of party, dismisses these social and economic strategic warnings as strategically unimportant (Femia and Werrell 2016).

The sad thing is that none of this horrible humanitarian crisis in Syria would have been inevitable if the White House had had officials who could put aside their "realism" school of international relations and open their eyes to the connection between economics and conflict. Imagine what it would have been like if a more attentive, prescient and creative US government had pursued a depoliticized "preventive defense" with Syrian farmers as partners before the drought. Former Secretary of Defense William Perry used to call this preventive defense. Using this method, you anticipate a problem and take actions to avoid it rather than standing flat-footed, then reacting with far more costly and less effective military responses to an economic problem.

It would have been relatively easy for experts from the foreign office of the US Department of Agriculture, US military civil affairs units, the US Army Corps of Engineers and the FAO to work in a depoliticized way to build robust resiliency programs in this earlier part of the lost decade. In a sense, the White House asked the wrong questions. Instead of saying, "How do we contain Syria?" it should have imagined this negative scenario and asked, "How do we help Syrian farmers build resiliency programs to protect the Syrian people? How do we help the Syrian people keep such a nightmare from occurring in the first place?".

Iraq: Another Economic Mistake

Iraq is another place where the United States was tested. This time it was against Iran for power and influence. In November 2016, the Iraqi government started to celebrate its military victory over the Islamic State of Iraq and Syria (ISIS). Investors once again saw reasons for optimism about Iraq's economic prospects. The stock market was on the upswing. Fitch gave the economy another vote of confidence in March when it upgraded Iraq's economic outlook to stable, mostly based on the country's improving public finances. And its Gross Domestic Product (GDP) growth was also booming. The defeat of ISIS prompted rare optimism and a window of opportunity for social inclusion and shared prosperity among all Iraqis (Hansen 2017).

But before long, that window of opportunity was squandered. Just eight months after Baghdad and the Western allies celebrated a hard-won victory over ISIS, the most serious anti-government protests to hit the country in years swept across its oil-rich south. Baghdad struggled to contain the protests. Demonstrators laid siege to government buildings, ports and oil companies.

The demonstrators faced serious electricity and water shortages as well as high unemployment. The demonstrators were fed up with their incompetent political leaders. They demanded sweeping reform to dismantle the corruption and mismanagement that has crippled the Iraqi economy, despite its oil wealth. The protests undermined hopes that elections in May 2018 would be a turning point and put the embattled country on the road to social inclusion and shared prosperity after decades of conflict (England 2018).

Interestingly enough, the Iraqi people have experienced a big build-up for a big letdown before. 2014 also seemed like the best of times. The war was over in Iraq, and investors were pouring money into its economy. Corporate earnings were booming. International investors were confident that economic rewards in Iraq outweighed any possible risks.

Iraq's economy looked promising in so many ways. In February 2014, its oil production surged to its highest level in over 30 years, and its oil exports hit a post-2003 high. It had one of the hottest commercial markets in the world. Asia Cell Communications had the biggest initial public offering (IPO) in the Middle East since 2008 (DeWraver and Albazzaz 2013).

The future also looked bright. With the second-largest proven oil reserves in the Organization of the Petroleum Exporting Countries (OPEC), the International Energy Agency predicted Iraq's oil output would double by the end of the decade. Its oil output was expected to grow by 600–700%. The Economist Intelligence Unit (EIU) predicted that Iraq's GDP growth would reach 8% in 2014 and almost 9% by 2020. No other country in the world had this kind of growth trajectory (Johnson 2014).

However, this was all a false dawn (FT View 2014). What did the optimists miss? US military doctrine says that phase three in a campaign plan is civilian-led stabilization. It follows two combat arms phases (first, seize the initiative, and second, dominate the enemy). Stabilization "consolidates the gains achieved through offense and defense in order to achieve sustainable outcomes that align with US national interests and strategic goals. In other words, successful post-conflict stabilization assures that the US wins the war, rather than a series of battles" US military doctrine also includes Reconstruction (phase four) (Flavin 2018; Brown 2009).

Hans Binnendijk and Stuart Johnson correctly argue that there has been a widening gap between the scale-down of combat operations and the start of stabilization and reconstruction operations (Binnendijk and Johnson 2004).

Bad things happen in this gap. The most dangerous events included ISIS filling the gap in Iraq. More recently, the Turkish military filled the gap against the Kurds in October 2019 when President Trump told US soldiers to cut and run from Syria. Once again, ISIS was free to begin another resurgence of terrorism.

Optimists thought that the fall of the divisive Prime Minister Maliki and the rise of a seemingly more inclusive Prime Minister Abadi was a hopeful sign. But it was too little, too late. Maliki's repression of the Sunnis was deep-rooted and contributed to the rapid expansion of ISIS. The initial ISIS military success and the subsequent and belated military response of the US and its allies are now well known.

The Rise and Fall of America's New Silk Road

Once in a blue moon and almost by mistake, someone like a George Marshall or a Dave Petraeus will come along to connect the economic and military dots at a high level. Unfortunately, the economic/security connection is not institutionalized. So, when the Dave Petraeus leaves the picture, foreign policy problems once again become militarized, and America looks for hammers to pound nails. In short, too many American strategists don't understand the importance of connecting economics and security.

Take Afghanistan, for instance. President Obama announced in 2011 that the lion's share of the 150,000 US troops would come home, leaving only about 10–15% of the troops behind (CNN 2011). There would also be a simultaneous and proportional 10–15% reduction of about 130,000 military contracts. US military doctrine says that civilians will take the lead in stabilization and reconstruction operations. But Hans Binnendijk and Stuart Johnson correctly argue that there is now a widening gap between the scale-down of combat operations and the start of stabilization and reconstruction operations (Binnendijk and Johnson 2004).

Bad things happen in this gap. So, when President Obama announced the US drawdown in Afghanistan, the US Treasury did a study that predicted how much of a negative impact there would be on the overall Afghan economy when 85–90% of the war economy went away. The best-case scenario was a 13% cut in Afghan Gross Domestic Product (GDP), which equates to the US Great Depression. The worst-case scenario was a 41% cut in Afghan GDP. Either way, the number of jobless people was expected to soar (Rosenberger 2011).

So, what happens militarily if nothing is done to fill this gap between combat power and stabilization and reconstruction? In East Timor, violence rose when the United Nations (UN) peacekeepers played cut and run (The New York Times 2012).

In Iraq, the world saw the rise of ISIS after the US played cut and run, leaving the country without making any serious stabilization and reconstruction efforts (NPR 2015). But, in Iraq, at least, oil production is not located near war zones, even if low oil prices recently reduced profit margins.

In Afghanistan, the good news is that there is lots of potential wealth. In fact, the US Geological Survey says that the country has a trillion dollars of potential mineral wealth (NBC News 2014).

Unfortunately, Afghanistan is struggling to turn this potential wealth into actual wealth. Former MIT Professor Walt Rostow would say that it lacks preconditions for economic takeoff (The Economist 2003).

If land-locked Afghanistan had good infrastructure, it could still have market access. But Afghan infrastructure has been largely destroyed because of continuous wars. In this regard, only 7% of the roads are paved. Therefore, this inadequate infrastructure equates to poor market access. When it rains, the roads turn to mud or flood. So, there is little incentive to increase production.

Therefore, Afghanistan faced a double whammy. On the cyclical side, the war economy was collapsing. On the structural side, it struggled with inadequate infrastructure and poor market access. With these combined, Afghanistan is faced depression economics. This is why it needed a Keynesian economic strategy to fill the gap in aggregate demand (or at least soften the blow).

Weak Trade Diversification

The International Monetary Fund (IMF) research shows that Afghanistan's external trade contributes little to economic growth. Its share of international trade is insignificant, and it has been running a trade deficit for the last two decades. Why the trade imbalance? IMF says that the country compares poorly with its neighbors, especially with respect to the low value of its official exports, which have remained below 10% of GDP since 2012, while imports are dominated by foreign-financed security spending and aid-related imports (Gitton and Muzaffari 2017).

In addition, IMF research highlights Afghanistan's low degree of trade diversification, which makes it more vulnerable to external shocks, like the current balance of payments crisis. Its external trade is over-concentrated in agricultural products (nearly 45% of exports). This reflects the small manufacturing base and a recent decline in specialized products, such as carpets and textiles, owing to global competition. And its heavy export concentration (to Pakistan and India) expose the country to demand shocks in those markets (Gitton and Muzaffari 2017).

IMF also cites several country-specific factors that have held back trade performance. Constant conflict over the last 35 years, a landlocked geography and infrastructure and institutional gaps have combined to thwart trade expansion. World Bank data documents the costs of Afghanistan's being landlocked. The World Bank's Logistics Performance Index (LPI) shows that on average, the delays and costs for both importing and exporting are far higher for landlocked countries than their coastal neighbors (Gitton and Muzaffari 2017).

Johns Hopkins Professor Fred Starr had been advocating a New Silk Road for decades (Starr 2007). His concept was transcontinental in scope and ran from China to Europe via Central and South Asia. But the US government did not push very hard for it. If they were going to support it, conventional wisdom says that the US State Department would be the logical place to do so, and the Policy Planning Council at there would ordinarily be the logical office in which to develop policy.

With this in mind, Mike Gfoeller (the Political Advisor), Lewis Elbinger (the Deputy Political Advisor) and I (the Chief Economist) at CENTCOM flew from Tampa to Washington, D.C. and met with the Policy Planning Council at STATE.[2] Unfortunately, the diplomats in policy planning wanted no part of the New Silk Road. It would be too difficult, with too many moving parts. Richard Holbrooke, the Special Representative for Afghanistan and Pakistan (SRAP), wanted no part of it either.

Since nobody else in Washington wanted to develop policy for a New Silk Road, General Petraeus signed off on CENTCOM doing policy formulation for a New Silk Road to close the gap in aggregate demand. A "tiger team" (or interagency task force) including Deputy CENTCOM Commander John Allen, Rear Admiral Jeffrey Harley, Ambassador Michael Gfoeller, myself, Dr. Adib Farhadi, Colonel Ted Hodgson, Dr. Martin Hanratty, Dr. Richard Ponzio, Lewis Elbinger and many other experts was created to develop a New Silk Road Strategy and Plan. Twenty economic development projects were designed and planned (Rosenberger 2011).

The hard infrastructure included transport, mining, energy and telecom projects. Transport included completing the Afghan Ring Road, the Afghan North–South Road Corridor, the Afghan East–West Road Corridor, the Kabul-Jalalabad-Peshawar Expressway, the Salang Tunnel, the Northern Rail Corridor and commercial aviation. Mining included the Aynak Copper Mine and the Hajigak Iron Ore Mine. Energy included the Turkmenistan, Afghanistan, Pakistan and India (TAPI) gas pipeline; the Central Asia-South Asia Electricity Project (CASA 1000); and the Sheberghan Gas Fired Thermal Power Facility. The Fiber Optic Ring was the telecom project.

To prove that there were no white elephant projects, the tiger team had a rigorous evaluation system. There were macroeconomic indicators, such as job creation, per capita income and government revenue. The cost benefit analyses only included

[2]At this time, Michael Gfoeller was the Political Advisor at CENTCOM, his Deputy was Lewis Elbinger and I was Chief Economist. The three of us attended that frustrating and unsuccessful meeting with the Deputy Director of the Policy Planning Council at the US State Department.

infrastructure projects in which there was a strong benefit to cost ratio. All others were rejected.

The Task Force's predictions of successful infrastructure projects also had demanding risk assessments. Oxford Analytica was rejected because their work only included economic risk. Instead, the task force opted for interdisciplinary company, which included political, military and economic algorithms. Soft infrastructure projects were also evaluated, including (a) legal, policy and regulatory reforms, (b) cross-border economic zones and (c) harmonizing regional customs.

The task force also developed plans to give the Afghan private sector persuasive reasons to stay in Afghanistan and enticing reasons for the Afghan diaspora to return.[3] Afghan rural communities lacked soft infrastructure to facilitate education and job creation. They lacked affordable electrical and internet services, resources and access to digital education libraries and entrepreneurship/mentorship programs.

The task force's strategic enablers involved building soft infrastructure in rural communities to create sustainable resources to support education and foster entrepreneurship. The plans for the small business center included a solar-powered convenience store and a micro-grid enhanced with a conferencing E-station. This would enable video teleconferencing and access to educational E-library centers and training videos to foster entrepreneurship.

Step by step, the CENTCOM task force was successful in getting 35 US government agencies onboard. While the policy planning council at the US State Department continued to oppose the New Silk Road Initiative, we were building at CENTCOM, we were successful at getting Marc Grossman, the new and more powerful Special Representative for Afghanistan and Pakistan (or SRAP), onboard. The CENTCOM task force enjoyed a declaratory policy victory when US Secretary of State Hillary Clinton embraced its New Silk Road initiative in a speech she gave in Chennai, India on 20 July 2011 (Rosenberger 2017).

With all of its persuasive feasibility plans done, the task force was ready to pursue an implementation plan. General Petraeus was the ideal choice to help get Wall Street onboard with the task force's finance plans and the media on board to mobilize public opinion. Petraeus holds a PhD in International Relations from Princeton University, and he taught economics at West Point.

The Strategic Mistake

But the task force suffered a fatal defeat when President Obama, who never understood the strategic value of its New Silk Road initiative, rushed to judgment and sent General Petraeus to "save the day" in Afghanistan.

This ill-advised decision meant that Petraeus would get bogged down with tactical military matters and have little time for the long view and the strategic New Silk Road Initiative. I personally believe that no Hollywood producer would ever miscast his

[3]For a graphic of these rural small business centers, see https://thinkrenewables.com/sbiz-centre/.

military commanders the way Obama did. Like George Patton, General Mattis was a "legendary warrior" who should have gone to Afghanistan. "Mad-man" Mattis would have excelled as a battlefield commander in Afghanistan. Instead, Obama had Mattis replace Petraeus at CENTCOM.

Why did this change of command matter? While General Petraeus was a grand strategist and passionate supporter of the New Silk Road at CENTCOM, General Mattis did not understand either the economic or strategic importance of the CENTCOM New Silk Road. Therefore, he had no problem and throwing away all the work the New Silk Road task force had done at CENTCOM. He bottomed out the funding for the New Silk Road at CENTCOM and tried to eliminate the Chief Economist position that Admiral William "Fox" Fallon had insightfully created to link economics and security. General Lloyd Austen ignored General Mattis's advice and maintained the Chief Economist position after Mattis retired. But the self-inflicted damage to America's version of the New Silk Road at CENTCOM was fatal. Instead of either competing with China or collaborating with it in the region, the fall of CENTCOM's New Silk Road Initiative meant that the US retreated economically from the region. In short, President Obama and the powers that be made a huge strategic mistake and squandered another window of economic opportunity.

Why not just let the US State Department run the New Silk Road Initiative? As we've seen, the Policy Planning Council at STATE never wanted the New Silk Road. Secretary of State Colin Powell created the Coordinator for Reconstruction and Stabilization (C/RS) position in 2004 to address these matters in theory.[4] But it never had enough resources to go from policy to implementation. The Bureau of Conflict and Stabilization (CSO) replaced C/RS in 2011, but it had similar resource shortfalls.[5]

Instead of beefing up its budget, it appears that President Trump now intends to take CSO to the chopping block. The Task Force for Business & Stability Operations (TF/BSO) in OSD was created in 2006 to develop pilot projects.[6] But TF/BSO and USAID had turf battles, and TF/BSO was bottomed out in November of 2014.

Economic White Flag to China

With no plans to fill the gap between (a) combat power and (b) stabilization and reconstruction, a muscle-bound America continues to win operational military victories, lose the peace and surrender US Silk Road Strategy to China. The New Silk Road task force members at CENTCOM may have felt partially vindicated when Beijing announced in 2013 that it would initiate its own scaled up version of the New Silk

[4]For more on C/RS, see https://2001-2009.state.gov/s/crs/c12936.htm.

[5]For more on CSO see https://www.state.gov/j/cso/.

[6]See www.rand.org/content/dam/rand/pubs/research_reports/RR1200/RR1243/RAND_RR1243.pdf.

Road, which they called the One Belt One Road Initiative (or BRI). The BRI is China's massive new trade corridor project. The good news is that China is serious.

The bad news is that much of China's New Silk Road goes around Afghanistan. CENTCOM's idea was to build infrastructure through Afghanistan, which, in turn, would strengthen the Afghan economy and foster transcontinental shared security. In any event, when America ill-advisedly abandoned its New Silk Road, China happily initiated its own multi-trillion version of the New Silk Road.

In late April 2019, the Chinese hosted their second international forum for their One Belt, One Road Initiative (BRI)—their version of the New Silk Road. 36 heads of state or government attended the forum. Enthusiasm was especially apparent in four regions of the world. In Southeast Asia, China secured the attendance of top leaders from 9 out of the 10 ASEAN member states. In Central Asia, four out of the five regional countries sent their top leaders. In Europe, 12 of the 36 heads of state or government attended, including those of Russia and Azerbaijan. And in Africa, the five top leaders of Djibouti, Egypt, Ethiopia, Kenya and Mozambique also attended.

In the years ahead, historians may well ask: How was China able to turn the tables on America? Just 8 years earlier, Eurasian strategists had been bullish on America's New Silk Road strategy and plans. That perception was punctuated by US Secretary of State Hillary Clinton's "New Silk Road" speech on 20 July 2011 in Chennai, India.

Back then, China was on the sidelines. In contrast, General Dave Petraeus and his interagency task force at the US Central Command (CENTCOM) were front and center, creating an inclusive New Silk Road strategy with plans for infrastructure that would go through Afghanistan and turn enemies into friends and aid into trade. All of this was intended to promote shared prosperity and collective security.

But behind the scenes, American foreign policy was starting to change. General Petraeus left CENTCOM and became the commander of the International Security Assistance Force (ISAF) in July 2010. General James Mattis replaced Petraeus and was the CENTCOM Commander from August 2010 until March 2013. General Mattis felt that the US State Department should run the New Silk Road. So, by 2013, Mattis had zeroed out the funding for the New Silk Road Task Force which General Petraeus had successfully created at CENTCOM. China turned the tables on America by filling this gap with its own version of the New Silk Road, the BRI. In contrast to America, which surrendered its New Silk Road strategy, Beijing is serious. The World Economic Forum says China is committing itself to a $8 trillion BRI. In the so-called Great Power Competition with China, America is giving the Eurasian playing field to China and retreating to Fortress America.

How was China so successful in turning the tables on America? And why is America so reluctant to implement its version of the New Silk Road plan? The short answer is that China understands how to connect economics and security in its foreign policy. In contrast, American foreign policy keeps economics and security in separate silos. For the most part, America waives the economic white flag and tries to solve conflicts in the world either militarily or with economic coercion. The Iranian nuclear agreement was an exception—but the Trump foreign policy team threw away all that brilliant work as well.

Conclusion

The U.S. foreign policy record is clear. A militarized U.S. foreign policy has marginalized U.S. economic policy. The day before the Syrian civil war, the U.S. said Syria was "stable" and "immune from Arab Spring." What did the U.S. miss? If a Syrian ballistic missile is launched somewhere in the Mideast, shared early warning systems alert affected nations while the missile is airborne. But if a country like Syria grossly mismanages its natural resources and makes itself increasingly vulnerable to a massive drought, food insecurity, migration and social unrest, the U.S. "realists" dismiss these economic warnings as strategically unimportant.

After years of socialism Syrian President Assad embraced free market reforms in 2009. That created a strategic opportunity for America to engage him with economic interdependence and shared prosperity. But the U.S. squandered this opportunity. After two successful combat arms phases of the war plan in Iraq in 2014, the U.S. failed to execute the stabilization and reconstruction phases of the war plan. ISIS terrorist group filled the gap.

To be fair, General Dave Petraeus tried shared prosperity with his New Silk Road Initiative. But after General Petraeus was sent to Afghanistan to be the Battlefield Commander, "the U.S." cut the funding for his New Silk Road at CENTCOM and threw away this singular attempt at shared prosperity as well. China turned the tables on the U.S. with its own multi-trillion-dollar version of the New Silk Road Initiative which they call the Belt and Road Initiative (or BRI). This U.S. marginalization of economics was a huge strategic mistake and another lost opportunity for ending the 18-year Afghan war.

Harry Truman used to say the only thing new in this world is the history you don't know. After World War II Jean Monnet's European Coal and Steel Community created social inclusion and shared prosperity … and turned previous French and German enemies into friends—not a bad way to keep the peace in Europe for 70 years. 18 years of trying to kill our way out of Afghanistan has failed. This militarized approach only deals with the symptoms of conflict and squanders opportunities for peace. The time for change is now. The U.S. should rebalance U.S. foreign policy with a greater emphasis on shared prosperity to reduce the negative social and economic demands for violence.

The Challenge of Reconciliation

Now let's move forward to September 2019. President Donald Trump says 18 years of fighting in Afghanistan is long enough. He says he wants to declare victory and cut and run from Afghanistan. That means that all US soldiers will hopefully be home well before what Trump plans to be his re-election in 2020.

Trump picked Dr. Zalmay Khalilzad, an excellent former US Ambassador to Afghanistan, as his US Special Representative for Afghan Reconciliation to help make this happen. The problem is that the Taliban sees President Ashraf Ghani's Afghan government as illegitimate and therefore refuses to meet with it. So, Khalilzad must fill the gap and meet with the Taliban himself.

Khalilzad's talks with the Taliban have been relatively successful. In the most recent round of negotiations, the two sides say they are close to agreement on two issues: A timetable for US troops to leave the country and what the Taliban needs to do to prevent international terrorists (like ISIS) from operating on Afghan soil.

However, Khalilzad says that two other issues are important to the US. First, there must be a ceasefire, and second, there must be an intra-Afghan political settlement. It took over a year before much progress was reached on these two issues.

Finally, in late February 2020 Khalilzad and the Taliban reached an agreement in Qatar. What are the four takeaways?

- First, a gradual U.S. troop withdrawal will start.
- Second, the U.S. has agreed to withdraw U.S. troops from Afghanistan in exchange for assurances by the Taliban that it will deny sanctuary to terrorist groups like Al Qaeda.
- Third, the U.S. now has about 12,000 troops in the country, down from about 100,000 at the peak of the war nearly a decade ago. They are supported by several thousand other troops from NATO allies.
- Fourth, the two sides have agreed to a gradual, conditions-based withdrawal over 14 months. In the first phase, about 5,000 troops are to leave within 135 days. During the gradual withdrawal, the Taliban and the Afghan government would have to work out a more concrete power-sharing settlement. That time frame would give the government the cover of American military protection while negotiating (Mashal 2020a, b).

Needless to say, lots of obstacles remain, with mistrust and uncertainty remaining on all sides. Mujib Mashal notes that corruption is still rampant, Afghan institutions are feeble, and the economy is heavily dependent on U.S. and other international aid. In addition, American efforts to instill an Afghan democratic system are at risk if the Taliban becomes dominant again (Mashal 2020a, b).

Is a Lasting Peace Problematic?

Therefore, the $64,000 question remains: Will the agreement last? First, economic reconciliation has been tabled. Unless economic reconciliation is front and center, Taliban fighters will become unemployed and the fighting will almost certainly resume. Therefore, there must be a robust job creation plan if there is to be a lasting peace.

Second, the World Bank research notes that a lasting peace is difficult because the US, the Ashraf Ghani government in Kabul and the Taliban lack the means to lock into an agreement. For starters, the Taliban cannot guarantee that if it accepts a lasting ceasefire, its more extreme members will not form a splinter group and keep fighting.

Conversely, the Ashraf Ghani government cannot make binding commitments that the Taliban can trust once they disarm. That's not surprising. Results from other similar civil wars show that many of the agreements that end civil wars are unstable and collapse into renewed fighting.

The World Bank also notes that the difficulty of negotiating a lasting peace is due to the inability of a third party to credibly commit to the peace. Credible guarantees offered by a third party committed to enforcing the terms of an Afghan peace treaty are painfully lacking.

Could the US provide a credible US guarantee for a settlement? Does it have a credible and well-established self-interest in preserving the peace with enough US military resources? Is Khalilzad in a position to provide costly signals of US commitment by deploying US troops to provide credible guarantees?

A credible US commitment is problematic for two reasons. First, Trump reneged on the nuclear agreement with Iran. Why should the Taliban trust him for a credible commitment in Afghanistan? Second, given his interest in permanently cutting and running from Afghanistan, it seems unlikely that he would want to deploy US troops for peacekeeping after they leave Afghanistan. That being the case, without such a credible US commitment and without an Afghan government seen as legitimate by the Taliban, there is no apparent incentive for the Taliban to disarm. And since disarmament is typically a precondition for the implementation of peace, these glaring omissions at least appear to create a significant obstacle to Khalilzad's successful negotiation with the Taliban.

If an external US military solution appears unlikely while Trump is president, is there another way to make the peace self-enforcing? Some analysts say that integrating part of the Taliban into the national army might be such a solution. The problem is that the Taliban would lose its bargaining power in relation to the Afghan government once its fighters demobilize. If the Afghan government retained its military, it could easily renege on its promise after the Taliban disarmed. It would have both the power and the incentive to either re-negotiate the agreement or defect from it unilaterally.

Reducing Post-conflict Risk

In peacetime, US politicians are fond of using the bumper sticker "peace through strength" when they speak to their base. "Peace through strength" is also music to the ears of the military industrial complex that President Eisenhower used to warn us about.

But what happens in the immediate post-conflict world? How does the government reduce the risk of fighting again? For starters, the pressure from the military industrial complex for high military spending is certainly present here as well. However, research findings from the World Bank indicate that high levels of military spending in the post-conflict world significantly increase the risk of reversion to war.

For starters, an important post-conflict problem is that neither side trusts the other. Thus, the more the government spends on the military, the more the rebel organization feels it must prepare for the renewal of conflict. Such mutual military escalation can easily trigger incidents that re-ignite the conflict.

Research indicates that high military spending may well increase the risk of reversion to conflict by inadvertently sending signals to the rebels that the government lacks confidence in the persistence of peace. Therefore, more Afghan government military spending is destabilizing. In addition to the problems resulting from higher military spending, the government could forego the opportunity to realize a peace dividend from reduced spending. Thus, getting to peace is difficult.

But even if peace is initially re-established with peacekeepers like the United Nations, research shows that it is often fragile. World Bank research indicates that countries like Afghanistan face two major risks:

First, when peace is re-established, it is often fragile. The 18-year Afghan civil war has developed a momentum of its own. Powerful forces tend to lock it into a syndrome of further conflict. Paul Collier calls these perpetuating factors the conflict trap. In other words, it does not take much to upset the apple cart. In fact, the best predictor of whether Afghanistan will be in a civil war next year is whether it is now in civil war.

Second, the likelihood of fighting happening again turns on whether or not the country inherits a severe economic and social decline from the period of conflict. Empirical research by the World Bank shows a striking pattern: Civil war is heavily concentrated in poor countries, like Afghanistan. In other words, poverty increases the likelihood of civil war. Thus, the World Bank's central argument is the failure of economic development is a root cause of conflict.

Economic Root Causes of the Afghan Conflict

Now that we've understood how costly the Afghan civil war has been, several other questions come to mind. First, why was Afghanistan so prone to violence 18 years ago? And second, why has the war lasted so long? If US strategists knew the answers to these two questions, it stands to reason that they might have a better chance of achieving political reconciliation and durable peace in Afghanistan.

Some scholars look at each civil war as totally unique and distinctive, with its own personalities and events. Most of us would agree that any all-embracing, general theory of civil war would therefore be patently ridiculous. That said, when one looks beyond the personalities and events, important patterns emerge. In fact, some of these patterns are surprisingly strong, which suggests that some characteristics tend to make a country prone to civil war. If so, what made a country prone to civil war?

Many people think they already know the initial circumstances or root causes of civil war. Those on the political right tend to assume that it is due to longstanding ethnic and religious hatreds. Those in the political center tend to assume that it is due to a lack of democracy and that violence occurs where opportunities for the peaceful resolution of political disputes are lacking. Some scholars on the political left point to a deep-rooted legacy of colonialism.

While each political group speculates, none of these narratives sits comfortably with the statistical evidence. Empirically, the World Bank's most striking pattern is that civil war is heavily concentrated in the poorest countries. Admittedly, war causes poverty. But more importantly, poverty increases the likelihood of civil war.

The World Bank's central argument is that the root cause of the Afghan conflict and other such civil wars is the failure of economic development. Countries with low, stagnant and unequally distributed per capita incomes, and that have remained dependent on primary commodities for their exports face dangerously high risks of prolonged conflict. In the absence of economic development, neither good political institutions, nor ethnic and religious homogeneity nor high military spending provide significant defenses against large-scale violence.

Now, let's turn to our second question. If the average civil war lasts 7 years, why has the 18-year war in Afghanistan lasted so long? IMF found that a wide range of poverty related indicators worsened during the Afghan conflict. These show that per capita income fell, food production dropped, exports growth declined and their external debt increased as a percentage of GDP.

These economic forces in Afghanistan contributed in large measure to what is known as the conflict trap: Once a country like Afghanistan has stumbled into conflict, powerful forces, perpetuating economic forces, act as a conflict trap and tend to lock it into a syndrome of further conflict.

The Primacy of Economic Reconciliation

How does Afghanistan get out of the conflict trap? The same way the French and Germans finally got out of a bilateral conflict trap after World War 2. President Harry S. Truman once said, "The only thing new in the world is the history you do not know." If one thinks back to the Treaty of Versailles, it's clear that victors of World War I came up with a punitive peace. French Prime Minister George Clemenceau wanted revenge, to make Germany pay for the wartime damage it (and the Central Powers) had done against Allied Powers, like France, during the war. Clemenceau got his economic coercion. Heavy reparations were used to keep the German economy down and punish Germany for its bad behavior. Clemenceau assumed that a weak economy would also keep the German military weak. How did Clemenceau's economic coercion work? A collective German resentment of these economic sanctions arguably aided the increase in the socio-economic demand for violence and, in turn, contributed to the rise of Hitler and World War II (Wilde 2019).

Professor Ulrich Krotz at the European University Institute in Italy points out that relations between France and Germany have gone through three grand periods since 1871: "hereditary enmity" (up to 1945), "reconciliation" (1945–63) and the "special relationship" embodied in a cooperation called Franco-German Friendship (since 1963). Given the fact that the French and Germans had fought each other for centuries and had developed this hereditary enmity towards each other,[7] was Clemenceau's economic coercion against Germany inevitable? After World War II, the statesman Jean Monnet argued that it was not. Jon Meacham, author of The Soul of America, would argue that people can and do change if you give them a chance to find their better angels (Meacham 2018).

In this regard, Monnet looked for common ground and social inclusion between the French and German people. In his strategic vision, French and German businessmen would work together and build a European Coal and Steel Community. Common bonds among businessmen would spread to the rest of the French and German populations and help turn longstanding enemies into friends.

Did Monnet appear naïve and hopelessly idealistic? Possibly. But appearances can be deceiving. Monnet showed that a web of economic interdependence and social inclusion could lead to shared prosperity and a more durable peace than more short-sighted economic coercion. In other words, shared prosperity would reduce the demand for violence and thus lower mutual threat perceptions—not a bad way to keep the peace in Europe for 70 years.

Of course, World War I and World War II were wars fought among nation states. That raises the question: Are the approaches of Monnet and Marshall regarding shared prosperity rather than economic coercion, still relevant when it comes to ending civil wars in more recent times? Yes. It doesn't matter whether French and Germans are reducing the demand for violence between two nation states via shared prosperity or whether Pashtuns and Tajiks are reducing the demand for violence inside one Afghan nation state via shared prosperity. The logic of Joseph Nye and Robert Keohane remains the same (O'Keohane and Nye 1989). When previously warring groups create shared prosperity via economic interdependence, they develop a vested interest in the status quo and mutual threat perceptions are mitigated. In this case, interdependence refers to situations where states (France and Germany) or actors (Pashtuns and Tajiks inside Afghanistan) are determined by external events in a reciprocal relationship with other states or actors, jointly limiting their autonomy. It is created through the expansion of international transactions, insofar as the costs associated with them constrain political activity. In the original New Silk Road Initiative which General David Petraeus and his task force at the U.S. Central Command created, the task force did a detailed cost benefit analysis that showed that while these relationships impose costs, the benefits may exceed them (Rosenberger 2017).

[7]The relationship between France and Germany, since 1871, according to Ulrich Krotz [1], has three grand periods: "hereditary enmity" (up to 1945), "reconciliation" (1945–63) and the "special relationship" embodied in a cooperation called Franco-German Friendship (since 1963). See Ulrich Krotz.

References

ACSAD, Drought vulnerability in the Arab Region: case study; drought in Syria—ten years of scarce water (2000–2010). ISDR and the Arab Center for the Studies of Arid Zones and Dry Lands (2011)

H. Binnendijk, S.E. Johnson, *Transforming for Stabilization and Reconstruction Operations* (NDU Press, Washington, DC, 2004)

R.D. Blackwill, J.M. Harris, *War by Other Means: Geo-Economics and Statecraft* (Belknap Press of Harvard University, Cambridge, 2016)

S.W. Bowen, *Hard Lessons: The Iraq Reconstruction Experience* (CreateSpace Independent Publishing Platform, 2009)

S. Brown, *Hard Lessons: The Iraq Reconstruction Experience* (CreateSpace Independent Publishing Platform (2 February 2009)

CNN, Obama announces Afghanistan troop withdrawal plan. CNN (2011), https://www.cnn.com/2011/POLITICS/06/22/afghanistan.troops.drawdown/index.html

P. Collier et al., *Breaking the Conflict Trap: Civil War and Development Policy* (Oxford University Press, Washington, DC, 2003)

B. Daragahi, International companies aim to set up shop in Iraq despite violence. Financial Times(2013), https://www.ft.com/content/84213d46-68a7-11e3-bb3e-00144feabdc0

F. de Chatel, The role of drought and climate change in the Syrian uprising: untangling the triggers of the revolution. Middle East. Stud. **50**(4), 521–535 (2014)

M. DeWraver, A. Albazzaz, Guest post: Iraq on the brink? Markets tell a different story. Financial Times (2013), https://www.ft.com/content/a062b5f1-6c61-3bb6-bbb5-e1f73859477d

J.R. Deni, *The Real Rebalancing: American Diplomacy and the Tragedy of President Obama's Foreign Policy* (The Strategic Studies Institute, US Army War College, 2015)

A. England, Syria's next step to reform. Financial Times (2008), https://www.ft.com/content/c6232b46-8fcc-11dd-9890-0000779fd18c

A. England, Iraq's political unrest stoke instability fears. Financial Times (2018), https://www.ft.com/content/40cdc0e4-89e7-11e8-b18d-0181731a0340

F. Femia, C. Werrell, Syria: climate change, drought, and social unrest. Climate and Security (2016), https://climateandsecurity.org/2012/02/29/syria-climate-change-drought-and-social-unrest/

Financial Times, FT view: the nightmare emerging in Iraq. Financial Times (2014), https://www.ft.com/content/235e8af2-f161-11e3-9161-00144feabdc0

B. Flavin, *Stabilization: A New Approach to Whole of Government Operational Planning and Execution* (US Army Peacekeeping and Stability Operations Institute, 2018)

S. Foltyn, Iraq's disillusioned give up on political leaders. Financial Times (2018), https://www.ft.com/content/00488e80-9970-11e8-9702-5946bae86e6d

J.L. Gelvin, *The New Middle East* (Oxford University Press, New York, NY, 2018)

P. Gitton, M. Muzaffari, Afghanistan's integration in regional trade: a stocktaking. International Monetary Fund (2017), https://www.google.com/url?sa=t&rct=j&q=&esrc=s&source=web&cd=1&ved=2ahUKEwivqrbchfroAhXnmuAKHVYpBgAQFjAAegQIBhAB&url=https%3A%2F%2Fwww.imf.org%2F~%2Fmedia%2FFiles%2FPublications%2FCR%2F2017%2Fcr17378.ashx&usg=AOvVaw005YSRojqReiK2EgA-CNK7

P.H. Gleick, *Water, Drought, Climate Change, and Conflict in Syria* (Pacific Institute, Oakland, CA, 2013)

R. Haas, *A World in Disarray* (Penguin Books, New York, 2017)

A. Hansen, Turning fragile optimism into stability in Iraq. National Democratic Institute (2017)

S. Johnson, Iraq: can a dusty war zone become a golden opportunity? Financial Times (2014), https://www.ft.com/content/fb32e390-2a80-11e3-ade3-00144feab7de

R.O. Keohane, J.S. Nye, *Power and Interdependence: World Politics in Transition* (Little Brown, Boston, 1989)

D. Lamothe, Mattis arrives in Afghanistan, says some in Taliban may be willing to pursue peace. The Washington Post (2018), https://www.washingtonpost.com/news/checkpoint/wp/2018/03/13/pentagon-chief-mattis-arrives-in-afghanistan-discusses-reconciliation-with-the-taliban/

H. Lang, D. Benaim, Cautious optimism in Iraq. US News and World Report (2017), https://www.americanprogress.org/issues/ext/2017/11/08/442571/cautious-optimism-iraq/

M. Mashal, Four takeaways from the U.S. deal with the Taliban. The New York Times (2020a), https://www.nytimes.com/2020/02/29/world/asia/us-taliban-afghanistan.html

M. Mashal, Taliban and U.S. strike deal to withdraw American troops from Afghanistan. The New York Times (2020b), https://www.nytimes.com/2020/02/29/world/asia/us-taliban-deal.html

J. Meacham, *The Soul of America: The Battle for Our Better Angels* (Random House, New York, 2018)

NBC News, Rare earth: Afghanistan sits on $1 trillion in minerals. NBC News (2014), https://www.nbcnews.com/science/science-news/rare-earth-afghanistan-sits-1-trillion-minerals-n196861

G. Nabhan, as cited by Francesco Femia and Caitlin Werrell, "Syria: Climate Change, Drought, and Social Unrest." The Center for Climate and Security (2013)

National Public Radio, Fact Check: Did Obama withdraw from Iraq too soon, allowing ISIS to grow? NPR (2015), https://www.npr.org/2015/12/19/459850716/fact-check-did-obama-withdraw-from-iraq-too-soon-allowing-isis-to-grow

C.A. Ohlers, A revised strategy for post-war stabilization and reconstruction. Orbis **63**(1), 2019 (2006)

M. Perry, How climate change and failed policies have contributed to conflict in Syria. Sustainable Food Trust (2016)

B. Plumer, Drought helped cause Syria's war. Will climate change bring more like it? The Washington Post (2013), https://www.washingtonpost.com

L. Rosenberger, Economic transition in Afghanistan: how to soften a hard landing. The Strategic Studies Institute. US Army War College (2011)

L. Rosenberger, The U.S. national security arena needs more than warriors. Financial Times (2016), https://www.ft.com/content/956595a2-c06a-11e6-81c2-f57d90f6741a

L. Rosenberger, The rise and fall of America's New Silk Road Strategy. Econo-Monitor, Roubini Global Economics (2017), https://defence.pk/pdf/threads/the-rise-and-fall-of-americas-new-silk-road-strategy.495583/

S. Saleeby, Sowing the seeds of dissent: economic grievances and the Syrian social contract's unraveling. Jadaliyya (2012), www.jadaliyya.com/pages/index/4383/sowing-the-seeds-of-dissent_economic-grievances-an

M. Salman, W. Mualla, The utilization of water resources for Agriculture in Syria: analysis of current situation and future challenges, in *Proceedings of the International Seminar on Water Issues of the World Federation of Scientists*, Erice, Sicily, Italy, IPTRID (2003)

M. Solh, Tackling the drought in Syria. Nat. Middle East (2010). https://doi.org/10.1038/nmiddleeast.2010.206

F. Starr, *The New Silk Roads: Transport and Trade in Greater Central Asia* (Central Asia Caucasus Institute Silk Road Studies Program, 2007)

H. Starr, *Henry Kissinger: Perceptions of International Politics* (The University Press of Kentucky, Lexington, 2015)

The Economist, Obituary: Walt Rostow. The Economist (2003), https://www.economist.com/obituary/2003/02/20/walt-rostow

The New York Times, Peacekeepers exit East Timor. The New York Times (2012), https://www.nytimes.com/2013/01/01/world/asia/in-east-timor-un-ends-peacekeeping-role.html

UN Food and Agriculture Organization, Syrian Arab Republic joint rapid food security needs assessment (JRFSNA). FAO Report (2012), https://reliefweb.int/report/syrian-arab-republic/joint-rapid-food-security-needs-assessment-jrfsna-june-2012

US Department of Defense, Military Support to Stabilization, "Transition and Reconstruction Operations Joint Operations Concept", Version 2.0, Dec 2006, https://www.jcs.mil/Portals/36/Documents/Doctrine/concepts/joc_sstro.pdf?ver=2017-12-28-162022-680

N. Van Dam, *Destroying a Nation: The Civil War in Syria* (IB Taurus and Company, London, 2017)
Y. Wada, L.P.H. Van Beek, M.F.P. Bierkens, Nonsustainable groundwater sustaining irrigation: a global assessment. Water Resour. Res. **48**, W00L06 (2012). https://doi.org/10.1029/2011WR010562
C. Ward, S. Ruckstuhl, *Water Scarcity, Climate Change and Conflict in the Middle East: Securing Livelihoods, Building Peace* (IB Taurus, New York, 2017)
C. Wieland et al., *The Syrian Uprising: Dynamics of an Insurgency* (University of St. Andrews Centre for Syrian Studies, Fife, Scotland, 2013)
C. Wieland, *Series on Syria from Reform to Revolt, Volume 1—Political Economy and International Relations*, ed. by R. Hinnebusch, T. Zintl, Chapter 11 (Syracuse University Press, Syracuse, 2015)
R. Wilde, How the treaty of Versailles contributed to Hitler's rise. Thought Company (2019)
C. Wright, Investing in Iraq: a frontier too far? Financial Times (2014), https://www.ft.com/content/9c26c6c9-4380-3192-9b19-60dfd591ec12
S.R. Zimmerman, D. Egal, I. Blum, *Task Force for Business and Stability Operations* (RAND Corporations, 2016), www.rand.org/content/dam/rand/pubs/research_reports/RR1200/RR1243/RAND_RR1243.pdf
R. Zoellick, The currency of power. Foreign Policy (2012), https://foreignpolicy.com/2012/10/08/the-currency-of-power/

Global Cooperation Over Great Power Competition in Afghanistan

Barnett R. Rubin

Abstract Like much of the world's political discourse, debate about Afghanistan is stuck in a fast-disappearing past. The coronavirus pandemic exposes the fatuity of the U.S. administration's elevation of "great power competition" as a goal to be pursued rather than an obstacle to be overcome for the sake of the cooperation we need to save our lives. The adoption of great power competition as the lodestar of US foreign policy also creates hurdles for the task of stabilizing Afghanistan, which requires international cooperation.

Keywords Great power competition · Afghanistan · International cooperation · Peace process · Russia · China

Introduction

Like much of the world's political discourse, debate about Afghanistan is stuck in a fast-disappearing past. The coronavirus pandemic exposes the fatuity of the U.S. administration's elevation of "great power competition" as a goal to be pursued rather than an obstacle to be overcome for the sake of the cooperation we need to save our lives. The adoption of great power competition as the lodestar of US foreign policy also creates hurdles for the task of stabilizing Afghanistan, which requires international cooperation.

The administration has set out its doctrine in the *U.S. National Security Strategy* issued by the President through the national security council and the *U.S. National Defense Strategy* issued by the Department of Defense. The recognition that U.S. national interests and values compete with and are sometimes opposed to the national interests of other states, as well as non-state actors, is hardly remarkable. Nor is the contention that in order to defend its interests and values the U.S. must strengthen its economy and maintain a military establishment to deter threats and, if necessary, fight. What is new is that these two strategy documents hardly even mention common

B. R. Rubin (✉)
New York University, New York, USA
e-mail: barnettrrubin@gmail.com

© The Author(s), under exclusive license to Springer Nature Switzerland AG 2021
A. Farhadi et al. (eds.), *The Great Power Competition Volume 1*,
https://doi.org/10.1007/978-3-030-64473-4_15

interests among states that can be realized only by overcoming competition and antagonism to establish cooperation. Cooperation on "areas of mutual interest" with Russia and China is literally an afterthought, mentioned only after a litany of malign behavior (National Security Strategy 2018). The NSS mentions pandemics once, in the context of biological warfare and never refers to climate change. Instead it laments that "climate policies will continue to shape the global energy system." U.S. leadership, the administration argues, is "indispensable" not to coordinating joint action to meet the single greatest threat to humanity, but to "countering an anti-growth energy agenda that is detrimental to U.S. economic and energy interests."

Ignoring pandemics and denying climate change are extreme examples of the danger that the Trump administration's worldview posed to national, and, indeed human security. In Afghanistan stabilization and counter-terrorism depend on the ability of the U.S. to mold and lead international cooperation. A political settlement in Afghanistan can endure only if the major powers with stakes in the outcome cooperate to support it. Those states include the U.S., Pakistan, Iran, Russia (the close alignment of these two increases their clout), China (its relationship with Pakistan multiplies its importance), and India. Central Asian and Persian Gulf states can contribute to the settlement process and its implementation and durability as well.

For Pakistan, China, Russia, and Iran, the most important question about any political settlement in Afghanistan is, will it provide for stability in their national interest or the common interests of the region, or stability that strengthens the capacity of the U.S. to threaten their interests, especially by keeping troops in Afghanistan and perpetuating a long-term monopoly on defense supply?

India, Central Asia, Saudi Arabia, and the UAE hope that the U.S. will stay on in Afghanistan to serve their own interests, for India against Pakistan, for Saudi Arabia against Iran and Qatar, and for Central Asia to balance Russia and China. The Arab states' main interest in a political settlement is to free U.S. resources in the region to support their efforts against Iran and remove an incentive for U.S.-Iran cooperation. Qatar has expressed a similar interest in the past in explaining its support for a political settlement.

South Asia Strategy

The *South Asia strategy* for Afghanistan announced by President Trump in August 2017, while predating the National Security Strategy, nonetheless bears the imprint of the same thinking. Its main international preoccupation is with Pakistan, denouncing "Pakistan's safe havens for terrorist organizations, the Taliban, and other groups that pose a threat to the region and beyond" (The White House 2020). Denouncing those safe havens was also the policy of the Obama administration, articulated by Secretary of State Hillary Clinton during her visit to Pakistan in October 2011, when she *said*, "You can't keep snakes in your backyard and expect them only to bite your neighbors. Eventually, those snakes are going to turn on whoever has them in their backyard."

She added, "We are working to establish concrete steps to address the planning and execution of attacks inside Pakistan and inside Afghanistan that cross the border."

Nowhere, however does the South Asia strategy address or even identify the reasons for Pakistan's behavior. Its sole remedy consists of threats and a strategic partnership with India, which Trump asked "to help us more with Afghanistan" (The White House 2020). The Trump administration's pursuit of other policies with respect to trade and Iran that have clashed with India's national interests have been only one reason that this additional help from India has not materialized. The South Asia strategy does not even mention China, Russia, Iran, Turkey, Saudi Arabia, Qatar, Turkey, or any of the states of Central Asia. The President's remarks did not refer to common interests about terrorism or anything else. Hence even before the NSS, Pakistan, Russia, Iran, and China suspected that the main U.S. goal in Afghanistan was not the partly common objective of stabilization to prevent the establishment of terrorist safe havens (subject to different definitions of terrorism), but pursuit of its primary strategic objectives.

U.S. objectives in accord with the NSS and South Asia strategy pose fundamental and sometimes existential threats to Afghanistan's neighbors:

- Pakistan considers a US military and intelligence presence in Afghanistan aligned with India as a potential existential threat, including to its nuclear weapons, though it does not say so explicitly in bilateral channels. It also faces the potential threat of India, with U.S. support, trying to fill any vacuum left by a U.S. withdrawal.
- Russia believes that the U.S. wishes to surround and destabilize Russia and, ultimately, replace its government through "color revolutions" (The Color Revolutions 2012). To the extent that a political settlement in Afghanistan provides the US with assets to pursue such policies, such as military and intelligence bases in Afghanistan or Central Asia, Russia will oppose it.
- Iran is convinced that the aim of U.S. policy is to overthrow the Islamic Republic, as it overthrew Mossadegh, and replace it with either a regime that will be subservient or weak, or chaos. Despite Iran's strong interest in a stable Afghanistan, stopping the existential threat from the U.S. takes precedence.
- China does not consider the U.S. presence in Afghanistan an existential threat, but rather as a potential challenge that it can manage. China considers Uighur separatism an existential threat and has therefore imposed an unprecedented surveillance regime on Xinjiang (Human Rights Watch 2019). China does not believe that the U.S. is supporting Uighur separatist fighters in Afghanistan and is therefore less opposed than Russia and Iran to the presence of U.S. forces. Nonetheless the U.S.'s increasingly forceful opposition to the Belt and Road Initiative, refusal to join the Asian Infrastructure Investment Bank (AIIB), launch of a trade war against Beijing, and efforts to build up its military presence in the Pacific and other regions around China increasingly lead the Chinese leadership's view of the U.S. to converge with that of Russia and Iran and to doubt the possibility of cooperating with the U.S. in Afghanistan.
- The shift by the US to supplying the ANDSF with American-made weapons and weapons systems, replacing the Soviet/Russian systems that Afghans have long

used, is also interpreted as a sign that the U.S. intends to maintain a long-term position in Afghanistan and military alignment with its security forces. Such defense supply relationships have proven durable even when stressed by political change, as in U.S. relations with Pakistan and Saudi Arabia.

Pakistan, Russia, Iran, and China do not believe that the U.S. regards the fight against terrorism as extending to terrorism directed against them. To varying degrees they suspect that a U.S. presence in Afghanistan will not be used in a common fight against terrorism, but selectively to oppose global or anti-American terrorism while supporting or facilitating violence or other destabilizing activities aimed at rivals or enemies of the U.S.

These views underlie the strategic convergence on Afghanistan of China, Russia, Pakistan, and Iran. Russia's "Moscow Process" aims to achieve a political settlement in Afghanistan that stabilizes the country while removing the U.S. presence that threatens Afghanistan's neighbors. Russia launched the Moscow Process as an alternative to a U.S.-led process, which Russia believes aims to stabilize Afghanistan on the basis of the U.S.-Afghan Bilateral Security Agreement, thereby making Afghanistan into the U.S.'s aircraft carrier in Central and South Asia.

Gaining Regional Support

None of these governments is necessarily engaged in a zero-sum struggle with the U.S. in Afghanistan, though they might be if their worst suspicions about Washington were true. The precondition for gaining these countries' active support for a political settlement in Afghanistan is persuading them through concrete policies that the U.S. does not aim to pursue its competition or opposition to them through a presence in Afghanistan. The converse is also true. Some steps the U.S. can take alone, with the Afghan government, or bilaterally. These steps will be most effective if situated in a multilateral process not under U.S. direct sponsorship.

The following unilateral or bilateral steps would help demonstrate such a commitment:

- Purchase of military equipment, including used and reconditioned MI-17 helicopters and spare parts, from the large stocks that both Russia and China have, combined with cooperation on training of security forces, including helicopter pilots. This is a relatively inexpensive way to meet the ANDSF need for helicopters for logistics and medical evacuation and requires much less training that using US models.
- Intelligence sharing, confidence building measures, and some measures of cooperation in the fight against IS in Afghanistan, including measures of transparency to refute suspicions about U.S. logistical backing for militants there. Consider engaging Chinese intelligence to address Russian and Iranian concerns or acceding to the Russian request for a UN investigation.

- Continuing to seek cooperation with China on Pakistan's involvement in the war. The U.S.'s historically close relationship to Pakistan has yielded little positive results in Afghanistan over the past 19 years. The U.S. and China have convergent interests and potentially complementary capacities in Afghanistan, though U.S. and Afghan hopes that China would rein in Pakistan in the short term have been disappointed, and China increasingly doubts U.S. interest in cooperation. Such cooperation, however, still suits the interests of Afghanistan and the U.S., unless the U.S. is committed to using its position in Afghanistan to weaken China in its own backyard.
- Actively support efforts at building connectivity of Afghanistan to its surrounding regions and within those regions, even while pressing for improvements in the Belt and Road Initiative. This would demonstrate U.S. interest in creating public goods in the region that establish the conditions for win–win outcomes for conflict. The U.S. should not take the lead in financing or design, but it can be much more involved than hitherto. The EU is also a potential partner, as it has adopted its first official policy on Euro-Asian connectivity and is considering how to link it with its Afghanistan and Central Asian policies. These policies could include:
  - Assuring that waivers on sanctions for Chabahar are implemented effectively so that compliance officers do not still prevent companies from bidding on or participating in contracts for Chabahar's construction, expansion, or linkage to ground routes. Even better would be exempting Afghanistan from compliance with all anti-Iran measures, such as ending petroleum purchases. India's joint venture with Afghanistan, Iran, and Japan to develop the port of Chabahar is essential for opening Afghanistan to the regional economies and ending its dependence on Pakistan for transit to international markets. It also strengthens the Indian and Japanese naval capacities in the Indian Ocean to balance China's expanding presence between Hong Kong and Djibouti. U.S. interest in Afghanistan would dictate active support for Chabahar and stronger Iran-Afghanistan relations, not just a waiver of sanctions.
  - Cooperating with Uzbekistan and other Central Asian states to improve Afghanistan's North–South linkages in ways that would benefit the Russian and Chinese economies, including linking Central Asia to Chabahar through the railway across North Afghanistan proposed by Uzbekistan, China, and India.
  - Joining the Asian Infrastructure Investment Bank with a generous capital contribution and making BUILD (a US program designed to compete with the BRI) a well-funded reality, signaling that the U.S. wishes to cooperate with China and others on developing much-needed infrastructure in Central Asia despite its reservations about the bilateral deals that form much of the basis for BRI. AIIB now operates according to international standards for development banks.
  - Support selected efforts to connect Afghanistan to the Belt and Road Initiative, including the China-Pakistan Economic Corridor (CPEC), most especially China-India cooperation on projects related to Afghanistan. These could

> include both major infrastructure projects and even cooperation on training of the ANDSF. This way China can signal to Pakistan that it considers India's role in Afghanistan as legitimate and ease Pakistan's anxieties about it. CPEC appears to some as cementing an economic alliance with Pakistan, but China's outreach to India to cooperate on a railroad linking Central Asia to Iran and Chabahar through Afghanistan illustrates China's long-term goal of relatively impartial hegemony in cooperation with regional powers.

Before a recent downturn in their relations, China has been willing to cooperate with India in Afghanistan, as it does on Myanmar under its China-India plus one doctrine. India (rightly) suspects that such offers are meant to lure Delhi into participating indirectly in the Belt and Road, but this should not be a doctrinaire reason for either India or the U.S. to oppose such cooperation when it serves our interests in Afghanistan. Joint China-India investments in connectivity and other public goods enable China to signal to Pakistan that it must accept an Indian role in Afghanistan. They would also integrate Afghanistan with the regional rising powers whose economic support and partnership it will need permanently.

Uzbekistan has started to develop a role in support of the peace process including on the political side. Its activities complement Turkmenistan's longstanding aspiration to serve as a neutral site for dialogue and Kazakhstan's promotion of regional connectivity and cooperation. Uzbekistan plays an important role in the Shanghai Cooperation Organization (SCO) and is relatively trusted by China, Russia, Iran, and Pakistan without any hostility toward the U.S., while its location makes it a lynchpin of Afghanistan's northward regional integration. It is worthwhile to engage further to understand how much it might be able to do and to keep it in mind as a facilitator of relations with China, Russia, and the SCO.

These measures have become more, rather than less urgent, since the Feb 29 signing of the agreement between the U.S. and the Taliban.

The combination of the delays in implementation of key provisions of the US-Taliban agreement and the persistent dispute over the outcome of the September 2019 Afghanistan presidential elections led US Secretary of State Mike Pompeo to rush to Kabul on March 23, 2020. Pompeo's failure to broker an agreement between presidential claimants Ashraf Ghani and Abdullah Abdullah was also a setback for the peace process, which requires a unified delegation to represent the Islamic Republic of Afghanistan in negotiations with the Taliban. Noting that "this leadership failure poses a direct threat to U.S. national interests," the State Department then announced that the U.S. was "reducing assistance by $1 billion this year,… is prepared to reduce by another $1 billion in 2021, … and will also reconsider our pledges to future donor conferences for Afghanistan" (United States Department of State 2020a; b). Economic regression in the U.S. will only reinforce this decision. The potential collapse of financial inflows magnifies the risks. While some may hope that frustrating the peace process will prolong the US's stay in Afghanistan, that option is fast disappearing as the COVID-19 pandemic ravages the US economy and society.

Even without Pompeo's threatened (though ultimately unimplemented) reduction of aid to pressure the Afghan political elite, the impact on Afghanistan of COVID-19 in the U.S. may be comparable to or exceed that of the breakup of the Soviet Union at the end of 1991. As it did then, the crisis combines a fragile peace process with a financial crisis. It was not mainly the withdrawal of Soviet troops in 1989 that led to the collapse of the Najibullah government. The withdrawal of troops enabled the mujahidin resistance to consolidate its hold on the provinces and the regime's supply lines, but it was the collapse of the foreign aid that paid for about half of the government budget that forced the resignation of President Najibullah in favor of an as yet unformed interim government. Najibullah could not withstand the mutinies of unpaid military units that defected to the resistance. Avoiding a comparable outcome today requires not only planning but a change in mentalities by all parties to the conflict.

Economic Crisis, Failure of Cooperation, and the Soviet Collapse

In 1988, the Afghan government budget was financed 44% by borrowing from the central bank, 26% by foreign aid, 6% by sales of natural gas, and 24% from other sources, primarily customs duties and sales from state monopolies (Rubin 1995). The withdrawal of Soviet troops ended gas exports in 1989, and by the fall of 1991 the government relied almost entirely on Soviet-bloc foreign aid and borrowing from the central bank. Two thirds of the foreign aid came in the form of commodities, mainly foodstuffs. The Afghan urban population depended on wheat from Siberia and Kazakhstan.

An agreement between the US and the USSR's dying remnant in September 1991 led to the cutoff of aid to the Afghan government, called the "Kabul regime" by the US and the Afghan mujahidin, as it is by the Taliban today. A third of the government's finances, including support for food supplies, disappeared. By then, according to President Najibullah, food prices had "gone up in a horrifying manner."[1] Regional powers, mainly Russia, Iran, and Pakistan, continued to support militias of both the Kabul government and anti—government mujahidin. Unpaid troops either deserted their posts or defected to those militias or mujahidin, capturing customs posts and any other assets that could produce income.

Aid Dependence and Crisis Today

Today Afghanistan is more dependent on foreign aid that it was at the end of the Soviet period. Whereas in 1988, 26% of the Afghan government budget was financed by aid, in 2018 foreign aid accounted for over 75% of the government expenditures, including off-budget military expenditures by the United States (Mashal 2019). The

[1] FOA.

cut in aid announced by Pompeo would include major reductions to support for the security sector. Foreign aid amounts to over 21% of total GDP, making Afghanistan the eighth most aid-dependent country in the world, behind Liberia, the Central African Republic, Sierra Leone, Somalia, and three island micro-states.

While Soviet aid went entirely to the government, most foreign aid since 2001 has been delivered by international and foreign agencies outside the government budget. From 2002 to 2010 over 80% of total aid was delivered outside the government. The proportion of this "external budget" in the country's public sector declined rapidly after the departure of most foreign troops by the end of 2014, as the foreign presence outside of Kabul dwindled, and today only about 20% of aid commitments go to the external budget.

Afghanistan has made progress in domestic financing of the government budget, but it no longer has recourse to deficit financing. The estimated 14% of GDP collected as government revenue in 2018 financed about 43% of the budget. Afghanistan's constitution, however, effectively forbids deficit financing by requiring that the central bank be "independent" (Article 12). The government has operated under a "no overdraft" rule since 2002.

Despite massive capital flight by the private sector, by the end of 2019 the influx of foreign capital over the past two decades left the central bank with reserves of about *$7.5 billion*—equal to about 40% of GDP or 10 months of imports. This high level of foreign exchange reserves, however, was offset by a current account deficit (omitting official transfers) estimated at 19% of GDP, and years of capital flight. As long ago as 2012 the Central Bank *imposed controls* on the export of currency. In the *third quarter of 2019*, foreign direct investment declined by 88% from the previous year, while the outflow of net portfolio investment increased by 82%, strong indicators of accelerating capital flight. If foreign aid is cut, reserves would be drained rapidly, triggering collapse of the exchange rate and further acceleration of private-sector capital flight. Pressure on the government to print more money regardless of the constitutional ban on deficit spending could become irresistible, leading to dramatic rises in food prices and the collapse of the real value of Afghans' wages.

Loss of capital inflows resulting from both aid and the operations of the international military and civilian presence would create pressures for deficit spending financed by money creation, regardless of constitutional requirements, and would be accompanied by dramatic rises in food prices just as wages would be collapsing. Food prices jumped in Afghanistan in the second week of March 2020 when Pakistan closed the border out of concern over the COVID-19 pandemic (Tolo News 2020). President Ghani had to appear on television to reassure the nation that there were no food shortages. Pakistan reopened the main border points on April 6, stabilizing food prices, though they remained higher than before March 14 (Voice of America 2020).

Pandemic in the Donors

The bulk of both foreign aid and operational capital inflows have come from the U.S., where in late March 2020 about 25% of the population (including me) lives under lockdown orders, a number that is sure to increase. Unemployment may reach 10% by April and keep growing. The economy may contract by 25% in this quarter alone. Government will spend two trillion dollars, about 10% of GDP, on domestic relief, in the next few months and will probably allocate more when Congress is back in session. This is the context in which Secretary Pompeo has made clear that the U.S. will not spend badly needed resources on a country whose leaders cannot cooperate with each other.

The amount of money needed to keep Afghanistan and a few similarly dependent countries solvent is miniscule compared to the vast funds that need to be injected into the U.S.'s contracting domestic economy, but pressure for "America First" financial retrenchment will be nearly irresistible. Afghans and their neighbors must plan for the rapid decline of aid and other capital inflows.

Consequences could include inflation, scarcity of food and other essential commodities, further acceleration of unemployment especially among educated youth, and massive population flows. Population flows already include an estimated 15,000 daily returnees from Iran, where the combination of U.S. sanctions, economic mismanagement, and the discriminatory rationing of healthcare is leading to mass exodus of Afghans. Thousands more are likely to attempt to flee Afghanistan for safety in the Persian Gulf, Central Asia, or elsewhere in the face of increasingly closed borders. The proportion of migrants infected with COVID-19 will increase exponentially. All of these crises are regional or global and require cooperation with neighbors of Afghanistan with whom the U.S. is engaged in varying levels of competition or even confrontation.

Peace Process on Rewind

As long as U.S. and NATO troops are still in Afghanistan, Afghan military and security services are likely to receive foreign funding, but the converse is also true: reductions in funding of the ANDSF are likely to accelerate with troop withdrawals. Continuation of that funding will depend on both agreement between Ghani and Abdullah and the implementation of the framework for a peace process envisaged in the U.S.-Taliban agreement. Given the likely economic consequences of the pandemic in the U.S., there is no possibility of a long-term US commitment to Afghanistan. A political agreement on the outcome of the presidential election and a peace process constitute the only alternative to collapse. Only a peace agreement supported by a regional consensus could begin to align state capacities and finances by reducing the security demands on the Afghan state.

The Taliban implemented a seven-day "reduction in violence" (RIV) during the week before the signing of the agreement. The start of intra-Afghan negotiations on ceasefire and transition within ten days after the February 29 signing was key to maintaining the momentum of the slowdown in bloodshed. However, the Afghan government's opposition to prisoner releases negotiated between the US and the Taliban forced delays. With the help of Qatar and the U.S., two weeks after the March 10 deadline for starting intra-Afghan negotiations Taliban and Afghan government technical teams held a video-conference on prisoner releases. On his way back from Kabul Pompeo stopped in Doha, where he met Taliban deputy leader Mullah Beradar at the Udeid air base that coordinates most U.S. bombing of Taliban positions in Afghanistan. The leaders reaffirmed commitment to the agreement. Pompeo stated that the Taliban were carrying out their obligations and emphasized the importance of proceeding with prisoner releases so as to start intra-Afghan negotiations.

On March 31, a technical delegation of the Taliban arrived in Kabul with the assistance of International Committee of the Red Cross (ICRC) to finalize the process of prisoners swap (Al Jazeera 2020). The process faced further delays and near breakdown as the Taliban insisted on releasing key commanders and the Afghan government insisted on gradual release of Taliban members who have not been convicted of grave crimes such as carrying out massive suicide attacks. The first tranche of 100 Taliban prisoners were released from Kabul prisons and in exchange the Taliban handed over 20 prisoners in their custody to ICRC in Kandahar (AP News 2020). It is only a matter of time until the pandemic starts spreading in overcrowded detention centers. Deaths from disease of inmates slated for release could damage the cooperation needed for prisoner exchange and increase the risk of either releasing prisoners or keeping them in captivity.

Every day that passes provides a chance for trust-destroying killings, like the Taliban attack that killed as many as 27 sleeping Afghan police and soldiers on a base in Zabul province in southern Afghanistan on the March 19 Nawruz holiday (Reuters 2020). The U.S. and Afghan government have responded by resuming aerial bombardment. The Afghan government was also slow to name an inclusive negotiating team.

The pandemic undermines the NSS's contention that a military buildup necessarily results in a position of strength, as did, of course, the entire military history of the U.S. in Afghanistan. The pandemic will inevitably have a direct impact on the withdrawal of U.S. troops from Afghanistan, which is the main interest of both the Taliban and President Trump. The firing of Captain Capt. Brett E. Crozier of the S.S. Theodore Roosevelt on April 2 after his warnings about the spread of Covid-19 among the 5,000-man crew of his battleship focused attention on the impact of the pandemic on military readiness. U.S. Afghanistan Commander Austin Miller *announced* that both new deployments and some departures are being delayed by quarantines and other measures. On March 24 Operation Resolute Support *announced* that four recently arrived service members had tested positive for COVID-19 and were in isolation (The Hill 2020). An additional 1500 recent arrivals, including 38 with symptoms, were still in preventive isolation. Congressional clamor over the safety of the forces mounted. The health and safety of U.S. forces are likely to take first priority, which risks resulting in a troop withdrawal that will be neither sufficiently compliant with the agreement to build the confidence of the Taliban nor be used effectively by the U.S. in peace negotiations.

Maintaining aid to sustain basic funding to the Afghan security forces and administration even as troops withdraw will be essential to preventing social collapse. Such a collapse would mean that neither the government, the Taliban, nor any faction could win any but worthless victories. The pressures on the U.S. government may lead to withdrawal and retrenchment, while the pressures on the Afghan government, opposition, and Taliban to protect loyalists are leading them to pursue increasingly meaningless political and military advantage. Left to their own devices all may defend themselves until they collapse in exhaustion.

Models of Regional and International Cooperation

These events signal the urgency of anchoring the U.S.-led process in a more sustainable and permanent international and regional framework. The current U.S.-Russia-China framework, which Pakistan has also joined, could provide a starting point for regional diplomacy in support of the Doha agreement and the Intra-Afghan Negotiations. Eventually it will need to develop into a more formal and inclusive framework capable of sustaining stability and the struggle against terrorism after the departure of the U.S. The history of attempts to resolve the conflict in Afghanistan provides a menus of examples of attempts to do so:

At all periods agreements were supported with an appropriate resolution of the UN Security Council and measures for monitoring and/or implementation.

In the 1980s the personal representative of the UN Secretary-General convened the Geneva Talks on Afghanistan. Diego Cordovez, the personal representative of the Secretary General (PRSG) shuttled in proximity talks in Geneva between the representatives of Pakistan and the Democratic Republic of Afghanistan (the "Soviet-installed Kabul regime"). The U.S. and U.S.S.R. acted as respective guarantors for the two parties. Pakistan was presumed to represent or at least control the mujahidin leadership. The knottiest problems were resolved in direct U.S.-USSR negotiations. The agreement was mostly unimplemented, because the nature and incentives of the parties changed too rapidly to be captured in a stable agreement.

In the 1990s during the "civil war period," the PRSG convened the six immediate neighbors of Afghanistan (Pakistan, Iran, Turkmenistan, Uzbekistan, Tajikistan, and China) plus two (Russia and the U.S.). Most of the discussions ended up as everybody versus Pakistan and led nowhere. They agreed on an arms embargo, but nothing was implemented. In the summer of 1999, a meeting of the six plus two in Tashkent including the Taliban ostensibly reached agreement on a process, but it was followed in short order by a Taliban offensive ostentatiously supported by Pakistan, and PRSG Lakhdar Brahimi suspended his mission. Starting in 1999 the UNSC passed a series of resolutions on Afghanistan, but these dealt with counter-terrorism, and largely overrode concern with any peace process.

In 2001 during the U.S. invasion of Afghanistan, the UN Special Representative of the Secretary General (SRSG), Brahimi, convened the Bonn Talks on Afghanistan. The formal negotiation sessions included only the UN and Afghans, but international observers had access to areas of the hotel where they could meet the UN and Afghan participants. The most important observers were from the U.S., Iran, Russia, and India. Pakistan attended but was marginalized. The Germans had a role as host. The U.S., Iran, and Russia all had reach-back to the field in Afghanistan and worked with the UN to shape the outcome, which finally expressed a consensus among them. The Bonn Agreement led to the authorization of the UN Assistance Mission in Afghanistan (UNAMA) led by the SRSG and requested Security Council authorization of the International Security Assistance Force (ISAF), but it did not provide any formal regional framework.

A year after the Bonn Conference, in December 2002, the Transitional Administration of Afghanistan, China, Iran, Pakistan, Turkmenistan, Uzbekistan, and Tajikistan issued a declaration on "Good Neighborly Relations," promising bilateral relations on the basis of mutual respect for sovereignty and territorial integrity, but it had no mechanism for follow-up, monitoring, or implementation except for being circulated as a UN Security Council document and being referred to in the preambles of subsequent declarations.

On December 4–5, 2005, representatives of eleven countries met in Kabul to establish the Regional Economic Cooperation Council for Afghanistan (RECCA) (RECCA 2020). Today members are Afghanistan, Azerbaijan, China, Egypt, Georgia, India, Iran, Iraq, Kazakhstan, Kyrgyzstan, Kuwait, Pakistan, Russia, Saudi Arabia, Tajikistan, Turkey, and Turkmenistan. RECCA has held seven international conferences and several thematic meetings. While it is not a political forum, it encountered a number of the same obstacles as multilateral political forums on Afghanistan. Regional countries largely viewed the formation of an Afghan-centered regional entity as an effort by the U.S. to marginalize existing regional organizations notably the Shanghai Cooperation Organization (SCO) (supported by Russia and China) and the Economic Cooperation Organization (ECO, preferred by Iran).

In November 2011, Afghanistan, Turkey, and the UN hosted an international conference on "Regional Security and Cooperation for a Secure and Stable Afghanistan." The conference launched the "Istanbul Process," which has also become known as the "Heart of Asia (HoA) Process." Its members were Afghanistan, its six immediate neighbors, India, Russia, China, the Central Asian Republics except Uzbekistan (which has since joined), Turkey, Azerbaijan, Saudi Arabia, and the UAE. It also included 28 supporting countries and international organizations. For the U.S. this conference formed part of the "diplomatic surge" for Afghanistan launched by Secretary of State Hillary Clinton in her speech to the Asia Society in February 2011. The "surge" also included direct talks with the Taliban, the launch of the (unfunded) "New Silk Road" Strategy, and the Second Bonn Conference in December 2011. It took energetic work by Turkey to convince Iran to participate in a meeting it saw as a U.S. initiative to exercise regional hegemony, to convince Pakistan to participate in a regional process on Afghanistan with India, and to gain the reluctant adherence of Russia and China, which saw in the effort another attempt to marginalize the SCO.

Both Turkey and Afghanistan envisaged the Istanbul Process as a more formal organization modeled on a Balkan Stability Pact with a secretariat in Kabul, but other members rejected that model. The HoA process did help normalize joint participation of Pakistan and India, each of which attended ministerial meetings hosted by the other. It seemed to gain momentum in 2013 when, as part of its new "look West" policy, China volunteered to assume the rotating chairmanship for the 2014 ministerial. At the October 2014 Beijing ministerial, Afghanistan and China with U.S. support proposed to establish a regional forum on reconciliation in Afghanistan within the framework of the Heart of Asia Process. Russia, however, blocked the proposal, complaining that it had not been consulted and did not know what the purpose or mode of operation of the proposed forum would be.

Alongside these multilateral efforts were innumerable "minilateral" efforts, many in the format of Afghanistan-Pakistan plus one. The U.S., UK, Turkey, and China have all convened such processes as well as a US-China-Afghanistan trilateral, the first and so-far only US-China plus one trilateral. During the 2011–2013 diplomatic surge, the U.S.-Pakistan-Afghanistan trilateral became institutionalized as the "Core Group," implying that a broader grouping would eventually form around it. In 2009–2010, when the Taliban were engaged in outreach (first genuine, then ISI-staged) to Saudi Arabia, Afghanistan suggested an Afghanistan-Saudi-Pakistan trilateral in Saudi Arabia to propose negotiations to the Taliban. Saudi Arabia instead suggested having such a meeting in Pakistan, which Afghanistan rejected. In January 2019 Uzbekistan hosted a trilateral with Afghanistan and India.

At the fifth HOA conference held in Islamabad in December 2015, US-China-Afghanistan and US-China-Pakistan trilaterals led to the formation of the Quadrilateral Coordination Group (QCG) including the U.S., China, Pakistan, and Afghanistan (United States Department of State 2020a; b). The U.S. and Afghanistan hoped that China would persuade or pressure Pakistan to compel the Taliban to negotiate with the Afghan government, China welcomed the opportunity to join the highest level of efforts for Afghan peace alongside the U.S., and Pakistan hoped that China would balance the pressure it was experiencing from the U.S. President Ghani personally briefed the ambassadors of Russia, Iran, and India on the QCG's activities, but this did not allay their misgivings at not being included. The QCG met largely at the senior official level and resulted in a meeting in Muree, Pakistan, in July 2015. That meeting was organized as a bilateral encounter between the Afghan government and the Taliban under the chairmanship of Pakistan, with the U.S. and China as observers. Senior members of the Taliban participated, in evident breach of a policy laid down by Mullah Omar against talking to the Afghan government before settling bilateral issues involving the U.S. first. The meeting was visibly dominated and controlled by the ISI, however, and the Taliban who participated were known ISI agents of influence within the Taliban rather than genuine representatives of the movement. This suited both Afghanistan and Pakistan, neither of which wanted to deal with the Taliban as a genuine political-military organization. Afghanistan was then depicting the insurgency as composed of "Taliban groups," and Pakistan wanted to control the terms of any settlement. The Pakistan Army extracted a statement from Taliban "deputy leader" Akhtar Muhammad Mansur that these Taliban were authorized to attend the meeting, but they nonetheless did not represent the Taliban as an organization or movement. Their presence caused a revolt in Quetta, with senior Taliban leaders asking if Mullah Omar had authorized this change in policy, which eventuated in the revelation that the movement's founder had died over two years earlier, in April 2013 (Rasmussen 2015). Despite this setback, preparations were underway for a second meeting in May 2016 when the U.S. killed Mansur in a drone attack as he was driving back to his home in Kuchlak, Balochistan after visiting Iran (Ahmad and Landay 2016; Reuters 2020). China concluded that the U.S. was not seriously interested in reconciliation, and the QCG did not meet again. China continued in a trilateral with Afghanistan and Pakistan.

Russian Presidential Special Envoy on Afghanistan Zamir Kabulov responded to the Istanbul Process by proposing a smaller, less official grouping that he thought would be more effective and independent of the United States: the six plus one. This included six states with significant influence over and stakes in Afghanistan (Russia, China, U.S., Pakistan, Iran, and India) plus Afghanistan, along with the UN. At least one meeting took place in this format in the Chinese permanent mission to the UN in Geneva in April or May 2014, but the Afghan government and U.S. did not support a process over which they had so little control.

The Shanghai Cooperation Organization founded a Contact Group on Afghanistan in 2005, when its heads of states meeting in Almaty, Kazakhstan, called for the withdrawal of foreign troops from Central Asia. It remained moribund for several years. Afghanistan became an observer at the SCO in 2012. In June 2017, when India and Pakistan became full members of the SCO, the organization re-launched an expanded contact group including Russia, China, Central Asian states, Afghanistan, Pakistan, and India, which held its first meeting in Moscow that October. The second meeting took place in Beijing in May 2018 (Tolo News 2018). President Xi Jinping strongly endorsed the SCO Contact Group at the SCO Summit in Qingdao in June 2018 (Tolo News 2018).

In March 2018 the government of Uzbekistan, transformed since Shavkat Mirziyoyev became president after the death of President Islam Karimov in September 2016, proclaimed its willingness to support the Afghan peace process by convening the Tashkent Conference on Afghanistan: Peace Process, Security Cooperation and Regional Connectivity. The participants, both neighbors and donors, included Afghanistan, China, France, Germany, India, Iran, Italy, Japan, Kazakhstan, Kyrgyzstan, Pakistan, Qatar, Russia, Saudi Arabia, Tajikistan, Turkey, Turkmenistan, United Arab Emirates, United Kingdom, United States, Uzbekistan, EU and UN. The declaration supported an inclusive peace process based on direct talks between the Afghan government and the Taliban and offered Uzbekistan's assistance.

Iran has sponsored several regional meetings on Afghanistan. Afghanistan's High Peace Council chief, Omar Duadzai, visited Iran in March 2019, to seek Iran's "technical support and coordination" in the peace process (BBC News 2019). In September 2019, it hosted a meeting of the national security advisors of Iran, Russia, China, India and Afghanistan "to explore ways of promoting peace and the fight against terrorism in the region." The absence of Pakistan was notable; Iran has accused Pakistan of supporting or enabling Saudi Arabia and the UAE to support

Baloch separatists based in Pakistan. Most recently it hosted an inter-parliamentary meeting on fighting terrorism in Afghanistan on December 7 with parliamentarians from Iran, Afghanistan, Pakistan, Turkey, Russia, and China. On April 13, 2020, the Institute for Political and International Studies, the think tank of Iran's foreign ministry, convened an international conference on Afghanistan using Skype. Participants included scholars and diplomats from Russia, China, the US, Pakistan, Iran, Uzbekistan, Norway, Germany, and the EU.

Following on the Russian military intervention in Syria in 2015, Russian diplomacy entered a new stage. According to Moscow's analysis, U.S. failure in Afghanistan was now inevitable, and Russia began to position itself to broker a regional solution leading to the complete withdrawal of the U.S. and the replacement of the U.S.-dominated Afghan government. The most novel parts of the initiative were Russia's engagement with the Taliban and rapprochement with Pakistan, reversing long-standing positions and strategic alignments.

In December 2016, Moscow hosted a trilateral with Pakistan and China, despite Afghan protests at not being invited (Radio Free Europe 2016). The meeting reportedly focused on the danger from IS and the need for "national reconciliation," with the Taliban seen as potential anti-IS allies. In February 2017 Iran, Afghanistan, and India joined the consultations in Moscow. In April 2017 all five Central Asian states also participated, while the U.S. declined an invitation. According to the Russian foreign ministry, these meetings focused on reaching consensus on how to engage the Taliban and oppose IS.

At the April meeting the Afghan representative announced that the next meeting would take place in Kabul, leading to the Kabul Process (see below). In August 2018 Russia announced that the Taliban had agreed to attend the next round of the Moscow Process in September, but Afghanistan refused to participate, and the meeting was rescheduled for November 2018 (Al Jazeera 2018). Afghanistan agreed to send a delegation of the High Peace Council when told by Kabulov that in that case Russia would invite a delegation of senior Afghan political leaders to participate (Radio Free Europe 2019). The U.S. decided to send an observer from the Moscow embassy. India sent two very senior retired ambassadors, making the meeting the first public encounter (other than hostage negotiations) between India and the Taliban.

In an attempt to preempt the Moscow Process and assert leadership of the international as well as domestic tracks of the peace process, the Afghan government convened the Kabul Process for Peace and Security Cooperation in June 2017. It invited 27 countries and organizations, in an attempt to place Afghanistan in a comprehensive international framework led by the Afghan government (and including its powerful friend, the U.S.) rather than a regional framework. The final statement quoted from the UN Global Counter Terrorism Strategy, that states (i.e. Pakistan) must "take appropriate practical measures to ensure that our respective territories are not used for terrorist installations or training camps, or for the preparation or organization of terrorist acts intended to be committed by other states or their citizens." President Ghani insisted that "success can only be achieved through a unified and close coordination of efforts under the Afghan Government's stewardship of the peace process" through the Kabul Process. At the second meeting of the Kabul Process, on February 28, 2018, President Ghani outlined a wide-ranging offer of peace to the Taliban. The Taliban rejected the concept of a process led by an Afghan government they considered illegitimate.[2] Defining the goal as a political settlement leading to the withdrawal of U.S. troops from Afghanistan and the region, Kabulov set about building a consensus among the region, starting with the newly convergent policies of Russia, China, Iran and Pakistan (Gul 2017). According to Kabulov, peace would come from the region, not from the U.S. The region would make peace with the U.S. if it could, which would be preferable, but without the U.S., if it must.[3] When Russia finally invited the United States to participate in the Moscow process, in April 2017, it declined the invitation. President Ghani tried to transform the Moscow process into a "Kabul process" at a meeting in Kabul in June 2017. The Moscow process, however, continued to advance, as both the Taliban and most of Afghanistan's political opposition agreed to participate.

In line with its elevation of the importance of great-power competition, the U.S. had at first tried to block the Moscow process. In August 2018, before the process led by Khalilzad had truly gotten off the ground, Moscow had convinced the Taliban to send an official delegation to Moscow for the next meeting, to which Russia also invited the Afghan government. U.S. pressure to marginalize Russia, however, led President Ghani not to attend officially but instead to send members of the High Peace Council.

As Khalilzad took over negotiations with the Taliban, he also set the stage for the alignment of Russian and U.S. policies in support of a political settlement. In September 2018, Khalilzad told the Russian ambassador in Washington that the U.S. would be open to participating in the Moscow process. Kabulov regarded Khalilzad as one of the architects of the United States' permanent presence in Afghanistan and agreed to meet him in Moscow later in September with the greatest skepticism, but the two found common ground. Russia and the United States agreed for the first time that the goal of a peace process was to produce an agreement that would stabilize Afghanistan and lead to the departure of U.S. military forces (US Embassy and Consulates in Russia 2018). The same ambiguity, as always, remained around the question of a residual counterterrorist force, but this was now presented as a subject for discussion rather than a red line. Khalilzad authorized working-level U.S. participation in the November 2018 session of the Moscow process, which included Taliban representatives (U.S. Department of State 2018).

As Khalilzad's negotiations with the Taliban progressed, he also increasingly coordinated with Kabulov. In February 2019, when the Council of Afghan Society, an association of Afghan expatriates in Russia, invited Taliban, along with representatives of Afghanistan's "constitutional coalition" for discussions on a settlement, the U.S. and Russia coordinated their response (The Diplomat 2019; Radio Free Europe 2019). Rather than back up the Ghani government's opposition to the meeting, the

---

[2]For an overview of the Moscow process, see Ekaterina Stepanova, *Russia's Afghan Policy in the Regional and Russia-West Contexts* (Paris, Institut français des relations internationales: May 2018),: 13–7.

[3]Personal interview, Moscow.

U.S. remained silent. Russia did not force a U.S. response by sending official representatives to the meeting—Russian Foreign Minister Sergey Lavrov declined an invitation to speak, and it remained a purely Afghan meeting. The Afghans who participated, as well as their Russian hosts, spoke favorably of turning the Moscow format into the framework for the intra-Afghan talks or negotiations that would result if the U.S.-Taliban talks in Doha reached agreement on a timetable for troop withdrawal and counterterrorism guarantees by the Taliban.

Global Consensus

Khalilzad followed up on the February meeting by trying to consolidate a global consensus in support of the process. On March 21, 2019, he invited to Washington the Russian, Chinese, and EU special envoys to Afghanistan. The meeting led to a joint US-Russia-China statement in support of the peace process, and a separate U.S.-EU declaration. Russia objected to including the EU at this stage. Russia agreed to try to use its convening capacity to supplement the Doha process with intra-Afghan dialogue. Foreign Minister Lavrov visited Doha to offer Russian support of the process.

The consensus on how a political settlement would meet the common security needs of both Russia and the U.S. was further laid out in a joint statement issued by Russia, China and the U.S. after a consultation among their Afghanistan envoys in Moscow in April 2019 (U.S. Embassy in Afghanistan 2019). The eight agreed points presented the new counterterrorism context of the agreement and provided a political road map more consistent with the Moscow format than the Kabul process. In particular, it distinguished an "Afghan-led, Afghan-owned peace process" from one led by the Afghan government, the role of which was more circumscribed than in Ghani's proposals. The U.S. agreed to an unprecedented joint statement with its two great-power rivals on the withdrawal of U.S. troops from Afghanistan. While the language did not call for the withdrawal of "all" troops, it did not explicitly exempt counterterrorism forces from the withdrawal (U.S. Embassy in Afghanistan 2019). Of particular relevance are the following excerpts from the text:

> *On the Peace Process:*
>
> The three sides encourage the Afghan Taliban to participate in peace talks with a broad, representative Afghan delegation that *includes the government* [emphasis added] as soon as possible.
>
> *On Counter-Terrorism:*
>
> The three sides support the Afghan government efforts to combat international terrorism and extremist organizations in Afghanistan. They take note of the Afghan Taliban's commitment to: fight ISIS and cut ties with Al-Qaeda, ETIM, and other international terrorist groups; ensure the areas they control will not be used to threaten any other country; and call on them to prevent terrorist recruiting, training, and fundraising, and expel any known terrorists.
>
> *On Troop Withdrawal:*

> The three sides call for an orderly and responsible withdrawal of foreign troops from Afghanistan as part of the overall peace process.
>
> The next stage was an attempt to bring regional players into U.S.-Russia-China process. To this end, when the three powers met again in Beijing on July 11, 2019, they invited Pakistan and Iran to join. Iran declined to attend, citing both U.S.-Iran tensions and its reluctance to join a process in which the great powers arrogated authority to themselves. Pakistan joined, however, and the four powers issued a joint statement calling on the "relevant parties to grasp the opportunity for peace and immediately start intra-Afghan negotiations between the Taliban, Afghan government, and other Afghans...as soon as possible" (U.S. Department of State 2019).

The U.S. could tell Russia that a peace process requires talks among the U.S., Afghan government, and Taliban, with Pakistan involvement, but it also needs a broader regional process.

It is also important that the process have authoritative institutional leadership that can convene smaller meetings of whatever combination may be deemed necessary. To institutionalize the process the U.S. could propose to the P-5, but in particular to Russia and China, that the UN Secretary-General appoint a special envoy or personal representative to convene a standing conference on peace and security in Afghanistan composed of the states in the Moscow process plus major donors, like the group of supporters in the HoA Process. That Special Envoy could co-convene the regional group with the chair of the SCO Contact Group and/or the OIC, given the overlap in membership. The SESG could also convene or act as an observer in the intra-Afghan negotiations or the slightly enlarged formats including the US, Pakistan, and perhaps the Arab States of the Persian Gulf, and act as a liaison to the regional group and standing conference.

Desperate Mediation?

Without illusions of return to normalcy, parties to the conflict would have the alternative of a humanitarian cessation of hostilities. They could pursue a common interest in cooperation to maintain order and minimal social services, above all public health. No one in today's Afghanistan celebrates the fall of the Najibullah government, the collapse of the state, and the outbreak of war as any kind of victory, but the pressures on all sides to pursue such a "victory" are as strong now as they were then. We do not have the time to wait for cooperation to evolve. The U.S., which has become accustomed to playing the leading role in everything concerning Afghanistan, cannot mediate cooperation in a conflict to which it is a party. Appeals for peace from an enemy are indistinguishable from demands for surrender.

Only authoritative mediation may enable the parties to overcome the imperatives of conflict to cooperate in their common interest. On March 24 UN Secretary-General Antonio Guterres called for an "immediate global ceasefire" to focus on "the true fight of our lives" (UN News 2020). To implement this appeal the Secretary-General should appoint a personal representative for Afghanistan with a strong background in humanitarian affairs. Even a chance of success would require the US and its partners to maintain consistent aid to the Afghan state even as troops withdraw.

Otherwise others may step in. Russia can restart its paused "Moscow process" unilaterally. China, through its ally Pakistan, might see little reason to rely further on a collapsing U.S. and the Afghans it supports. That and even grimmer scenarios may lie in store unless the parties to the conflict cede to the cooperation imperative imposed on them by the coronavirus and the delays in the peace process. If the U.S. as it claims, supports the UN Secretary-General's initiative for a humanitarian cease-fire, it should urge the UN to convene the relevant powers (virtually if necessary) in an urgent effort to prevent the collapse of Afghanistan.

References

AP News, Afghan Taliban confirms release of 1st government prisoners (2020). *AP News*. https://apnews.com/054d1c9054ebd3b08553e6345a8d82ae

J. Ahmad, J. Landay, U.S. says late Taliban leader was planning attacks on Americans (2016). *Reuters*. https://www.reuters.com/article/us-usa-afghanistan-taliban-idUSKCN0YC0P6. Accessed 15 Apr 2020

Al Jazeera, Afghanistan peace conference kicks off in Moscow. *Al Jazeera* (2018). https://www.aljazeera.com/news/2018/11/afghanistan-peace-conference-kicks-moscow-181109083021481.html. Accessed 15 Apr 2020

Al Jazeera, Afghanistan: Taliban team in Kabul for prisoner exchange process.*Al Jazeera* (2020). https://www.aljazeera.com/news/2020/03/afghanistan-taliban-team-kabul-prisoner-exchange-process-200331164500379.html. Accessed 16 Apr 2020

BBC News, مذاکره با طالبان؛ شورای عالی صلح افغانستان خواهان حمایت و همکاری ایران است. *BBC News*. صفحه افغانستان (2019). https://www.bbc.com/persian/afghanistan-47594724

K. Gannon, R. Faiez, Taliban to take part in 'Intra-Afghan' talks in Moscow. *The Diplomat* (2019). https://thediplomat.com/2019/02/taliban-to-take-part-in-intra-afghan-talks-in-moscow/

A. Gul, Russia to host wider regional conference on Afghanistan. *Voice of America* (2017). https://www.voanews.com/east-asia-pacific/russia-host-wider-regional-conference-afghanistan

Human Rights Watch, China: how mass surveillance works in Xinjiang. Human Rights Watch (2019). https://www.hrw.org/video-photos/interactive/2019/05/02/china-how-mass-surveillance-works-xinjiang

R Kheel, Four coalition troops in Afghanistan test positive for coronavirus. *The Hill* (2020). https://thehill.com/policy/defense/489197-four-coalition-troops-in-afghanistan-test-positive-for-coronavirus. Accessed 15 Apr 2020

R. Legvold, The color revolutions. *Foreign Affairs* (2012). https://www.foreignaffairs.com/reviews/capsule-review/color-revolutions

M. Mashal, Afghanistan needs billions in aid even after a peace deal, World Bank says. *The New York Times* (2019). https://www.nytimes.com/2019/12/05/world/asia/afghanistan-aid-world-bank.html

Ministry of Foreign Affairs Islamic Republic of Afghanistan, Deepening connectivity and expanding trade through investment in infrastructure and improving synergy. RECCA (2020). https://recca.af. Accessed 15 Apr 2020

National Security Strategy, The White House (2018). https://www.whitehouse.gov/wp-content/uploads/2017/12/NSS-Final-12-18-2017-0905.pdf. Accessed 15 Apr 2020

Radio Free Europe, Kabul protests being left out of Moscow meeting on peace in Afghanistan. *Radio Free Europe/Radio Liberty* (2016). https://www.rferl.org/a/kabul-protests-left-out-moscow-meeting-russia-china-pakistan-peace-in-afghanistan-taliban/28201061.html. Accessed 15 Apr 2020

Radio Free Europe, Taliban to join Moscow talks without Afghan government officials. *Radio Free Europe* (2019). https://www.rferl.org/a/taliban-to-take-part-in-intra-afghan-talks-without-government-officials/29750099.html
S.E. Rasmussen, Taliban officially announce death of Mullah Omar. *The Guardian* (2015).https://www.theguardian.com/world/2015/jul/30/taliban-officially-announces-death-of-mullah-omar
B.R. Rubin, *The Fragmentation of Afghanistan: State Formation and Collapse in the International System* (Yale University Press, New Haven, 1995).
A.Q. Sediqi, Afghan Military Base Stormed, 27 Security Personnel Killed. *Reuters* (2020).https://www.reuters.com/article/us-afghanistan-attack/afghan-military-base-stormed-27-security-personnel-killed-idUSKBN21728X. Accessed 15 Apr 2020
Tolo News, China's leader proposes SCO-Afghanistan contact group. *TOLOnews* (2018). https://tolonews.com/afghanistan/china's-leader-proposes-sco-afghanistan-contact-group. Accessed 18 Dec 2020
Tolo News, Religious leaders condemn price-gouging, hoarding. *TOLO News* (2020). https://tolonews.com/afghanistan/religious-scholars-condemn-price-gouging-hoarding. Accessed 18 Dec 2020
UN News, COVID-19: UN chief calls for global ceasefire to focus on 'the true fight of our lives.' *UN News* (2020). https://news.un.org/en/story/2020/03/1059972
United States Department of State, Department press briefing—November 7, 2018. U.S. Department of State (2018). https://www.state.gov/briefings/department-press-briefing-november-7-2018/
United States Department of State, Four-party joint statement on Afghan Peace Process. U.S. Department of State (2019). https://www.state.gov/four-party-joint-statement-on-afghan-peace-process/
United States Department of State, Joint press release of the quadrilateral coordination group on Afghan peace and reconciliation. U.S. Department of State 2009–2017 (2020a). https://www.state.gov/r/pa/prs/ps/2016/01/251105.htm. Accessed 15 Apr 2020
United States Department of State, On the political impasse in Afghanistan. United States Department of State (2020b). https://www.state.gov/on-the-political-impasse-in-afghanistan/. Accessed 15 Apr 2020
United States Embassy in Afghanistan, Joint statement on trilateral meeting on Afghan peace process. U.S. Embassy in Afghanistan (2019). https://af.usembassy.gov/joint-statement-on-trilateral-meeting-on-afghan-peace-process/
United States Embassy and Consulates in Russia, Special representative for Afghanistan reconciliation Ambassador Khalilzad meets with MFA counterparts in Moscow. *US Embassy and Consulates in Russia* (2018). https://ru.usembassy.gov/special-representative-for-afghanistan-reconciliation-ambassador-zalmay-khalilzad-meets-with-mfa-counterparts-in-moscow/
Voice of America, Pakistan to reopen border for Afghan visitors stranded due to COVID-19 closure. *Voice of America* (2020). https://www.voanews.com/south-central-asia/pakistan-reopen-border-afghan-visitors-stranded-due-covid-19-closure. Accessed 15 Apr 2020
The White House. Remarks by President Trump on the Strategy in Afghanistan and South Asia. The White House (2020). https://www.whitehouse.gov/briefings-statements/remarks-president-trump-strategy-afghanistan-south-asia/. Accessed 15 Apr 2020

Barnett R. Rubin is a Senior Fellow at the Center on International Cooperation of New York University. He is also a non-resident fellow of the Quincy Institute for Responsible Statecraft. He has taught at Yale and Columbia Universities, headed the Center for Preventive Action at the Council on Foreign Relations, and served as senior advisor to both the US Special Representative for Afghanistan and Pakistan (2009–2013) and the UN Special Representative of the Secretary-General for Afghanistan (2001–2002). His most recent book, *Afghanistan: What Everyone Needs to Know*, was published by Oxford University Press in July 2020. He thanks Nematullah Bizhan, Adib Farhadi, Scott Guggenheim, Leah Zamore, Per Albert Ilsaas, Roland Kobia, and others who cannot be named for their assistance.

Great Power Competition in Information Environment in the CASA Region

Sean Ryan

Abstract This paper focuses on analyzing the information environment as it pertains to what is now called, "The Great Power Competition" between Russia, China, and the United States in the Central and Southern Asian states region. This paper asks and answers critical foundational questions in an effort to refine the information landscape for future analysis.

Keywords Great power competition · Information environment · CASA region · United states · Russia · China

Introduction

This paper focuses on analyzing the information environment as it pertains to what is now called, "The Great Power Competition" between Russia, China, and the United States in the Central and Southern Asian states region. Several questions arose when considering the three framing questions presented by organizers to focus this discussion. To begin some of the most critical of these foundational questions are presented below in an effort to refine the information landscape for future analysis.

Foundational Questions

Any discussion on Great Power Competition should start with at least a cursory overview of each power's strategic goals or objectives. This proved to be one of the more simplistic questions at hand. Beyond the documented objectives of each "great power" and associated assumptions, there exists a myriad of questions that center on simple, or what should be simple, definitions that shape this discussion. These questions include the following:

S. Ryan (✉)
West Liberty University, Gary E. West College of Business Main Hall 256 208 University Drive, West Liberty WV26074, US
e-mail: sean.ryan@westliberty.edu

© The Author(s), under exclusive license to Springer Nature Switzerland AG 2021
A. Farhadi et al. (eds.), *The Great Power Competition Volume 1*,
https://doi.org/10.1007/978-3-030-64473-4_16

- What is the CASA region?
- What countries are included, for example, defines issues like what natural resources are resident in the region that the competing powers will struggle to attain? What ethnicities and languages come into play? What relationships inherently exist due to geographic proximity that competing powers must strive to either work with or challenge? All of these cascading questions factor into how the information environment may serve the great powers in this part of the world.
- How we define the CASA region heavily influences synchronization of messaging across the US government. Conflicting definitions, especially between the Departments of State and Defense, highlight critical challenges for the US Government. For example, Department of State and Department of Defense define the CASA region quite differently.

Department of State regional boundaries are presented in Appendix A reflecting a region spanning from Sri Lanka in the south to Kazakhstan in the north. Department of Defense Combatant Command boundaries depicted in Appendix B are defined in Chapter 6, United States Code Title 10 (Feickert, 2012; USC10, 2019). While federal law defines each set of boundaries, these two US Government agencies each maintain key strategic relationships and conduct their own discrete activities and messaging. Competing with China and Russia in the information environment implies the need to synchronize efforts across US Government agencies. This is made significantly more difficult when each agency defines the CASA region differently.

- What does the information environment include?

This question is foundational to any meaningful discussion on great power competition. Internal debates within the Department of Defense amplify the importance of answering this question. Does the information environment include just social media, the traditional news media, publicly available information, or all information? For the purposes of strategic analysis, it is most prudent for analysts to view the information environment in the broadest terms.

Consistent with the question defining the information environment is the question of how does the information environment impact the great power competition. In terms of how the information environment impacts the broader great power competition it is important to recognize that leaders at all levels of business and government rely on information to make decisions (Grynkewich et al., 2018). Therefore, it is essential to recognize that information and activities are intrinsically linked. Great power competition is rooted in multiple disciplines including economic, diplomacy, security, and information. It is important to consider all messaging and activities through an informational lens, which runs contrary to common interpretations of traditional military doctrine (JCOIE, 2018). Recent Chinese information efforts highlight how text messaging can be leveraged to spread information and disinformation similar to how social media has been used previously (Wong et al., 2020).

- Is there actually a state of competition between China, Russia, and the US?

This paper will not address this particular question as there is, in my opinion, no great value in answering it. Whether or not the respective powers consider themselves to be competing does not change the fact that each nation, as a function of their respective activities, stands as an obstacle in the path of every other 'great power's efforts to achieve national objectives.

Strategic Perspectives

Populations and Economic Factors

United States strategic interests frequently become the center of discussions regarding the CASA region. Terrorism, namely the September 11th, 2001 attacks deepened US involvement in the region. While we cannot overlook the re-emergence of terrorist sanctuaries as a critical national interest, there are broader perspectives to consider.

Consider the affected populations and economies. Approximately 1.8 Billion people or one fourth of the world's population reside in the CASA region. Expand that view to include the Great Powers and Iran, as regional influencers, and that perspective swells to nearly 3.8 Billion people or roughly one half of the world's population (CIA, 2019). Table 1 depicts the affected populations in the CASA region and Great Power Competition.

Population figures are not inherent justifications for national interests. When economies are combined to include the directly impacted populations, a different landscape takes shape. Whereas Afghanistan is commonly referred to as one of the poorest countries in the world, the US Geologic Survey estimated that more than $900 Billion in natural resources exist there (USGS, 2006; 2011). NBC (2014) reported that the Chinese state owned mining enterprise, China Metallurgical Group, signed a $3 Billion contract with the Afghan Government to exploit the Mes Anyak copper deposit, for example.

India is also involved in capitalizing on Afghan metal and mineral reserves. With strategic iron ore deposits to exploit, the Afghan Government signed another contract possibly worth $10.8 Billion with companies from regional neighbor India (Reuters, 2013). These mineral deposits are critical to manufacturing. Beyond Afghanistan, significant resources exist in exploitable quantities throughout the CASA region.

While the CASA may be fraught with social instability, poverty, and political turmoil, there are significant economic opportunities there. The natural resources that exist in the CASA region, including Afghanistan, possess strategic implications for the world's largest—and most resource hungry—economies. A key question with strategic ramifications centers on who controls those resources and how the related information is shared or restricted.

Table 1 Affected population and select economic data

Country	Population	Percentage of CASA region	Select economic data GNP	Comparative economic data
Afghanistan	34,940,837	1.9%		
Bangladesh	159,453,001	8.8%	11.5 T BDT 2018	
Bhutan	766,397	0.0%		
India	1,296,834,042	71.2%	145.2B INR 2019	2.5 T USD 2018
Kazakhstan	18,744,548	1.0%	59.6B KZT 2018	
Kyrgyzstan	5,849,296	0.3%		
Nepal	29,717,587	1.6%		
Pakistan	207,862,518	11.4%	13.8 M PKR	
Sri Lanka	22,576,592	1.2%	14B LKR 2018	
Tajikistan	8,604,882	0.5%		
Turkmenistan	5,411,012	0.3%		
Uzbekistan	30,023,709	1.6%		
CASA Region Population	**1,820,784,421**	**100%**		
China	1,384,688,986		896,8B HML 2018	14 T USD 2018
Russia	142,122,776			
United States	329,256,465		19.5 T USD 2019	20.4 T USD 2018
Iran	83,024,745			
Other significant Country Populations	1,939,092,972			
Total Involved Populations	**3,759,877,393**			

Sources CIA Factbook, 2018, Tradingeconomics.com, IMF as cited by WEF

Information Environment

The CASA region is the scene of strategic struggles dating to before the reign of Alexander the Great in the fourth century BC. The phrase, "Graveyard of Empires," has been applied to Afghanistan for decades if not generations. In a country where central control is questionable, or entirely elusive, strategic decisions become increasingly complex especially in an era of increasing economic interdependence.

Who are prominent users of information and who are the likely targets of influence? What factors or technologies shape the information environment in the CASA region? In an era when corporate entities control information access to billions of people, influence once attributed to nation states is now heavily affected—if not outright controlled—by wealthy billionaires and corporations. Who are the ultimate beneficiaries of information and influence?

Answering the above question about who are the likely influence targets and beneficiaries deals with a complex topic. Formal governance may reside in the central government and extend to provincial governors, but that represents only a beginning. Tribes and clans control or heavily influence territory containing strategic mineral deposits. Formal and informal leaders within these tribes and clans become key strategic influence targets. Similarly, the tribes and clans controlling or influencing access to those same natural resources will factor into the decision calculus and subsequent negotiations related to those who desire those natural resources.

The Haqqani family provides an example of the importance of the information environment in the Great Power Competition. The Haqqanis earned fame in the 1980s as warriors and freedom fighters against the Soviets. An especially powerful clan, the Haqqani's are active members of the Zadran tribe (Prusher, 2002). Prusher explained that the Zadran's tribal area extends throughout Paktia, Paktika, and Khost provinices in Southeastern Afghanistan and parts of Waziristan in Pakistan. The three provinces in Afghanistan are home to more than 1.5 million Afghans, most of whom are ethnic Pashtun (NPS, n.d.). This sociocultural demographic becomes particularly important when considering exploitation of the Anyak copper, chromite, and cobalt deposits that were identified by the USGS (2011).

Can the China Metallurgical Group afford to negotiate with only the central Afghan Government for exploitation of mineral deposits? Security must be a major concern considering the degree of resistance that prevails against the central Government and foreign involvement. The Haqqani's maintain significant influence over the security conditions in these mineral rich provinces. Consequently, how foreign entities negotiate with leaders from the Haqqani family and the Zadran tribe to ensure continued security becomes a genuinely important topic. Given the extremely rugged terrain in this remote are in Southeastern Afghanistan and Waziristan, the information domain provides important options for shaping attitudes.

Cell phones, social media and face-to-face communications combine to present a variety of avenues for information exchange. Statistica reported that cell phone usage in Afghanistan peaked in 2017 at nearly 24 million users (2019). The Chinese recently demonstrated the efficacy of cell phone SMS text messaging for information dissemination (Wong et al. 2020). Facebook reportedly reaches approximately 2.5 billion people worldwide with nearly 3.5 million users in Afghanistan (Statistica 2019). Many other social media applications similarly offer mechanisms for communicating, sharing information, and spreading influence. Together, this myriad of technological platforms offer a complex information environment.

While social media applications are constantly being developed, Facebook currently factors into the CASA region on a broad scale. Three out of the world's 15 leading countries in terms of the numbers of Facebook users are in the CASA region (Statistica 2019). India is the world's leading Facebook user with more than 260 million users, more than the United States. Bangladesh and Pakistan ranked twelfth and thirteenth, respectively. While Afghanistan has 3.5 million uses, this represents more than 10% of the population. By comparison, the US has 180 million Facebook users. Table 2 depicts select data on Facebook users by country.

Table 2 Select facebook data

Country	Facebook users (Millions)
India	260
Bangladesh	37
Pakistan	33
Afghanistan	3.5
United States	180

Facebook data derived from Statistica.com

The penetration of Facebook may be driven from economic considerations. Extensive business ties between India and the United States and nearly 20 years of US presence in Afghanistan certainly impacted the prevalence of US-based social media platforms. Will prominent use of social media shift from US-based applications to Chinese-based applications in the coming years? Social media usage will provide a key indicator on how people are networking and, by extension, how people are being influenced.

The above social media data conveys a picture that may surprise some people. There are nearly twice as many Facebook users in the CASA region than in the United States! Interviews, negotiations, video teleconferenced conversations are possible through Facebook and Facebook Messenger. Other networking and social media applications provide similar capabilities.

It will be difficult to acquire reliable data on Chinese applications due to how tightly the Chinese Government controls such information. However, it is likely that Chinese applications like Renren, Weibo, and QQ will grow in popularity throughout the CASA region as the United States effectively withdraws over time. Statistica (2020) reported that five of the top ten social media applications, as measured by numbers of users, are based in China. Table 3 depicts those top 10 applications and

Table 3 Top 10 global social media applications

	Application	Users (in Millions)
1	Facebook	2,449
2	YouTube	2,000
3	WhatsApp	1,600
4	Facebook Messenger	1,300
5	Weixin/WeChat	1,151
6	Instagram	1,000
7	Douyin/Tik Tok	800
8	QQ	731
9	Qzone	517
10	Weibo	497

Data derived from Statistica.com as of January 2020

the number of users as of January 2020. It is a matter of practicality to the people in the CASA region.

There are many avenues for influencing key decision makers. Decision makers may occupy formal positions in the Afghan Government or be local leaders who can influence security. Facebook and other social media applications offer vehicles for connecting people virtually. Text messaging is already used extensively in Afghan business.

Ultimately, the information environment provides practical and redundant means for networking people together. As Ibarra and Andrews (1992) concluded, networking factors into influence and power dynamics. The ability to communicate directly and closely impacts perceptions, which in turn affects influence. Social networks will inherently adapt to the evolving demands in the CASA region. As business considerations evolve, financial and economic factors will assume primacy over other concerns. The importance of shaping the information environment will continue to grow. The ability to communicate in real time, face to face, even if through virtual means will become increasingly important as new, dynamic relationships take shape.

Individual perceptions and decision making can be targeted and affected through social media. Whereas most of the preceding discussion focused on the CASA region, the domestic populations in each of the Great Power competitors are also primary influence targets. Russian involvement in the 2016 US Presidential election process demonstrated that much.

Strategic Objectives

China

China is clearly executing a strategy aimed at increasing its global position and influence. Increasing its power and influence through economic expansion places China at odds with the US and Russia globally, as well as with its regional neighbor, India (Chandran 2017; CSIS 2019). China's Belt and Road Initiative, also referred to in circles as the new Silk Road, is central to economic expansion. The BARI (also referred to as OBOR) offers access to expanded markets for Chinese goods as well as improving access to raw materials and natural resources.

Increasing access to materials and markets moves China ahead in its drive to increase its advantages through economic development and innovation (CSIS, 2019). While India continues to resist BARI progress, the remaining central Asian states stand to benefit—or, be exploited economically. India refused to attend BARI conferences and openly contests China's efforts to market the benefits of the Belts and Roads Initiative (Chandran, 2017). CSIS asserted that China is using the information environment to advance its initiatives (2019). This is clearly the case given the rhetoric surrounding the latest conference.

China is demonstrating increasing sophistication in its economic and information strategy. While China benefited greatly from intellectual property theft in recent decades, President Xi attempted to allay fears of continued economic predation by announcing penalties for theft of intellectual property (Bloomberg, 2018). Whether or not companies, corporations, and governments are willing to accept Xi's efforts to market a level playing field, it is clear that China wants to avoid open military conflict at this point in time (Chandran, 2017). It is equally clear that China aims to reassert itself as a world power and views current times as a "New Era" (Chen, 2017). The concept of an era is particularly important in Chinese culture as it frames what is strategically possible.

The China-Pakistan Economic Corridor and the Shanghai Cooperation Organization (SCO) represent increasingly sophisticated efforts to access natural resources and expand influence throughout the CASA region (Khalid et al., 2017). Control and shaping of information related to Chinese initiatives plays an important role in Xi's efforts to market China's evolving economic expansion. India's resistance to Xi's efforts to sell the OBORI demonstrate an awareness, at least by India, that there are potential risks of undo strategic influence from long-term predatory debt that can be incurred from Chinese loans.

Chinese actions support stated strategic objectives of increasing its power and influence through economic expansion and innovation (Araya, 2019; Chen, 2017; CSIS, 2019; Khalid et al., 2017). Acquiring natural resources and expanding markets remains a central effort in Xi's strategy. China's geographic proximity to the CASA region provides it with a decisive strategic advantage. When the China-Pakistan Economic Corridor is completed, China will clearly possess advantages over the US and Russia in terms of influencing Pakistani decision making.

The COVID-19 crisis is also providing China with information opportunities. Chinese communist leaders are working to control information, most recently about how they handled the COVID-19 outbreak (Rosenberger, 2020). Three major lines of effort recently demonstrated include the following:

- Controlling information from China;
- Promoting China in a favorable light by deflecting Chinese shortcomings and emphasizing other countries' failings while highlighting the positive aspects of Chinese partnership; and,
- Attempting to sow discord and division in American civil society. This last line of operation using social media and disinformation. These formerly Russian tactics were displayed during and since the 2016 US presidential elections.

A fourth line of effort focused on promoting Chinese nationalism is highly likely given recent efforts in the wake of the COVID-19 outbreak. Pei (2020) highlighted how Chinese leaders studied strategic failures experienced by Russia late in the Cold War. Xi's regime currently confronts political and social resistance from ethnic and religious minorities as well as from pro-democracy elements. Thus far, Xi's responses to these resistance elements can be characterized as classic totalitarian population control measures, such as the use of force to intimidate potentially vocal adversaries, extensive use of re-education camps, and close monitoring of dissident activities.

Tight controls over electronic information related to the COVID-19 crisis were also observed (Pei, 2020).

Extensive efforts along all four lines of effort should be anticipated if any conclusions can be drawn from the initial Chinese information efforts in the wake of the COVID-19 crisis. Recent Chinese tactics using broadcast text messaging are a new innovation revealed by Wong et al., (2020) that will likely accompany efforts in social media. These efforts will not be confined to the CASA region.

Instead, Chinese information offensives related to the COVID-19 outbreak will focus on separating the US from our European and Asian allies and creating or increasing distrust between the US population and the US Government. Chinese use of bots to exploit social media as a means of amplifying distrust and discord and sowing seeds of doubt were already identified in Italy (Wong et al., 2020). It is obvious that the Chinese are attempting to capitalize on the inherent ambiguity and fear associated with the novel COVID-19 virus. While COVID-19-related messaging may only be tactical in nature, it will certainly align with Xi's strategic plans.

Russia

Russian strategic objectives and behaviors are pragmatically rooted. Wanting to reestablish Russia as a premier world power, Russian leaders do not want military conflict with the United States. Rather, Russian goals center around security, power, influence, and economic development. Russia is adamant about security, maintaining status as Eurasian power, regaining the mantle of global power and advancing economic development. Russia wants to reestablish the buffer zones to the west and south that she existed under the Soviet Union. The collapse of the Soviet Union and fragmentation of the empire left Russia's borders exposed and NATO expansion is clearly perceived as a threat. Russia wants to rebuild a buffer from Syria to Afghanistan through security agreements, economic arrangements, and troop deployments. Rebuilding the western buffers is more difficult. Belarus and Kaliningrad remain, but NATO expansion in contravention of the handshake deal on removing nukes from Kazakhstan and Ukraine really touches on Russian insecurities (Shleifer and Treisman, 2011).

Gurganus and Rumer (2019) view Russia somewhat differently as being openly activist, working to expand global influence. Using Syria, Crimea, and Ukraine as examples Gurganus and Rumer make a compelling argument that Russia retains, and refined, its capabilities in operations short of war, including information activities and political warfare (2019). Their perspectives are consistent with the reporting from the Director of National Intelligence that concluded Russia interfered with the 2016 US Presidential election (DNI, 2017). Placed against the backdrop of NATO expansion and economic factors, it can be argued that Russia is seizing economic opportunities while working to strengthen security geographically. Regardless of its motivations behind these specific operations, Russia clearly demonstrated capabilities and the willingness to act in the information domain.

Shleifer and Treisman (2011) provide perspectives based on perceptions of practicality. Shiefer and Treisman contend that Russia is acting out of pragmatic reasons. Global price drops in natural gas place Russia in a potentially compromising position. According to Shiefer and Treisman, Russia is focused on three primary objectives:

- securing economic growth,
- fostering friendly regimes in other former Soviet states, and
- preventing terrorism at home.

What Shiefer and Treisman fail to explain in their analysis is why Russia would expend the energy and assume the risk of social media attacks against the United States.

CSIS analyst, Olga Oliker (2016), contends that Russian leaders seek to increase Russia's influence and power, including economic growth. Economically, Russian leaders seek to grow their GDP into one of the world's largest economies. And, Russia will continue to view any threat to their return to status as a world power as a threat. Their demonstrated capabilities in information and political warfare are likely to serve as centerpieces of an activist foreign policy (Gurganus and Rumer 2019; Olkier 2016).

It is obvious that there is debate among policy analysts about the fine points of Russia's strategic direction. All analysts share some perspectives. Economic motivations make compelling arguments, especially considering the current size of Russia's economy, Regardless, the opportunities to expand markets and increase influence in the wake of US campaigns in the CASA region place Russia in a prime position to aggressively compete in the information domain.

Conference Questions

Three central shaping questions provided at the beginning of this conference focused analysis of the information environment in the CASA region in terms of great power competition. At the time of the conference, the negotiated peace settlement was viewed as an upcoming pivotal event. Perceptions of the US commitment to credible and reputable Afghan governance will heavily influence individual perspectives about the US and will shape future US relations in the CASA region (Ryan, 2017). The three shaping questions provided by the conference were:

- What risks do the US face in the Information Environment as it relates to Central Asia?
- In what ways will the Great Powers influence the region in the Information Environment after a negotiated peace?
- How will they attempt to undermine or amplify peace and stability efforts?

The current conflict in Afghanistan stands as a shaping event regardless of how the CASA region is defined and what interests each great power has in the region. The United States became ensnared in Afghanistan after the attacks against the United

States on September 11, 2001. It does not matter what occurred prior to the trigger event, or what tactical, operational, political, or strategic objectives the US held when US Special Forces invaded Afghanistan in 2001. Today, nearly 20 years later, the US and numerous other countries are still ensnared there in conflict while negotiations are underway for a brokered end to the conflict as we have come to know it. Stability in Afghanistan is a central issue to great power engagement in the region.

What Risks Do the US Face in the Information Environment as It Relates to Central Asia?

After negotiating a peace agreement with Taliban leaders in the face of a fractured Afghan government, the United States is at risk of being discredited. Credibility is an element of influence. Ibarra and Andrews (1993) demonstrated that perceptions and influence are shaped by proximity, social interaction and network centrality. This concept clearly cuts across the information domain and physical geography. As neighboring states, it is reasonable to conclude that Russia and China can degrade perceptions of US power in the CASA region.

Having avoided direct overt involvement in Afghanistan stabilization and security development in the post-9/11 era, Russia and China are positioned to capitalize on the United States disengagement. Regional organizations, like the Shanghai Cooperation Organization, provide ample opportunities to influence key decision makers in Afghanistan and other CASA countries. Similarly, economic initiatives open the door to increasing their positions in the CASA region while diminishing the relative position of the United States.

Stability benefits all great powers in terms of mitigating potential terrorist sanctuaries and opening access to natural resources. Clearly the US is disadvantaged geographically in terms of physical presence depending on how transition occurs. A poor transition leaves the US without presence, or with minimal presence, in the region making it increasingly difficult to compete. It is reasonable to expect to see the Afghan conflict exploited and distance between the US and the region highlighted. These factors will serve to divide the American people internally and to divide the US from regional allies externally.

Further, the legacy of US experiences in Afghanistan and the CASA region offer Russia a unique opportunity. Focusing on building agreements with the former Soviet states in Central Asia, Russia will exploit every opportunity to distance the US from potential regional allies thus increasing the sense of security. Combining economic and security opportunities into pragmatic agreements will be a centerpiece of Russian information activities in the CASA region.

In What Ways Will the Great Powers Influence the Region in the Information Environment After a Negotiated Peace?

In short, China and Russia will use the information environment to pursue sustainable economic advantages over the United States. Both Russia and China are pragmatic in their regional relationships. Russia's activist policies and China's non-interventionist approach converge in the likelihood that each power is likely to engage regional power brokers. Engagements will be multi-faceted leveraging diplomatic, economic, security, and informational elements. Whether contacts within the CASA region are elected national leaders, tribal/clan elders, or warlords, will be less relevant than what they can offer to a prospective partner. It is reasonable to assume that the COVID-19 pandemic will factor heavily into messaging for at least the next year.

China

China will continue to extend the Belts or Roads Initiative (OBORI) (Araya, 2019). Emphasizing economic opportunities, China will highlight the difficulties and negative perceptions associated with American and Coalition presence. Chinese efforts will primarily equate with an intense marketing effort aimed at promoting positional advantage over the United States.

The entire regional dynamic will inherently shift as the US role in Afghanistan changes. Changing roles will ultimately result in changing expectations and perceptions, which emphasizes the importance of considering influence and persuasion as a series of psycho-social processes (Lutz and Kakkar, 1976; Solomon et al., 1985). As social exchange theory suggests, people will make decisions about relationships based on their ongoing cost–benefit analysis (Cropanzano and Mitchell, 2005; Thibault and Kelley, 1959). A reduced or realigned role by the US will open the door to interpretations by Afghan leaders that may favor China.

China will seek to promote themselves over the US and Russia in Afghanistan as a regional partner and global power. China will emphasize that it does not place 'strings' on economic aid and support. Along with the potential promised from the OBORI, China will leverage perceptions enhanced by its increasing economic presence in ports throughout the region. Additionally, China will seek to marginalize India; particularly leveraging Pakistan through the China Pakistan Economic Corridor announced in 2015 (Chandran 2017; Khalid et al., 2017). Chandran (2017) also highlights the potential for China to create influence by offering loans that create long-term debt situations.

China also possesses opportunities to generate influence and shape perspectives through the Shanghai Cooperation Organization (SCO). SCO involves China, Kazakhstan, Kyrgyzstan, Russia, Tajikistan, and Uzbekistan, India and Pakistan. Afghanistan holds observer status, while the SCO denied observer status for the United States (Khalid, Qureshi and Hassan, 2017). Considered collectively, China's activities reflect a posture aimed at establishing regional economic and informational dominance.

Russia

Russia will seek to further marginalize the US in the region now that the peace agreement was formalized in Afghanistan. Russia's motivations in the CASA region are multifold, although the economic factors dominate their decision calculus (Gurganus and Rumer, 2019; Oliker, 2016).

While improving their relationships with their former states, like Kazakhstan, fits into Russian strategic calculus, and solidifying markets for petroleum and natural gas in the far East will mitigate deteriorating revenues from Europe, (Shleifer and Treisman, 2011) Russia will attempt to supplant US influence in the region in the wake of what can be portrayed as a US withdrawal. Russia may also choose to compliment direct efforts to rebuild or deepen relations with former Soviet states. Russia may attempt to use external organizations to alienate the US from regional partners. The Shanghai Cooperation Organization is uniquely positioned to serve such a purpose due to its membership and that the US is excluded (Khalid et al., 2017). Khalid et al., allege that the SCO already contributed to a distancing between the US and Uzbekistan.

Another likely information activity by Russia will target US domestic perspectives. Applying Rational Choice Theory and Vaughan's Amoral Calculator Model, the US should expect Russia to capitalize on the formalized peace agreement and the timing of a US Presidential election year (March and Olsen, 1975; Vaughan, 1998). Whereas they consider attacks through the cyber domain to be violations of their sovereignty, Russian leaders will use those capabilities to incite division within the United States.

More specifically Russia may attempt to sow discontent and division within the US population by leveraging social media to emphasize any number of factors related to the CASA region. Weaknesses that may undermine long-term peace, perceived injustice to the Afghan government and people, risks to the Afghan people, and an overarching sense of futility in what has become America's longest war are potential themes in social media campaigns. Elicitation of these themes may increase as the US election grows closer.

How will they attempt to undermine or amplify peace and stability efforts?

Russian and China each benefited from the US and NATO allies becoming embroiled in the sustained conflict in Afghanistan. Aside from the extreme costs incurred by the US and her allies in terms of blood and treasure, beyond the financial burdens, Russia and China were able to capitalize on the growing perspectives that the US is meddling in Asian affairs. Over time, both countries moved from passive observers to active participants in non-military terms while the US bore the burden for security.

Strategically the US' wavering commitment and tenuous situation in Afghanistan serves to weaken regional views of the United States as a partner. China's efforts to securing strategic mineral deposits, for example, were made possible in part by the US-led Coalition efforts to improve security in Afghanistan. From these standpoints, neither China nor Russia want an unstable Afghanistan.

Neither China nor Russia are likely to attempt to sabotage US peace efforts in Afghanistan. Rather, they are likely to allow those efforts to fail under their own weight. Once US efforts are seen to be failing, and possibly alienating potential Afghan allies, then each country will most likely move to secure stable relationships with those Afghan leaders who will emerge as influential among the dominant tribes.

Conversely, both China and Russia recognize the complexity of the situation in Afghanistan. Involving themselves in that situation, even informationally, carries inherent risks. There exists too much risk and not enough reward for either China or Russia to engage informationally regarding the peace and stability efforts. There are so many factions and parochial interests that any involvement or interventions will risk inciting aggression from one faction or another.

Strategic Framework

It is important to note that Peter Ziehan, (2020), a geopolitical strategist, highlights how the United States was the primary architect of the post-World War Two international order. Accepting this point anchors further analysis. Ziehan also emphasizes that neither Russia nor China are currently capable of fulfilling the role that the US fills in terms of securing international shipping and trade. The United States' deep water Navy provides security for international shipping as it projects power.

US analysts cannot assume that Russia and China will adhere to international norms established by past or present US initiatives. Any analysis of potential Chinese or Russian activities should start in a framework unbounded by international norms. Whether or not a particular activity yields some favorable advantage for Russia or China should become the standard against which activities should be measured.

US strategic planners should consider a holistic strategy model when assessing strategic advantage or risk in the CASA region when it comes to great power competition. Strategy analysts may find the traditional Ends-Ways-Means model depicted in Joint Publication 5–0 to fall short in terms of assessing strategic position. Figure 1 depicts the classic US Army strategy model (Dorff, 2001), which will be familiar to most strategic planners in the United States Government.

Dorff (2001) described the three major elements of strategy traditionally included in national security strategy development, Ends-Ways-Means. Also shown are three relationships between the major elements. These relationships can be used to evaluate a strategy in terms of viability. In summary, a feasible strategy will include adequate resources allocated to achieve defined objectives through acceptable methods over time. If any of these major elements is ill-defined or deficient, or if a relationship is not wholly sufficient, then the strategy may be out of balance or contingent upon additional authorizations. Such conditions place a strategy at increased risk.

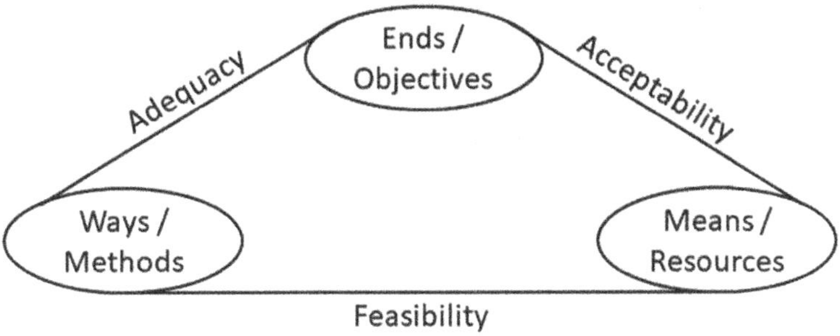

Fig. 1 Strategy Model. Derived from US Army War College Guide to Strategy, by Dorff, R., 2001

This model accounts for cultural variations. The acceptability of specific methods, or ways, is dependent on values and cultural norms. Consequently, this model applies in all cultural contexts, and organizational contexts.

It is important to note that some recent versions of the Joint Publication 5–0 refer to suitability rather than feasibility to describe these general criteria. Suitability and feasibility are not synonymous. Suitability, in fact, can confound the discrete nature of the relationships described by feasibility, acceptability and adequacy. For that reason, acceptability, feasibility and adequacy are suggested for accurately describing the relationships between the major elements and to serve as top level evaluative criteria in the traditional Ends-Ways-Means strategy model.

Extending the traditional model to include the will of the decision maker to achieve long term strategic objectives becomes especially relevant in the discussion of Great Power Competition. Determination, commitment, desire, or the will of the decision maker is particularly important due to the long term nature of strategy. This third dimension in the strategy model can become a discriminator in terms of the efficacy of a strategy.

Figure 2 extends the traditional strategy model to include an all-too-often assumed variable: the will of the ultimate decision maker to carry out a strategy.

This enhanced strategy model removes an implicit assumption that the strategy will remain supported over time. While this may seem trivial it becomes a critical differentiator over time. Whereas the United States generally produces a new National Security Strategy during each new presidency, and elects a new President every four to eight years, China already took steps to ensure strategic continuity by including key strategic objectives in the Chinese Constitution (Chen, 2017). United States term limits, while purposeful, places China in a position of strategic advantage when it comes to implementing strategic plans, which are inherently long term.

Fig. 2 Enhanced Strategy Model. Building upon the US Army Strategy Model depicted in Fig. 1, this depiction includes the will of the decision maker in maintaining, resourcing, and executing a strategy

Summary

There are several questions that need to be addressed to further analyze the information environment in the context of the Great Power Competition. As these foundational questions are addressed, the strategic objectives of China and Russia focus on factors that include economics, power, and influence. Objective analysis of strategies must account for demonstrated actions and messaging—words and deeds.

Whereas China seeks sustainable competitive advantage in terms of economic dominance and innovation, it wants to avoid direct military confrontation with the United States. Economic advantage offers China influence and power. Chinese strategists ensured the long-term sanctity of strategic objectives by embedding those objectives in the Chinese Constitution. Concurrently, Chinese information practitioners are studying and learning from US and Russian actions and are continuing to adapt and innovate in the wake of the COVID-19 pandemic.

Russia also seeks to improve power and influence through economic expansion. Russian activities, however, are more complex involving tactical objectives focused on creating space between Russia and opposing NATO activities. It is consequently more challenging to differentiate between Russian activism and their long-term strategic goals. In the CASA region, Russian objectives will include strengthening relationships with their former vassal states. Russian leaders clearly seek to rebuild a stronger economic foundation and to reclaim a valid position as a world power.

Leveraging the information environment is essential to both China and Russia for attaining their objectives. While the perceived end of hostilities in Afghanistan may be seen as a pivotal event, as signaled by the signed peace agreement between select Taliban leaders and the United States, social and political fractures in the Afghan state and diverse international perspectives on the ramifications of the peace agreement increase risk to US credibility. Russian activities through social media and information environment will focus on exploiting such vulnerabilities.

The US should expect that Russia, and possibly China, may leverage US activities in the CASA region to sow internal discord, especially in this election year. Both China and Russia benefit from stability in the CASA region. It is doubtful that either state would attempt to undermine peace initiatives. However, each state can reasonably be expected to pursue economic advantage in the region in the wake of US influence.

Finally, it is important to identify implicit assumptions when assessing opposing strategies in the Great Power Competition. Implicit assumptions create increased risks in strategies. One implicit assumption that exists in the traditional ends-ways-means strategy model focuses on the role of the decision maker / implementer. The importance of the will to sustain and implement a strategy cannot be overstated; to address this shortcoming in the traditional model an enhanced model was offered that captures the will of the ultimate decision maker. It is especially important for strategic analysts to consider Great Power Competition from unbounded perspectives.

Appendix A: Department of State Bureau of South and Central Asian Affairs Map

Retrieved from Department of State website, March 22, 2020 (DoS, 2020).

Appendix B: DoD and USCENTCOM Unified Command Plan Boundaries

US Central Command Area of Responsibility, Retrieved from: https://www.centcom.mil/AREA-OF-RESPONSIBILITY/

References

D. Araya, China's grand strategy. AI and Big Data. Forbes (2019). https://www.forbes.com/sites/danielaraya/2019/01/14/chinas-grand-strategy/#2ffb35121f18

N. Chandran, China has designs to spread money and influence across the world—India isn't having it. CNBC (2017). https://www.cnbc.com/2017/05/15/china-has-designs-to-spread-money-and-influence-across-the-world--india-isnt-having-it.html.

D. Chen, China has a new grand strategy and the west should be ready. The Diplomat. (2017) https://thediplomat.com/2017/10/china-has-a-new-grand-strategy-and-the-west-should-be-ready/

R. Cropanzano, M.S. Mitchell, Social exchange theory: an interdisciplinary review. J. Manag. **31**(6), 874–900 (2005)

Department of Defense, Joint Concept for Operating in the Information Environment (JCOIE) (Washington, DC, 2018)

Department of State, Countries and regions: south and central Asian region (2020). https://2009-2017.state.gov/p/sca/index.htm

Director, National Intelligence (DNI), Assessing Russian activities and intentions in recent US elections: the analytic process and cyber incident attribution. Report ICA 2017-01D (Washington, DC, 2017)

J. Donati, Indian firms seek to renegotiate $10.8 billion Afghan iron ore deal: Kabul official. Reuters (2013). https://www.reuters.com/article/us-afghanistan-mines/indian-firms-seek-to-renegotiate-10-8-billion-afghan-iron-ore-deal-kabul-official-idUSBRE99S08J20131029

R. Dorff et al., A primer in strategy development. US Army War College Guide to Strategy. In: J.R. Cerami, J.F. Holcomb (eds.) (Strategic Studies Institute, Carlisle Barracks, 2001)

A. Feickert, The unified command plan and combatant commands: background and issues for congress. Report 42,077, Congressional Research Service, Washington, DC (2012, July 17)

A.G. Grynkewich, H. Cabayan, R. C. Jones, S.K. Thomson, S. Meredith III, G.S. Seese, D. DiEulis, What do others think and how do we know what they are thinking (Joint Chiefs of Staff, Washington DC, United States, 2018)

J. Gurganus, E. Rumer, Russia's global intentions in perspective. Carnegie Endowment (2019). https://carnegieendowment.org/2019/02/20/russia-s-global-ambitions-in-perspective-pub-78067

H. Ibarra, S.B. Andrews, Power, social influence, and sense making: effects of network centrality and proximity on employee perceptions. Admin. Sci. Quart. 277–303 (1993)

Q. Khalid, S. Hassan, US Afghan strategy: policy responses of China and Pakistan (2001–2017). Glob. Soc. Sci. Rev.**2**(2), 68–84 (2017)

J.G. March, J.P. Olsen, The uncertainty of the past: organizational learning under ambiguity. Eur. J. Polit. Res. **3**(2), 147–171 (1975).

Naval Post Graduate School (NPS), Afghanistan. Program for Culture and Conflict Studies. https://nps.edu/web/ccs/afghanistan1

O. Oliker, Unpacking Russia's new national security strategy. CSIS (2016)

M. Pei, China's coming upheaval. Competition, the coronavirus, and the weakness of Xi Jinping. Foreign Affairs (2020)

S. Peters, V. King, T. Mack et al., Summaries of important areas for mineral investment and production opportunities of nonfuel minerals in Afghanistan, vol. 1. Open-File Report 2011–1204 USGS Afghanistan Project Product No. 199. US Geologic Survey. Washington, DC (2011). https://pubs.usgs.gov/of/2011/1204/pdf/Front_matter_vol_I.pdf, ps://www.csis.org/analysis/unpacking-russias-new-national-security-strategy

I. Prusher,Afghan power brokers: playing the tribal loyalty card. Christian Science Monitor. 2002-06-10. Archived from the original on March 2, 2008 (2002). https://web.archive.org/web/20080302080521/http:/www.csmonitor.com/2002/0610/p01s03d-wosc.html

L. Rosenberger, China's coronavirus information offensive: Beijing is using new methods to spin the pandemic to its advantage. Foreign Affairs (2020).https://www.foreignaffairs.com/articles/china/2020-04-22/chinas-coronavirus-information-offensive?utm_medium=newsletters&utm_

source=twofa&utm_campaign=China%E2%80%99s%20Coronavirus%20Information%20Offensive&utm_content=20200424&utm_term=FA%20This%20Week%20-%2011 2017
S.R. Ryan, Rapport development and native language use between US advisors and Afghan counterparts (2017). https://scholarworks.waldenu.edu/cgi/viewcontent.cgi?article=4643&context=dissertations
A. Shleifer, D. Treisman, Why Moscow says no: a question of Russian interests, not psychology. Foreign Affairs (2011). https://nrs.harvard.edu/urn-3:HUL.InstRepos:27867127
M.R. Solomon, C. Surprenant, J.A. Czepiel, E.G. Gutman, A role theory perspective on dyadic interactions: the service encounter. J. Market. **49**(1), 99–111 (1985)
Statistica, Global social networks ranked by number of users (2019). https://www.statista.com/statistics/272014/global-social-networks-ranked-by-number-of-users/
Statistica, Top 15 countries based on number of facebook users (2019). https://www.statista.com/statistics/268136/top-15-countries-based-on-number-of-facebook-users/
Statistica, Number of mobile cellular subscriptions in Afghanistan (2019). https://www.statista.com/statistics/497074/number-of-mobile-cellular-subscriptions-in-afghanistan/
J. Thibault, H. Kelley, The social psychology of groups (Wiley, New York, 1959)
United States Code, United States Code, Title 10 (2019). Retrieved from: https://uscode.house.gov/view.xhtml?path=/prelim@title10&edition=prelim
United States Department of Defense, US Central Command Area of Responsibility (2020), Retrieved from: https://www.centcom.mil/AREA-OF-RESPONSIBILITY/
United States Government, CIA World Fact Book (2020). https://www.cia.gov/library/publications/the-world-factbook/
D. Vaughan, Rational choice, situated action, and the social control of organizations. Law Soc. Rev. 23–61 (1998). Retrieved from: https://citeseerx.ist.psu.edu/viewdoc/download?doi=10.1.1.459.5688&rep=rep1&type=pdf
World Economic Forum, World's Biggest Economies in 2018 (2018). https://www.weforum.org/agenda/2018/04/the-worlds-biggest-economies-in-2018/
P. Ziehan, Disunited nations: the scramble for power in an ungoverned world. (Harper Collins Publishers, New York, 2020)

Central Asia Great Power Competition: A 200-Year History Continues

Mitchell Shivers

Abstract This chapter considers the Central Asia region (Kazakhstan, Kyrgyzstan, Tajikistan, Uzbekistan, and Turkmenistan) and its potential participation in any continuing Great Power Competition (GPC). It summarizes a part of the long history of external involvement in the affairs of Central Asian states, considers economic factors which influence relations with non-regional actors, comments on the present contests underway, and posits possible scenarios for future cooperation, competition and influence among inter/intra-regional players.

Keywords Great power competition · CASA region · Central Asia · Great game · Economy · United States · Great powers

Introduction

This chapter considers the Central Asia region (Kazakhstan, Kyrgyzstan, Tajikistan, Uzbekistan, and Turkmenistan) and its potential participation in any continuing Great Power Competition (GPC). It summarizes a part of the long history of external involvement in the affairs of Central Asian states, considers economic factors which influence relations with non-regional actors, comments on the present contests underway, and posits possible scenarios for future cooperation, competition and influence among inter/intra-regional players.

Since the early 1800s, Central Asian nations have become adept at play in the "great game." The 1900s and 2000s ushered in large shifts of influence amongst external players, but Central Asia's subordination to others remained. Autocratic, poor, landlocked, generally arid, mountainous, stubbornly resistant to the rule of law with a persistent legacy of Soviet style command economic management, the Region lags in development and continues to provide a fertile environment for extra-regional parties seeking to shape circumstances there to their own advantage.

M. Shivers (✉)
Hughes+Shivers, LLC, 2700 Woodley Road, NW, Washington 20008, DC, USA
e-mail: shiversm@gmail.com

© The Author(s), under exclusive license to Springer Nature Switzerland AG 2021

A. Farhadi et al. (eds.), *The Great Power Competition Volume 1*,
https://doi.org/10.1007/978-3-030-64473-4_17

Central Asia's strategic location and significant natural resources offer some prospect for greater cooperation amongst intra-regional players, but also invite interference by other, non-regional actors seeking advantages for themselves.

Central Asia's History of External Interference

The Nineteenth Century

The Great Power Competition is anything but new to Central Asia. Since the early Nineteenth Century, great powers have jostled and gamed with one another diplomatically and politically using the mountains, deserts and steppes of Central Asia as the pitch. Indeed, a 100 year span during the nineteenth century is called the Great Game. This time describes great power contests in Central Asia between the Russian and British empires as each sought to project its self-interests over the region.

The immediate period after the decline of Napoleon's First Empire ushered in a time of increased Russian and British global expansion. Once regarded as inhospitable due to its rough geography, climate and warring tribes, Central Asia eventually invited Russo-British attention due to its strategic position as a crossroads and for its terrain's various possible military uses. The British wished to control the strategic military approaches to their invaluable India, and Russia wanted to advance its control of neighboring lands to build its own defenses, to settle disputes with local tribes, expand trade and to thwart any potential advance of the British. Central Asia was their chessboard.

Eugeny Sergeev (2013), notes the six key, driving events that put the Great game in motion: the cessation of the Caucasian War (1817–1864); the Sepoy Mutiny (1857); the Second Opium War (1856–1860); the Anglo-Persian War (1856–1857); the end of British and the beginning of Russian territorial expansion in relation to processes of industrialization in the mid-1800s; and the American Civil War which sent Russia and Britain in search of alternative sources of cotton."

These events stemmed from the pursuit of the perceived economic interests of the powers involved. The Great Powers sought influence over territories to enhance their own economic circumstances by increasing their sources of raw materials and expanding the markets for their finished goods. The decline of mercantilism's role in national policies, and the movement toward freer trade, as evidenced by the repeal of the Corn Laws in Great Britain in 1846,[1] contributed to the desire for new foreign markets. Industrialization of the great powers, however, was key.

Great Britain's first Industrial Revolution began in the late 1700s. Its huge effect on the British economy accelerated into the first half of the nineteenth century. The move away from de-centrally produced, handcrafted goods to the mass production of standardized products significantly increased the nation's potential economic output.

[1] The Corn Laws were trade barriers on food and grain meant to restrict such imports in Great Britain from1815 to 1846.

However, Britain needed much larger stocks of raw materials, among them cotton, to supply its rapidly growing factories. India, what would later come to be known as the jewel of the British empire, provided the solution.

India became essential to Great Britain's rise to economic dominance in the nineteenth century. While the India of the 1700s was already relatively advanced and, as a whole, stronger economically than Great Britain, circumstances would change by the beginning of the 1800s. Professor Angus Maddison (2006) estimates that in 1700, the Indian subcontinent's economy was nine times that of Great Britain. Indeed, at the time, India had the largest economy. As the century progressed, India's advantage declined as industrialization in Great Britain had greater effect. By 1820, India's GDP advantage over Great Britain had shrunk to only three times larger (Maddison 2006).

Through self-advantaged trade Great Britain was enjoying huge economic benefits from its trade, and a growing political relationship with distant India. Great Britain's earliest dealings with India were limited to trade and were largely managed by the East India Company, a public–private organization that used the British Crown's authority. The decline of the Mughal Empire in India enabled the East India Company to increasingly assert control over the Indian sub-continent. By the early 1800s, the East India Company had nearly twice the number of soldiers in India than the British army did. These East India Company presidency armies, compromised of the Bengal Army, Madras Army and Bombay Army, were funded by the Company's large accumulation of wealth and were manned by Indian sepoys.

In 1800 Russia was a large global power. The short rule of Tsar Paul I from 1796 to 1801, witnessed Russia join the British and Austrians to oppose Napoleon. That war ended in 1800. Nevertheless, after peace, Paul I, now annoyed with the British, schemed with Napoleon to invade British India (Van der Oye 2014).

Following Paul I's assassination in 1801, his son, Alexander I, ascended to the throne and ruled from 1801–1825. He quickly abandoned the joint plan to attack India. Initially perceived to be a domestic reformer, Alexander I's interest soon turned to foreign policy. While he is most remembered for his defeat of Napoleon's invasion of Russia, he is also recalled for his frequently shifting international allegiances.

In 1807, Napoleon once again promoted the idea of a joint Russo-Franco invasion of India to Alexander I, but, once again, it was by Russia without its involvement. As before, Napoleon had hoped to strike at Great Britain, on land, in faraway India (Hauner 1990).

No longer allies with Great Britain, Napoleon now wished to achieve a victory over Great Britain by avoiding France's likely naval defeat, instead opting for a potential land victory in India where France's chances, allied with Russia, were much better.

Still, Alexander I's shifting alliances played into widespread British fears that he would someday wish to advance on India. Those fears arose against the backdrop of Great Britain's rise as a global economic powerhouse and its supplanting of China as the globe's economic leader (Maddison 2006). Great Britain's economic dependence on India made them wary of any other nation's plans to disrupt their envious growth and growing power.

The 1812 Russian victory over France was initially met with public acclaim in Britain. However, the acclaim quickly dissipated. Lord Castlereagh, the British foreign secretary, warily viewed Russian motives since they had regularly demonstrated a desire to expand their influence and territory. Brendan Simms (2013) noted, "...in addition to his suspicions of France, Castlereagh was wary of the hegemonic pretensions of other powers—especially tsarist Russia, which bestrode Europe like a colossus by the time of the 1814–15 Congress of Vienna, the peace conference held at the end of the Napoleonic Wars"(Simms 2013).

M.S. Anderson (1956) noted that the campaign of 1812 came at a moment when the British public opinion toward Russia was particularly volatile and uncertain. The British, once aligned with tsarist Russia, had begun to emphasize the inherent differences between the two empires, paying particular attention to the British public's preference for their individual rights.

In 1817, General Sir Robert Thomas Wilson's *A Sketch of the Military and Political Power of Russia, In the Year 1817*, famously alarmed the British public about his perception of Russia's intentions. Wilson, when serving in a diplomatic post in Moscow, had been the first to inform the British of Napoleon's defeat at Waterloo. In his *Sketch*, a best seller, he challenged any British generosity toward Russia's motives abroad, and claimed that Russia was a direct threat to British India. The first seeds of the British phobia of Russia had been sown (Hopkirk 2017).

It is against this British suspicion of the Russians, and Russia's history of imperial expansion, that the Great Game began to unfold. The competition in Central Asia, between the two great global powers ebbed and flowed as it continued for the balance of the century. Such diplomatic and political competition was, at times, intense. Indeed, David Fromkin (1980) stated, "By the last quarter of the nineteenth century, it was a common assumption in Europe that the next great war—the inevitable war—was going to be the final showdown between Britain and Russia."

Fromkin (1980) continued by saying that by the middle of the nineteenth century, Great Britain had valid reasons for standing in opposition of Russian expansion into Asia. Fromkin noted that a Russian expansion would:

> (1) upset the balance of power by making Russia much stronger than the other European powers; (2) it would culminate in a Russian invasion of British India; (3) it would encourage India to revolt against Britain; (4) it would cause the Islamic regimes of Asia to collapse, which in turn would lead to the outbreak of a general war between the European powers in order to determine which of them would get what share of the valuable spoils; (5) it would strengthen a country and a regime that were the chief enemies of popular political freedom in the world; (6) it would strengthen a people whom Britons hated; (7) it threatened to disrupt the profitable British trade with Asia; (8) it would strengthen the sort of protectionist, closed economic society which free-trading Britain morally disapproved of; and (9) it would threaten the line of naval communications upon which Britain's commercial and political position in the world depended. To these the British Foreign Secretary and Prime Minister Lord Salisbury added a tenth toward the end of the century, when he observed that England would have to stop Russia from acquiring Constantinople because, having made such an issue of it for so long, England would lose her reputation as a formidable power if she finally yielded the point (Fromkin 1980).

In 1800, the East India Company controlled 70 million acres on the Indian subcontinent and over 90 million inhabitants (The Economist 2011).That such a prize could be at risk was unacceptable to the British.

It was inevitable that Afghanistan, positioned at the buffer space between British India and the Central Asia, would become key in the competition between Russia and Britain. Afghanistan represented the most likely route of advance for Russian troops if they were to pursue a move on the East India Company's bounty. When Russia secured its principal territorial objectives against Persia in several wars ending in 1828, it undermined British influence in Persia and also brought Russia ever closer to India.

Throughout the nineteenth century, the most powerful nations continued to expand their territory and the numbers of peoples under their control. The Russian Empire expanded overland to the Pacific and south toward Persia, Afghanistan and India. The British Empire eventually stretched from Western-most Europe to India to the South of China and Australia. Almost every European nation colonized parts of Africa, and much of Central and South America were also under colonial rule.

Much of this expansion was in search of maintaining the critical economic status quo among nations as well as gaining ground on a competitor, and then maybe, conquest. Among many of the war-prone nineteenth century European states, keeping up with one's competitors was an essential part of survival tactics.

The Twentieth Century

By the beginning of the twentieth century, Russia and Great Britian had still not been to war with each other. While differences remained, the threat of Russia invading India seemed more remote. Change was underway. Russia lost a war with Japan, and Russian revolutionaries achieved a constitution in 1906 that limited the monarch's absolute power. The rise of Germany was looming over Europe, when, in 1907, Britain and Russia attempted to resolve three Asian matters when they agreed the Anglo-Russian Entente of 1907 in St. Petersburg. The Entente, really three agreements by the two colonial rivals, dealt with Persia, Afghanistan and Tibet, effectively making them protectorates. Persia was divided into three parts, Afghanistan was agreed to be "outside the sphere of Russian influence, and Tibet's business would be handled by China. That neither Russia nor Britain consulted with the territories concerned was not unusual for the times" due to the colonialist mindset (Finn 2017).

In 1912, the Republic of China was founded when, in the prior year, the Xinhai Revolution overthrew the Qing dynasty. Two years later, Russia and Britain would enter WWI together as allies in the Triple Entente, the possibility of which was enabled by the 1907 Entente.

As the twentieth century began to unfold, revealing a new set of rivalries, conflicts, and rising powers, the attention began to shift from Central Asia. Russia, Great Britain, Persia, India and China all were experiencing more pressing and more distant problems that required their urgent attention. Central Asia became the sole charge of the Soviet Union and it's attention and energy was largely focused elsewhere. With the inevitable coming of the Great War, everyone's focus was on Europe and less so on Central Asia.

The five Soviet Republics in Central Asia, Kazakhstan, Tajikistan, Uzbekistan, Kyrgyzstan, and Turkmenistan, together Soviet Central Asia, were formed in the 1920s and 1930s that. Their boundaries were alleged to be imperfect attempts to create Stalin inspired formulaic ethnic republics, but the peoples inhabiting the lands were more demographically diverse, making such a goal unachievable. Others, notably, historian Arne Haugen, have challenged that notion.

Throughout much of the twentieth century Central Asia stayed mostly out of the West's eye. As an integral part of the Union of Soviet Socialist Republics from 1918 until 1991, it was difficult for westerners to travel to, and interchange with, the Central Asia republics. Reporting in Central Asia was largely confined to state-controlled less reliable news reports, and Central Asians had little interchange with westerners.

Overall, Central Asia's experience within the Soviet Union was mixed.

Alexander Cooley (2012) writing for the Asia Society online notes that there were three features in Soviet empire rule:

> Like other colonial powers, or imperial Russia before, the Soviets really took Central Asia and turned it into a site for extraction of raw materials, for example cotton in Uzbekistan, oil from Azerbaijan and mining in northern Kazakhstan. Cotton production in Uzbekistan has dire environmental consequences, such as the shrinking coastline of the Ural Sea. In addition, other ill effects like from pesticides that contributed to the alarming rate of cancer in the region. In northern Kazakhstan, there were nuclear test sites.
>
> Strategy of divide and conquer installed by the Soviets in the region [*sic*]. The Soviets did not try to compartmentalize individual ethnic groups like other colonial powers. Instead, they did the opposite by dividing ethnicity and creating large minorities with an administrative unit. The Soviet concern in the 1920s and 1930s was that there might be a unified movement (pan-Islamic movements or pan-Turkic movements) that would lead the various people to oppose against Soviet rule. So they deliberately drew up these large republics with the aim of having a dominant ethnic group, but not too dominant, to play them off one against each other.
>
> The last similarity is the ideology of a superior civilization. It was true that the Soviets wanted to Sovietized [*sic*] the region, but they didn't really manage in the end. On the other hand, the Soviet authority did not really care as long as these Central Asian regions did not come out of line.

World War II had a huge impact on Soviet Central Asia, like elsewhere. Large numbers of Soviet refugees fled the fighting in the west and resettled to Central Asia. Such movement, fleeing from the German front, was either officially sanctioned or very informal. Officially, Soviet leaders resettled millions through deportations and evacuations. Deportations sought to resettle populations thought likely to support Germany, while evacuations sought to move citizens toward the Russian rear and safety. Industry was moved from militarily vulnerable locations to Central Asia where it employed relocated Soviet citizens. This meant that millions of Soviet citizens, of German, Finnish and other European backgrounds, took up residence in the region. These mass movements had the effect of creating hardships for all involved, but the voluntary and involuntary relocation of Soviet citizens/residents and industry to

Central Asia contributed significantly to the Soviet victory over Germany by enabling war production to continue.

The subsequent Cold War ushered in a relatively quiet "world stage" period for Central Asia. Fiona Hill (2002) wrote that "Before 1991, the states of Central Asia were marginal backwaters, republics of the Soviet Union that played no major role in the Cold War relationship between the USSR and the United States, or in Soviet Union's relationship with the principal regional powers of Turkey, Iran, and China."

The Office of the Historian, Department of State, United States of America (2020), counters the Soviet invasion of Afghanistan at the end of 1979, intent on installing a friendly socialist government on its southern border, "… was a watershed event of the Cold War." It marked the only invasion by the Soviet Union of a country outside the Eastern Bloc during the Cold War and was met with widespread international condemnation.

The Soviets, at times numbering a peak force of 115,000, occupied Afghanistan for nine years. The war cost almost 15, 000 Russian soldiers their lives. The brutal, long war also killed about 1 million Afghan civilians, 90,000 Mujahedeen fighters, and another 18,000 Afghan forces (Taylor 2014).

After the Soviet invasion came a virulent civil war that resulted in a protracted contest that killed many civilians and destroyed much of Kabul's infrastructure. The 1991 collapse of the Soviet Union, accelerated by the Soviet experience in Afghanistan, drastically changed the dynamics on the ground. The Soviet backed government, still receiving Soviet aid, could no longer count on its critical assistance. The end of an era had come for the Soviets and their puppet state.

Beginning in November 1994, the Taliban, supported by Pakistan's ISI, began to assert itself by force and in September 1996 entered Kabul, thereafter controlling a large part of Afghanistan. The Taliban, a totalitarian fundamentalist Islamic based political ideology, and its military arm, were ostensibly aimed at settling Afghanistan's civil war and restoring peace to the nation. Ultimately, the Taliban controlled about three-quarters of Afghanistan's territory. Their cruel, barbaric ways led to condemnation from many foreign governments, and the Taliban's national government, the Islamic Emirate of Afghanistan, was only ever recognized by three nations (Saudi Arabia, the United Arab Emirates and Pakistan). Pakistan was notable in its strong support for the Taliban regime and many observers judged the Taliban to be a proxy for Pakistan's interest in Afghanistan as a strategic land reserve in case it one day went to war with India.

Indirectly, it was Pakistan's interference with Afghanistan's internal affairs, by its support of the Taliban that eventually led to the longest war in American history. Ironically, it was the Soviet interference with Afghanistan that fundamentally altered Pakistan's calculus toward Afghanistan. The Soviets, as once-removed Pakistani neighbors, were thought tolerable by Islamabad. That was not the case when the Soviet militarily occupied Afghanistan controlling its government, and were then no longer once removed. After the Soviet retreat from Afghanistan, Pakistan resolved to have a "user friendly" regime installed as Afghanistan's next government (Weinbaum 1991).

The Taliban experience with governing Afghanistan proved to be short lived. When, after 9/11 in 2001, the Taliban refused to surrender or assist in the capture of al Qaeda's leader, Osama bin Laden, they were toppled by a small U.S. contingent of para-military CIA operators and soldiers. Driven into safe havens in Pakistan, the Taliban regrouped, slowly at first, and then launched a protracted insurgency against Afghanistan, and U.S.-led NATO forces.

After 9/11, the United States dramatically increased its focus and engagement with Central Asian states. The sudden, much larger presence of the United States in the region, in turn, drew increased attention from Afghanistan's immediate neighbors: China, Tajikistan, Uzbekistan, Turkmenistan, Iran and Pakistan.

There were a number of reasons for the United States' increased focus on Central Asia. First, Afghanistan, as a land-locked nation, is dependent on neighboring countries for land and air access for all imported goods. Second, in addition to principal logistic routes, the U.S. needed backup land and air access in case its primary pathways for land and air transport were threatened. Third, Afghanistan's borders are very porous and facilitated the transport of the goods of criminal enterprises, via Central Asia, the proceeds of which were increasingly facilitating terrorist activities. Fourth, Central Asian cooperation would bolster the U.S. military capabilities by providing a transit base for aircraft, temporary basing for transit troops and through transport of non-lethal cargoes. Fifth, the common pursuit of regionally based terrorists was essential. Sixth, the earlier Taliban regime had provided safe haven for Central Asian insurgent groups.

After their independence from the Soviet Union in 1991, and before 2001, the so-called former Soviet "stans" (Kazakhstan, Tajikistan, Uzbekistan, Kyrgyzstan, and Turkmenistan, but excluding Afghanistan), the United States engaged Central Asian states in a range of relatively minor political, economic, and military programs. In particular, military–to-military cooperation increased, and denuclearization of the region succeeded. American corporations also expressed greater interest in Central Asian hydrocarbon extraction and set up joint venture partnerships with various public and private enterprises in the region. Still, American general interests in Central Asia remained checked by Russia's continuing dominating influence. That all changed quickly on 9/11. By the end of 2001, the United States enjoyed military basing rights at two of the three Central Asian neighbors of Afghanistan (Hill 2002).

The increased U.S. engagement with Central Asia has proven a mixed bag. The reasons for such are understandable. The authoritarian governments in the region are wary of American democratic ideals, Russia continues to control important economic levers in Central Asia, the Russian language remains *lingua franca* and America's war in Afghanistan has gone poorly. Importantly, as a super power America must ration its energies, and remain focused on truly strategic matters. Central Asia, with its long ties to Russia and its close proximity to an ascendant, now more globally oriented, China, with little directly at stake for the U.S., represents a poor payback for any overly ambitious American diplomatic efforts.

Central Asia and Modern Great Power Competition

The Players

Three great powers, the United States, Russia and China, are in competition, of some sort, in Central Asia. Of these, only Russia and China have sufficient reasons and the wherewithal to be ongoing, long-term contestants. The U.S.' strategic interests are sufficiently remote from the region to render Central Asia to be, at least for now, at best a second-rate priority. Given that the two of the world's current great powers neighbor Central Asia, it's hard to imagine the United States being anything but a mentor and spoiler.

The 2019 Trump administration peace initiative in Afghanistan, if successful, promises few, if any, remaining U.S. troops in the war-torn nation. As the American military profile lessens in Afghanistan, so will its policymakers', legislators' and citizens' interest. If the old precept, "follow the money," holds true, then the attention of American citizens, and their leaders, will dramatically shift away from Central Asia and Afghanistan. The United States' military experience in the Middle East and Afghanistan has given it both a wariness and weariness about its battles there, and an accumulation of doubts about the wisdom of its spending on grand projects in attempts to positively alter faraway outcomes.

Some 2,400 American lives have been lost in Afghanistan, with all too little to show for their sacrifice. The U.S. public now supports less military engagement abroad, and gives greater attention to the numerous domestic challenges. President Trump's 2016 election campaign tapped into that frustration and upended conventional U.S. politics. While the United States' next game plan in the region has yet to materialize, the current retrenchment suggests that far fewer resources will be devoted to new Central Asian initiatives.

However, this could be subject to change with evolving events and circumstances. First, if there is a new peace and stability in Afghanistan that creates more favorable regional economic conditions with increased intra and inter-regional trade. If authoritarian Central Asian governments further liberalize and place greater emphasis on the rule of law, human rights and anti-corruption efforts. If China's ambitious Belt and Road Initiative (BRI) stumbles and leaves Central Asian nations uneasy. Or if energy prices leave Russia unable to leverage its inherent historical and economic advantages in the region leaving Central Asia to seek assistance elsewhere.

China's modern post-Mao rise has been remarkable, and it has profoundly changed the Great Power competitive landscape in Central Asia. China's broad economic reforms, introduced in the mid-70s, were sweeping. A. Doak Barnett (1986), writing in *Foreign Policy* just a decade after Mao Tse Tung's death noted, "the Chinese have moved rapidly from ideological dogmatism toward eclectic pragmatism, from extreme totalitarianism toward liberalized authoritarianism, from a command economy toward "market socialism," and from autarkic isolationism toward international interdependence." As late as 1990, Chinese relationships with Central Asian states were close to non-existent. China's "Go West" economic policy, launched in 2001 to benefit its lagging western provinces, kick-started China's greater interest in trade with Central Asia.

China's still growing economic might in Central Asia today is considerable and is only likely to increase. In *2013* Chinese trade with Central Asia topped US$50 Billion, far surpassing Russia's trade (Rumer et al. 2016). In 2018, five years later, U.S. trade with Central Asia totaled only US$1.1 Billion of exports from the U.S. and less than US$1.5 Billion in imports to the United States from Central Asia. In contrast with today's China, Russian—Central Asian trade in 2018 was US$18.9 Billion in Russian exports, and US$7.3 Billion in imports to Russia (Trading Economics 2020). Clearly, Central Asia's authoritarian leaders, still guiding large swaths of slowly decentralizing command and control enterprises, can see their "bread is buttered" more and more by China.

Given their proximity, China and Central Asia would be expected to have a large trading relationship, and as indicated above, they now do. What is remarkable is the speed by which the trading relationship has grown in recent years. As recently as the mid-1990s, just after the Soviet Union's collapse, China's trade with Central Asia was only between US$350–US$750 million (Rumer et al. 2016). It is not clear whether increasing trade relationships always translate into mutually beneficial broader ties between nations. Countries may very well be trading with one another grudgingly, out of some reluctant necessity, to satisfy a need not easily filled elsewhere, or just based upon close proximity. Alberto Behar and Anthony Venables writing in the *Handbook for Transportation Economics,* state, "Trade flows depend on characteristics of the source and destination countries, such as their economic size as reflected in income. It also depends on 'between country' characteristics such as their policy towards each other and their cultural affinity" (2011). Behar and Venables (2011) go on to write, "GDP and distance typically account for 70% of the cross-country variation in trade, but other variables are also significant. Other geographical characteristics include having a common border (neighbours trade more) and country area (large countries trade less); islands trade more, but landlocked countries trade less."

While Russia and China both share borders with Central Asia, China's new and ascendant economic might now dominates the shared neighbors' landscape. Indeed, even Russia is more and more connected economically with China. Sino-Russo trade exceeded US$100 Billion for the first time in 2018, but Russia's exports with China are narrow. According to Yohei Ishikawa (2019), "Oil and other minerals accounted for 76% of Russian exports to China, with wood and paper products responsible for another 8%. After a new oil pipeline from East Siberia to the Asia–Pacific region went online in 2009, Russia's resource exports to China soared. In 2018, higher oil prices boosted the value of these exports." The tremendous fall in oil prices, and the simultaneous downturn in both economies, during the COVID-19 health and economic crisis will surely reverse that increased trade. Both nations desire to broaden the range of exports traded between the two countries, but, for now, China, Russia's largest export partner, accounts for 25% of Russia's oil exports.

Ishikawa (2019) goes on to write, "This has Russian political and business leaders worried over the prospect of their country becoming an "ancillary supplier" of resources for China. Specifically, they fret that the bilateral trade structure will give

China huge pricing power over Russia and lead to increasingly wider economic inequity between underdeveloped East Siberia and the growing northeastern parts of China—a situation that could create tension along the border in the Russian Far East." Other nations have expressed anxiety over the uneven benefits of trading with China, as well. For example, today Chinese tariffs on U.S. made goods remain substantially higher than the Chinese tariffs on other world suppliers (Brown and Lovely 2020). That, and the control China's state-owned enterprises exert of imports, remain one reason the large U.S. trade deficit with China continues.

Chinese willingness, even at the official state level, to steal foreign technology also makes technology savvy nations wary of trade with the growing power. Chinese built military carrier-based aircraft, stolen from Russia, is a case in point. Attempting to accelerate its aircraft development for its first naval carrier, and save money, China copied Russian aircraft only to be sadly disappointed by its performance failures (Peck 2019). U.S. companies and institutions have long complained about Chinese pressure to share confidential proprietary data with Chinese joint venture partners. In other cases, outright Chinese espionage has taken place. William Evanina, director of the U.S. National Counterintelligence and Security Center, speaking at a Center for Strategic and International Studies conference, said that China's trade secret thefts cost the United States "anywhere from US$300 billion to US$600 billion" every year" (Reuters 2020).

China's rapidly growing economic power, forecast to become the world's top economy, fuels an ever-expanding Chinese influence around the globe. This is especially so in the regions immediately surrounding China, as neighbors, and near neighbors, ponder policies aimed at maximizing the benefits of their Chinese trade while not falling completely under China's sphere. Unlike the United States, which China plans to replace as the world's super power, China rules through an authoritarian regime. Authoritarian regimes may be less likely to resolve domestic and international disputes through peaceful, lawful means. On the other hand, for better or for worse, such regimes may have a greater consistency in their policies over time.

Earlier, the British used their 100-year colonization of India to fuel their expanding industrial base and provide a ready market for their factories. Today, China is faced with an underemployed workforce, hidden by officially inflated employment numbers (Feng et al. 2015).

China's Uncertain Rise to Supremacy

China's decades long fast rates of economic growth, and the sheer size of its population, presently at a billion more than the United States' population, suggests to many that China's rise to lasting economic global supremacy is without question. Others point to demographic trends underway in China that challenge that assumption (Wang 2010). Forecasts point to a rapidly ageing China which, before long, project too few workers supporting too large a population (Xi 2019). The consequence is economically troubling for China. Indeed, earlier Chinese policies, like its infamous

One Child Policy, in effect from 1979 to 2015, now make China's population growth skewed to a rapid, disproportionate increase in ageing that its economic overtake of the United States, once broadly accepted, is now in question (at least, in *real* GDP terms) and China is likely to be surpassed by India, mid-century, as the globe's most populace nation.

China's demographics, and their implications for China's economy, matter since much of its rise in external influence is premised on its economic clout. If the United States remains the world's preeminent economic power, that suggests the status quo, that nations will continue to be led, at least in part, by American leadership in global affairs. If China's worrying demographic trends materialize themselves as true, and its economic supremacy never materializes, then American preeminence will likely remain unchanged.

As the economic rivalry for global leadership between the U.S. and China plays out, its direct effect on their relationship with Central Asia is uncertain. However, China is likely to remain the biggest external influence on Central Asia due to its combination of: proximity to Central Asia; its huge GDP (eleven times that of Russia's) and its desire to find ever-larger markets for imports and exports. Russia enjoys strong historical ties to Central Asia, shares the world's second longest international border with Kazakhstan and has strong security, economic and energy links to the region. America has no on-going compelling interest in Central Asia, so, absent any new significant development, U.S. interest in Central Asia is likely to wane post a withdrawal of American forces in Afghanistan.

Like the United States, Russia has recently made its own pivot to the East yet again. Given its habitually fractured relationships with the West, Russia now seeks to further enhance its partnership with China by building an ever-closer alliance. Such an alliance might mean that Russia is now more willing to take a subordinate role to China with respect to Central Asia.

Writing for the Carnegie Endowment for International Peace, Arkady Dubnov (2018) states that, "China holds a highly significant place on Russia's agenda in Central Asia. Beijing has become an indispensable strategic partner for the Kremlin as it seeks alternatives to the West, and it is an indispensable economic partner for the region's countries." Practically, Russia now depends on China to deliver economic benefits to Central Asia that Russia itself cannot provide.

Both Russia and China have a distinct foreign policy advantage in Central Asia aside from their proximity to the region. As authoritarian governments, each has demonstrated a disinterest in reforming the internal governance methods of other nations, and each Central Asian nation is, to varying degrees authoritarian, as well. That has always placed the United States, with its strong bent on moving other non-democratic nations toward democracy, at a regional disadvantage. The U. S. promotion of human rights is seldom well received in Central Asia capitals. While China and Russia are mute on the subject, American foreign aid and military assistance is often tied to governance reforms unwelcomed by the leaders of Central Asia.

China's Belt and Road Initiative (BRI), while not specifically aimed solely at Central Asia, will engage the region in many substantive ways. BRI's stated aim is to increase China's connectivity to the world, perhaps so, but it is also meant to address the slower domestic growth implied by China's ageing demographics. Known, too, as One Belt, One Road the enormous 2013 announced infrastructure initiative straddles continents and engages 60% of the world's population. It is a US$ 1 Trillion plus multi-pronged effort by the Chinese to reshape large swaths of world commerce. In many ways it is a modern day equivalent of America's post World War II Marshall Plan, but even more. While the BRI has been described as 12 times larger than the Marshall Plan (in constant dollars), the BRI will be stretched over a 3.5 times longer time period. Bigger in scale, the BRI nevertheless shares a common objective with America's earlier Marshall Plan.

The United States was the world's largest manufacturer at the end of World War II. Given the expected sharp decline for goods in the immediate aftermath of the war, new markets were needed for America's excess manufacturing capacity. Now the same holds true for China's forecast slowing due to its maturing economy and the adverse effects of its ageing population. Each nation was confronted by conditions, fixed in the short term, at least, that would slow its economic trajectory. The United States embarked on the Marshall Plan in 1948 to rebuild Western Europe, in part to help halt the spread of communism, but as importantly to keep factories at home operating at high capacity. China, through the BRI, publically seeks "world connectivity", but it can enlist the younger populations of the many third world nations along the BRI route, as substitutes for its ageing workforce, thereby ensuring its growth continues with full factories and ready consumers. Interestingly, unlike America's rival at the time of the Marshall Plan, the Soviet Union, China's rivals today, Russia and America, will also be able to exploit China's investment in the BRI.

It's noteworthy that Chinese President, Xi Jinping, first announced the BRI in Kazakhstan in 2013 during a state visit. With the world still in a painfully slow recovery from the 2008 Great Financial Crisis, China was eager to build bridges, literal and figurative, ensuring its rapid growth and ascendancy would continue. It was fitting that Xi, speaking in the world's largest landlocked nation, would announce such an ambitious program to connect so many countries. It was a geopolitical masterpiece! Rail, road and pipelines are featured aspects of the BRI.

The Central Asian Economies

The sizable population of Uzbekistan at 33.4 million inhabitants (number one amongst Central Asian nations) and Kazakhstan's considerable landmass of 1 million square miles (the world's 9th largest nation) set them apart among nations in Central Asia (NationMaster 2020). Kazakhstan represents 60% of the region's GDP (Zarakhovich 2006). Notably, Central Asia is a rather small region of only 74 million inhabitants, ranking fifth among Asian sub-regions, and is entirely comprised of landlocked nations (Worldometer 2020).

Prior to their independence from the Soviet Union in 1991, the Central Asian states functioned as economic subordinates within the U.S.S.R. Their development supported overall Soviet Union goals. Prior to gaining independence the new, separate Central Asian states achieved uneven degrees of economic development. During the Soviet years, power grids, road systems and water resources were planned by Moscow regionally with less regard given to the sovereign boundaries constructed after independence. Accordingly, some states had unusual influence over other regional nations' critical infrastructure. Kyrgyzstan, for example, controlled key regional electricity generating sites and water distribution, while some *intra*-state road transit relied on passage through a neighboring state.

It is notable that the former Central Asian Soviet republics achieved their independence more through the U.S.S.R.'s collapse than their own desire for nationhood. The region's newly found independent status reflected the ending of the Soviet regime, and not a hard fought-for extended struggle toward a separate standing. Indeed, some suggest there was a real reluctance on the part of the Central Asian states to separate when the opportunity presented itself (Olcott 1998).

Since the nineteenth century, and prior to independence, the Russian exploitation of Central Asia was quite acceptable to most of its inhabitants. After all, compared to modest nearby nations, the Central Asian states enjoyed relative stability, a reasonable standard of living and the opportunities afforded to them as part of a major world power. The Soviet Union's authoritarian ways seemed of little concern to the citizens of Central Asia. After independence the first leaders of the new states, unexpectedly thrust into their positions, had previously enjoyed leadership positions in the former order. Among these leaders' first challenges was determining each nation's economic possibilities given the regional construct left by the Soviets.

The first leaders of the post-independence era Central Asia inherited economies that were relics of the old Soviet system. The shock of sudden independence took its toll. Initially per capita incomes fell drastically and the private sector composed less than half of the still state-owned dominant part of the economy. Foreign remittances, long an important part of income in Central Asia, fell sharply as Russia's economy staggered.

Sparsely populated and landlocked, Central Asia has a combined landmass the size of Europe, but is largely inhospitable and is not conducive to economic activities. The region's west contains the world's largest inland sea, the Caspian, but enjoys no further access to the seas beyond Eurasia. During the twentieth century the discovery of significant natural resources of gas, oil, gold, uranium, coal, hydropower and solar potential have given promise to a brighter future for the region. Nor does this discovery grant Central Asia to energy self-sufficiency, it creates opportunities for the Central Asian states to become significant energy exporters if it can build the pathways needed to deliver these resources to foreign markets.

To take full advantage of its energy potential Central Asia will need to attract foreign direct investment (FDI) to efficiently extract minerals and hydrocarbons, build plants and develop the expensive transport infrastructure that is required to reach foreign markets. Certainly, building better access to foreign markets will be a major undertaking that will require regional, if not, global, cooperation. Presently, Central Asia simply doesn't have the internal capital needed for these ambitious undertakings.

Kazakhstan's Tengizchevroil joint venture provides an example of the region's need for FDI. Soon after Kazakhstan's independence, its new leader, Nursultan Nazarbayev, struck a 40-year deal in 1993 creating a joint venture between the U.S.'s Chevron (50%) and Exxon Mobil (25%), Kazakhstan's KazMunayGas (20%), and Russia's Lukoil (5%), to develop the nation's largest oil field, Tengiz, as well as the nearby Korolev field. Over 25 years later, these fields are among the largest and most lucrative in the world, capable of producing some 540,000 barrels per day. There a, mostly foreign invested, US$37 billion project is underway to expand production capacity by 2022 to 1 million barrels per day (Todd 2019). At current estimates, Tengiz might produce some 1 million barrels per day for thirty years into the future.

The pay-offs to Kazakhstan have been huge. Recently, oil and gas revenues are estimated to have contributed 50% of Kazakhstan's current annual budget. Importantly, the energy field is also an important employer in Kazakhstan. It's estimated that in addition to Tengizchevroil's 48,000 employees; Kazakhstan's oil service industry employs another 170,000 in well paying jobs. Though majority owned by foreign firms, Tengizchevroil, has committed to transferring, highly desirable, and marketable skills to Kazakhstanis.

Kazakhstan utilized FDI to develop their capital-intensive oil fields on the Caspian, and other Central Asian nations will need to do much the same for their development. That will require additional concrete steps to further improve each country's business climate to ensure that the West's foreign investors have a good expectation that their investments will be honored. It won't necessarily be the same hurdle for Chinese and Russian investors, state sponsored or otherwise, since the command control legacy systems of both nations will act in a more concerted way to protect the national commercial interests of both those countries.

Central Asia's initial embrace of a movement to market economies after independence has been slow but steady for the most part. Established in 2003, the World Bank's *Doing Business* survey, annually ranks 190 nations on their ease of doing business, and is a useful gauge for transparency, rule of law and other business friendly measures. While four Central Asian nations are currently measurably improving in their international standing for their business climate, Turkmenistan remains resistant to these much-needed and desirable reforms. Recently, the World Bank singled out Kazakhstan, Uzbekistan and Kyrgyzstan for their significant gains in addressing previous hurdles to doing business. Kazakhstan now ranks 25th in this much followed world ranking. Its improvement is part of its plan to become a top-30 world economy by 2050 (The World Bank 2020).

In general, the region's nations must compete aggressively, and fairly, for the scarce private global foreign direct investment available to emerging markets like Central Asia. Adopting welcoming investment climates that encourage good business practices are essential. That will ensure each nation, and each of its enterprises, are able to choose the terms for such investments wisely. Nation's that unduly restrict business opportunities through opaque practices that create uneven playing fields might advantage the few, but such practices cannot serve the many. Attracting Russian and Chinese investments, with their statist influences, may prove to be a lower hurdle than attracting the more demanding private investors from the West.

Corruption of the national public sectors in Central Asia has been a problem since independence. To a degree, some of the current corruption has undoubtedly been inherited from their old Soviet regimes. Still, some thirty years on since independence, the corruption persists. Transparency International's annual Corruption Perception Index (CPI) ranks 180 nations' public sectors for the perception of the corruptness. The Central Asia nations, although steadily improving, albeit slowly, rank poorly. In 2019, Kazakhstan, the best-ranked Central Asian state in the CPI, ranked 113 (1 = highest). Turkmenistan, ranked lowest of the Central Asian states, ranked 165 of 180 nations in 2019 (Transparency International 2019).

While the CPI rankings are not without criticism, they're a reasonable approximation, and comparison, of globally perceived public sector corruption (The Guardian 2013).

The corruption level of a given nation's government officials is problematic, in particular, to United States businesses. American companies, and all non-U.S. corporations issuing securities in the United States, are legally obliged to adhere to the United States' very strict Foreign Corrupt Practices Act ((FCPA), passed in 1977. Aimed at eliminating the private sector's bribery of foreign leaders in order to obtain business, FCPA imposes harsh felony criminal penalties on guilty corporations and their executives (U.S. Department of Justice 2017). Most other countries, and their corporations, operate without such legal penalties for foreign bribery. Those state owned, or private, enterprises unburdened by FCPA, and willing to bribe officials in nations tolerant of corruption, enjoy a huge advantage in bidding for large state sponsored projects. Accordingly, U.S. and other companies subject to FCPA typically operate at a disadvantage and refrain from attempting to operate in highly corrupt nations.

Today, a surplus in the global supply of oil and gas, Central Asia's own way of doing business under its authoritarian leaders, coupled with the close proximity of tolerant Russia and China, each with their own interests in the region, along with Central Asia's distance from America, all conspire to limit American interest in the region.

What's Ahead? Uncertainty

Central Asia's future is, perhaps, now less predictable than it was during its post-independence period. The slowly evolving region continues to have a great need for external capital to pursue development, but there is slack global demand for its natural resources, diminished potential for Russia to play its desired supporting role, and America's waning interest in ambitious foreign initiatives, and especially in Central Asian region. On the plus side, Central Asia's overall infrastructure has improved, as has its inter-connectivity by rail and road with China and the European Union.

A still ascendant China continues to have the ways and means to remain heavily involved in the region, but its ambitions may have now been checked by recent events regarding the coronavirus pandemic, which may have a lasting impact on its economy by shifting global supply chains away from its factories. However, there is China's ageing population that could motivate China to become even more engaged with the region. It is just too soon to say how these matters will play out. Given Russia's current economic issues, and the United States' likely complete withdrawal from the region, Central Asia, for the time being, may not be able to foster very much Great Power Competition. China may achieve regional hegemony by default.

Among the possibilities: Russia persists with its regional engagement because of its historic role, its soft power influence (for example, Russian is still the official language of three Central Asian states) and its desire to maintain its self-proclaimed sphere of influence over the region for security purposes, but it reluctantly accepts subordinate status to China. China vigorously asserts itself as the dominant hegemon of the region, and America accepts a lesser role of mentor, coach and occasional spoiler to other Great Powers in Central Asia.

Russia, unable to still economically assert itself with any dominance, will likely continue its mix of military and economic support, promoting cohesion and pro-Russia relations and an emphasis on deterring any Western entreaties. While harmonized trade is important to the region and necessary, Russia's leadership in this realm alone will not be sufficient to achieve its own regional goals. Without economic clout, Russia's only other levers are really too "soft" to be consequential. An example, the Moscow-led mutual defense pact, the Collective Security Treaty Organization (CSTO) includes Kazakhstan, Tajikistan and the Kyrgyz Republic, but it has largely failed to advance key Russian interests. Indeed, both China and the United States have both had military *bases* in the region since the creation of the CSTO.

American military planning has, for three years since the publishing of the 2018 National Defense Strategy, shifted its attention to Great Power Competition elsewhere, with only glancing reference to Central Asia as a contested area. Central Command (CENTCOM), with Central Asia in its area of operations, now largely concentrates on its Middle East goals in its annual "posture" statements to Congress. Little is said of Central Asia. Last year, CENTCOM stated, "We recognize the U.S. is rightly shifting its resources toward Europe and East Asia to balance great power competition, but remain mindful that the CENTCOM AOR represents a geopolitical crossroads and a principal zone for that competition as well. Of the five major threats identified in the NDS, four—competition with China; competition with Russia; Iran's rogue, malign activities; and combatting VEOs—reside or are contested on a significant scale in the CENTCOM AOR every day (Votel 2019). Even with its caveat, CENTCOM doesn't sound convincing about its plans for any meaningful Central Asian engagements.

China's immediate role in the region will revolve around its unfolding Belt and Road Initiative (BRI). On paper, and in the abstract, at least, the BRI appears very attractive to all, but are the terms attached to this grand lending and grant initiative more onerous than recipients might be willing to accept? Will there be host nation resistance to the various burdens and obligations imposed by China as it seeks to fashion its national interests. Will the BRI's temptations prove too much for Central Asia? Already, Western nations, including the United States and Australia, are laying the groundwork to obliquely but actively oppose any onerous terms associated with the BRI, thus thwarting some of the BRI's goals.

A new "Blue Dot Network" established by the United States, Japan and Australia, and soliciting the involvement of others, seeks to, "…serve as a globally recognized seal of approval for major infrastructure projects, letting people know that projects are sustainable and not exploitative." Most observers believe this collective effort, launched on the sidelines of the 35th ASEAN Summit, is clearly directed at pushing back against China's BRI. If the West is unwilling to match China's largesse, is it at least attempting to make China's deal making more costly? Are more such Western challenges likely as the Great Powers consider their possibilities in Central Asia in light of BRI?

Of course, numerous possibilities for Central Asia may evolve after the global pandemic COVID-19 crisis of 2019–2020. The full extent of the pandemic's injury to world order is not yet understood, nor, indeed, is its long-term ramifications.

Today, Central Asia is better placed to more effectively deal with the future. As Richard Pomfret (2019) notes, Central Asia's achievements since independence have been noteworthy. Having overcome the initial significant economic challenges of hyperinflation, widespread unemployment and budget shortfalls, each Central Asian state went on to make successful transitions to market economies. They did so while avoiding not uncommon pitfalls like breaking up or warring between themselves. Each nation, some more, some less, has made progress in democratizing and structurally improving their economic diversification and, save one, improving its business practices. Central Asia's growing rail, road and electronic inter-connectivity, will at long-last enable the region to escape much of the burden of being land-locked. Even so, a continuing over-reliance on natural resources for exports will be a drag on the region's economies in the short run.

So, at least in the near term, these Great Power competitors, China, Russia and the United States, will devote themselves mostly to other arenas, arenas with greater impact globally than the once useful "pitch" provided by Central Asia. Given the continuing needs of Central Asia, its strategic location neighboring two of the Great Powers, and the various counter-balancing interests of the Great Powers, it seems likely GPC will continue in the region, but for now, at a reduced level. Exactly how Central Asia will navigate its passage through this next period is anyone's guess, but we can expect a wide variety of lesser contests to continue.

References

M. Anderson, British public opinion and the Russian campaign of 1812. Slavonic East Eur. Rev. **34**(83), 408–425 (1956)

A.D. Barnett, Ten years after Mao. Foreign Policy. (1986). https://www.foreignaffairs.com/articles/asia/1986-09-01/ten-years-after-mao. Accessed. 2 April 2020

A. Behar, A.J. Venables, Transport Costs And International Trade. In: André de Palma, Robin Lindsey, Emile Quinet, Roger Vickerman (Eds.). *A Handbook of Transport Economics*(Edward Elgar Publishing, North Hampton, 2011)

C. Brown, M. Lovely,Trump's phase one deal relies on China's state-owned enterprises. Peterson Institute for International Economics. (2020).https://www.piie.com/blogs/trade-and-investment-policy-watch/trumps-phase-one-deal-relies-chinas-state-owned-enterprises. Accessed 31 Mar 2020

A. Cooley, Central Asia: a political history from the 19th century to the present. The Asia Society. Foreign Affairs. (2012). https://www.foreignaffairs.com/articles/central-asia/2012-08-07/new-great-game-central-asia . Accessed 10 Mar 2010

A. Dubnov, Reflecting on a quarter century of Russia's relations With Central Asia. Carnegie Endowment for International Peace. (2018). https://carnegieendowment.org/2018/04/19/reflecting-on-quarter-century-of-russia-s-relations-with-central-asia-pub-76117. Accessed 17 Mar 2020

The Economist. The company that ruled the waves. The Economist. (2011). https://www.economist.com/christmas-specials/2011/12/17/the-company-that-ruled-the-waves. Accessed 20 Mar 2020.

S. Feng, Y. Hu, R. Moffitt, Long run trends in unemployment and labor force participation in China. National Bureau of Economic Research. (2015).https://www.nber.org/papers/w21460.pdf. Accessed 1 Apr 2020

T. Finn, Anglo-Russian entente 1907. Gov UK Blog History of Government. (2017). https://history.blog.gov.uk. Accessed 23 Mar 2020

D. Fromkin, The great game in Asia. Foreign Affairs. (1980). https://www.foreignaffairs.com/articles/south-asia/1980-03-01/great-game-asia

The Guardian. Datablog: is Transparency International's measure of corruption still valid. The Guardian. (2013). https://www.theguardian.com/global-development/poverty-matters/2013/dec/03/transparency-international-measure-corruption-valid

M. Hauner, *What Is Asia to Us?* (Unwin Hyman, London, 1990).

F. Hill, The United States and Russia in Central Asia: Uzbekistan, Tajikistan, Afghanistan, Pakistan, and Iran. On the Record, Brookings Institution. (2002).https://www.brookings.edu/on-the-record/the-united-states-and-russia-in-central-asia-uzbekistan-tajikistan-afghanistan-pakistan-and-iran/. Accessed 16 Mar 2020

P. Hopkirk, *The great game: the struggle for empire in Central Asia* (Kodansha International, Tokyo, 2017).Accessed 12 Feb 2020

Y. Ishikawa, Russia's cozy economic ties with China are not all they seem. Nikkei Asian Rev. (2019) https://asia.nikkei.com/Spotlight/Comment/Russia-s-cozy-economic-ties-with-China-are-not-all-they-seem. Accessed 20 Mar 2020

A. Maddison, *The world economy: volume 1: a millennial perspective and volume 2: historical statistics*(Development Centre Studies, OECD Publishing, 2006). https://doi.org/10.1787/9789264022621-en. Accessed 14 Mar 2020

NationMaster. Countries compared. NationMaster. (2020). https://www.nationmaster.com/country-info/stats/Geography/Land-area/Square-miles

Office of the Historian, The Soviet invasion of Afghanistan and the U.S. response, 1978–1980. U.S. Department of State. (2020). https://history.state.gov/milestones/1977-1980/soviet-invasion-afghanistan. Accessed 19 Mar 2020

M. Olcott, Central Asia: confronting independence (James A. Baker III Institute for Public Policy, Rice University, 1998).https://large.stanford.edu/publications/power/references/baker/studies/assets/docs/UnlockingtheAssets_CentralAsia_ConfrontingIndependence.pdf. Accessed 3 Apr 2020

M. Peck, Russia is angry that China stole and copied its jet fighter designs. The National Interest. (2019). https://nationalinterest.org/blog/buzz/russia-angry-china-stole-and-copied-its-jet-fighter-designs-80351. Accessed 1 Apr 2020

R. Pomfret, *The Central Asian Economies in the Twenty-First Century* (Princeton University Press, Princeton, 2019).

Reuters, China theft of technology is biggest law enforcement threat to US, FBI says. The Guardian. (2020). https://www.theguardian.com/world/2020/feb/06/china-technology-theft-fbi-biggest-threat

E. Rumer, R. Sokolsky, P. Stronski, U.S. policy toward Central Asia 3.0. Carnegie Endowment for International Peace. (2016).https://carnegieendowment.org/2016/01/25/u.s.-policy-toward-central-asia-3.0-pub-62556. Accessed 20 Feb 2020

E. Sergeev, *The Great Game, 1856–1907: Russo-British Relations in Central and East Asia* (Johns Hopkins Press, Baltimore, 2013).

B. Simms, Castlereagh's catechism. Foreign Affairs. (2013).https://www.foreignaffairs.com/reviews/review-essay/2013-03-01/castlereaghs-catechism

A. Taylor, The Soviet war in Afghanistan, 1979–1989. The Atlantic. (2014).https://www.theatlantic.com/photo/2014/08/the-soviet-war-in-afghanistan-1979-1989/100786/. Accessed 18 Mar 2020

F. Todd, Unpacking Chevron's Tengiz expansion project: will it go live by 2022? NS Energy. (2019).https://www.nsenergybusiness.com/features/chevron-tengiz-expansion-project-2022/. Accessed 1 Apr 2020

Trading Economics. Russia exports by country. Trading Econom. (2020).https://tradingeconomics.com/russia/exports-by-country. Accessed 31 Mar 2020

Transparency International, 2019 corruption perception index. Transparency Int. (2019). https://www.transparency.org/cpi2019?/news/feature/cpi-2019

U.S. Department of Justice, Foreign corrupt practices act. U.S. Department of Justice. (2017). https://www.justice.gov/criminal-fraud/foreign-corrupt-practices-act

D.S. Van der Oye, Paul's great game: Russia's plan to invade British India. Central Asian Survey **33**(2), 143–152 (2014). https://doi.org/10.1080/02634937.2014.909672

J. Votel, Great power competition: the current and future challenges in the Middle East. Testimony, Senate Armed Service Committee. (2019).https://www.centcom.mil/Portals/6/Documents/Transcripts/Votel020519.pdf Accessed 9Apr 2020

F. Wang, China's population destiny: the looming crisis. Brookings Institution. (2010).https://www.brookings.edu/articles/chinas-population-destiny-the-looming-crisis/ Accessed 1 Apr 2020

M. Weinbaum, Pakistan and Afghanistan: the strategic relationship. Asian Surv. **31**(6), 496–511 (1991). https://doi.org/10.2307/2645079

The World Bank, Doing Business 2020. The World Bank. (2020). https://www.doingbusiness.org. Accessed 4 Apr 2020

Worldometer. Central Asia population. Worldometer. (2020). https://www.worldometers.info/world-population/central-asia-population/

F. Xi, Why ageing China won't overtake the US economy as the world's biggest–now or in the future. South China Morning Post. (2019).https://www.scmp.com/comment/insight-opinion/article/3003524/ageing-china-simply-cannot-overtake-us-economy-worlds. Accessed 1 Apr 2020

Y. Zarakhovich, Kazakhstan comes on strong. Time Magazine. (2006). https://content.time.com/time/world/article/0,8599,1539999,00.html. Accessed 30 Mar 2020

China and the United States in the Middle East: Between Dependency and Rivalry

Michael K. Singh

Abstract While China is becoming more active in the Middle East economically, diplomatically, and militarily, its policies in the region are often difficult for Western policymakers to predict or understand. Beijing views the Middle East through three lenses—its direct interests in the region, Xi Jinping's signature Belt and Road Initiative, and US-China relations—and its policies are the result of the complex interplay of the three. China's strategy emphasizes maximizing economic benefits while minimizing commitments in the region, while incrementally seeking to make strategic gains, bolster China's regional and global profile, and challenge American predominance. The future of China's policy depends not just on these factors, however, but on the approach the United States chooses to take to on Beijing overall as well as on the unanswered question of what Middle East strategy Washington will pursue to complement its broader strategy of great-power competition. The United States current posture of heavy presence but strategic diffidence in the region offers China the best of both worlds—a continued American security umbrella over a region increasingly interested in diversifying its great-power relationships.

Keywords China · United States · Great power competition · Belt and road · Middle East strategy

Introduction

From the mid-2000s onward, the United States has gradually come to see China as its foremost peer and rival. This view has been shared by policymakers of both major political parties. It was Hillary Clinton, then a US Senator from New York, who wrote in 2011 of the need for a "pivot" to Asia, and of America's "Pacific Century" (Clinton 2011).The pivot was later rebranded a "rebalancing" to avoid the impression that the US was turning away from other regions of the globe as it sought to increase

M. K. Singh (✉)
Washington Institute, Washington, USA
e-mail: msingh@washingtoninstitute.org

© The Author(s), under exclusive license to Springer Nature Switzerland AG 2021

A. Farhadi et al. (eds.), *The Great Power Competition Volume 1*,
https://doi.org/10.1007/978-3-030-64473-4_18

its focus on Asia, but otherwise continued apace during President Barack Obama's second term as president.

Despite his otherwise strenuous efforts to distance himself from the initiatives of his predecessor, President Donald Trump embraced and extended this focus on Asia, which saw its apotheosis in the National Defense Strategy (NDS) issued in 2018. The introduction to that document, penned by Secretary of Defense James Mattis, asserted bluntly that "Inter-state strategic competition, not terrorism, is now the primary concern in U.S. national security" (US Department of Defense 2018). The strategy represented a conscious effort to turn the ship of state away from two decades of continuous focus on counter-terrorism and the Middle East. It was also intended as a clear-eyed recognition of the reality that the threat posed by revisionist, nuclear-armed powers far outweighed that of the non-state actors whose attacks had drawn the United States into most of its 21st-century conflicts.

However, the embrace of "strategic competition" clashed with another reality: the United States, whatever its aspiration, remained mired in those same conflicts. US troops were engaged in hostilities in Afghanistan, Iraq, Syria, and Africa, supported by scarce air, naval, and so-called "ISR" (intelligence, reconnaissance, and surveillance) assets. The NDS and a companion document issued by the White House, the National Security Strategy, described the approach the Trump Administration hoped to take. But of course, the United States was not starting with a blank page, and these documents offered little indication of how the world's sole superpower intended to transition from a strategy emphasizing conflicts on the periphery to one prioritizing great-power competition.

Nor did these strategy documents indicate what this profound strategic shift meant for US policies in the periphery. Both note the importance, for example, of preventing Iran from acquiring nuclear weapons and of defeating terrorist organizations, these are appendages to the broader strategy of great-power competition rather than integrated into it. There is little indication of how this strategy might play out in the Middle East – whether it means, for example, that the US will maintain longstanding objectives but devote fewer resources to accomplishing them, or whether Middle East issues will now be viewed through the lens of great-power competition, or whether the region will be forsaken altogether. Nor does the NSS or NDS provide meaningful guidance regarding how the US sees Chinese (or for that matter, Russian) involvement in the Middle East—potential American losses in a zero-sum game for influence, a welcome diversion of a rival's resources, or something in between?

This omission stems from US strategy pronouncements defining US objectives—and, just as important, the American understanding of Chinese objectives—only in the broadest terms. For example, the NSS states that "China…aspire(s) to project power worldwide, but they interact most with their neighbors," and that Beijing aims "to persuade other states to heed its political and security agenda" (The White House 2017). US aims are described with less specificity still. This leaves unanswered whether American policymakers view their objectives with respect to great-power competitors as essentially regional in nature—say, preventing China from establishing hegemony in East Asia—or whether they see strategic competition as a global struggle ala the Cold War.

Understanding American and Chinese national aims with greater precision is vital to determining how great-power competition will unfold in the Middle East and other regions outside the immediate environs of the two powers—how and to what extent each will engage in the region's affairs, and whether the relationship between the two will be one of cooperation, dependency, or rivalry. The focus of this essay will be the latter—how Chinese policy in the Middle East is shaped by its interests in the region and beyond.

Three Lenses for Chinese Policy in the Middle East

Understanding American policy in the Middle East requires not simply a familiarity with US interests there—the understanding of which changes little from one presidential administration to the next—but also of broader domestic and geopolitical dynamics that shape the overall formulation of national security policy. The same is true for China and other powers. For understanding Chinese policy in the Middle East, three lenses are especially important—first, China's interests in the Middle East; second, China's efforts to project power westward through the Middle East toward Europe and Africa under the rubric of the "Belt and Road" Initiatives (BRI); and third, China's complex relationship with the United States, which remains the dominant external power in the Middle East.

The role that China plays in the Middle East today, and how that role will develop in the future, is the result of the interaction of these dynamics, together with other factors such as domestic politics and China's history. This interaction can lead to outcomes that are often counterintuitive for American policymakers. For example, two of China's most valued relationships in the region are with countries bitterly opposed to one another, Israel and Iran. The former offers China both a destination for infrastructure investment and, to the increasing chagrin of US officials, a potential source for high-tech partnerships. The latter, however, is a natural partner in Chinese efforts to counter American predominance in global governance, and one whose relative international diplomatic and economic isolation makes it particularly, if reluctantly, dependent on Beijing.

Likewise, despite China's partnership with Iran, its willingness to help Tehran resist US sanctions has generally been limited. This owes in large part to the relatively greater priority Beijing places on preserving good trade relations with the United States; for this reason, predictions that US sanctions will fail due to Chinese defiance have frequently proven erroneous, because they misapprehend Beijing's priorities. The same tensions played out in public view when the United States invited China to participate in the Manama Conference in June 2019. The conference, promoted endorsed by the Trump administration as a meeting to promote investment and economic development in the Palestinian territories, was boycotted by the Palestinians, upset with the United States' recognition of Jerusalem as Israel's capital. China, together with Russia, initially announced it too would decline to attend, consistent with Beijing's efforts to take a balanced approach to the Israelis and Palestinians. Yet China quickly reversed course, ultimately sending an observer to the conference—loath, perhaps, to pick a fight with Washington over a peripheral issue, as well as to miss out on the investment opportunities that might arise.

Chinese Interests in the Middle East

From an American perspective, what is most striking about Chinese interests in the Middle East is the degree to which they overlap with those of the United States, in marked contrast to the sharp divergence in, say, American and Russian interests. For Beijing, a number of interests in particular appear to predominate—chief among them energy, other trade and investment, freedom of navigation, and counter-terrorism.

Energy

The roles occupied by the United States and China in the global energy landscape have changed dramatically in recent years. The change in the American role was vividly illustrated by President Trump's involvement in April 2020 in a deal struck between OPEC and Russia to cut back on oil production in an effort to *raise* prices (Faucon et al. 2020). This represented a sea change from the position the United States—unwittingly on the brink of a shale revolution—found itself in just over a decade earlier, when then-US Treasury Secretary Hank Paulson pleaded with oil producers to increase production and open up their oil sectors to greater investment (Associated Press 2008).

China's role in global energy markets has undergone a perhaps less stark yet still enormously significant transformation. China has grown to become one of the world's top consumers of oil, second only to the United States (Energy Information Administration 2020a; b). And though China became an oil importer only in the mid-1990s—providing a major impetus to the expansion of its ties with states in the Middle East—those imports have since skyrocketed, driven by increasing demand and declining domestic production. China become the world's largest importer of oil in 2017, surpassing the United States, which in late 2019 became a net exporter of oil (Energy Information Administration 2020a; b). According to the BP Energy Outlook, these trends are likely to continue even if the COVID19 pandemic and rising US-China trade tensions, among other factors, depress demand for and trade in oil; China's oil demand will surpass that of the United States, which will become an increasingly major supplier, and China will continue to lead global growth in oil imports until eventually being overtaken by India in that regard (BP 2019). China is also the world's second-largest importer of liquefied natural gas (LNG), and led the world in LNG import growth in 2018. Qatar is China's second-largest source of LNG imports; Iran is a potential supplier as well, though US sanctions currently obstruct Iran's LNG exports (International Gas Union 2019).

It is widely assumed that China's growing role as an energy consumer and importer necessarily means increased interdependency between China and the Middle East's oil states. The reality is more complex. While China is a major customer for producers such as Saudi Arabia and Iraq, they also count among their major export destinations Beijing's regional rivals in Asia, such as Japan, South Korea, India, and Taiwan. China itself has become somewhat less dependent on Middle Eastern suppliers. This is partly due to China's growing energy relationship with Russia, which surpassed Saudi Arabia as China's top oil supplier in 2016, and in 2017 OPEC's overall share of Chinese oil imports dipped to 56%, down from a high of 67% in 2012, according to the US Energy Information Administration (Energy Information Administration 2018). Yet it is also, ironically, partly the result of the United States' growing energy independence, which has freed up supplies in West Africa (Angola was China's third-largest supplier of oil in 2018) and South America (Brazil, Venezuela, and Colombia are all significant Chinese suppliers). Within the Middle East, China has diversified its energy relationships, importing significant quantities of oil from Saudi Arabia, Iraq, Oman, Iran, Kuwait, the UAE, and Libya in 2018.

China's energy relations with the Middle East go well beyond the oil trade, however. Energy is so central to Beijing's regional relationships that in the "'1 + 2 + 3′ cooperation pattern" outlined in the Arab Policy Paper released by the Chinese government in 2016, energy cooperation both comprises the first pillar, and elements of the third, which focuses on nuclear and "new" energy in addition to satellites (State Council of the People's Republic of China 2016). James Dorsey of the S. Rajaratnam School of International Studies at Singapore's Nanyang Technological University has argued that Chinese firms have "replaced US oil majors as major investors in Middle Eastern and North African national oil companies," citing Saudi Aramco mulling the sale of a 5% stake to a consortium of Chinese state-owned firms, and Abu Dhabi's ADNOC's sale of a significant share to Chinese government-held entities (Dorsey 2019a; b). China has also been active in the nuclear, solar, wind, and hydroelectric sectors in the Middle East, consistent with the third pillar of its "1 + 2 + 3" approach (Fulton 2020). The growing Chinese role in regional energy investment—perhaps even more than its role as a consumer of Middle Eastern oil—raises the regional stakes for Beijing.

Non-Energy Trade and Investment

While energy may be at the core of China's economic interests in the Middle East, it is far from the entire story. For China, the Middle East is both a source of and a destination for investment and trade. According to the American Enterprise Institute, Chinese investment in the Middle East—Arab states, Iran, Turkey, and Israel—was a combined $140.5 billion from 2005–2019 (American Enterprise Institute 2019). Overall, this amount is not especially impressive—it is less than the amount China invested in the sub-Saharan Africa, and roughly on par with its investments in South America. More noteworthy is the trend. In 2018, for example, China invested $28.11

billion in the Middle East, more than in any region except Europe. In a similar vein, nearly $20 billion of the roughly $25 billion in Chinese investment and construction in Egypt since 2005 has come since 2013, as has roughly $22 billion of the $30 billion it has invested in the UAE over that time period (Molavi 2019). Dorsey notes that China's investment is broad in scope, writing, "In just one slice of the vast Eurasian interests of Chinese companies, China's ambassador to Saudi Arabia, Li Chengwen, noted in 2013 that 140 Chinese companies were involved in contracts worth $18 billion in Saudi Arabia's construction, telecommunications, infrastructure, and petrochemical sectors" (Dorsey 2019a; b).The Middle East has also been a growing destination for Chinese exports, which increased by 67% in nominal terms from 2010 to 2019, far exceeding the overall rate of growth in Chinese exports globally over that time period, and far exceeding the 22% increase in American exports to the region (International Monetary Fund 2020). Finally, the region is a destination not only for Chinese trade and investment, but for its workers. Approximately 550,000 nationals were living in the Middle East in 2014, with 200,000 in the UAE alone (Dorsey 2019a; b).

The economic dependence between China and the Middle East runs in both directions. In 2019, the Middle East sent $132 billion in exports to China, more than to any other single destination, far outstripping the United States ($41 billion) and nearly as much as to the entire European Union ($160 billion). Yet while Chinese investment in the region seems ubiquitous, Chinese imports heavily favor energy producers. For example, Oman sent China nearly 40% of all its exports in 2019, and enjoyed a nearly $16 billion trade surplus with Beijing. Egypt, on the other hand, exported a mere $358 million to China that year while importing from China goods and services worth over $12 billion, making China responsible for nearly one-third of Egypt's overall trade deficit. (International Monetary Fund 2020) One important source of Chinese support for Middle Eastern economies comes in the form of tourism. China supplanted the United States as the largest source of outbound tourists and tourism revenue in 2014 (Chaziza 2019). For example, the number of Chinese tourists arriving in Egypt grew by 280% from 2014 to 2018, with China's overall share of tourism in Egypt rising from 0.6 to 2.07% in that period. For comparison, US tourist arrivals in Egypt rose 86% in that same period, with the US market share rising from 1.56 to 2.53%. Over the same time period, Chinese tourism to Israel grew by 225%, while US tourism grew 49% (UN World Tourism Organization 2020).

While analysts and officials in the United States debate whether the overall trends of Chinese trade and investment in the Middle East, and vice versa, should be of concern, two areas of Chinese economic activity in the region merit particular attention: arms and technology. Chinese arms sales in the Middle East pale in comparison to those of the United States and Russia, yet have grown quickly in the past decade, increasing 82% during the period 2014–2019 compared to the previous five years. Furthermore, the systems sold by China raise particular concerns—Beijing has, for example, been energetically promoting the export of armed UAVs, selling them over the past decade to Algeria, Egypt, Iraq, Jordan, Saudi Arabia, and the UAE (Stockholm International Peace Research Institute 2020). In addition, China has sold significant quantities of anti-ship cruise missiles to Iran, among others; these types of weapons have been responsible for attacks by Lebanese Hezbollah and Yemen's Houthi rebels, both of which receive arms from Iran. China even reportedly built a factory in Iran in 2010 to manufacture anti-ship cruise missiles (UPI 2010).

Chinese investment in high technology has focused on Israel, considered a global leader in high-tech startups. According to a recent report by the RAND Corporation, Chinese venture capital investment in Israel increased from $500 million in 2014 to $1 billion in 2016, and Israeli tech-startups received $325 million in Chinese investment in the first three quarters of 2018 alone, an increase of over one-third from the previous year. The report indicates that China has focused on areas such as artificial intelligence, robotics, and information technology—all notable for their potential application to military challenges and for being focus areas for Beijing's "Made in China 2025" plan (Efron et al. 2020). While China-Israel technology ties are notable for the friction they have caused in the US-Israel relationship, they are not the only case of Chinese technological investment in the Middle East raising eyebrows in Washington. As of September 2019, Chinese 5G provider Huawei had signed contracts with eleven telecommunications firms in the GCC, including in Saudi Arabia, Oman, the UAE, Kuwait, and Bahrain, prompting official US complaints (Halligan 2019).

Freedom of Navigation

China's economic dependence on the Middle East, and particularly the region's entanglement in Chinese energy security, also lends Beijing a keen interest in freedom of navigation in the region's seas. About 21% of petroleum liquids consumed globally and one-quarter of the global liquefied natural gas trade pass through the Strait of Hormuz, and in 2018, 76% of the crude oil and condensates passing through the Strait was bound for Asian markets. Options for bypassing the Strait of Hormuz are quite limited—Saudi Arabia and the UAE have pipelines circumventing the chokepoint that together can carry 6.5 million barrels per day of oil, just 31% of the volume that passes through the Strait (Energy Information Administration 2019a; b).

Also of vital importance to Chinese energy and economic security are the Suez Canal and Bab el-Mandeb strait. These waterways are not as vital to Beijing from an energy security perspective, yet they are still important. According to the US Energy Information Administration, 9% of seaborne-traded petroleum passes through the latter, including, in 2018, 2.6 million barrels per day that is destined for Asian markets (Energy Information Administration 2019a; b). Adding to the Suez Canala and Bab el-Mandeb's centrality to Chinese economic interests is the fact that 60% of Chinese exports traverse these waterways, which has led to heavy Chinese investment in the canal's ports and terminals (Dorsey 2019a; b).

Counterterrorism

At a meeting of judicial chiefs of Shanghai Cooperation Organization countries in 2006, Xiao Yang, chief of China's Supreme People's Court, exhorted his peers to combat what he termed the "three evil forces" of terrorism, extremism, and ethnic separatism. Beijing views the Middle East as a source of all three.

China's conception of terrorism is both like and unlike that of the United States, for reasons the "three evils" formulation suggests. Like the United States, China has been the victim of domestic terrorist attacks that it worries have links to international terrorist organizations. These have included bomb attacks in Urumqi, the capital of China's restive Xinjiang region, in April and May 2014, and knife attacks in Kunming and Guangzhou in March and May of 2014, respectively. In 2015, a Chinese hostage was murdered by ISIS in Syria. Thousands of Chinese nationals—exact estimates vary, and even China's Syria envoy has indicated he does not know the number with any precision—have reportedly traveled to Syria to join groups such as ISIS, Hurras ad-Din, and others. Chinese officials, like their counterparts in Europe and elsewhere, worry about the security consequences of the return of these foreign terrorist fighters (Reuters 2018a; b).

Yet for China, unlike the United States, worries about terrorism and extremism are inextricably tied to concerns about ethnic rights and the internal stability of China itself. Alireza Nader and Andrew Scobell quote Chinese scholar Li Weijian of the Shanghai Institutes of International Studies as writing, "As the strategic extension of China's western border region, the trends governing the situation in the Middle East and the region's pan-nationalisms and extremist religious ideological trends have a direct influence on China's security and stability" (Nader and Scobell 2016). This also means that, despite its professed concern about terrorism, China tends not to share the blanket US opposition to Islamic extremist groups like Hezbollah or Hamas, provided that they do not give succor to those Beijing regards as threats to China's internal security. Nor, unlike the United States and most researchers of extremism, does China view political liberalization, justice, or democracy as part of the answer, for obvious reasons. Instead, China emphasizes economic development as the key to overcoming extremism and radicalization. Rolland (2017) quotes Chinese scholar Gan Junxian as asserting, for example, that "increased economic exchanges and trade, enhanced living standards in Central Asia, and cultural exchanges to strengthen trust between people will eliminate the basis for fundamentalism and terrorism."

Other Interests

One could reasonably enumerate any number of other Chinese interests in the Middle East, all deriving from the basic desire of states to safeguard their security, prosperity, and system of government and way of life. Where China differs from the United States and other Western countries, however, is in its readiness to link issues that are of deep concern to Beijing but otherwise unrelated to the Middle East—for example, the matter of international recognition of its "One China" policy—to its regional relations. Indeed, China devoted a paragraph of its Arab Policy Paper, issued on the eve of President Xi Jinping's visit to the Middle East in 2016, to the matter of Taiwan. The paper states that:

> The Taiwan question concerns the core interests of China. The one China principle is the important basis for China to establish and develop relations with Arab states and regional organizations. Arab states and regional organizations have always been committed to the one China principle, refrained from having any official relations or official exchanges with Taiwan, and supported China in peaceful development of cross-Straits relations and the great cause of national reunification. China appreciates all these (State Council of the People's Republic of China 2016).

Nor is Beijing's concern limited to this single issue. In his remarks at the 2019 Manama Dialogue, delivered in proficient Arabic, Chinese Middle East envoy Li (2019) devoted nearly half of his speech to a stern rebuke of the Japanese Defense Minister's earlier comments on the South China Sea. While there is little indication that Beijing has sought to promote its stance on the South China Sea as actively as it has the "One China" policy, regional governments are doubtless aware that China views as global interests that others might perceive as strictly Asian affairs. There are few parallels to such linkages in international affairs; the Trump administration suggested that it would tie foreign assistance to states' UN voting records on Israel, for example, but such initiatives tend to be short-lived and sporadically enforced.

The Belt and Road Initiative

The second lens through which Chinese policy in the Middle East should be viewed is the Belt and Road Initiative (BRI), Xi Jinping's signature foreign policy strategy. In the Chinese formulation, the BRI aims to foster connectivity in order to bring about the realization of what Xi termed a "community of common destiny"—that is, to more closely bind far-flung regions of the world to China. While the BRI is most frequently associated in Western opinion with infrastructure investment, Xi described five forms of connectivity the initiatives aim to advance: political, transportation, trade and investment, financial integration including greater use of the Chinese currency (renminbi), and people-to-people exchanges (Scobell et al. 2018).

The BRI consists of two, or in some descriptions three, separate pillars: the Silk Road Economic Belt, which focuses on overland routes from Western China through Central Asia and beyond; the Maritime Silk Road, which focuses on maritime corridors from the South China Sea to the Indian Ocean and beyond; and a "New Digital Silk Road," focusing on telecommunications and information connectivity (Kliman and Grace 2018). So ambitious is the BRI that it is more appropriate to ask which regions and countries China does not include in it rather than which ones it does;

it encompasses all of Eurasia, including Russia, Central Asia, and the Middle East; Southeast Asia; all of South Asia apart from India; much of Eastern Europe; parts of Africa, especially coastal East Africa; and even countries in South America. The BRI speaks to the breadth of Chinese economic, political, and security activities, as well as to the depth of its ambitions.

In explaining how the BRI emerged and what drives it, Rolland (2017) describes it "as an attempt to set the direction for China to achieve its ambitions as a preponderant regional power, in the context of mounting challenges in both the economic and strategic domains." She states that the BRI emerged as a result of two developments that led the Chinese leadership to rethink their approach to achieving China's rise: the global financial crisis of 2008, which hampered Chinese economic growth; and the US "rebalance" to Asia, viewed in Beijing as an effort to contain China. Rolland asserts that the BRI has both economic drivers as well as unstated strategic drivers. The former aims not only to promote Chinese economic growth generally, but dispose of excess industrial capacity, boost China's state-owned enterprises, and encourage greater international use of the renminbi. The latter focuses on fostering development in Western China and Central Asia in order to promote internal stability in China, promote China's energy security, boost China's diplomatic profile and influence, and counter the perceived US effort at containment (Rolland 2017).

For all these hoped-for benefits, Chinese analysts also acknowledge that the BRI brings with it economic and strategic risks. China's westward expansion has exposed it to increased threats of terrorism, which claimed the lives of fourteen Chinese nationals in Afghanistan and Pakistan in 2004, as well as war, which put at risk tens of thousands of Chinese citizens in Libya and Yemen in 2011–2012. Chinese analysts also have expressed worry about the BRI becoming a lightning rod for extremists, increasing the risk to Chinese workers. Finally, Chinese scholars have recognized the risk that BRI could place China in greater competition with other major powers, including not only the United States but also India, Japan, and others (Wuthnow 2017). China's growing economic footprint has also resulted in a degree of backlash in the region itself, with an Iranian parliamentarian complaining in 2010 of the impact of Chinese imports on domestic producers, and instances of violence against Chinese workers in Algeria and Iraq (Chen 2011).

Many of the activities now included under the umbrella of the BRI are not new; certainly China was active in infrastructure construction and port development in the Middle East before the strategy was articulated. Nor are the drivers of the BRI entirely distinct from the formulation of Chinese interests in the Middle East described above; building pipelines advances both transport connectivity as well as enhancing Chinese energy security, for example. Why then should the BRI be considered a distinct lens through which to view Chinese policy in the Middle East? If the interests enumerated in the previous section describe the ways in which Beijing sees the Middle East directly affecting its prosperity and security, the BRI treats the Middle East as a piece in a broader puzzle—not simply as a destination for Chinese diplomats, investment, etc. in its own right, but as a link in a chain, and a particularly important one given its geographic centrality. Thus, while projects such as the development of Oman's Duqm port or Israel's Ashdod port might serve Chinese economic interests in their own right, they take on particular importance in the context of BRI given their role in facilitating the westward projection of Chinese power and influence. As Rolland notes, "BRI is…not merely a list of revamped construction projects but a grand strategy that serves China's vision for itself as the uncontested leading power in the region" (Rolland 2017).

US-China Relations

The United States and China are increasingly viewed as rivals if not adversaries, and for good reason: from a US perspective, China is the only other country in the world likely to match the United States in economic and military power, and appears determined to use that power in ways inimical to American interests. Yet the day-to-day reality of the US-China relationship is far more complex, as demonstrated by the frequent warm words and reciprocal visits between American and Chinese leaders as well as the interdependence of the US and Chinese economies. This conflict—between rivalry and interdependence—is the central one for US-Chinese relations.

In the Middle East, China is confronted with the unavoidable reality that the United States is the region's most powerful external actor, and has been not just since the end of the Cold War, but arguably since Egyptian President Anwar Sadat expelled Soviet military advisors from Egypt in 1972. China's relationship with the United States in the region is characterized by three tensions:

(1) First, China aims to dethrone the United States, but not supplant it. As noted above, Beijing perceives the United States as a declining power, and the American rebalance to Asia as an effort to contain and marginalize China in regional and global affairs which must be countered for China's rise to continue. However, China is neither capable nor necessarily interested in replacing the United States as the world's sole superpower. Furthermore, China remains highly interdependent on the US, especially economically; the unavoidable reality for Beijing is that its relationship with Washington is—so far, at least—more important than almost any interest it possesses in the Middle East.

(2) Second, China aspires to be seen in the world and within China as a great power, but wishes to cast itself as an entirely different kind of power than those that preceded it, especially the United States. China characterizes Western powers' approach to the developing world as exploitative and militaristic, and seeks to position itself as a purportedly benevolent and peaceable alternative. Similarly, China has criticized the American tendency to seek and form alliances and engage in interventions in the Middle East and elsewhere, and has publicly espoused a doctrine of "non-interference" and sought to cultivate friendly relations with all states while eschewing formal alliances with any, save North Korea. This doctrine is captured by Mao Zedong's "five principles of peaceful coexistence," namely mutual respect for sovereignty and territorial integrity, mutual non-aggression, non-interference in other states' internal affairs, equality and

mutual benefit, and peaceful coexistence (Ministry of Foreign Affairs of the People's Republic of China 2014).

(3) Third, while China is critical of the US approach to the Middle East—especially military interventions in Iraq, Syria, and elsewhere—it remains dependent on an American security umbrella there. It is the US Fifth Fleet, not the PLA Navy, that secures oil shipments bound for China through the Strait of Hormuz, and American security cooperation with the Iraqi military is vital for safeguarding Chinese energy investments there. This dependence is why China, unlike other US rivals, has not cheered the withdrawal of US forces from Syria, Iraq, and Afghanistan; indeed, the prospect of American disengagement poses a difficult challenge to Chinese strategy in the region.

As Western analysts of China have debated whether the growing rivalry between China and the United States is merely a geopolitical struggle or also an ideological one, the notion that a "China model" exists as an alternative to the "Washington consensus" has been promoted by Beijing and has sparked worries in Washington. One Middle East analyst described the China model as "balancing economic development, state modernization, and political control…for countries that want to carefully manage their economic and political transformation" (Hoyakem 2009).

Yet there are substantial questions whether a China model truly exists, much less whether it is applicable to the states of the Middle East. Kamrava (2018) concludes, for example, that while the notion of achieving economic growth while avoiding political liberalization has an undeniable appeal for authoritarian states, Middle Eastern regimes generally lack the "necessary ingredients and preconditions" —notably state capacity—that enabled China's success.

Nevertheless, a narrative about the risk of the China model spreading across the region will likely persist due to a combination of factors—discontent in the Middle East with American policies and the market- and democracy-oriented Washington Consensus; anxieties in the West about losing ground ideologically to China in regions such as the Middle East; and a desire by Middle Eastern governments to curry Beijing's favor and import the tools of political control it seems to employ so effectively. Despite these factors, there is little evidence that US policy faces a challenge, whether from within the Middle East or from China itself, from the adoption of a China model by regional governments. Indeed, the bigger challenge may come from within Washington itself, where questioning of the US-led Middle East status quo may be most acute.

China's Challenges and Opportunities

In pursuing its interests and objectives in the Middle East, China—like all external powers before it—faces numerous obstacles. Yet Beijing, in contrast to the United States and Europe, also perceives opportunity in the region.

Economic Instability

In the first quarter of 2020, the Middle East faced what arguably constituted a triple economic crisis: the economic impact of national measures to contain the outbreak's spread, such as closing retail businesses and factories; the further impact of declining external sources of income and support, including tourism, remittances, portfolio inflows, and aid, as the rest of the world grappled with the outbreak; and the crash in oil prices stemming from a Saudi-Russian price war and the crash in demand resulting from the pandemic. The crisis demonstrated just how quickly economic fortunes can change, as prior to the pandemic the Middle East was experiencing modest growth and some countries, such as Egypt, were seen as the cream of the emerging-markets crop from the perspective of foreign investors.

For China—focused as it has been on energy and (often debt-financed) infrastructure projects, this instability also represented a multifaceted crisis. Low oil prices would ordinarily be greeted enthusiastically in Beijing, but, like other major oil consumers, China was in no position to capitalize as the decline in prices was driven as much by the evisceration of demand as the glut in supply. Furthermore, the decline in prices leaves energy investments less profitable, and pressure on regional finances likely promises a reduction in spending for the foreseeable future on the sort of mega-projects that have attracted Chinese involvement, many of which could be left dormant in the wake of the coronavirus. It is possible that the crisis may direct particular ire toward China, both because debt burdens are likely to mount and Beijing may hesitate to offer relief, and because blame for the pandemic's global spread may be placed at the doorstep of the Chinese government, which is widely regarded as having misled the world about the initial outbreak in Wuhan. The crisis may also underscore the need for regional governments to enact measures to encourage the growth of robust private sectors, which could prove a disadvantage for China's economic engagement in the region.

Nevertheless, Beijing may also perceive potential advantages in the region's economic instability. Amid economic crisis, distressed governments may seek to raise funds through the sale of assets, offering China an opportunity to expand its regional portfolio in line with BRI objectives. Beijing, and the Asian Infrastructure Investment Bank (AIIB) which it dominates, may also act as a lender of last resort if Western states prove unable or unwilling to help, and if traditional regional benefactors such as Saudi Arabia, the UAE, Qatar, and Kuwait must struggle with their own financial squeezes. This could be particularly the case for states like Oman, which is already heavily dependent on China for trade, is viewed by China as an important BRI node, and which has resisted the political conditions associated with financial aid from its neighbors.

Political Instability

While economics may come first in China's relations with the Middle East, political instability in the Middle East is a greater challenge than economic turbulence because China has far fewer tools with which to address the former, especially compared to competitors like the United States and Russia. Amid the "Arab Spring" in 2011, Beijing sought to maintain friendly relations with both sides of just about every regional conflict, cultivating cordial ties both with Muammar Qadhafi and Libyan rebels, for example (despite its professed prioritization of sovereignty and non-interference), just as China had been among the first governments to recognize Hamas rule in Gaza the decade prior. Yet Beijing was caught flat-footed nevertheless—it had to scramble to mount its first-ever expeditionary naval operation in 2011 to evacuate 30,000 Chinese nationals from Libya, having reportedly been previously unaware that so many of its citizens resided there. It later had to mount a smaller such operation in Yemen. While both operations were ultimately celebrated in China, they demonstrated clearly the mismatch between Beijing's regional posture and its vulnerabilities there—a mismatch China has since strived to overcome, in no small part by opening its first overseas naval base in Djibouti.

Since the Arab Spring, Beijing's effort to be friends with everyone has faced increasing headwinds, and not just in countries experiencing civil wars. Since 2011, the Middle East has witnessed a rise in intraregional rivalry, as the old US-led order has decayed, and states have instead organized themselves into ad hoc blocs and taken more activist stances on conflicts within the region. In just a decade, for example, GCC states have intervened in Bahrain, Yemen, Libya, and the Horn of Africa, Turkey has intervened in Syria, Libya, and the Horn of Africa, and Iran has intervened in Syria, Iraq, Lebanon, Yemen, and elsewhere.

Avoiding taking sides in this environment has proven challenging, and has relegated China to the sidelines, despite its great-power aspirations. With regard to Syria, for example, China is widely viewed as a junior partner to Russia given the fact that Beijing has cast its veto multiple times in support of Moscow's position, breaking from its previous practice of using its veto power largely for matters closer to home. Beyond these votes, China has exercised little diplomatic muscle and no military power; while this quiescence may be in keeping with China's desire to distinguish itself as a different kind of power, the contrast between its profile and that of the US, Russia, and even European powers such as the UK, France, and Germany is stark and demonstrates the difficulty of marrying non-interference with geopolitical relevance. It also demonstrates that regional actors, whatever their complaints about American, European, or Russian policies, continue to turn to these powers for assistance and mediation rather than to Beijing, reflecting perhaps China's brief modern history in the region and a simple lack of familiarity and well-developed relationships.

Finally, China's preferred approach to tackling one of the region's most pressing problems—extremism—has proven wanting. As noted above, Beijing asserts that economic development, poverty reduction, and cultural exchanges are sufficient to prevent radicalization and discontent. However, copious research since the 9/11 attacks demonstrates conclusively that this is not the case. Extremism has multiple drivers, including political exclusion, corruption, abusive security forces, and ideological appeal—factors which Chinese initiatives avoid addressing, and to which they sometimes contribute in countries receiving Chinese aid and investment. As Kliman and Grace (2018) note in their report on the BRI, China—despite adopting an anti-bribery law in 2011—risks exacerbating corruption in the countries in which it operates, and its export and promotion of digital surveillance risks heightening a sense of injustice and oppression as it helps to solidify government control of local populations.

Changing Role of the United States

Under successive administrations, the United States has made clear that it views its engagement in the Middle East as excessive and often counterproductive, and desires, if not to withdraw, certainly to reduce the resources and effort devoted to the region relative to others. This Middle East fatigue has several causes—the desire to "rebalance" to Asia, a goal shared by Republicans and Democrats alike; a broad and growing sense that American interventions in the Middle East have paid scant strategic dividends at best; and even a view that the US has fewer national interests at stake in the region than in the past, largely due to the United States' increasing energy independence.

This desire to step back from the Middle East was first expressed by President Barack Obama, whose initial presidential campaign in 2008 focused heavily on criticism of the US-led war in Iraq. However, it has had its apotheosis under President Trump, who has spoken in characteristically blunt terms about the futility of US engagement in the Middle East, referring to Syria as "blood-stained sand" not worth fighting over, and the US invasion of Iraq in 2003 as "the worst decision ever made in the history of our country" (Trump 2019). Both Obama and Trump directed particular criticism toward China—Obama described Beijing as a "free rider" in Iraq, and Trump suggested that China and others "should be protecting their own ships" in the Gulf. President Trump has backed up his words with action: withdrawing some US forces from Syria and pushing to withdraw them from Afghanistan; acquiescing to a Turkish incursion into territories in northeast Syria occupied by US Kurdish allies; and declining to respond to Iranian attacks on internationally-flagged oil tankers and the Aramco oil processing facilities at Abqaiq and Khurais.

The American drive to step back from the Middle East has, unsurprisingly, unsettled and often angered US allies in the region. Since the Cold War, the United States has been the preponderant power in the region, and the Middle East's de facto security architecture has been characterized not by regional multilateral organizations or alliances, but by a central US role as security guarantor, and strong bilateral security relationships between the United States and allies who often lack meaningful ties with one another. The erratic nature of American engagement and rhetoric on the Middle East since the mid-2000s has meant that US disengagement has often been perceived as more thoroughgoing that it has actually been; if prior to 9/11 the US had a strong commitment to the region but a modest presence, the US posture is now the reverse, an uncertain commitment despite an enduring presence. This uncertainty is exacerbated by the fact that the US has refrained from articulating what a new, less resource- and effort-intensive Middle East strategy might look like.

American diffidence toward the Middle East has also presented Beijing with a conundrum. On one hand, China's approach to the region has long assumed, and tacitly depended upon, a US security umbrella. There are three reasons for this. First, while the United States became heavily involved in the Middle East after World War II in part to contest Soviet influence in the region, China's involvement there began during a period—the 1990s—when the United States was already far and away the region's dominant power. Second, China has long lacked the capacity to project diplomatic or military power to the Middle East, even had it desired to do so. Finally, China as noted above has sought to style itself as a different kind of great power, one which eschews the military interventions practiced by peers; this is far easier to pull off when someone else is shouldering the security burden, which if unmet would significantly increase the risks to Chinese interests in the region.

Despite this dependence on the US security umbrella in the Middle East, China has—like other states—been critical of American policies in the region as contributing to instability. As a revisionist power, Beijing also, as noted above, broadly aims to reduce American dominance of international affairs and increase its own role. Thus, while US disengagement from the region poses a serious challenge to China's approach to the Middle East, it also opens an opportunity for China—albeit one it may not want—or be able to afford—to take. This conundrum is reflected in the positions Chinese officials and analysts have taken on US policies in the region. For example, China strongly opposed the US intervention in Iraq in 2003, yet also later opposed the withdrawal of American forces (Singh 2016).

China's Middle East Strategy

A Chinese strategy in the Middle East must be inferred, as China, like the United States, tends to issue aspirational documents rather than propounding on its regional strategies. These statements, such as the "Five Principles of Peaceful Coexistence," the "Arab Policy Paper" with its "1 + 2 + 3" framework, and so on, are useful for understanding China's approach to the region, but do not together constitute anything approaching a strategy. At the core of China's implicit strategy for the region are two realities: first, that a state's foreign policy, as Chinese ambassador to the United States Cui Tiankai acknowledged in 2015, inevitably follows its investment; in other words, where merchants and traders go, diplomats and soldiers follow (Tiankai 2015). Second, while China is a revisionist power and interested in reshaping global affairs to enhance its own role and diminish that of the United States, the Middle East is not Beijing's preferred arena to wage this struggle—it is not looking for a fight with

Washington in the region, and likely assesses that the reverse is also true, at least for now.

The result is a strategy which seeks to build on what Beijing deems to have worked for it so far in the Middle East—practicing "non-interference," remaining friends to all and avoiding taking sides in regional disputes, and looking to reap economic and political benefits where they are available—while incrementally expanding its regional role and challenge to the United States. Several distinct elements of this strategy can be identified from China's recent policies.

Lead with Economic Benefits

China's economic relationships in the region are fundamentally driven by mutual need—China needs energy and uses for excess capacity and capital, and states in the Middle East need investment, infrastructure, and low-cost goods that the region's under-developed private sector do not provide. These economic relationships provide most states in the region with a clear stake in partnership with Beijing, and, far more than any geopolitical motivation, mean that regional states would resist being forced to choose between the United States and China if it came to that.

However, China's economic entanglement in the Middle East also pays other strategic dividends—it provides a form of leverage that China has not hesitated to use in other contexts, for example in withholding rare earth exports to Japan in 2010 and barring fruit imports from the Philippines in 2012. China has also used tourism as a political tool in its relations with Taiwan, and Mordechai Chaziza has argued that Beijing has also done so with regard to Turkey and Israel (Chaziza 2019). Similarly, during a 2017 summit in Beijing with Turkish President Recep Tayyip Erdogan, Xi Jinping was quoted by China's Foreign Ministry as telling his counterpart, "In order to promote even greater development of relations, China and Turkey must respect and give consideration to each other's core concerns, and deepen security and counter-terrorism cooperation." This was widely viewed as Xi telling Erdogan that economic benefits would be based on better Turkish cooperation on the Uighur issue (Reuters 2017).

China's economic focus in the region carries other potential benefits. Kliman and Grace have argued, for example, that Chinese investments will give Beijing leverage in instances where projects are debt-financed, will reshape the commercial playing field in favor of Chinese firms and against those from the West, and enhance Chinese influence over international commercial standards. In addition, by inviting third-country banks to participating in the financing of Chinese-led projects, Beijing's economic risks can be shared, and other states can come to acquire a stake in its success in the region (Kliman and Grace 2018).

Everyone's Friend, no One's Ally

In the early years after the establishment of the People's Republic of China, Beijing acted in sympathy with anti-colonial or "liberation" movements, for example by cultivating ties with Nasser's Egypt in 1956 and supporting Yasser Arafat's Palestine Liberation Organization (PLO). In the ensuing years, however, China's approach to the Middle East went from ideological to outwardly cautious and conventional. In 2017, for example, China proposed a "four-point plan" for Israeli-Palestinian peace (Lederer 2017):

(1) Advancing the two-state solution based on 1967 borders with East Jerusalem as the capital of a new Palestinian state;
(2) Upholding "the concept of common, comprehensive, cooperative and sustainable security," immediately ending Israeli settlement building, taking immediate measures to prevent violence against civilians, and calling for an early resumption of peace talks;
(3) Coordinating international efforts to put forward "peace-promoting measures that entail joint participation at an early date;"
(4) Promoting peace through development and cooperation between the Palestinians and Israel.

As a plan, it seemed designed more to avoid offense than take bold stands and, more than anything, to broadcast China's aspiration to a more prominent status in the region. In doing so, it followed a well-established pattern for Beijing—seeking to enjoy the benefits of friendship with all, without the burdens of alliance to any of the region's players. Jonathan Fulton has described China's approach as "fence-sitting"—that is, avoiding entanglement in regional conflicts and competitions while formalizing relationships that highlight mutual benefit and minimize commitment. China has signed "strategic partnership" or somewhat enhanced "comprehensive strategic partnership" agreements with Algeria, Djibouti, Egypt, Iran, Iraq, Jordan, Kuwait, Morocco, Oman, Qatar, Saudi Arabia, Turkey, and the UAE since 2014.

While it would be easy to dismiss such vehicles as diplomatic niceties, Fulton argues that China uses them with purpose. First, they enable China to explicitly demonstrate balance in its regional ties; for example, Xi Jinping in his first visit to the region initially stopped in Riyadh to sign China's partnership agreement with Saudi Arabia, then immediately flew to Tehran to sign a comparable agreement with Iran. Second, the accords nevertheless broadcast Chinese priorities in the region—Fulton notes that Beijing hands out few "comprehensive strategic partnership agreements," which in the Middle East it has concluded with Egypt, Iran, Saudi Arabia, and the UAE. When Xi flew to Abu Dhabi in 2018, he upgraded China's relationship with the UAE to a comprehensive strategic partnership; when Qatar's Emir Tamim visited Beijing in 2019, no such upgrade was awarded (Fulton 2020).

China has counted implicitly on regional counterparts' desire for the economic benefits of good relations with China and, frankly, on low expectations to maintain this approach to a fractious region. Yet analysts debate whether Beijing will be able to sustain it. As James Dorsey notes, Xi's visit to Riyadh in 2016, though outwardly warm and successful, was in fact characterized by serious differences over Syria, Iran, and Islamism (Dorsey 2019a; b). In a similar vein, then-Arab League Secretary General Amre Moussa publicly rebuked China in 2010 when Beijing refused to endorse a League resolution declaring East Jerusalem the Palestinian capital (Chen 2011). Dorsey argues that it is only a matter of time before the contradiction between China's growing interests in the Middle East and its effort to remain neutral in the region's conflicts becomes unmanageable, either because the gradual exit of the United States forces China to step up, or because a conflict inimical to Beijing's interests—say, between Israel and Iran—finally arrives (Dorsey 2019a; b).

Quietly Seek Strategic Gains

China's avowed policy of non-interference in the Middle East can lead the casual observer to see Beijing as lacking anything beyond an economic interest—paired, perhaps, with a general desire for prestige—in the region, or even as lacking in confidence in its dealings there. However, the reality is more complex: behind the veneer of this relatively innocuous approach, Beijing quietly targets and pursues strategic gains in the region with determination.

Two clear examples of this pursuit of strategic gains are China's keen interest in infrastructure projects that further its BRI objectives, as well as the role technology plays in its regional relations. As to the former, China has energetically invested in the construction of ports in key locations—such as Duqm on Oman's Arabian Sea coast, Jizan port in Saudi Arabia, Khalifa port in the UAE, Ain Sokhna in Djibouti, and Ashdod on Israel's Mediterranean coast—industrial parks and other projects in proximity to those ports, and rail lines, such as one from Western China to Iran initiated by Xi Jinping on his 2016 visit there or another envisioned from Ashdod to Eilat, and oil and gas pipelines from the UAE, to Iraq, to Iran. Such projects predate the BRI, which served more as a rebranding and rationalization of such activities than a new approach to the region. On the surface such projects are both economically beneficial to the region and consistent with US objectives such as regional stability and economic development; yet on another level, they connect to similar Chinese-led projects further east, such as ports in Pakistan and Southeast Asia, that taken together boost Beijing's ability to project power westward and could serve in the future as a foundation for a more robust Chinese presence.

Regarding technology, Efron et al. (2020) have noted that China's high-technology investments in Israel tend to focus on areas such as artificial intelligence, robotics, and information technology, which are also areas of focus for Beijing's "Made in China 2025" plan. The plan's aim is to bolster China's role as a high-tech manufacturing leader by fostering indigenous innovation, in part through technology transfer facilitated by foreign investment. Efron et al. (2020) cites a Mercator Institute report in stating that "Chinese companies are using overseas investments to speed up China's technological catch-up and to leapfrog stages of technological development," which, although not problematic by itself, is objectionable because this activity is 'partly supported and guided by the state' as part of 'an overarching political' program that aims to 'generate large-scale technology transfer.'" China's focus is not on Israel only; Beijing has taken the same approach to Silicon Valley in the United States, prompting the US government to tighten rules on foreign investment to limit China's potential gains, and has sought to promote Chinese telecommunications providers in the Middle East as it has in other regions (Northam 2018).

One of the clearest ways in which China has sought strategic gains in the region, however, is its cultivation of relations with Iran. Superficially, Beijing has balanced its relations with Iran with those it enjoys with Tehran's regional rivals, notably Saudi Arabia—Xi was careful to visit both on his first regional trip, and accorded to both the same level of partnership. Yet there can be little doubt that Iran plays a more central role in China's regional policy than any other state; RAND Corp. has identified Iran as the "pivotal state" for China within the Middle East, in the same category as Venezuela and Pakistan, for example (Scobell and Nader 2016).

While Iran and China enjoy ancient historical and civilizational links which retain more relevance today than perhaps appreciated by Western policymakers, the drivers behind the Iran-China relationship are more contemporary. Iran offers China an appealing combination of geopolitical affinity, insofar as both are revisionist states looking to challenge American predominance in international affairs, and Iran is indeed the only state on the vital Gulf littoral not friendly to the United States; geographic proximity, especially insofar as Iran presents the possibility of providing China with energy imports via overland routes; and economic appeal, given that Western firms tend to steer clear of the Iranian market for fear of political risk.

The China-Iran relationship has netted important gains for Tehran, as Beijing has assisted Iran's military development—particularly its nuclear, missile, and satellite programs—and has used its permanent seat on the UN Security Council to dilute and delay sanctions targeting Iran. For China, Iran is something of a long-term bet; Beijing infrequently touts its relationship with Iran but is only willing to stick its neck out so far for it—declining, for example, to flout US sanctions in recent years—but appears to be counting on the relationship to pay dividends over time as Iran emerges from international isolation and China's Mideast role expands. Scholar of Sino-Iranian relations John Garver describes China as seeking "a strategic partnership with Iran that will serve as a major element of Chinese influence in the post-American-dominated West Asia…without antagonizing the United States upon which China's modernization depends…a difficult task requiring considerable diplomatic skill." Garver suggests that China seeks to "manage the contradictions" between the US and Iran to its own advantage (Garver 2016).

Incrementally Boost Chinese Role, Challenge US Role

China is not looking for a fight with the United States in the Middle East; it is cognizant that it lacks the capacity, either in resources or partnerships, to prevail, and equally aware that the United States has considerable leverage it could employ against Beijing in the region, such as influence over oil flows. Moreover, China has benefited enormously from the US-led order in the Middle East as noted above, and does not appear to have yet reached a conclusion that it is no longer beneficial to Beijing's interests on balance, and indeed may never do so. Yet these regional considerations cannot be divorced entirely from China's global aspirations, which demand that it work to present itself internationally and domestically as a great power, and challenge American dominance of global affairs. It does so in the Middle East, albeit incrementally and relatively quietly.

In its practice of diplomacy in the region, China has preferred to create or promote structures that minimize the American role or exclude it entirely, and encourage regional partners to engage with them. Examples of this abound. Chinese scholar Sun (2019) has characterized this as a "whole-of-region diplomacy" that aims to strengthen south-south relations and bolster the role of developing countries in international relations. Sun describes this diplomatic engagement as following three patterns: first, the creation of bilateral mechanisms such as the China-Arab States Cooperation Forum and the China-GCC Strategic Dialogue, which he describes as "1 + N" diplomacy; second, engagement with existing regional mechanisms such as the Organization of the Islamic Conference and the Union of the Arab Maghreb, which he describes as "N + 1" diplomacy; and third, the inclusion of regional states in multilateral organizations that China dominates, such as the Shanghai Cooperation Organization, Conference on Interaction and Confidence-Building Measures in Asia, and the Asian Infrastructure Investment Bank. Beijing supplements these formal mechanisms with increasingly energetic, if not fruitful, diplomatic initiatives in the region. It has proposed an Israeli-Palestinian peace initiative, appointed envoys for Syria and "Middle East issues," and has hosted numerous regional delegations in Beijing in an effort to position itself as an international mediator in the style of the United States, EU, and Russia.

China has also used its permanent seat on the UN Security Council to advance Middle East aims in unprecedented ways in recent years. While China has used its veto power less frequently than any other member of the P5, its use of the veto has accelerated in recent years, with eleven of China's fourteen vetoes (excluding one cast prior to 1971, when the Chinese UN Security Council seat was held by the Republic of China) coming since 2000. Prior to 2011, China had never vetoed a resolution related to the Middle East, despite the region's over-representation in UN business; since 2011, eight of China's nine vetoes have concerned Syria. This development may reflect, in part, China's burgeoning partnership with a fellow revisionist power, Russia. However, China has taken pains—though few in the region have noticed—to distinguish itself from Russia by abstaining on numerous occasions on which Russia has vetoed Syria-related resolutions. It may also reflect greater determination to use the tools at Beijing's disposal to frustrate American aims, and together with Moscow to defeat what China sees as a Western-led regime change effort in Syria (Calabrese 2019).

In addition to traditional diplomacy, China has increasingly employed soft power, including information operations, in the Middle East, as it has done around the world. Most observers point to language, education, and tourism as China's major tools of soft power in the Middle East, and with ample reason. China has set up Confucius Institutes in Bahrain, Egypt, Iran, Israel, Jordan, Lebanon, Turkey, and the UAE, and has made a conscious effort to encourage regional governments to promote Chinese-language instruction in their countries and send students to China for exchange visits. As Chinese tourism has grown, as detailed above, China has required hotels, for example in the UAE, to train staff in Mandarin and provide Chinese-language newspapers and television channels in order to qualify as "China Ready" (Fulton 2020). China has also engaged in disinformation campaigns in the Middle East targeting the United States, including for example an Arabic-language video produced by the state-run China Global Television Network alleging an American origin of the COVID-19 pandemic (Wong et al. 2020). The Alliance for Securing Democracy has also noted the predilection of Chinese official media for amplifying Iranian government-produced propaganda (Brandt and Schafer 2020).

Finally, China has been boosting its military presence in the region, focusing on activities that demonstrate presence and capability that also enhance Beijing's ability to protect its citizens and investments in the region. These entered their present, more energetic phase in 2008, when the PLA Navy deployed vessels to conduct counter-piracy missions in the region and protect commercial vessels. This deployment enabled numerous port calls; indeed, the PLA Navy conducted more port visits, 45, in the Middle East than in any other region from 2009 to 2015 (Scobell et al. 2018). The port calls are just one example of China's military diplomacy; Chinese fighter jets have landed for refueling in Turkey and Iran, and China has dispatched and received dozens of high-level military delegations with the region. In 2017, perhaps as a result of the PLA Navy's need to evacuate Chinese nationals from Libya and Yemen during the Arab Spring, China opened its first naval base in the region in 2017, in Djibouti.

While these military activities have been relatively routine, even innocuous, in comparison with the activities of other external powers, they nevertheless appear designed in part to challenge the United States. For example, in December 2019, the PLA Navy joined the Russian and Iranian navies in joint exercises in the Gulf of Oman, at a time the United States was seeking to recruit partners to join in an anti-Iran maritime security coalition in the Gulf (which Chinese officials indicated they might join in August 2019, illustrating the mixed approach Beijing takes to the US in the region). Beijing downplayed the event, describing it as a "normal military exchange…not necessarily connected with the regional situation," but it was the first exercise of its kind whose political symbolism exceeded its military significance (Reuters 2019). China has also reportedly engaged in intelligence cooperation with the Syrian regime and may even have deployed Special Forces personnel to Tartous, enabled by a 2015 law authorizing overseas counterterrorism operations, inching toward involvement in a conflict which would pit it against the United States and Europe (Clarke and Saltskog 2019). In Djibouti, the US Defense Department in 2018 accused Chinese nationals of targeting American military aircraft with lasers, a charge that the Chinese government denied (Reuters 2018a; b).

The Way Forward

The overall trajectory of Chinese policy toward the Middle East seems certain—Beijing is gradually investing greater economic, diplomatic, and military resources there. At present, China would appear to devote proportionately few resources to the region given its strategic importance to Beijing; this imbalance will inevitably be rectified over time. The precise policies China pursues in the Middle East will be the result of the interaction of the three lenses described above—its interests, the Belt and Road Initiative, and the US-China relationship, as well as other factors such as China's economic fortunes and its domestic politics.

But perhaps even more, Chinese policy—as well as that of other outside actors—will be profoundly affected by the United States' own strategy, both toward the Middle East and toward China. Two fundamental questions are especially important: First, will the US pursue a zero-sum approach toward China, or do policymakers view some level of cooperation with Beijing as necessary or desirable? Second, what will the US strategy in the Middle East be as it shifts toward a focus on "Great Power Competition"? The United States starts with an overwhelming preponderance in the Middle East, which both attracts criticism yet also to a great extent crowds out powers like Russia and China. Going forward, Washington faces a spectrum of choices—whether it wishes to, as it did during the Cold War, define as a vital interest excluding its rivals from the region; whether it wishes to seek to cooperate with those rivals in the Middle East, whether to diffuse tensions or to share burdens; or even whether it wishes to force those rivals into greater commitments in the Middle East, forcing them to expend resources on peripheral matters, as many advocates of great-power competition strategy believe the United States has done for the past two decades.

If the United States chooses to continue competing for influence in the Middle East, it will face two difficult strategic choices. First, Washington must choose whether to continue its longstanding efforts to encourage economic and political reform, or to downplay or discontinue these efforts in order to avoid pushing authoritarian allies toward less demanding partners like Moscow and Beijing. The temptation to adopt a "go-along-to-get-along" approach is strong, both because it is easier, and because the US track record in promoting reform is poor. However, doing so may mean tolerating systems characterized by the maladies—for example, corruption and abusive security forces—that create opportunities for the very adversaries the United States seek to counter, whether great-power rivals or extremists.

Second, the United States will need to choose how it asks allies to relate to great-power rivals in the Middle East, especially China—whether, that is, to seek to force partners to choose between the United States and China, as the Trump administration does in its Africa strategy, or to accept their increasing ties with China as not necessarily incompatible with their partnerships with the United States. The choice is not straightforward—while partners will resist making a choice, they also value the benefits that come with their affiliation with the United States and may be wary of Beijing's—or for that matter Moscow's—suitability as an alternative.

There are, of course, many options on the spectrum between these two extremes. The Center for a New American Security issued numerous recommendations for the United States to respond to China's BRI, focused on coordinating with allies on matters such as defense, commerce, and technology, and addressing the problems such as corruption or vacuums such as infrastructure financing which create opportunities for China (Kliman and Grace 2018). In conjunction with this approach, the United States could choose to pursue cooperation with Beijing in the Middle East. Yet Washington should keep its expectations of such efforts low; those cooperative efforts that have succeeded are either modest, such as joint counter-piracy patrols, or over-hyped, such as cooperation on the Iran nuclear deal, which only materialized when the United States decided to make major concessions to the Iranian and Chinese positions. More robust cooperation would not necessarily require greater trust between Washington and Beijing—indeed, joint efforts in the region could be a mechanism for developing greater confidence in the bilateral relationship—but certainly would require a decision by Beijing to prioritize such confidence-building over the competing desire to undermine the United States.

Should the United States choose not to compete for influence in the region, it should not assume that the result will be a Chinese quagmire in the region. Such a decision would almost certainly force China to develop—or accelerate the development of—power-projection capabilities similar to those for which they currently depend on the United States. These capabilities would, it must be noted, be transferrable to other regions—just like those of the United States—not necessarily trapped in the Middle East. And Chinese moves to secure their own energy security would necessarily also give Beijing more control over that of others, including primarily its neighbors in East Asia who also depend heavily on Middle Eastern energy imports. But it seems highly unlikely that China would replicate American strategy in the region, which is the legacy of a Cold War competition with the Soviet Union that was both geopolitical and ideological. Instead, China would be more apt to leave the region to its own devices, narrowly defining Chinese interests and working where possible through what it regarded as capable regional partners, chiefly Iran. The United States should also avoid the mistaken belief that Chinese conduct in the Middle East will not have implications for the rest of the world—Washington and its democratic allies may not wish or need to contest every piece of territory in the world, but commercial and diplomatic standards cannot be easily defended in one place if they are forsaken in another.

Perhaps the least advantageous approach for the United States, however, would be to avoid answering these key strategic questions about the Middle East and China. Continuing the status quo—characterized by a large US presence but uncertain US commitment—would be a high-cost, high-risk, low-reward approach. The perception that the United States is hesitating leads partners to hedge—accelerating, perhaps, the development of their relations with Moscow and Beijing—and encourages adversaries to believe that the US is susceptible to being pushed out, placing US personnel at risk. It also seems likely to lead China to decide that it needs to step up its own diplomatic and military involvement in the region, increasing the chances of increased US-Chinese tensions there should Washington decide it does not wish to depart after all. The choice to compete with China in the Middle East or not is largely in Washington's hands; deferring or avoiding that choice means, in reality, delegating it to Beijing.

References

American Enterprise Institute, China global investment tracker. AEI.Org (2019). https://www.aei.org/china-global-investment-tracker/. Accessed 18 Apr 2020

Associated Press, Paulson: no quick fix for oil prices. *The New York Times* (2008). https://www.nytimes.com/2008/06/02/business/02oil.html. Accessed 28 Apr 2020

J. Brandt, B. Schafer, Five things to know about Beijing's disinformation approach. Securing Democracy (2020). https://securingdemocracy.gmfus.org/five-things-to-know-about-beijings-disinformation-approach/. Accessed 24 Apr 2020

British Petroleum, Energy outlook 2019 (2019). https://www.bp.com/en/global/corporate/energy-economics/energy-outlook.html. Accessed 28 Apr 2020

J. Calabrese, Chine and Syria: in war and reconstruction. Middle East Institute (2019). https://www.mei.edu/publications/china-and-syria-war-and-reconstruction. Accessed 30 Apr 2020

M. Chaziza, China in the Middle East: tourism as a steath weapon. Middle East Q. **26**(4), 1–8 (2019)

J. Chen, *The Emergence of China in the Middle East* (National Defense University Press, Washington, DC, 2011).

C. Clarke, M. Saltskog, The U.S. withdrawal from Syria is an opportunity for China. Lawfare (2019). https://www.lawfareblog.com/us-withdrawal-syria-opportunity-china. Accessed 27 Apr 2020

H. Clinton, America's Pacific century. *Foreign Policy* (2011). https://foreignpolicy.com/2011/10/11/americas-pacific-century/

J. Dorsey, *China and the Middle East* (Palgrave McMillan, Cham, 2019a)

J. Dorsey, It's when not if China's Middle Eastern tightrope snaps—analysis. Eurasia Rev. (2019b). https://www.eurasiareview.com/12112019-its-when-not-if-chinas-middle-eastern-tightrope-snaps-analysis/. Accessed 23 Apr 2020

S. Efron, K. Schwindt, E. Haskel, *Chinese Investment in Israeli Technology and Infrastructure* (RAND Corp, Santa Monica, 2020).

Energy Information Administration, China surpassed the United States as the world's largest crude oil importer in 2017 (2018). https://www.eia.gov/todayinenergy/detail.php?id=34812. Accessed 28 Apr 2020

Energy Information Administration, The Bab-el Mandeb Strait is a strategic route for oil and natural gas shipments (2019a). https://www.eia.gov/todayinenergy/detail.php?id=41073. Accessed 20 March 2020

Energy Information Administration, The Strait of Hormuz is the world's most important oil transit chokepoint (2019b). https://www.eia.gov/todayinenergy/detail.php?id=39932. Accessed 20 March 2020

Energy Information Administration, Despite the U.S. becoming a net petroleum exporter, most regions are still net importers (2020a). https://www.eia.gov/todayinenergy/detail.php?id=42735#. Accessed 28 Apr 2020

Energy Information Administration, What countries are the top producers and consumers of oil? (2020b). https://www.eia.gov/tools/faqs/faq.php?id=709&t=6. Accessed 28 Apr 2020

B. Faucon, S. Said, T. Puko, U.S., Saudi Arabia, Russia lead pact for record cuts in oil output. Wall Street J. (2020). https://www.wsj.com/articles/opecallies-look-to-resolve-saudi-mexico-standoff-and-seal-broader-oil-deal-11586695794. Accessed 07 Jan 2020

J. Fulton, China's challenge to US dominance, in *China's Great Game in the Middle East*, ed. by C. Lons (European Countil on Foreign Relations, 2020)

J. Garver, China and Iran: expanding cooperation, in *Toward Well-Oiled Relations*, ed. by N. Horesch (Palgrave-MacMillan, Basingstoke, 2016)

N. Halligan, Huawei remains strong in the Middle East despite US restrictions. Arabian Business (2019). https://www.arabianbusiness.com/technology/428314-huawei-remains-strong-in-middle-east-despite-us-restrictions

E. Hokayem, They've come a long way in 60 years, and so have we. The National (2009)

International Gas Union, 2019 World LNG report (2019). https://www.igu.org/app/uploads-wp/2019/06/IGU-Annual-Report-2019_23.pdf

International Monetary Fund, Direction of trade statistics. IMF Data (2020). https://data.imf.org/?sk=9D6028D4-F14A-464C-A2F2-59B2CD424B85

M. Kamrava, The China model and the Middle East, in *The Red Star and the Crescent*, ed. by J. Reardon-Anderson (Oxford University Press, New York, 2018)

D. Kliman, A. Grace, *Power Play: Addressing China's Belt and Road Strategy* (Center for a New American Security, Washington, DC, 2018).

E. M. Lederer, China urges support for new Israel-Palestinian peace plan. The Associated Press (2017). https://apnews.com/article/8eb70c17ecfd4621a40467999a842d20. Accessed 11 Jan 2020

C. Li, *Remarks at the Manama Dialogue* (Manama, IISS, 2019).

Ministry of Foreign Affairs of the People's Republic of China, The five principles of peaceful coexistence (2014). https://www.fmprc.gov.cn/mfa_eng/wjb_663304/zwjg_665342/zwbd_665378/t1179045.shtml. Accessed 22 Apr 2020

A. Molavi, China's global investments are declining everywhere except for one region. Foreign Policy (2019). https://foreignpolicy.com/2019/05/16/chinas-global-investments-are-declining-everywhere-except-for-one-region/

A. Nader, A. Scobell, *China in the Middle East: The Wary Dragon* (RAND Corporation, Santa Monica, 2016).

J. Northam, China makes a big play in Silicon Valley. NPR (2018). https://www.npr.org/2018/10/07/654339389/china-makes-a-big-play-in-silicon-valley. Accessed 3 Apr 2020

Reuters, China's Xi calls for greater counter-terrorism cooperation with Turkey (2017). https://www.reuters.com/article/us-china-silkroad-turkey/chinas-xi-calls-for-greater-counter-terrorism-cooperation-with-turkey-idUSKBN18A01D. Accessed 24 Apr 2020

Reuters, China denies U.S. accusation of lasers pointed at planes in Djibouti (2018a). https://www.reuters.com/article/us-usa-china-djibouti/china-denies-u-s-accusation-of-lasers-pointed-at-planes-in-djibouti-idUSKBN1I429M. Accessed 23 Apr 2020

Reuters, China envoy says no accurate figure on Uygurs fighting in Syria. *South China Morning Post* (2018b). https://www.scmp.com/news/china/diplomacy-defence/article/2160465/china-envoy-says-no-accurate-figure-uygurs-fighting. Accessed 21 March 2020

Reuters, Russia, China, Iran start joint naval drills in Indian Ocean (2019). https://www.reuters.com/article/us-iran-military-russia-china/russia-china-iran-start-joint-naval-drills-in-indian-ocean-idUSKBN1YV0IB. Accessed 24 Apr 2020

N. Rolland, China's Eurasian century? National Bureau of Asian Research (2017). https://www.nbr.org/wp-content/uploads/pdfs/programs/chinas_eurasian_century_ch3.pdf

A. Scobell, B. Lin, H. Shatz, M. Johnson, *At the Dawn of Belt and Road* (RAND Corporation, Santa Monica, 2018).

A. Scobell, A. Nader, *China in the Middle East: The Wary Dragon* (RAND Corporation, Santa Monica, 2016).

M. Singh, Chinese policy in the Middle East in the wake of the Arab uprisings, in *Toward well-oiled relations?*, ed. by N. Horesh (Palgrave MacMillan, Basingstoke, 2016)

State Council of the People's Republic of China, China's Arab policy Paper (2016). https://english.www.gov.cn/archive/publications/2016/01/13/content_281475271412746.htm. Accessed 22 Apr 2020

Stockholm International Peace Research Institute, SIRPRI arms transfer database (2020). https://www.sipri.org/databases/armstransfers

D. Sun, China's whole-of-region diplomacy in the Middle East. China Q. Int. Strateg. Stud. **5**(1), 49–64 (2019)

The White House, A new national security strategy for a new era (2017). https://www.whitehouse.gov/articles/new-national-security-strategy-new-era/. Accessed 23 Apr 2020

C. Tiankai, US-China CEO bilateral investment dialogue [Interview] (2015). Accessed 12 Feb 2015

D. J. Trump, Trump on Syria: 'Let someone else fight over this long blood-stained sand'. NBC News. (2019). https://www.nbcnews.com/now/video/trump-onsyria-let-someone-else-fight-over-this-long-blood-stained-sand-71910469988. Accessed 07 Jan 2020

UN World Tourism Organization, Yearbook of tourism statistics, data 2014–2018 (2020). https://www.e-unwto.org/doi/book/10.18111/9789284421442

UPI, China opens missile plant in Iran (2010). https://www.upi.com/Defense-News/2010/04/23/China-opens-missile-plant-in-Iran/82791272037022/. Accessed 23 March 2020

US Department of Defense, What is the National Defense Strategy (2018). https://www.defense.gov/Explore/Features/story/Article/1656414/what-is-the-national-defense-strategy/. Accessed 23 Apr 2020

E. Wong, M. Rosenberg, J. Barnes, Chinese agents helped spread messages that sowed virus panic in U.S., officials say. *The New York Times* (2020). https://www.nytimes.com/2020/04/22/us/politics/coronavirus-china-disinformation.html

J. Wuthnow, *Chinese Perspectives on the Belt and Road Initiative* (National Defense University Press, Washington, DC, 2017).

How US Policy Toward Iran Has Undermined US Interests in the Middle East

Barbara Slavin

Abstract US foreign policy for the past three decades—with a few notable exceptions—has failed to achieve a less hostile relationship with the Islamic Republic of Iran that could have avoided a debacle in Iraq, prevented or mitigated other regional conflicts, shored up the transatlantic alliance and allowed the United States to manage a long-sought "pivot" to Asia to address a rising China. Many observers insist that Iran's unique system—built on an Islamic theocracy and resistance to "US imperialism"—is not capable of fully reconciling with the US. But this paper will argue that there have been several opportunities to achieve a mutually beneficial détente and that the US decision to withdraw unilaterally in 2018 from the Joint Comprehensive Plan of Action (JCPOA) wrecked a chance at reaching that outcome while boosting Russian and Chinese power in the Middle East. The US exit from the JCPOA and the subsequent imposition of an oil embargo on Iran has backfired, leading to growing tensions that initially resulted in the US sending more troops to the region, rather than achieving the drawdown and greater burden sharing sought by successive US presidents (Gibbons-Neff, How U.S. Troops Are Preparing for the Worst in the Middle East. The New York Times 2020). The US now faces the prospect of being forced to withdraw from Iraq in a humiliating fashion. Meanwhile, overuse of economic sanctions threatens to undermine them as a tool of diplomacy and less risky substitute for kinetic action. The sanctions have also contributed to the rising death toll from the novel coronavirus pandemic by weakening economies and reducing the ability of governments such as Iran's to respond to this humanitarian crisis.

Keywords US · Policy · Middle East · US interests · Iran · China · Great power competition · JCPOA

B. Slavin (✉)
The Atlantic Council, 2929 28th street NW, Washington, DC 20008, USA
e-mail: barbaraslavin31@gmail.com

© The Author(s), under exclusive license to Springer Nature Switzerland AG 2021
A. Farhadi et al. (eds.), *The Great Power Competition Volume 1*,
https://doi.org/10.1007/978-3-030-64473-4_19

Background

The US and Iran, once close security partners, experienced a nasty divorce after the 1979 Islamic Revolution. The decisive blow was the decision by revolutionary leader Ayatollah Ruhollah Khomeini to condone and prolong the captivity of 52 Americans taken hostage in the US Embassy in Tehran on Nov. 4, 1979. Khomeini supported the student hostage holders in a successful effort to outflank leftist factions within his own revolutionary movement and to prevent the US from interfering in Iranian politics on the side of pro-Western politicians. The hostages were not released for 444 days, until Jimmy Carter had left the White House and Ronald Reagan became president.

The Reagan administration, worried that Iran would fall under Soviet influence, made covert overtures to Tehran that led to the 1986 Iran-Contra scandal in which it was revealed that the US had sold weapons to Iran and used the proceeds illegally to fund anti-Communist guerrillas in Central America. At the same time, however, the US provided crucial intelligence to Iraq that helped prevent it from being defeated in the 1980–88 Iran-Iraq war. In the last years of the war, the Reagan administration intervened more directly. It escorted Kuwaiti tankers down the Persian Gulf, which led to hostilities between the US Navy and Iran's Revolutionary Guard naval forces, several dozen casualties and severe damage to the Iranian side. The US also shot down an Iranian civilian airliner, killing all 290 on board, mistaking it for a hostile Iranian F-14. Iran accepted a cease-fire in the war shortly thereafter.

The first real opportunity for an improvement in ties came under Reagan's successor, George H. W. Bush, and followed the end of the Iran-Iraq war in 1988 and Khomeini's death in 1989. President Ali Akbar Hashemi Rafsanjani, who had engineered the rise of Ayatollah Ali Khamenei to succeed Khomeini as Supreme Leader, was initially the most powerful political figure in Iran. Rafsanjani, who had traveled across the United States before the revolution to visit a brother studying at the University of California at Berkeley, lacked Khomeini's knee-jerk anti-Americanism. The new president's top priority was reconstructing Iran following the Iran-Iraq war and he sought Western help. He was encouraged by President Bush's remark in his 1989 inaugural address that "goodwill begets goodwill"—a reference to Iranian assistance in freeing remaining Americans held captive by pro-Iranian Lebanese Shi'ite militants in Beirut.[1] Iran remained neutral in the US-led 1991 Gulf War, which expelled Iraqi invaders from Kuwait, and saw new benefit in reducing hostility toward the US during the unipolar period created by the collapse of the Soviet Union.

However, Iran also continued acts of terrorism in Europe—killing Kurdish and other dissidents—that confirmed its pariah reputation in the West. Israel, which had

[1] Inaugural Address of George Bush, January 20, 1989. "To the world, too, we offer new engagement and a renewed vow: We will stay strong to protect the peace. The 'offered hand' is a reluctant fist; but once made, strong, and can be used with great effect. There are today Americans who are held against their will in foreign lands, and Americans who are unaccounted for. Assistance can be shown here, and will be long remembered. Good will begets good will. Good faith can be a spiral that endlessly moves on." https://avalon.law.yale.edu/20th_century/bush.asp.

seen Saddam Hussein's Iraq as the greater threat prior to the 1991 Gulf War, began lobbying Washington against any improved ties with Tehran following Iraq's expulsion from Kuwait and subjection to severe UN sanctions. The Bush administration did not invite Iran to participate in a major post-war conference in Madrid on Middle East peace and abandoned efforts to improve ties with Tehran following several failed attempts (Slavin 2007). Iran, in turn, provided support to Palestinian militant groups that carried out terrorist bombings in Israel and sought to wreck chances for Israeli-Palestinian implementation of the 1993 Oslo accords.

The Clinton administration embraced a policy of "dual containment" of both Iraq and Iran despite major differences between the two countries' policies and systems of government. Under pressure from Israel and its US supporters, Clinton in 1995 barred a lucrative deal between Iran and the American oil company, Conoco, placed an embargo on most US trade with Iran and in 1996 signed into law the Iran-Libya Sanctions Act, the first US secondary sanctions measure that sought to penalize *foreign* companies investing in Iran's oil and gas sector. In possible retaliation, Iran-backed Saudi Shi'ite militants blew up an American barracks in Khobar Towers, Saudi Arabia, killing 19 Americans and wounding nearly 400.

A slight thaw occurred in Clinton's second term, after the election of Mohammad Khatami as Iran's president in 1997. Khatami, a mild-mannered intellectual and former director of Iran's national library, called for a "dialogue of civilizations" with the US and encouraged exchanges of athletes, academics and artists (Khatami 1998). Iran-linked attacks on Americans in the Middle East ceased. The Clinton administration eased sanctions on the export of food and medicine to Iran and other countries on the State Department's list of state sponsors of terrorism. Iran, in turn, was allowed to sell carpets, pistachios and caviar to the US.

Secretary of State Madeleine Albright, in a major speech in 2000, apologized for the US role in overthrowing Iranian Prime Minister Mohammed Mossadegh in 1953 and re-installing the Shah on the throne—an enduring source of Iranian grievance—and expressed regret that the US had sided with Saddam in the Iran-Iraq war (State 2000). Iran and the US participated in a UN-led process among Afghanistan's neighbors to deal with the consequences of the Taliban takeover of that country in 1996. However, Khatami faced domestic resistance to his reform initiatives and failed to reciprocate some US overtures.

Wasting the Unipolar Moment

The terrorist attacks of Sept. 11, 2001 presented the most opportune moment for US-Iran rapprochement after the 1979 revolution. The Iranian government and people expressed sympathy and solidarity with the US following the attacks—unlike other Muslim majority nations—and shared US antipathy toward the Sunni Muslim fundamentalists responsible. However, President George W. Bush adopted a policy of war on "all terrorism of global reach" including groups supported by Iran such as Hezbollah that were not involved in 9/11 (House 2009). His hawkish administration minimized cooperation with Iran in Afghanistan and Bush notably included Iran in an "Axis of Evil" with Saddam's Iraq and North Korea in Bush's 2002 State of the Union Address (House 2002).[2] Bush also rebuffed a 2003 offer from Iran, transmitted by the Swiss ambassador in Tehran, for talks on a "grand bargain" that could have addressed other issues of concern, including Iran's support for Palestinian and Lebanese militants.[3] His decision to invade Iraq without Iranian cooperation and to replace its Sunni dictatorship with a religious and ethnic-based system opened Iraq to deep Iranian penetration, encouraged the growth of Sunni fundamentalist militancy and upset the regional balance of power between Iran and Saudi Arabia. Iran, threatened with "regime change" by influential members of the Bush administration, sent back Iraqi fighters it had nurtured during the Iran-Iraq war and trained new militias to fight Iraqi Sunnis and American and allied forces. Iran provided improvised explosive devices to these militias that, according to the Pentagon, killed more than 600 American servicemen in Iraq from 2003–2011 (Bump 2020).

The administration of Barack Obama saw a chance to change the dysfunctional dynamic in which the US remained estranged from the most populous and well-educated country in the Middle East. Back channel talks on a resolution of a dispute with Iran over its advancing nuclear program bore fruit after the 2013 election of Iranian President Hassan Rouhani, a former national security adviser and nuclear negotiator. Equipped with a skilled diplomatic team led by US-educated Foreign Minister Javad Zarif, the Rouhani administration obtained the support of Supreme Leader Ayatollah Ali Khamenei for a historic compromise that traded verifiable, if time-limited, curbs on Iran's nuclear program for significant sanctions relief. While the JCPOA did not address other US differences with Iran, it provided a foundation for further high-level discussions.

Implementation of the JCPOA had barely begun when Donald J. Trump was elected US president in November 2016. Trump had campaigned on a promise to scrap the JCPOA and only grudgingly remained in the deal until May 2018. His decision to quit despite Iranian compliance and to try a year later to impose a total embargo on the export of Iranian oil incentivized Iran to undertake a series of aggressive actions in the Persian Gulf against US Arab allies and against Americans in Iraq and to begin to move outside the limits of the JCPOA in the nuclear arena. Domestically, hardline forces that had never embraced reconciliation with the West became even more powerful and cracked down brutally on Iranian protesters as economic conditions worsened (Fassihi and Specia 2019). Iran strengthened ties with Russia and China, pivoting firmly toward authoritarian powers that also seek to contain and if possible, supplant US influence in the Middle East and beyond.

[2]George W. Bush State of the Union Speech, January 29, 2002. ("States like these, and their terrorist allies, constitute an axis of evil, arming to threaten the peace of the world. By seeking weapons of mass destruction, these regimes pose a grave and growing danger. They could provide these arms to terrorists, giving them the means to match their hatred. They could attack our allies or attempt to blackmail the United States. In any of these cases, the price of indifference would be catastrophic.")

[3]Slavin, "Bitter Friends, Bosom Enemies," Appendix ("Iran's 2003 Offer to the United States).

Rising US-Iran Tensions in Iraq

In January 2020, the US and Iran came to the brink of war after rocket attacks, allegedly by Iran-backed Iraqi militias, killed an American contractor in Kirkuk and the US responded by attacking Iraqi militia bases, killing more than two dozen militiamen (Rubin et al. 2020). Members of these groups retaliated by besieging the US Embassy in Baghdad on New Year's Eve; even though the Iraqi government under caretaker prime minister Adel Abdul Mahdi succeeded in convincing the attackers to disperse, the Trump administration escalated the situation on January 3 by assassinating Qasem Soleimani, the long-time head of the Quds Force, in a drone attack outside Baghdad Airport, and the deputy head of Iraq's popular militia forces, Abu Mahdi al-Muhandis (Crowley et al. 2020). Iran then retaliated with missile strikes that injured 100 Americans on two bases in Iraq but did not result in any fatalities (Rubin et al. 2020). Iran, on high alert, accidentally shot down a Ukrainian airliner that had just taken off from Tehran, killing all 176 people on board. For three days, the Iranian government did not admit culpability (2020b).

The tragedy—which led to protests in Iran—as well as the killing of Soleimani, which had also sparked huge demonstrations of mourning and nationalism, paused US-Iran hostilities but did not end them. A new cycle erupted in March 2020, leading to two more American and a number of Iraqi deaths (Loveluck and Ryan 2020). Iran's leadership has vowed to expel the US from Iraq and the region at large and Iraqi politicians are struggling to reconcile the demands of their powerful neighbor for a US exit with their desire to keep ties to the US and defend Iraqi sovereignty (Fassihi and Hubbard 2020). As of this writing, many in the region are bracing for more attacks, the US has consolidated its forces in Iraq on two bases—in Irbil and outside Baghdad—and the long-term presence of US troops in Iraq to fight ISIS and keep a foothold in the country remains in jeopardy. The US and Iraq have agreed to negotiate a new Strategic Framework Agreement, replacing one reached in 2008, to set conditions for any continuing US military presence (Presse 2020).

In hindsight, successive US administrations, with the slight exception of the second Clinton presidency and Barack Obama's tenure, wasted a period of relative American global dominance in which the US could have narrowed the breach with Iran and focused on other priorities. Instead, American foreign policy practitioners in both the Republican and Democratic parties catered too heavily to the anti-Iran views of traditional US allies in the Middle East and their US supporters. A more balanced US policy, ironically, might have better served the long-term interests of Israel and of Iran's Arab rivals as well as the United States by diminishing Iran's motivation to act against them directly and through partners. Unfortunately, the US wound up encouraging aggressive behavior on the part of both Iran and its adversaries, further destabilizing the region and undermining US influence.

Cementing Regional Ties with Russia and China

US policies have strengthened powerful elements in Iran, Russia and China in their desire to create an alternative to the rules-based order crafted by the US and its allies in the aftermath of World War II. All three countries see themselves as important historic powers with a right to spheres of influence and resent what they regard as a US sense of entitlement to interfere at will in their backyards and around the globe.

The rise of Russian and Chinese influence is particularly apparent in the Middle East, where erratic US statements and abrupt policy shifts have convinced even long-time US allies that they cannot count on American protection. While Donald Trump can be assured of a warm reception in Israel, Saudi Arabia and the United Arab Emirates, Russian President Vladimir Putin and Chinese leader Xi Jinping can visit any country in the Middle East and expect a polite if not effusive welcome. The US has hostile relations with Iran, increasingly testy ties with Turkey and has become largely irrelevant to the resolution of the Syrian civil war, the conflict in Yemen, the civil war in Libya and even the Israeli-Palestinian dispute. Meanwhile, the US decision to quit the JCPOA and impose ever more draconian sanctions on Iran has put Iraq—a country into which the US has poured so many lives and so much treasure—in an increasingly difficult position trying to placate its powerful neighbor while maintaining security and economic ties with the US.

Russia and China have played a more skillful regional game—shoring up relations with Iran along with its regional rivals.

The improvement in ties between Iran and Russia is especially striking given historic Iranian resentment of Russia for swallowing up, in the eighteenth and nineteenth centuries, portions of what had long been the Persian Empire. To this day, the 1828 Treaty of Turkmenchay, under which Persia ceded control of what is now Armenia, Georgia and parts of Azerbaijan to Russia, is a metaphor for Iranian defeat and humiliation. After World War II, Russia's successor, the Soviet Union, also briefly occupied Iran's remaining part of Azerbaijan. The US helped work to expel the Soviets, earning Iranian goodwill. Much of that goodwill was squandered in 1953, when the CIA backed successful efforts to oust Prime Minister Mohammad Mossadegh and put the Shah back on the throne.

Ayatollah Khomeini, the leader of the 1979 revolution that removed the Shah, espoused a policy of distancing Iran from both big nuclear powers, popularizing the slogan of "Neither East, nor West." However, after Khomeini's death in 1989, Iran turned to Russia for weapons and civilian nuclear power expertise. Most dramatically, Iran collaborated with Moscow to keep the regime of Bashar al-Assad in power in Syria after the 2011 uprising there, going so far as to allow Russian planes to refuel at an Iranian airbase (Gibbons-Neff 2016). While the interests of Russia and Iran in Syria are not identical—Iran dislikes Russia's apparent acceptance of Israeli freedom of maneuver in Syria against Iranian targets—both are still working assiduously to help the Assad regime—with Russian air power and Iranian and Shi'ite militias trained by Iran and by Hezbollah—regain control over remaining Syrian territory held by Sunni fighters and Kurdish groups. Russia and Iran are part of the so-called

Astana process, along with Turkey, that seeks to exclude the United States from shaping a political solution to the Syria war.

Iran is looking forward to the lifting of a UN embargo on the sale of conventional arms to Iran in October 2020 to buy advanced fighter planes and missile defenses from Russia. Blocked by US sanctions from purchasing American civilian aircraft—a program to buy $18 billion worth of Boeings under the JCPOA collapsed after Trump came into office—Iran also plans on purchasing Russian Tu-204 passenger jets and aims to produce major components of the aircraft in Iran (Esfandiary and Tabatabai, 2019). The Trump administration failed to prevent the arms embargo from ending, given that the US was no longer a party to the JCPOA. Russia, which chaired the UN Security Council in October, clearly has a strong interest in seeing the embargo end. As permanent members of the Security Council, Russia and China have shielded Iran from further punishment by the international community for its missile program and regional activities and blocked US attempts to hold Iran and Russia to account for their intervention in Syria.

From Silk Road to Belt and Road

Iran's relations with China date to the Silk Road more than 100 years before the birth of Christ and in the modern era, to 1971, before the Iranian revolution. While the Shah's Iran turned largely to the US to arm his military, China provided weapons to Iran after the revolution during the Iran-Iraq war and also gave assistance to Iran's nuclear program in the 1990s.

China became Iran's major trading partner in the mid-2000s as US secondary sanctions frightened off European, Japanese and South Korean businesses and for a time, China was the largest purchaser of Iranian oil (Slavin 2011). While Beijing has cut back on trade and investment since the Trump administration withdrew from the JCPOA, it has never completely severed ties and has used its ability to create new companies not entangled with the US economy to continue to purchase oil and build infrastructure in Iran. China does not criticize Iran for its domestic repression, which mirrors its own, and showed solidarity with Iran after an outbreak of a deadly coronavirus in China spread to the Islamic Republic. Iran is envisioned as an important link in Beijing's Belt and Road initiative. At the same time, China has greatly increased trade and investment in Iran's Arab rivals, Saudi Arabia and the United Arab Emirates, including arms sales and joint military exercises (Gurol and Scita 2020).

In strategic terms, Iran has strengthened ties with both Russia and China. Iran has observer status in the Shanghai Cooperation Organization, which includes Russia and China. In December 2019, Iran, Russia and China conducted their first trilateral maritime military maneuvers in the northern part of the Indian Ocean (2019b). China bills itself as a force for stability in the region which avoids meddling in the internal affairs of other countries, while Russia's willingness to send its military and mercenaries to intervene in Middle Eastern civil wars suggest a wider ambition to augment and even supplant the US as a guarantor of regional security (Hinnant 2020). Neither China nor Russia yet has the blue water navy sufficient to assure freedom of navigation. But Beijing could well build on its economic ties to increase its naval presence in the region, which already includes a naval base in Djibouti.

Overuse of Sanctions Undermines their Efficacy

Sanctions have long been Washington's tool of choice in trying to counter and contain the Islamic Republic of Iran. Beginning in 1979 after the seizure of US diplomatic hostages, successive US administrations have imposed a variety of economic penalties, starting with asset freezes and arms embargos. What began as unilateral steps escalated under the Clinton administration, with the passage of the Iran-Libya Sanctions Act, to secondary sanctions and attempts to dissuade third countries from investing in Iran's oil and gas industry. These efforts grew under the Obama administration into a truly multilateral regime that included UN and European Union sanctions directed at Iran's oil exports and banks that were designed to pressure Iran to negotiate curbs on its nuclear program. Those sanctions were lifted as part of the JCPOA.

Under the Trump administration, the use of sanctions reached new—and counterproductive—heights. Withdrawing unilaterally from the JCPOA and re-imposing the sanctions lifted by the deal, the US sought to force other countries to penalize Tehran even when it remained in full compliance with the nuclear agreement. The Trump administration has gone even farther than the pre-JCPOA restrictions by refusing to grant waivers to third countries to continue to purchase limited amounts of Iranian oil.

US officials tout statistics showing that Iran's economy has severely contracted as proof that sanctions are working (Esfandiari 2020). However, Iran has become more aggressive regionally and more repressive at home, casting doubt on the efficacy of this tool if they are truly meant to alter Iranian policies and not simply at collective punishment. Ordinary Iranians have suffered greatly, both from inflation and shortages in imported medicines, while regime insiders appear to be doing just fine (Watch 2019). The sanctions look particularly punitive given the Covid-19 pandemic. Although the Trump administration notes that medicine and medical devices are not sanctioned, the reluctance of foreign banks to deal with Iran and the overall weakening of the Iranian economy as a result of sanctions made the country far less able to contend with an unprecedented medical emergency.

As of this writing, Iran has refused to return to negotiations with the US or other countries about its nuclear program or regional interventions and has announced that it will no longer observe limits on its nuclear activities (Sanger and Broad 2020). Iran is insisting that the US first offer sanctions relief before it will attend multilateral talks on nuclear or other issues. It has offered to return to full JCPOA compliance only if the incoming Biden administration does the same.

The Trump administration, sometimes egged on by Congress, has used sanctions against a variety of other countries including Russia, Venezuela, North Korea and China. These punishments are incentivizing global efforts to circumvent sanctions and threaten to create a coalition of the sanctioned as targets collaborate to continue trading with each other. Countries are increasingly using local currencies to avoid the dollar and resorting to barter arrangements. Russia, for example, has helped Venezuela evade US sanctions on its oil industry (Faiola and De Young 2020). North Korea is investing heavily in crypto-currencies and Iran is also exploring their increased use to circumvent the dollar and the Society for Worldwide Interbank Financial Telecommunication or SWIFT, which facilitates inter-bank transactions (Sanger 2020).

China, as pointed out earlier, has continued to purchase Iranian oil through a variety of means using local currencies and barter. Even Europe, despite its dependency on the US market, created a vehicle for continuing to trade with Iran called INSTEX (Instrument in Support of Trade Exchanges). While the mechanism has gotten off to a slow start, it is viewed by some European policymakers as a necessity to establish European economic sovereignty and to hedge against further US secondary sanctions directed at European trade with countries other than Iran. As of February 2020, nine European countries had become members of INSTEX (2019c). In late March 2020, INSTEX announced that it had carried out its first transaction with Iran, involving about a half million dollars worth of German medical goods related to blood treatment (Norman 2020).

Beyond such arrangements, Iran, which has been subjected to sanctions for much of its history, has perfected many means of smuggling oil and refined products across its borders and of masking the nationality and identity of oil tankers. According to TankerTrackers.com, Iranian oil exports doubled in January 2020 to 1 million barrels of oil a day, exported to East Asia and Syria (2020a). That exceeds the amount Iran is estimated to need to sell abroad to keep its economy afloat (Tavakol 2019). Exports dropped, however, due to reduced demand sparked by the Covid-19 crisis. Sanctions and the world economic crisis caused by the pandemic are also forcing Iran to diversify its economy and reduce reliance on oil earnings to fund its expenditures. Iran's budget for 2020 reduces dependency on the petroleum sector to less than 25 percent—the lowest it has been since before the revolution (Khajehpour 2019). The collapse of oil prices in the face of Covid-19 is likely to reduce this percentage even further.

A Policy of Strategic Confusion

As the Trump administration nears its end, US policy in the Middle East continues to suffer from confusion over goals and uncertainty over means. Is the aim a new and better JCPOA? Containing Iran? Regime change? The answer depends on who in the Trump administration is talking at any given period of time. This confusion contributes to Iranian paranoia and encourages Iranian aggression; it also incentivizes Iran's regional foes to hedge and look for security assistance from other quarters.

The Trump administration early on wholeheartedly embraced a right-wing Israeli government led by Prime Minister Benjamin Netanyahu and a Saudi government led by a headstrong young prince, Mohammed bin Salman (MBS). Crafted by Donald Trump's son-in-law, Jared Kushner, a diplomatic neophyte, the policy aimed at promoting an alliance between Israel and Saudi Arabia that would resolve the Palestinian issue in Israel's favor and weaken Iran's regional reach. Neither has occurred. MBS took a series of disastrous steps including backing a boycott of Qatar that fractured the Gulf Cooperation Council and pushed Qatar closer to Iran, briefly taking hostage the Lebanese prime minister, continuing a calamitous war in Yemen and ordering the brutal murder of a Saudi columnist for the Washington Post in one of Saudi Arabia's own consulates. He also initiated an oil price war with Russia at a particularly inopportune time, just as the Covid-19 pandemic was expanding exponentially in early 2020 and dramatically depressing demand. Although an agreement was reached in April 2020 to cut global production, the result of MBS's initial action was a collapse in oil prices that threatens Saudi aspirations to diversify their economy and provide jobs for millions of young Saudis (Krauss 2020; Blas and Pismennaya 2020).

Israel, meanwhile, pushed the Trump administration to move the US Embassy to Jerusalem, to recognize Israeli annexation of the Golan Heights and to accept Israeli settlements in the West Bank as "facts on the ground" (Jakes and Halbfinger 2019). However, Israel and the Palestinians are no closer to a peace agreement—Palestinians quickly rejected the "deal of the century" unveiled by President Trump in late January 2020—and many Israelis have lost faith in the US to contain Iranian expansion. Indeed, Israel stepped up operations in Syria and Iraq to target Iranian precision weapons and to establish a buffer zone north of the Golan Heights (Tibon and Landau 2020; Specia 2019).

One reason US allies feel insecure is that the Trump administration has shown itself extremely reluctant to risk American lives in support of regional friends. Trump abandoned Syrian Kurds after they helped the US defeat the ISIS caliphate, balked at the last minute at retaliating for the Iranian shoot-down of a US drone in June 2019, did not respond kinetically to a precision attack on Saudi Arabia's main oil facility three months later and gave conflicting commands over the presence of a small US military contingent in Syria. After numerous rockets lobbed at or near US personnel in Iraq, the US finally retaliated for the death of a US contractor with disproportionate strikes on an Iran-backed Iraqi militia late in 2019 that killed two dozen Iraqis. This led in turn to the attempted break-in of the US Embassy in Baghdad on New Year's Eve 2019 by an Iran-backed Iraqi militia, Kataib Hezbollah, and the US assassination by drone of Quds Force leader Soleimani, followed by Iranian missile strikes—without causing fatalities but leading to traumatic brain injury of 100 Americans—on two Iraqi bases where Americans were situated.

Afterwards, Trump asserted falsely that no Americans were hurt and that the Iranians were "standing down" in their confrontation with the US (2020c). Iran's Supreme Leader, however, made clear that Iran has no such intention and that its goal remains to expel US forces from Iraq and eventually the entire Middle East (Cunningham 2020). Tit-for-tat attacks resumed in March 2020 and there were worrisome reports at the time that hawks in the Trump administration were pushing for a massive attack on Iran-backed militias and even Iran proper—something that could lead to a wider war and push Iraqis to demand a complete US withdrawal (Mazzetti and Schmitt 2020). Indeed, the US has agreed to negotiate a new Strategic Framework Agreement with Iraq that could lead to such a withdrawal (2020e).

Tellingly, the US did not invoke Article 5 of the NATO treaty after the attacks in 2019 and 2020 on Americans in Iraq. This perhaps reflected recognition that US allies, including those involved in a NATO-led training and anti-ISIS mission in Iraq, might not want to become involved in a crisis instigated by the US escalation with Iran—as well as the result of NATO resentment at persistent Trump administration bashing of the alliance. Another sign of the damage Trump policy on Iran has done to transatlantic relations is the fact that more countries have signed on to a French plan to patrol the Persian Gulf than a US-led maritime coalition, Operation Sentinel (2019a).

The US could have enhanced security in the Middle East and managed a "pivot to Asia" with fewer negative consequences had it remained in and built on the JCPOA. By withdrawing unilaterally and seeking to block Iran from exporting any oil, the Trump administration squandered leverage to work for a more advantageous and broader agreement. Turning Iran, if not into an ally at least into a less hostile competitor, would have allowed the US to draw down its large deployments in the Middle East peacefully and in a dignified manner and regional allies to focus more on economic development than expensive arms purchases. It would also have facilitated continuing efforts to contain ISIS and other Sunni terrorist groups.

Is It Too Late to Repair the Damage?

With the US in the throes of both a pandemic and a presidential election year, it is unlikely that the Trump administration—or Iran—will show the necessary flexibility to change the trajectory of US-Iran relations. US officials continue to tout "maximum pressure" as a successful strategy despite its adverse consequences in the short and medium-term. There are, however, several steps the US could take to de-escalate tensions and improve the prospects for a less chaotic Middle East in 2021.

The US could emphasize to Iran through intermediaries that it seeks no wider conflict. It could embrace a French plan providing Iran with $15 billion in credits for future oil sales in return for Iran halting further steps out of the JCPOA. The US could also make clear to the Europeans that it will not sanction the Iranian counterpart to INSTEX or interfere in any other way with this mechanism as long as it is confined to humanitarian trade. It could approve other channels for the provision of food and medicine to Iranians. In that context, the inauguration in January 2020 of a Swiss channel for humanitarian trade was a good first step (2020d).

The incoming Biden administration could tone down hostile propaganda against Iran by the State Department, particularly in the Persian language, and confine its support for the Iranian people to general expressions upholding human rights, rather than comments that flirt with advocating regime change. The State Department could lift sanctions on Iranian Foreign Minister Zarif, Iran's chief diplomat, and ease restrictions on the movement of Iranian diplomats in New York. It could indicate a willingness to talk to Iranians within multilateral forums and stop seeking a photo op between presidents that is politically untenable for Iranian President Hassan Rouhani. The US should also renew efforts to free remaining Americans detained in Iran by offering to swap them for Iranians jailed in the US. Such swaps, beyond their inherent humanitarian benefit, can be important confidence building measures.

The administration should use the Covid-19 crisis to pause and if possible, suspend, sanctions on Iran that interfere with its ability to respond to this global health emergency. Merely repeating that US sanctions do not specifically forbid the supply of medicine and medical devices to Iran is disingenuous, given the damage the sanctions have done to the Iranian economy which left the country ill-prepared to deal with the virus.

The US should encourage its Arab friends to reduce tensions with Iran and quietly support conversations between them and Iran about regional security. This includes considering an Iranian proposal called "Hope"—for Hormuz Peace Endeavor—that builds on the Security Council resolution that ended the Iran-Iraq war and pledges, among other things, non-interference in the internal affairs of Persian Gulf neighbors (Haghirian and Zaccara 2019).

To make these changes would be to admit that "maximum pressure" has produced negative results for Iranians and for the US and its allies in terms of Iran's nuclear and regional posture. However, the US will never be able to focus on other key priorities while it remains mired in the Middle East and it will not be able to extricate itself from the region in a dignified manner without first reducing tensions with Tehran.

References

French Warship to Spearhead European Mission in Gulf From Next Year. Reuters via EURACTIV.com (2019a), https://www.euractiv.com/section/defence-and-security/news/french-warship-to-spearhead-european-mission-in-gulf-from-next-year/

Iran Launches Novel Drills With Russia, China Amid Tensions with US. Radio Farde (2019b), https://www.rferl.org/a/iran-russia-china-naval-maneuvers-persian-gulf-oman/30348593.html

Six More Countries Join Trump-Bustling Iran Barter Group. The Guardian (2019c), https://www.theguardian.com/world/2019/dec/01/six-more-countries-join-trump-busting-iran-barter-group.

Iran Exporting Million Barrels of Oil a Day Tracking Firm Data Indicates. bneIntelliNews (2020a), https://intellinews.com/iran-exporting-million-barrels-of-oil-a-day-tracking-firm-data-indicates-175930/

Plane Shot Down Because of Human Error, Iran Says. The New York Times (2020b)

Trump Says Iran is 'Standing Down,' Vows To Continue Pressure. National Public Radio (2020c), https://www.npr.org/2020/01/08/794404894/president-trump-to-deliver-statement-on-iran

U.S. Says First Shipments of Medicine to Iran Delivered via Swiss Humanitarian Channel. The New York Times (2020d), https://www.nytimes.com/reuters/2020/02/03/world/middleeast/30reuters-swiss-iran.html

US, Iraq to Hold Talks on Troop Future. Agence France-Presse (2020e), https://www.military.com/daily-news/2020/04/08/us-iraq-hold-talks-troop-future.html

Agence France Presse, US, Iraq to Hold Talks on Troop Future. Military.com (2020), https://www.military.com/daily-news/2020/04/08/us-iraq-hold-talks-troop-future.html.

J. Blas, E. Pismennaya, Saudis Boost Oil Output, Defying Trump's Plea To End Price War. Bloomberg (2020), https://finance.yahoo.com/news/saudi-arabia-resists-trump-attempt-094943623.html

P. Bump, Why the Administration Claims that Soleimani Killed Hundreds of Americans. TheWashington Post (2020), https://www.washingtonpost.com/politics/2020/01/08/why-administration-claims-that-soleimani-killed-hundreds-americans/

M. Crowley, F. Hassan, E. Scmitt, US Strike In Iraq Kills Qassim Soleimani, Commander of Iranian Forces. The New York Times (2020), https://www.nytimes.com/2020/01/02/world/middleeast/qassem-soleimani-iraq-iran-attack.html

E. Cunningham, In Rare Friday Sermon, Iran's Khameni says US Suffered Blow to 'Superpower Image'. The Washington Post (2020), https://www.washingtonpost.com/world/middle_east/in-rare-friday-sermon-irans-khamenei-says-us-suffered-blow-to-superpower-image/2020/01/17/76ec4bf0-389b-11ea-a1ff-c48c1d59a4a1_story.html

G. Esfandiari, Iran Had a Rough Year—And 2020 Is Likely To Be Equally Difficult. Radio Free Europe (2020), https://www.rferl.org/a/iran-rough-year-2020-predictions-difficult-economy/30355331.html

D. Esfandiary, A. Tabatabai, *Triple Axis: Iran's Relations with Russia and China* (IB Taurus, New York, 2019).

A. Faiola, K. De Young, In the US Embargo On Venezuelan Oil, Russia Is A Clear Winner. The Washington Post (2020), https://www.washingtonpost.com/world/the_americas/in-the-us-embargo-on-venezuelan-oil-russia-is-a-clear-winner/2020/02/06/c45ca39e-476e-11ea-91ab-ce439aa5c7c1_story.html

F. Fassihi, B. Hubbard, Iran's Supreme Leader Rebukes US in Rare Friday Sermon. The New York Times (2020), https://www.nytimes.com/2020/01/17/world/middleeast/iran-ayatollah-khamenei-Friday.html

F. Fassihi, M. Specia, Iran Used Firearms in Deadly Crackdown on Protestors, Officials Admit. The New York Times (2019), https://www.nytimes.com/2019/12/03/world/middleeast/iran-protest-crackdown.html

T. Gibbons-Neff, Why Russia Has Advanced Strike Figures at an Airbase In Iran. The Washington Post (2016), https://www.washingtonpost.com/news/checkpoint/wp/2016/08/19/why-russia-has-advanced-strike-fighters-at-an-air-base-in-iran/

T. Gibbons-Neff, How U.S. Troops Are Preparing for the Worst in the Middle East. The New York Times (2020), https://www.nytimes.com/2020/01/06/world/middleeast/troops-iran-iraq.html

J. Gurol, J. Scita, China's Persian Gulf Strategy: Keep Tehran and Riyadh Content. Atlantic Council (2020), https://atlanticcouncil.org/blogs/menasource/chinas-persian-gulf-strategy-keep-tehran-and-riyadh-content/

M. Haghiran, L. Zaccara, Making Sense of HOPE: Can Iran's Hormuz Peace Endeavor Succeed? The Atlantic Council IranSource Blog (2019), https://atlanticcouncil.org/blogs/iransource/making-sense-of-hope-can-irans-hormuz-peace-endeavor-succeed/

L. Hinnant, US Allies See Middle East Strategy Vacuum That Putin Can Fill. The Times of Israel (2020), https://www.timesofisrael.com/us-allies-see-middle-east-strategy-vacuum-that-putin-can-fill/

Human Rights Watch, Iran: Sanctions Threatening Health (2019)

L .Jakes, D.M. Halbfinger, In Shift, U.S. Says Israeli Settlements in West Bank Do Not Violate International Law. The New York Times (2019), https://www.nytimes.com/2019/11/18/world/middleeast/trump-israel-west-bank-settlements.html

B. Khajehpour, Analyzing Iran's New State Budget. Al-Monitor (2019), https://www.al-monitor.com/pulse/originals/2019/12/iran-new-budget-rouhani-sanctions-economy.html

S.M. Khatami, CNN-Transcript of Interview With Iranian President Khatami, in *CNN*, ed. by Amanpour, C. (1998)

C. Krauss, Oil Nations, Prodded by Trump, Reach Deal to Slash Production. The New York Times (2020), https://www.nytimes.com/2020/04/12/business/energy-environment/opec-russia-saudi-arabia-oil-coronavirus.html

L. Loveluck, M. Ryan, Militia Attacks On Americans in Iraq Are Becoming More Audacious. The Washington Post (2020), https://www.washingtonpost.com/world/middle_east/militia-attacks-on-americans-in-iraq-becoming-more-audacious-us-wrestling-with-how-to-respond/2020/03/27/7b31d76c-6d38-11ea-a156-0048b62cdb51_story.html

M. Mazzetti, E. Schmitt, Iraq Meets Warning from Top Commander. The New York Times (2020), https://www.nytimes.com/2020/03/27/world/middleeast/pentagon-iran-iraq-militias-coronavirus.html

L. Norman, EU Ramps Up Trade System With Iran Despire U.S. Threats. The Wall Street Journal (2020), https://www.wsj.com/articles/eu-ramps-up-trade-system-with-iran-despite-u-s-threats-11585661594

A.J. Rubin, F. Fassihi, E. Schmitt, V. Yee, Iran Fires on US Forces at 2 Bases in Iraq, Calling it Fierce Revenge. The New York Times (2020), https://www.nytimes.com/2020/01/07/world/middleeast/iran-fires-missiles-us.html

D. Sanger, North Korea's Internet Use Sergers, Thwarting Sanctions and Fueling Theft. The New York Times (2020), https://www.nytimes.com/2020/02/09/us/politics/north-korea-internet-sanctions.html

D.E. Sanger, W.J. Broad, Iran Challenges Trump, Announcing End of Nuclear Restrictions. The New York Times (2020), https://www.nytimes.com/2020/01/05/world/middleeast/trump-iran-nuclear-agreement.html

B. Slavin, Out of Sync: Iran and the United States, in *Bitter Friends, Bosom Enemies: Iran, the U.S. and the Twisted Path to Confrontation.* (St.Martin's Press, New York, 2007)

B. Slavin, Iran Turns to China, Barter to Survive Sanctions. Atlantic Council (2011), https://www.atlanticcouncil.org/in-depth-research-reports/issue-brief/iran-turns-to-china-barter-to-survive-sanctions-1/

M. Specia, Israel Launches Airstrikes in Syria to Target Iranian Forces. The New York Times(2019), https://www.nytimes.com/2019/11/20/world/middleeast/israel-syria-iran-strikes.html

M. Tavakol, Iran's Crude Oil Exports: What Minimum is Enough to Stay Afloat? The Atlantic Council Iran Source Blog (2019), https://www.atlanticcouncil.org/blogs/iransource/iran-s-crude-oil-exports-what-minimum-is-enough-to-stay-afloat/

A. Tibon, N. Landau, Trump's "Deal of the Century" Was Written in a Way So the Palestinians Would Reject It. Maybe That Was the Plan. Haaretz (2020), https://www.haaretz.com/israel-news/.premium-trump-s-deal-of-the-century-was-written-so-the-palestinians-would-reject-it-1.8443677

U.S. State Department, 3/17/000 Albright Remarks on American-Iranian Relations (2000), https://1997-2001.state.gov/statements/2000/000317.html
The White House, President Delivers State of the Union Address (2002), https://georgewbush-whitehouse.archives.gov/news/releases/2002/01/20020129-11.html
The White House, Selected Speeches of President George W. Bush 2001–2008 (2009)

Afghanistan and the Great Powers in Regional Geopolitics, Economics and Security

Marvin G. Weinbaum

Abstract Afghanistan occupies a point of convergence of great power interests in a new Great Game across a region dealing with such issues as terrorist and criminal activity, narcotics trafficking and porous borders. The great powers have also in recent years importantly converged in their trying to promote an Afghan peace process. A promising future for Afghanistan depends as well on a better-integrated region economically and substantial progress toward political reconciliation among the neighboring states. Afghanistan occupies a place in a region where rivalries play out usually to its detriment. At present, the security-driven agendas of its neighbors, particularly the enmity between India and Pakistan, strongly encumber the trans-regional movement of goods and services on which Afghanistan depends so heavily. In turn, the armed struggle in Afghanistan challenges the well being of the region. The country's long history of instability and armed conflict threatens the region's ability to realize its potentials for economic growth through economic connectivity and regional political cooperation. The appropriate means for resolving the Afghan conflict and regional disagreements is generally thought to be the exercise of diplomacy. Yet spoilers, domestic and foreign, have demonstrated their ability to impede negotiations with which they feel uncomfortable. Russia, China and the US, with their perceived stakes in the region and in pursuit of their own strategic objectives, have at times exacerbated the differences among the region's states. But these powers are also in a position to encourage integration and thus further regional peace and stability. In Afghanistan, the future lies ultimately with the Afghans themselves, but the actions of the regional powers and the international community, most of all the great powers, figure prominently in the country's economic survival and hopes for security.

Keywords Great Game · Great powers · Afghan conflict · Regional connectivity · Taliban · South asia · Istanbul process

M. G. Weinbaum (✉)
Director of Afghanistan and Pakistan Studies, Middle East Institute, 3001 Veazey Terrace Apt. 407, Washington, DC 20008, US
e-mail: mweinbaum@mei.edu

© The Author(s), under exclusive license to Springer Nature Switzerland AG 2021

A. Farhadi et al. (eds.), *The Great Power Competition Volume 1*,
https://doi.org/10.1007/978-3-030-64473-4_20

Introduction

Afghanistan is regularly described as the crossroads of the region, the link between South Asia, West and Central Asia. Its borders include two of the globe's major powers, Russia and China, and three major regional actors, Pakistan, Iran and India. Other players in the neighborhood, Saudi Arabia and the Gulf States, also bulk large in the region's politics and economics. A third global power, the United States, was at first drawn to the region as a more or less silent alliance partner in a Cold War containment strategy. The Soviet Union's Afghan invasion in 1979 thrust the US into a more active if still indirect involvement that after more than a decade in disengagement saw in the aftermath of September 2001 the US emerge as a pivotal actor in determining the future of Afghanistan.

The Afghan state has had to endure in these last two decades in a security-driven region and an increasing competitive international environment. It finds itself a point of convergence of great power interests in a multi-polar contest among the US, Russia and China in a new Great Game for political influence, markets and geo-strategic positioning (Harpviken and Tadjbakhsh 2016). Much of the present catalyst for this competitive environment has been the insurgency in Afghanistan and the American-led war on global terrorism.

Afghanistan is an arena in which regional rivalries also play out. The country has been caught up in India and Pakistan's enduring strategic struggle. Whether because of Pakistan's quest of an illusive strategic depth or fear of Indian machinations across the border, Afghanistan has hosted high stakes shadow combat. The Iran-Saudi enmity is also openly reflected in Afghanistan as they contest in promoting Shiite and Wahabbi brands of Islam among the population. A subtler struggle goes on between Iran and Pakistan in the competition for political influence in Kabul as well as for Afghanistan's commercial markets. The ability of any one of Afghanistan's neighbors to undermine Afghanistan's domestic stability is well established.

Afghanistan in turn poses challenges to the well being of the region. The country's long history of instability and armed conflict threatens the region's ability to realize its potential economic growth through economic connectivity and regional political cooperation. Afghanistan's ungoverned spaces, so feared by the West, have offered opportunities for insurgencies aimed at several of the region's states to find sanctuary and hatch their plans.

The Great Powers' Interest in Afghanistan

Afghanistan was until 1979 only marginally important to the great powers. Interest in Afghanistan by the US and its adversary the Soviet Union went only so far as their mutual concern that the other be kept from exercising a commanding influence in the country. Afghanistan occupied a break in the line of countries recruited to provide the US with the containment of the Soviet Union. But an unaligned Afghanistan was enough to ensure there was no further compromise of its strategy of encirclement. Moscow was most focused on its southern tier states and was determined that Afghan soil not be used to weaken its hold on its Muslim Central Asia. As a result, the US and Soviets, acting defensively, were each satisfied with having Afghanistan effectively detached from the tensions of the cold war. This began in the mid-1950s with a peaceful competition intended to balance the other's influence in providing the badly underdeveloped country largely infrastructural improvements. There was created a virtual condominium in which the US and Russia sometimes, as in the building of highways, complemented each other's investments. The aim of both was to use their own foreign assistance to balance the other's influence. In no other place on the globe did there exist such cooperation between these two adversaries. However, in one area the US ceded to the Russians an unchecked advantage, the training, equipping and socialization of the Afghan military. The Afghan World War II leadership had preferred this role for the US, but in deference to Pakistan, the US excluded Afghanistan from the Baghdad Pact, then CENTO. It was a decision that ultimately led to two communist coups against the Kabul government and shattered the idea that Afghanistan could stand aside from a Cold War competition to which others in the region were drawn.

For those who thought that Afghanistan had escaped being a battleground in the Cold War, that belief was to shatter with the Soviet invasion of Afghanistan in December 1979. Even while the incursion was not undertaken as part of a grand strategic plan, Soviet intentions for installing a favored Afghan president and his party morphed into an indefinite occupation as Islamic militants mounted attacks on the imposed regime. The Red Army's presence in the country was sufficient for the US to become more deeply involved, albeit indirectly, in an effort seen as ensuring that Russia did not move against CENTO member Pakistan.

With the agreement reached in Geneva for the Soviet Union to fully withdraw its forces by February 1989, the US quickly lost interest not only in Afghanistan but also in the region. In 1990, all aid to Pakistan ended, and with the breakup of the Soviet Union in 1992, the US disengagement was complete. In the US's absent from the region, there followed a period of conflict in Afghanistan in some ways more frightening than the jihad against the Soviet Union. Under conditions of warlordism and anarchy, years of fratricide among the victorious mujahideen parties gave rise to the Taliban and al-Qaeda.

The advent of 9/11 brought the US rushing back to Afghanistan and the region and into a commitment, however reluctantly, to the building and defense of the Afghan state. Two other developments kept the US regionally reengaged. Fighting global terrorism became the centerpiece in American security thinking. Although attention also would soon be drawn to Iraq and later to Syria and Iran, the Afghanistan-Pakistan theater remained of high concern, lest a Taliban insurgency allow the area to again become a launching pad for international terrorism. Secondly, 1998 saw Pakistan and India emerge as confirmed nuclear powers and later evidence that Pakistan had surreptitiously shared its nuclear knowhow with Libya, Iran and North Korea. Growing militant Islamic radicalism in Pakistan also was recognized as posing a threat to the security of Pakistan's nuclear arsenal. The export of Islamic radicalism from Afghanistan to Pakistan and the Central Asian states put Afghanistan squarely at the center of US concerns. Along with the dangers of terrorism and nuclear proliferation loomed the possibility that Pakistan and India's newly reignited Kashmir conflict could move toward a catastrophic nuclear exchange carrying regional and global consequences whose prevention required American intercession.

As much as the interests of the major powers diverged elsewhere, terrorism and proliferation in the region provided important areas of convergent interests. In 2001 the major actors as well as the regional powers were able to coordinate their efforts in bringing about the reformation of an Afghan state at the Bonn conference. In such areas as counter-terrorism, anti-money laundering, narcotics control, and in dealing with transnational organized crime, these powers have demonstrated an ability to work together. As discussed below, in recent years they have been aligned in trying to further an Afghan peace process. Even so, elements of competition are also still present in otherwise complementary efforts.

Regional Economic Integration

As measured by trade and other economic activities, South and Central Asia compose the least integrated region in the world. For all the logic of greater connectivity, rather than nations acting together in areas where they share common purpose, there are deeply embedded national rivalries, and entrenched economic and other elite interests have impeded change (Harpviken and Tadjbakhsh 2016). While terrorist networks, drug trafficking, and smuggling already connect the region, the movement of people, energy and ideas confront barriers erected by the regional powers' defensive hedging strategies (Harpviken and Tadjbakhsh 2016). This comes despite the creation over the last decade of regional organizations (see below) designed to break these barriers.

The benefits of cooperation among the region's countries would far outweigh any drawbacks. All would be in a better position to break down trade restraints and share development goals that include building a regional transportation infrastructure, harnessing untapped energy and accessing water resources. Economically, this approach makes sense by enabling them to reduce costs, pool their resources and to take advantage of their complementary capacities (Harpviken and Tadjbakhsh 2016). Improved coordination could overcome such problems as wasted resources and development assistance working at cross-purposes.

An inter-regional framework that furthers cooperation and coordination would also allow countries of South and Central Asia to pursue counterterrorism and counternarcotics and meet their other security challenges. Having such a framework would additionally allow the region's state actors to collectively and individually deal on stronger terms with external powers, including China, Russia, and Saudi Arabia. Most immediately, China's expansionist economic plans contained in its Belt and Road Initiative (BRI) could potentially overpower or marginalize Central Asian and South Asian steps to greater integration.

Within South Asia increased trade would on balance be in the interests of all its states. At present South Asia has anemic intra-regional trade, at least of the licit kind. At 5 percent of total trade, it is the lowest in the world (Nisha 2014). While informal trade is much higher, it denies governments needed taxes. Protectionist and politically motivated obstacles, most of them non-tariff, impede economic investment and growth. Freer trade within South Asia is also a necessary precondition for increased commerce with Central Asia. Connectivity could create greater space for private sectors to work together with the public sectors within and across borders. Given the incentives and appropriate legal frameworks, the private sector may help facilitate the expansion and transfer of administrative and technical skills. Furthermore, region-wide connectivity can bring better opportunities for drawing on human resources in order to improve civil services and vocational education. Yet it is necessary to be careful that hopes and plans do not too far overreach their capacity to succeed. To fail too often may discourage and threaten worthy projects. It could turn off the foreign and local investment and the popular support needed to be successful.

Afghanistan and Regional Interconnectivity

Afghanistan is pivotal to prospects for increased connectivity. At present, the armed struggle in Afghanistan along with the security driven agendas of its neighbors, particularly the rivalry between India and Pakistan, strongly encumbers trans-regional movement of goods and services. But in many respects because of its geographic location and wider reflections of its conflict, without sustained peace in Afghanistan economic growth regionally is effectively on hold.

At the same time, it is difficult to conceive of a promising economic future for Afghanistan without a better-integrated region. It is Afghanistan's challenge to transform itself from an at-risk landlocked country dependent on the willingness of neighbors to open their borders into a land bridge between South and Central Asia (Weinbaum and Humayoon 2009). The hope is that by facilitating Afghanistan's emergence as a hub of economic activity integrated within a regional framework, the Afghan economy can be strengthened and the regime as a whole may be better able to fend off insurgency. No country in the vicinity believes it would be well served by a Taliban victory. India and Pakistan have reason to be concerned about a radical Islamic government in Afghanistan that inspires and empowers Islamic militancy in their own countries. As the perceived threat from militant Islamists grows, the states of South Asia may find additional motives for cooperation among themselves and with Central Asian states. Uzbekistan, Turkmenistan, Tajikistan, and Kyrgyzstan fear security risks from Afghanistan once the United States has entirely withdrawn its forces and should the Taliban emerge victorious.

If Afghanistan is going to take its place as a catalyst for broad regional wellbeing, regional actors must step up their international economic and development assistance to Afghanistan. They can make the region more hospitable by following through on collaborative projects and lowering trade and border restrictions. They can also do more in sharing their development knowhow and personnel. But it may be further necessary that regional partners, without becoming directly involved in Afghanistan's political affairs, impress upon the Afghan government the need for domestic reforms. The implementation of projects, public or private, is obviously compromised by corruption that more broadly undermines and distorts economic growth. Also, the corrosive effects on development plans and the society of poppy cultivation have to be confronted.

Greater regional economic integration of the kind that can be crucial to Afghanistan cannot fully succeed without substantial progress toward political reconciliation among the countries. The mutual distrust among South and Central Asian countries and their separate bilateral conflicts figure strongly in retarding the region's overall economic growth and prosperity. Failure to reduce tensions that are for the most part security driven has impeded the economic and structural reforms necessary to increase trade, facilitate investment, and promote energy transfers and cultural exchanges. Economic growth and prosperity are impeded by unresolved political disputes, including Afghanistan's historic border disagreement with Pakistan, and Pakistan's decades of conflict with India over Kashmir, and more recently India's treatment of Muslims nationally. Suspicions and long standing border issues among the Central Asian states have a similar effect. Widespread distrust leading to defensive trade, energy and other economic policies constrains opportunities to enjoy the benefits of cooperation. These states have yet to appreciate their common stakes and entail the risk-taking that comes with adopting new security perspectives.

While the progress of South and Central Asian integration is put on hold by the prolonged conflict in Afghanistan, a number of these countries continue to take interest in the country for its natural resources and export markets. Iran and Pakistan compete commercially in Afghanistan. Iran and Saudi Arabia in their sectarian appeals compete in Afghanistan and Pakistan. China and India have both made bids for Afghanistan's mineral resources. China in particular sees security in Afghanistan as linked with the protection of its economic investments in Central and South Asia.

Russia, China and the US, with their perceived stakes in the region and in pursuit of their own strategic objectives have often exacerbated the differences among the region's states. But the major powers are also in a position to encourage integration and thus further regional peace and stability. In hopes of having the broad region take greater responsibility toward contributing to Afghanistan's economic growth and stability, the US promoted the convening of a meeting in Istanbul in 2011. In what would become the Istanbul Process, the gathering consisted of 14 countries in South and Central Asia as well as Saudi Arabia, the United Arab Emirates, and Turkey. The conference coincided with the US Silk Road initiative – from a concept first proposed by Professor Frederick Starr – aimed at furthering Afghanistan integration in its region through reviving traditional trade routes and building needed infrastructural links (Starr 2007). It also fit well with American policy determination to encourage greater regional commitment to Afghanistan (US Department of State 2017). But the conference quickly distanced itself from the US by denying it even official observer status. Becoming known as well the Heart of Asia, the Afghanistan-centered effort focused on fostering economic cooperation among regional states. It mandated for itself the introduction of confidence-building measures that would contribute to strengthening people-to-people contacts and the bringing about of greater cohesion in regional integration (Heart of Asia-Istanbul Process 2019).

Other organizations have taken up the growing efforts toward strengthening connectivity between South, Central and West Asia. As already noted, the proliferation of regional organizations and grouping—SAARC, SCO, RECCA, ECO, the Istanbul/Heart of Asia process and various quadrilateral and trilateral groupings are all testimony to the belief in diplomacy as the leading tool for expanding economic growth and resolving regional differences. The most pressing trade and environmental issues, and political dispute differences can best be resolved through consultation and cooperation. Without diplomacy there seems no hope of resolving such time-worn issues as the Kashmir dispute, the contested Durand Line, India-China border issues, differences over water rights and obligations between India and Pakistan and between Afghanistan and Iran. Multilateral cooperation can contribute to attempts to defang terrorist organizations in the region, including dealing with terrorist money laundering. Seldom can these programs be handled by one country alone.

Russia and China in their initiated Shanghai Cooperation Organization (SCO) can also claim regional integration as its goal. Originally, the organization founded in 1996 with the furtherance of regional security its declared purpose, had five members (ADB Asia Regional Information Center 2020). But it gradually broadened its scope of interest to encompass the region's development in sponsoring projects in transportation and energy and telecommunications, and discussing economic and cultural issues among others. Logically, with regional integration entering the agenda, the SCO invited Pakistan and India first as observer states and then as full members. As the Afghanistan conflict loomed larger in regional thinking, and cross border terrorism and narcotics became a growing concern to all the member states, Afghanistan's presence in meetings of SCO become indispensible. Not able to meet the full obligations of membership, Afghanistan has remained, however, in observer status.

Separately, China's has made the case that its Belt and Road Initiative is aimed at strengthening the economic capacity of the countries to which it is connected. At various meetings of regional actors over the last several years, concrete project proposals have defined the need for a regional transportation infrastructure, the harnessing of untapped energy, accessing water resources, and breaking down trade restrictions. But to date, inter-regional, multilateral dialogue has mostly resulted in mere wish lists of projects that are for the near future unrealizable.

Few projects or programs have reached the implementation phase. It is where even the best laid plans and development strategies to achieve connectivity have come up empty. The few successes are nearly all bilateral rather than multilateral. Some are unable to realize their potential because they were poorly conceived and fell prey to competing national objectives. The actions of non-state actors have also been an aggravating factor. Most countries have failed because they depended on government policies where they encounter problems of political will, administrative capacity, and endemic corruption. What all these efforts and others have in common is that they cannot be fully realized without an Afghanistan largely at peace.

Diplomacy in Afghanistan and the Region

Diplomacy is generally seen as the instrument of choice in the search for ways to resolve the Afghan conflict peacefully and realize the aspirations of the region. With no decisive military victory by either the Kabul government or insurgent forces in sight, there is a widespread belief that however long it takes the conflict must end with a negotiated political settlement. Putting aside the question of whether compromise with the Taliban is possible, successful diplomacy requires interlocutors who are credible representatives of their respective sides. It also requires that the main parties to the conflict be direct participants in the process. While trust is necessary, there must also be enforcement mechanisms to ensure that the parties honor their agreements.

The mantra sounded so often among present day peacemakers is that there must be an "Afghan-owned, Afghan-led" process. Yet there is no occasion when Afghans in major armed conflict at the national level when left to themselves have been able to settle their differences through political means. Following the Soviet withdrawal in 1989, and after the Mujahideen seizure of power in 1992, the results were hardly peaceful. There followed the birth of warlordism, anarchy, and a destructive civil war. Over the period from 1975 through the present, Afghans have been engaged in six identifiable civil wars, and in none have they been able to achieve political compromise. In every case the outcome was determined by one side eventually winning, the other losing (Weinbaum and Majidyar 2019).

When there have been diplomatic processes, they have required the strong agency of external actors. The 1988 Geneva Accord that after years of negotiations arranged for the departure of Soviet forces from Afghanistan would have been impossible without the patronage of the UN. The Afghan insurgents, the mujahideen, were never party to the negotiations or the agreement. It is generally agreed that the 2001 post conflict Bonn meeting that created the political framework for a new Afghan state might have broken up without the heavy pressure put on the Afghan delegates by Russia, Iran, the US and others. After a contentious presidential election in 2014, likely intra-elite violence in Afghanistan was headed off by US Secretary of State John Kerry's forcing on the political leaders the formation of a unity government. In the extensive negotiations in Doha between the US and Taliban, not only did the Taliban refuse to allow the Kabul government's participation, the agreement with the US did not guarantee Kabul's full recognition in subsequent intra-Afghan talks. In all probability, for progress in those talks, outside mediation will be essential.

The challenges of diplomacy and obstacles to an Afghan peace process are all too obvious. The Afghan conflict is complicated by the possibility that spoilers, domestic or foreign, have frequently demonstrated their ability to impede negotiations with which they are uncomfortable. Aside from other insurgent groups such as Islamic State and al-Qaeda which have never accepted the idea of a negotiated peace, it is uncertain whether the Taliban political delegation can speak for all the field commanders, especially once the terms for a settlement might be known. Much the same can be said about the absence of any consensus among Afghanistan's political elites and its various interest groups over what they would find acceptable, women's groups being only the most obvious. The Taliban also have a poor track record in delivering on promises and do not set a good example of tolerant, inclusive governance where they hold sway. Among the many other thorny issues, the difficulty in disarming or integrating Taliban fighters with government ones stands out. Nor is there seemingly a lining up of countries that would be willing and able to supervise or provide compliance with any agreement reached. But perhaps the most significant obstacle to successful diplomacy is the absence of a shared perspective among negotiators over a future Afghan state—is it to be a constitutional democracy or an Islamic Emirate?

At the regional level, there is high demand for diplomacy. A more fractionalized and fluid international system and fading historical memories have cleared the way for opening new channels of diplomacy as well as new alignments and partnerships. Changes globally that have loosened the structure of international relations are mirrored in South Asia. Older external alliances have been shaken and to some extent realigned. Some newer ones replaced them. Not too long ago a pariah held in deep suspicion, Russia has succeeded in reinserting itself diplomatically, especially in Afghanistan. This is occurring both in the security and economic realms. China with its BRI is poised to refashion much of the international economic, political and potential military relations over a wide swath of Asia. Once reluctant, China appears ready to go beyond just its economic interests in its concern over the domestic affairs and disputes of certain countries. All of these changes are of course occurring against the backdrop of a receding and uncertain US engagement globally.

Overall, the record of diplomacy in the region is at best only mixed. Although there is more multilateral engagement than ever and greater awareness of the mutual benefits in cooperation and more regional institutional support, most of the standing differences and antagonisms among countries in the region have hardly budged over time. Diplomacy has thus far failed to overcome long-standing issues that divide states in the region, and there are some backward steps as well. If allowed to progress, a strengthening of regional ties through commercial and energy interdependency may in time make some of these and other disputes more amenable to settlement. But even if at present there are few major diplomatic breakthroughs to point to, it is difficult to imagine progress in achieving solutions to the region's most pressing problems in the absence of skilled, dedicated diplomacy and the political will to employ it.

Afghan Scenarios and the Regional and Great Powers

With Afghanistan the focal point for so much conflict regionally, it is necessary to understand how the opportunities for the area's integration will be contingent on what future directions the Afghan state may take. The course of events in Afghanistan will inevitably have still broader consequences for the security and stability of countries in the region and beyond. In turn, its neighbors and, as history has shown, especially the great powers will be instrumental in shaping Afghanistan's future. It may be therefore instructive to identify various possible scenarios that Afghanistan may follow over the next several years. At least four broad alternative futures seem plausible, albeit of varying probability (Weinbaum and Majidyar 2019).

The first of these scenarios, and undoubtedly the brightest, depicts an Afghanistan where over the near future the Kabul government has after strenuous negotiations been able to reach a compromise agreement with the Taliban interlocutors. Other elements of the insurgency are either soon defeated or at least marginalized and only some isolated fighting continues. The Taliban has approved the general framework of a political order that while more Islamic in character also provides for real power sharing and preserves most of the social and economic gains over the last two decades. Despite criticism of the agreement from the various pro-government factions as well as from dissenting Taliban field commanders, the number of defections from either side has been minimal.

Although Scenario One is likely to be the product of intra-Afghan negotiations, it is difficult to imagine it occurring in absence of the strong role of outside actors facilitating its outcome. The high degree of involvement by the major powers along with the regional states reflects not only their stakes in Afghanistan but also a preference for this scenario over any other alternatives. None stand to profit from a fragmented Afghanistan in chaos or its being dominated by an Islamic regime potentially bent on facilitating the export of a radical ideology and regime change. For Pakistan, an Afghanistan feeling secure could mean a country with less security dependence on India. All the states in the region would presumably stand to gain economically from an Afghanistan at peace, united and prospering, allowing it to assume its place as the region's crossroads of commerce and transfer of energy.

Scenario Two assumes a large measure of continuity in the current conflict, with the Taliban making gradual gains in solidifying control over much of the Afghan countryside while the Afghan security forces succeed in maintaining their grip on population centers. The heavy toll on both combatants and civilians is unrelieved. The search by Afghans for a diplomatic means to end the conflict drags on with little evidence of progress and despite strong popular demands to see an early end to the fighting. Modest progress is registered in addressing corruption but public trust and confidence in the government remains wanting.

This second scenario assumes two conditions: first that US forces and those of its allies retain at least a small presence in the country in support of Afghan security forces (ANSF). A full withdrawal of foreign troops raises doubt that the ANSF can remain an effective fighting force. The great test would come with the loss of tactical air support for ground operations and troop evacuations, without which it would be difficult to protect provincial capitals and Kabul. The second assumption is that despite the bitter disputes among Afghanistan's power brokers there remains sufficient cohesion among the ruling elites that none will choose to break with and take up arms against the central authority. No small part of the ability to stave off the Taliban countrywide depends on there being a central government thought capable of delivering basic security.

Under conditions of continuity, the policies of regional powers and major actors are likely to remain in concert in their preference for a political solution to the conflict and in their willingness to facilitate intra-Afghan dialogue. This great and regional power consensus on Afghanistan might well dissolve in the absence of progress with intra-Afghan negotiations following a complete withdrawal of foreign troops. Armed with their hedging strategies, these countries would be likely to reassess their options, reactivating their relations with proxies and opening new lines of communication to Taliban leaders.

A third scenario envisions the Taliban's emergence as the preeminent authority in Afghanistan. It might be achieved in steps through a negotiated agreement or after relentless military pressure, and most probably a combination of the two. A contributing factor may well be a steady erosion of political stability brought on by the continued disunity among the country's political elites. As a variation of the means by which the Taliban come to power, a disintegrating Kabul government could result in power brokers informally striking separate deals with the Taliban. Their probably uncoordinated actions would come in anticipation of a Taliban political ascendency and the desire to preserve their own interests in a restored emirate.

An ascendant Taliban may be prepared, at least initially, to form a more inclusive government in a interim power-sharing arrangement. In contrast to its earlier emirate, the Taliban in Scenario Three might be more tolerant of non-Taliban groups and individuals. And possibly, having learned lessons from the past, the Taliban leadership would also be somewhat more flexible in the social sphere. Implementation of rules could also be expected to vary somewhat countrywide. But in any case, in this scenario governance would take place within the framework of the Taliban leadership's interpretations of Sharia law. And judging from its public statements, negotiating behavior, and governing practices in areas currently under its control, the Taliban has no intention of budging on long-held core principles.

Scenario Four foresees an outcome where rather than any single force capturing national power through military means or by political agreement that no force would be able to solidify power. Unlike the Taliban in the 1990s, neither the insurgents nor their opponents possess the unity of command that existed when Mullah Omar and Ahmed Shah Massoud could count on the allegiance of their Taliban and Northern Alliance fighters. Neither side today has the cohesion that with a collapse of the state would enable them to monopolize power. In all probability there would follow a bitter civil war over territory and resources leaving much of the country as ungoverned space. The present unity of purpose among Taliban elements in pursuit of defeating the sitting government will disappear with the spoils of victory as ambitious commanders clash over local control. Similarly, in this dissolving state, regional and ethnic aligned leaders and their militias would compete. With no viable economy and the ongoing violence, as many as a third of Afghanistan's population can be expected to flee. By contrast to when the communists held power in the 1980s and portions of the urban middle class remained in the country, Afghanistan's far larger middle class but especially its numerous young educated people will be among the first to seek refuge elsewhere–anywhere.

In Scenario 4, the major and regional powers would find a chaotic civil war the least desirable of outcomes. However much some may be dissatisfied with a weak or distracted Afghanistan, none would welcome the unpredictability that comes with anarchy across most of the country, or being dragged into the conflict backing various fighting elements expected to serve as proxies furthering their interests. There always exists the danger that through these proxies, regional and major powers could find themselves in direct confrontation.

It is difficult to imagine that whichever scenario unfolds, Afghanistan would for the foreseeable return to the kind of peace it experienced in the decades prior to 1978. The country was hardly idyllic, being economically and socially impoverished and burdened by corrupt and dysfunctional political institutions. But it was not plagued as in the years that followed by an uncontained culture of violence. Afghanistan is now filled with a large body of combatants who have known only conflict throughout their lives. Without a concerted effort to provide alternative livelihoods, they are unequipped for anything but for fighting. The country as never before has on its soil both light and heavy weaponry and the men with the training to use them. And even were an agreement somehow finally reached with the Taliban, violence might continue. The remaining insurgent challenges from Daish (Islamic State-Khorasan) and al Qaeda aside, the deep political and social divisions created after over four decades of civil war and the implant of criminal networks may mean that violence has become endemic in Afghanistan. The frequent charge that the conflict in Afghanistan is at its core a proxy war overlooks the very real fault lines that exist among the Afghans.

The Great and Regional Powers in Bringing Peace to Afghanistan

The strained relations among states in the region would appear to create a difficult environment for peace building (Maley 2009). For the time being, however, to a remarkable extent the regional powers have largely stood together in giving support to the initiation of an Afghan peace process. As has been stated, all would prefer a negotiated solution to end the conflict to any alternatives. While they have never possessed the leverage acting individually or working collectively that could impose a peaceful outcome in Afghanistan, a comprehensive peace would be impossible to sustain without their cooperation. The great powers are similarly limited in their ability to pressure the Afghan players but well positioned to facilitate peace efforts and, potentially, create impediments.

For all of their differences and the rising tensions elsewhere in the international arena, the interests of the US, Russia and China like those of most of Afghanistan's neighbors have converged on such issues as terrorist and criminal activity, narcotics trafficking and porous borders. Russia has been especially anxious to block the export of heroin through Central Asia and is of late alarmed about the emergence of the Islamic State as a terrorist force in Afghanistan. China has been for some time troubled that fighters with the insurgent Turkestan Islamic Movement (TIM) from northeast China have been able to find safe haven in Pakistan and more lately in Afghanistan. With its deep economic involvement in the region, China is also determined to protect from terrorist attacks its investments and workers, particularly in Pakistan's Balochistan province.

Together in promoting an Afghan peace process, the great and regional powers, whether intentionally or not, have with their intensive contacts with the Taliban accorded it a degree of legitimacy far beyond anything its leaders have previously experienced. As a consequence the great and regional powers have in effect simultaneously degraded the legitimacy of the Kabul regime. Nevertheless, the external actors have all managed to retain normal diplomatic relations with Kabul government. They are also united with the government in calling for a ceasefire or at least a sustained reduction of violence as necessary to reaching any agreement with the Taliban. Moreover, all the great and regional powers have operated under the assumption that in any settlement the Taliban would be absorbed into the current political order, albeit with a constitution incorporating more Islamic features. None have advocated the restoration of an Islamic emirate. The major powers and the regional actors seem to also recognize the possible implications of a breakdown of dialogue and the need to avoid state failure and a chaotic civil war.

The pace of American efforts to build a multi-national coalition backing its peace efforts picked up sharply in 2019 with US Special Representative Zalmay Khalilzad's traveling extensively across the region while negotiating with a Taliban delegation in Doha. Efforts to enlist great power cooperation were clearly on display with the convening of a trilateral meeting with Russian and Chinese representatives on April 25, 2019. The trilateral meeting found agreement in several key issues, ones meant to strengthen the US hand in Doha (Bone and Nijrabi 2019). The Taliban were called on to agree to open negotiations directly with the Kabul regime and other Afghan stakeholders, and to break all ties to al-Qaeda and other terrorist groups, insuring they would not be allowed to use Afghan soil to mount attacks elsewhere. It was in the belief that the Taliban were in accord with these points that American negotiators agreed in Doha in late February 2020 to a phased withdrawal of US troops from Afghanistan.

Many countries sought to exert their influence to draw the Taliban to the negotiating table. Although often overlapping and supplementary to US efforts primarily focused on reaching a bilateral agreement, several European countries, most notably Norway and Germany, have especially concentrated on getting an intra-Afghan dialogue off the ground. Pakistan and Russia, China and Qatar have shown strong readiness to provide venues for peace talks. Beginning in 2016 the Russians became active in a Moscow Process. It initially involved a meeting between Russian, Chinese and Pakistani diplomats. The following year it included representatives from the Kabul government, India and several Central Asian countries. In 2018 the Taliban was invited to attend as well as the Afghan delegation. Refusing to hold direct negotiations with Kabul's representatives, the Taliban failed to attend and the meeting was then called off. A high-profile meeting took place in 2019 that included a delegation of mainly Afghan politicians and civil society participants and several government officials not recognized as officially representing the Kabul government. Although the US was invited to join the Moscow process, Washington was unwilling to have representatives at the meetings, at least not in any official capacity. Meanwhile, talks continued in Doha between the US and Taliban and while the Kabul government was not pleased with being excluded, it otherwise welcomed the possibility of an agreement that might pave the way for an intra-Afghan dialogue.

Pakistan has generally been believed to be crucial to bringing the Taliban to the negotiating table. Few doubt that over nearly three decades Pakistan's security forces have maintained close ties to the senior Taliban command in Quetta and even closer ones with the Haqqani Network. Even so, Pakistan's influence with senior Taliban leadership is sometimes overstated. Pakistan's ISI, for more than 25 years mentor to the Taliban, finds itself unable to sway its leadership on issues they consider as core convictions.

Despite the overall cooperation among the great powers in backing a peace process, the strategic competition and suspicions that mark their global relationship can find reflection in the peace efforts. Russian efforts at peace making are often viewed in Washington as an attempt to upstage the US. The Russians have born some resentment over the lead role of the Americans in the peace process since they desired to be seen as major players. What had been for Russia for years in the wake of the Soviet military's departure in 1989 a deliberately low profile in Afghanistan has clearly changed with its determination to have a deeper imprint on the country's future. Moscow has been eager to take the lead in facilitating intra-Afghan talks, especially as the US has been preoccupied with reaching a bilateral agreement with the Taliban aimed at securing a US military withdrawal. Moscow has readily sought the role of mediator, a part it is also angling to play between Pakistan and India. Moscow has for many years been pushing for a regional security conference along the lines of the Organization for Security and Cooperation in Europe, the OSCE. That architecture could also be extended to cover Afghanistan.

Compared to other states in the region, China has long been thought to be a peripheral player in Afghanistan. As a rule, Chinese foreign policy has tended to steer largely clear of direct involvement in other countries' domestic affairs, preferring instead to be able to work productively with whatever regime is in power. Unlike Russia, Iran and India, China has never been a participant in past proxy wars in Afghanistan. Its development and humanitarian assistance has been relatively modest, especially given its largesse elsewhere. China has, however, shown itself over the last several years ready to assume a more active and visible role in Afghanistan. Above all, it is about fears of Muslim radicalism among the Uyghur populations in Xinjiang and TIP terrorism. China's desire to gain access to mineral resources in Afghanistan, specifically Ayak cooper deposits, also can account for its interest in Afghanistan's stability. Furthermore, a more peaceful Afghanistan can contribute to the success of China's BRI. Although for the time being at the periphery of that grand scheme, Afghanistan could provide a valuable extension of its network. This possibility seems more realistic with Beijing's plans to build a road across the Wakhan Corridor, that stretch of highly mountainous Afghan territory that links it geographically with China. Thus far China's activities in Afghanistan have had nothing like the polarizing effect that its policies have had in the Indian subcontinent. But while China has long believed that economic investment could take care of its security interests in a country like Afghanistan, it has found that those interests may be insufficiently protected where there is a power vacuum and where ethnic and sectarian conflicts bulk large (Imran 2019).

These developments take place against the backdrop of a US anxious to disengage militarily from Afghanistan. The prospect that the US could with its presence protect Chinese investments in Afghanistan is now past. In Beijing's evolved thinking, interest in helping to foster a peace in Afghanistan, is also linked with Indian involvement in the country and instability in the ever-important Gulf region. Its diplomatic engagement encouraging a negotiated end to the conflict began with its initiative in hosting in 2014 the first of several quadrilateral dialogues that brought together delegations from Washington, Kabul and Islamabad. China has continued to offer its good offices for facilitation of a peace process. Because China is viewed in many quarters as being a more disinterested as well as influential actor, it is easier for Beijing to offer itself as an honest broker.

At the same time that Afghanistan, Russia and China as well as other regional actors may be adjusting to American intentions to sharply reduce its military footprint in Afghanistan, these countries paradoxically harbor suspicions about the U.S.'s intensions to leave the region. According to this way of thinking, and despite the Doha agreement with the Taliban, the US is believed to have no intention of actually quitting the region but instead is still looking for a means to establish a permanent presence in Afghanistan in order to pursue a larger regional strategy. The US is seen as determined to retain a geostrategic foothold from which it is able to check Russian and Chinese regional ambitions, pen in Iran, maintain leverage over Pakistan, and seize Afghanistan's mineral resources. In its most conspiratorial form, the US is viewed as having deliberately prolonged the Afghan conflict and used its war on terrorism as a ruse to justify its staying on indefinitely.

And yet, despite the desire to have the US to withdraw its military from Afghanistan, governments in Russia, Iran, Pakistan and other countries are also quietly fearful of a too rapid exit of US and allied foreign troops, and a subsequent sharp reduction of financial backing for the Afghan government by the US and international donors. Their concern is that a too rapid US departure could throw the burden of a tumultuous Afghanistan in their laps, leaving the strong prospect of an outpouring of refugees and an anarchic civil war into which they could be drawn.

Looking Forward

While Afghanistan's future lies ultimately with the Afghans themselves, the actions of the regional powers and the international community, most of all the great powers, figure mightily in its economic survival and hopes for security. Without the facilitation of neighboring countries together with the commitments of international donors, there is little prospect of landlocked Afghanistan attaining sustainable economic growth. This in turn may have to wait on greater economic integration and political cooperation among regional powers. Afghanistan also has little chance of defending its sovereignty and territorial integrity if it becomes a football in a competition among regional powers or a trophy in a great game involving Russia, China and the US.

As with achieving a viable Afghan economy, an Afghan peace is probably impossible as well without a coordinated process of regional and international support. Regional and great powers can take the credit for their lead in promoting a political solution. Despite disappointments with aspects of the US-Taliban agreement, a host of European countries together with the EU were quick to lend their approval as an important first step. But as already noted, there are decided limits on outside powers' ability to help shape a peace process. Although the US and Afghanistan's other benefactors have considerable leverage with the Kabul government, its leaders have shown no hesitation in pushing back against pressures they claim impinge on the country's sovereignty or that threaten their individual and group interests. More limited still are foreign powers' ability to sway Taliban senior leaders who must build consensus internally and who resist compromising on what they interpret as Islamic doctrine. Where other countries have at times been more effective is in their exercise of spoiler power. Pakistan, the country outside of Afghanistan with the most at stake in a peace agreement, has been capable in undermining progress when distrustful of those participating or displeased with the direction of negotiations.

For all the effort and deep desire of the Afghan people for peace and the general preference of regional or great powers to see a diplomatic solution to the conflict that leaves Afghanistan a peaceful, stable, united and prospering country, the prospects for ending violence in the country with a grand bargain with insurgents seem dim for the foreseeable future. Even with a peace deal struck between the US and the Taliban supposedly opening a pathway toward intra-Afghan negotiations, it is at best but the first step in an undoubtedly long process certain to highlight at every step two sharply contrasting visions of an Afghan polity. One envisions a state with the pluralistic values of a liberal constitutional political system; the other is dedicated to the imposition of a Sharia-based political order in the form of an Islamic emirate. Historically, such "civilizational" differences have almost always compromised and are resolved with one side winning, the side other losing. In observing the manner of Taliban negotiating in the many months of intensive talks leading to the February 2020 accord with the US, its leaders' unwillingness to make any significant concessions augers badly for the success of dialogue among Afghanistan themselves. And while the US and its allies have resolved that there is no military solution to Afghan conflict, it is uncertain that the Taliban senior leadership taking a long view has reached the same conclusion.

Perhaps the best that can be expected over the next few years is that while peace negotiations may drag on, Afghanistan's centers of population continue to remain under largely government control. This will require, however, a continued strong US and international commitment financially and to some extent militarily. It is difficult to believe that in the absence of a ceasefire that the Afghan National Security Forces (ANSF) can long deny the Taliban's capture of heavily populated areas of the country without the retention of some US forces, particularly the continued availability of supportive airpower. Even then, the state's survival will need political elites to show greater capacity to unify, overcoming the deep wellspring of suspicion among them. This is a tall order in light of the almost inevitable sharp decline of foreign assistance in the wake of another disruptive national election.

With the US forces now slated to disengage from Afghanistan in whole or in part, American influence in the region will almost inevitably weaken. The result may well be the unraveling of Afghan security forces, leading to the dismantling of the Afghan state. But rather than a victorious Taliban regime such as that which followed in the 1990s, no force may be able to consolidate power for some time. Various ethnic and sectional militias are likely to contest for turf while a Taliban fractures over competing claims to top leadership. The result would be an unbounded, fragmented Afghan civil war in which an energized Islamic State and al-Qaeda are also players. There are bound to follow millions of refugees and a resulting major humanitarian crisis with global disruption. At the same time a largely ungoverned country can provide fertile ground for global terrorist groups to organize and plan.

The regional and the great power dynamics in the event of an anarchic Afghanistan are almost certain to be affected and perhaps be profound. The convergence of interests that has largely marked US, Russian, and Chinese relations in backing a peace process is likely to dissolve. All will feel threatened by a grossly destabilized Afghanistan and the opportunities it affords for export for anti-state Islamic forces. But their responses to this challenge may be very different and lead to tensions and potential confrontation, if only indirectly. The three great powers may well adopt their own coping strategies, aligning with different Afghan contenders in the armed conflict. Among the three, the United States may hold the weakest hand in influencing the ultimate outcome.

References

ADB Asia Regional Information Center, Cross border infrastructure. Shanghai Cooperation Organization (SCO) (2020). https://aric.adb.org/initiative/shanghai-cooperation-organization

M. Bonn, A.B. Nirjabi, Great powers in Afghanistan, between cooperation and competition. International Policy Digest (2019). https://intpolicydigest.org/2019/05/09/great-powers-in-afghanistan-between-cooperation-and-competition/

K.B. Harpviken, S. Tadjbakhsh, *A Rockb between Hard Places: Afghanistan as an Arena of Regional* (Hurst & Company, London, 2016).

Heart of Asia-Istanbul Process, About heart of Asia-Istanbul process (2019). https://www.hoa.gov.af/about-us/about-hoa-ip.html

S. Imran, Sino-US involvement in Afghanistan: implications for South Asian stability and security (2019). https://www.scribd.com/document/439479995/Sino-US-Involvement-in-Afghanistan-pdf

W. Maley, Afghanistan and its region. In: J.A. Their (ed.) The future of Afghanistan (The US Institute of Peace, Washington, DC, 2009)

B. Nisha, South Asia: on the cusp of regional connectivity. Huffington Post (2014, November 25). https://www.huffingtonpost.com/nisha-biswal/south-asia-on-the-cusp-of_b_6219228.html

F.S. Starr, *The New Silk Roads: Transport and Trade in Greater Central Asia* (Central Asia-Caucasus Institute and Silk Road Studies Program, Washington, DC, 2007).

F.Z. Sumar, Prospects for regional integration in central Asia. US Department of State (2014, October 28). https://www.state.gov/p/sca/rls/rmks/2014/233577.htm

M. Weinbaum, H.H. Humayoon, The intertwined destinies of Afghanistan and Pakistan. In: J.A. Thier (ed.), The future of Afghanistan (US, Institute of Peace, Washington, DC, 2009)

M. Weinbaum, K.A. Majidyar, Afghanistan's unending wars. In: Paul Salem, R. Ross Harrison (eds.), Escaping the conflict trap; civil wars in the Middle East (The Middle East Institute, Washington, DC, 2019)

US Department of State, US support for the new silk road (2017). https://2009-2017.state.gov/p/sca/ci/af/newsilkroad/index.htm

The manufacturer's authorised representative in the EU is Springer Nature Customer Service Centre GmbH, Europaplatz 3, 69115 Heidelberg, Germany. If you have any concerns regarding our products, please contact ProductSafety@springernature.com

Printed and bound by CPI Group (UK) Ltd, Croydon, CR0 4YY
30/09/2026
02225495-0012